Methods in Molecular Biology™

Series Editor
John M. Walker
School of Life Sciences
University of Hertfordshire
Hatfield, Hertfordshire, AL10 9AB, UK

For other titles published in this series, go to
www.springer.com/series/7651

RNA Therapeutics

Function, Design, and Delivery

Edited by

Mouldy Sioud

The Norwegian Radium Hospital, Oslo, Norway

Humana Press

Editor
Mouldy Sioud
Department of Immunology
Norwegian Radium Hospital
Institute for Cancer Research
0310 Montebello, Oslo
Norway
mosioud@ulrik.uio.no

ISSN 1064-3745 e-ISSN 1940-6029
ISBN 978-1-4939-6151-1 ISBN 978-1-60761-657-3 (eBook)
DOI 10.1007/978-1-60761-657-3
Springer New York Dordrecht Heidelberg London

Cover illustration: The cover image shows the distribution of GFP-positive NBT-12 cells subsequent to paxilin knockdown using small interfering RNAs (see chapter 13, Fig. 3B). Also, the image illustrates the secondary structure of a small regulatory RNA.

Printed on acid-free paper

Humana Press is part of Springer Science+Business Media (www.springer.com)

Preface

Besides being the physical link between DNA and proteins, RNA plays several other key roles, including RNA catalysis and gene regulation mediated mainly by non-coding small RNAs. This regulation can occur at some of the most important levels of genome function, such as chromatin structure, chromosome segregation, transcription, RNA processing, RNA stability, and translation.

The discovery of catalytic RNAs has opened up a wealth of opportunities to allow investigators to regulate gene expression post-transcriptionally using ribozymes and derivates. In addition to ribozymes, a new RNA-based strategy for gene expression in mammalian cells has recently been described. This strategy known as RNA interference (RNAi) has been shown to function in mammalian cells after introduction of small interfering RNAs (siRNAs) or expression of short hairpin RNAs (shRNAs) by RNA polymerase III promoters. Although much is known about the mechanisms of RNAi, there are a number of challenges that applications of this gene-silencing technology need to overcome, including the activation of innate immunity, off-target effects, and in vivo delivery. The present book on RNA is intended to cover some the most important functions of RNA, particularly small RNAs, and to facilitate the use of RNA agents in human therapies.

Since most, if not all, biological functions are mediated by structure on some level, and it is clear that natural RNA modifications play a vital role in such a structure, **Chapter 1** highlights the importance of this process in gene regulation. And **Chapter 2** describes a detailed method for analyzing naturally occurring RNA modifications. It should be noted that modified nucleotides occur in almost all classes of endogenous RNAs, and there are about 100 different base modifications known that may carry out several functions. Recent studies with chemically modified siRNAs have provided convincing evidence that 2′-ribose modifications can evade the recognition of siRNAs by immune receptors. **Chapter 3** elegantly describes the current strategies, allowing the separation of immune activation from gene silencing. Regarding siRNA chemical modifications, the major challenge is to identify the modification that eliminates siRNA unwanted effects without interfering with gene silencing.

Although much has been accomplished regarding RNA delivery to mammalian cells, obstacles regarding the in vivo delivery of siRNAs remain. Also, technologies that mediate targeted delivery of siRNAs are needed to improve their therapeutic efficacy and safety. **Chapters 4–7** describe the most current methods for delivery, including the use of innovative nanotechnologies for the delivery of oligonucleotides and siRNAs. Also, a large number of targeting strategies using antibodies, peptide, and RNA aptamers are described in **Chapter 7**.

To design an effective siRNA sequence, one must consider the base composition of the chosen site and whether the site will be accessible. **Chapter 8** describes a method for the identification of effective siRNA sequences. Also, **Chapters 9–11** offer valuable protocols for the design, the delivery, and assessments of shRNA activity on tumor growth and viral infection.

A recently established transfected cell array technology has opened new experimental dimensions in the field of functional genomics. This technology allows for the transfection of several thousands of different siRNAs in microarray format. Thus, the silencing of several genes in a spatially separated manner can be monitored. **Chapters 12** and **13** describe this new technology, which is expected to facilitate functional genomics and drug target validation.

As indicated above, eukaryotes produce several types of small RNAs that function in diverse pathways. These small RNAs are conventionally grouped on their origin into two classes: microRNAs (miRNAs) and siRNAs. MiRNAs are usually generated from the dsRNA region of the hairpin-shaped precursors, whereas siRNAs are derived from long double-stranded RNAs. **Chapter 14** highlights the biogenesis of miRNAs and shows that intron-derived miRNA can induce RNAi not only in vitro but also in adult mice. Given that several effector proteins are shared by siRNA and miRNA pathways, there is a potential danger of inhibiting the miRNA biogenesis by exogenously or endogenously expressed siRNAs. **Chapter 15** underlies the risk of this danger.

Currently, different methodologies are used to profile miRNA expression. These include Northern blotting with radiolabeled probes, cloning approaches, quantitative PCR-based amplification, and microarray-based expression profiling. A collection of four chapters describes these recent technologies (**Chapters 16–19**).

In the 1980s, the discovery that certain RNAs can perform catalysis has led to the development of a new class of therapeutic RNAs called *trans*-cleaving ribozymes. Such ribozymes bind mRNA through base-pairing interactions and subsequently cleave the bound target mRNA in vivo. During recent years, much progress has been made toward assessing the potential therapeutic utility of ribozymes, with the hammerhead, hairpin, ribonuclease P, and *Tetrahymena* ribozymes being the main focus of this translational research. By using the *Tetrahymena* group I intron-based *trans*-splicing ribozyme, in **Chapter 20,** Kim and colleagues describe the conversion of human telomerase reverse transcriptase-encoding RNA to therapeutic transgene herpes simplex virus thymidine kinase, and **Chapters 21 and 22** cover the recent therapeutic use of hairpin and ribonuclease P ribozymes.

In addition to RNA catalyst, small structured single-stranded RNAs or DNAs, also known as aptamers, are attractive tools in therapy. They are in vitro-evolved RNA structures that bind with high affinity to a given ligand. The high-affinity binders are usually selected from a random pool of nucleotides by a process called systemic evolution of ligands by exponential enrichment (SELEX). **Chapters 23 and 24** describe the use of cancer cells to select cell-specific RNA aptamers and the selection of aptamers binding to human S100 calcium-binding protein B, respectively. **Chapter 25** describes the design of 2′-modified small RNAs able to function as TLR-7/8 antagonists. These novel agents could be used to treat TLR-induced diseases such as SLE.

Cancer immunotherapy offers an attractive therapeutic addition, delivering treatment of high specificity, low toxicity, and prolonged activity. In the case of vaccines, several mechanisms operating under physiologic conditions to modulate immunity contribute to immunologic unresponsiveness seen in most patients. Critical to the initiation of immunity, dendritic cells (DC) are key participants in immune regulation. Therefore, modulation of DC-using RNAi is especially important in overcoming immune tolerance to tumor cells. In this respect, **Chapter 26** describes the modulation of DC function by dual siRNAs targeting indoleamine 2,3-dioxygenase (IDO) and simultaneously activating RIG1 protein. IDO is a key immunomodulatory enzyme that promotes peripheral immune tolerance by inhibiting T-cell activation and proliferation through tryptophan catabolism.

One approach to stimulate effective cytotoxic T-cell (CTL) responses in cancer patients would be to rebuild this process in vitro, that is, to isolate DCs from the patient and load them with tumor antigens in such a manner that the antigens will be processed and presented to T cells. This approach has been accomplished by incubating DCs with peptides and proteins or by transfecting the cells with DNA constructs. Transfecting DCs with mRNA-encoding antigens is yet another method to load DCs with antigens. Messenger RNAs can be isolated directly from tumor cells or synthesized in vitro from DNA constructs. **Chapter 27** describes an elegant protocol for the generation of immunogenic DC transfected with mRNAs.

The generation of therapeutic T-cell populations for adoptive immunotherapy of cancer requires extensive ex vivo cell processing, including the isolation or the creation of antigen-specific T cells and their subsequent propagation to clinically relevant numbers. However, novel antibody-based immunotherapeutic strategies that exploit chimeric immune receptors (CIR), expressed on the surface of transduced human peripheral blood mononuclear cells (PBMC), to redirect potent non-MHC-dependent cytotoxicity to tumor cells expressing a tumor-associated antigen or self-antigens (e.g., CD19, CD20, CD33) has been described. Of note, this strategy overcomes the problem of MHC restriction. **Chapter 28** describes the manufacturing of T cells expressing CIRs. These genetically modified T cells expressed the chimeric receptor and killed cancer cells. These preclinical data confirm the feasibility of this approach to manufacturing T-cell products.

Based on several observations, vaccine strategies will be unsuccessful unless they are coupled with a means of counteracting the suppressive influence caused by certain immune cells and tumor cells. **Chapter 29** describes the recent advances in breaking self-tolerance to tumor cells. Notably, bone marrow transplantation (BMT) is currently used in the treatment of a variety of diseases such as leukemia. However, significant complications still limit the efficacy of this treatment, including the occurrence of graft-versus-host disease (GvHD) and infections. **Chapter 30** highlights the challenges facing this treatment and describes the possibilities of overcoming GvHD as well as genetic manipulations of hematopoietic $CD34^+$ stem/progenitor cells using RNAi.

Topics covered in this volume will be of interest to researchers, clinicians, teachers, and biotech companies interested in RNA-based therapies. It is my hope that the readers will benefit from this collection of excellent chapters dealing with RNA as tools and therapies.

Finally, I would like to thank the authors for their contributions, Anne Dybwad for excellent editorial assistance, the series editor, John Walker, and all those involved in the production of the book.

Mouldy Sioud

Contents

Contributors

HILDE ALMÅSBAK • *Department of Immunology, Radiumhospitalet-Rikshospitalet, University Hospital, Oslo, Norway*

LUIS MARAT ALVAREZ-SALAS • *Laboratorio de Terapia Génica, Departamento de Genética y Biología Molecular, CINVESTAV, México D.F., México*

GUILLERMO AQUINO-JARQUIN • *Departamento de Fisiología, Biofísica y Neurociencias, CINVESTAV, México D.F., México*

PATRICK ARBUTHNOT • *Antiviral Gene Therapy Research Unit, Department of Molecular Medicine and Haematology, University of the Witwatersrand Medical School, Wits, South Africa*

YONG BAI • *Program in Comparative Biochemistry, University of California, Berkeley, CA, USA*

AUDE BONEHILL • *Laboratory of Molecular and Cellular Therapy, Department of Physiology – Immunology, Medical School of the Vrije Universiteit Brussel (VUB), Brussels, Belgium*

TORUNN BRULAND • *Department of Cancer Research and Molecular Medicine, Norwegian University of Science and Technology (NTNU), Trondheim, Norway*

LINA CEKAITE • *Department of Immunology, The Norwegian Radium Hospital, Oslo, Norway*

CHEN PU CHANG • *Department of Cell and Neurobiology, Keck School of Medicine, University of Southern California, Los Angeles, CA, USA*

HUI CHEN • *Department of Chemistry and Department of Physiology and Functional Genomics, Shands Cancer Center and Center for Research at the Bio/nano Interface, UF Genetics Institute and McKnight Brain Institute, University of Florida, Gainesville, FL, USA*

GARY A. CLAWSON • *Gittlen Cancer Research Foundation, Hershey Medical Center, Hershey, PA, USA; Departments of Pathology, Biochemistry and Molecular Biology, College of Medicine, Pennsylvania State University, Hershey, PA, USA*

DALIA LOPEZ COLON • *Department of Chemistry and Department of Physiology and Functional Genomics, Shands Cancer Center and Center for Research at the Bio/nano Interface, UF Genetics Institute and McKnight Brain Institute, University of Florida, Gainesville, FL, USA*

JURGEN CORTHALS • *Laboratory of Molecular and Cellular Therapy, Department of Physiology – Immunology, Medical School of the Vrije Universiteit Brussel (VUB), Brussels, Belgium*

OMID C. FAROKHZAD • *Laboratory of Nanomedicine and Biomaterials, Department of Anesthesia, Brigham and Women's Hospital, Harvard Medical School, Boston, MA, USA*

CHRISTINA SÆTEN FJELDBO • *Department of Cancer Research and Molecular Medicine, Norwegian University of Science and Technology (NTNU), Trondheim, Norway*

ANDERS FJOSE • *Department of Molecular Biology, University of Bergen, Bergen, Norway*

YNGVAR FLØISAND • *Department of Hematology, Rikshopitalet–Radiumhospitalet, University Hospital, Oslo, Norway*

SATOSHI FUJITA • *Research Institute for Cell Engineering (RICE), National Institute of Advanced Industrial Science and Technology (AIST), Tokyo, Japan*
WEIWEI GAO • *Laboratory of Nanomedicine and Biomaterials, Department of Anesthesia, Brigham and Women's Hospital, Harvard Medical School, Boston, MA, USA*
CLARA-CECILIE GÜNTHER • *Department of Mathematical Sciences, Norwegian University of Science and Technology (NTNU), Trondheim, Norway*
ELIZABETH H. GUTHRIE • *Department of Microbiology and Immunology, Lineberger Comprehensive Cancer Center, University of North Carolina, Chapel Hill, NC, USA*
KAZUMI HAKAMADA • *Research Institute for Cell Engineering (RICE), National Institute of Advanced Industrial Science and Technology (AIST), Tokyo, Japan*
CARLO HEIRMAN • *Laboratory of Molecular and Cellular Therapy, Department of Physiology – Immunology, Medical School of the Vrije Universiteit Brussel (VUB), Brussels, Belgium*
MAX TZE-HAN HUANG • *Department of Microbiology and Immunology, Lineberger Comprehensive Cancer Center, University of North Carolina, Chapel Hill, NC, USA*
JIN-SOOK JEONG • *Department of Pathology and Medical Research Center for Cancer Molecular Therapy, Dong-A University College of Medicine, Busan, Korea*
ATHENA KANTARTZIS • *Department of Biochemistry and Biophysics, University of Rochester Medical Center, Rochester, NY, USA*
JOHN KARIJOLICH • *Department of Biochemistry and Biophysics, University of Rochester Medical Center, Rochester, NY, USA*
IN-HOO KIM • *Molecular Imaging and Therapy Branch, Research Institute, National Cancer Center, Goyang, Korea*
ASTRID LÆGREID • *Department of Cancer Research and Molecular Medicine, Norwegian University of Science and Technology (NTNU), Trondheim, Norway*
HERMANN LAGE • *Charité Campus Mitte, Institute of Pathology, Berlin, Germany*
METTE LANGAAS • *Department of Mathematical Sciences, Norwegian University of Science and Technology (NTNU), Trondheim, Norway*
ROBERT LANGER • *Harvard-MIT Center of Cancer Nanotechnology Excellence, Massachusetts Institute of Technology, Cambridge, MA, USA*
SANG-JIN LEE • *Genitourinary Cancer Branch, Research Institute, National Cancer Center, Goyang, Korea*
SEONG-WOOK LEE • *Department of Molecular Biology, Institute of Nanosensor and Biotechnology, Dankook University, Yongin, Korea*
SHI-LUNG LIN • *Department of Cell and Neurobiology, Keck School of Medicine, University of Southern California, Los Angeles, CA, USA*
FENYONG LIU • *Program in Comparative Biochemistry, University of California, Berkeley, CA, USA*
MARIANNE LUNDBY • *Department of Cellular Therapy, Radiumhospitalet-Rikshospitalet, University Hospital, Oslo, Norway*
DAVID H. MATHEWS • *Department of Biochemistry and Biophysics and Department of Biostatistics and Computational Biology, University of Rochester Medical Center, Rochester, NY, USA*
LING MENG • *Department of Chemistry and Department of Physiology and Functional Genomics, Shands Cancer Center and Center for Research at the Bio/nano Interface, UF Genetics Institute and McKnight Brain Institute, University of Florida, Gainesville, FL, USA*

KRISTINE MISUND • *Department of Cancer Research and Molecular Medicine, Norwegian University of Science and Technology (NTNU), Trondheim, Norway*

JUN MIYAKE • *Research Institute for Cell Engineering (RICE), National Institute of Advanced Industrial Science and Technology (AIST), Tokyo, Japan*

MASATO MIYAKE • *Research Institute for Cell Engineering (RICE), National Institute of Advanced Industrial Science and Technology (AIST), Tokyo, Japan*

ANNE MOBERGSLIEN • *Department of Immunology, Institute for Cancer Research, Rikshospitalet – Radiumhospitalet, University Hospital, Oslo, Norway*

CHRIS B. MOORE • *Department of Microbiology and Immunology, University of North Carolina, Chapel Hill, NC 27599, USA*

AKIRA NAGASAKI • *Research Institute for Cell Engineering (RICE), National Institute of Advanced Industrial Science and Technology (AIST), Tokyo, Japan*

BART NEYNS • *Laboratory of Molecular and Cellular Therapy, Department of Physiology – Immunology, Medical School of the Vrije Universiteit Brussel (VUB), Brussels, Belgium*

MEGHAN O'DONOGHUE • *Department of Chemistry and Department of Physiology and Functional Genomics, Shands Cancer Center and Center for Research at the Bio/nano Interface, UF Genetics Institute and McKnight Brain Institute, University of Florida, Gainesville, FL, USA*

REIKO ONUKI-NAGASAKI • *Research Institute for Cell Engineering (RICE), National Institute of Advanced Industrial Science and Technology (AIST), Tokyo, Japan*

WEIHUA PAN • *Gittlen Cancer Research Foundation, Hershey Medical Center, Hershey, PA, USA; Departments of Pathology, Biochemistry and Molecular Biology, College of Medicine, Pennsylvania State University, Hershey, PA, USA*

ALEX RADOVIC-MORENO • *Harvard-MIT Center of Cancer Nanotechnology Excellence, Massachusetts Institute of Technology, Cambridge, MA, USA*

ANNE-MARIE RASMUSSEN • *Department of Immunology, Radiumhospitalet-Rikshospitalet, University Hospital, Oslo, Norway*

PAUL JAY FANNIN RIDER • *Program in Infectious Diseases and Immunity, School of Public Health, University of California, Berkeley, CA, USA*

SEUNGIL RO • *Department of Physiology and Cell Biology, Anderson Biomedical Science, University of Nevada School of Medicine, Reno, NV, USA*

RAMIRO ROJAS-HERNÁNDEZ • *Laboratorio de Terapia Génica, Departamento de Genética y Biología Molecular, CINVESTAV, México D.F., México*

SHEENA M. SAAYMAN • *Antiviral Gene Therapy Research Unit, Department of Molecular Medicine and Haematology, University of the Witwatersrand Medical School, Wits, Johannesburg, South Africa*

KWAME SEFAH • *Department of Chemistry and Department of Physiology and Functional Genomics, Shands Cancer Center and Center for Research at the Bio/nano Interface, UF Genetics Institute and McKnight Brain Institute, University of Florida, Gainesville, FL, USA*

JINJUN SHI • *Laboratory of Nanomedicine and Biomaterials, Department of Anesthesia, Brigham and Women's Hospital, Harvard Medical School, Boston, MA, USA*

MOULDY SIOUD • *Department of Immunology, Institute for Cancer Research, Rikshospitalet-Radiumhospitalet, University Hospital, Oslo, Norway*

RUI SONG • *Department of Physiology and Cell Biology, Anderson Biomedical Science, University of Nevada School of Medicine, Reno, NV, USA*

TONJE STRØMMEN STEIGEDAL • *Department of Cancer Research and Molecular Medicine, Norwegian University of Science and Technology (NTNU), Trondheim, Norway*
ULRIKE STEIN • *Max-Delbrück-Center for Molecular Medicine and Experimental Clinical Research Center, Charité Berlin, Berlin, Germany*
DAG R. SØRENSEN • *Department of Comparative Medicine, Radiumhospitalet-Rikshopitalet Universtity Hospital, Oslo, Norway*
WEIHONG TAN • *Department of Chemistry and Department of Physiology and Functional Genomics, Shands Cancer Center and Center for Research at the Bio/nano Interface, UF Genetics Institute and McKnight Brain Institute, University of Florida, Gainesville, FL, USA*
DEBRA TAXMAN • *Department of Microbiology and Immunology, Lineberger Comprehensive Cancer Center, University of North Carolina, Chapel Hill, NC, USA*
KRIS THIELEMANS • *Laboratory of Molecular and Cellular Therapy, Department of Physiology – Immunology, Medical School of the Vrije Universiteit Brussel (VUB), Brussels, Belgium*
LIV THOMMESEN • *Department of Cancer Research and Molecular Medicine, Norwegian University of Science and Technology (NTNU), Trondheim, Norway*
TARO Q.P. UYEDA • *Research Institute for Cell Engineering (RICE), National Institute of Advanced Industrial Science and Technology (AIST), Tokyo, Japan*
AN M. T. VAN NUFFEL • *Laboratory of Molecular and Cellular Therapy, Department of Physiology – Immunology, Medical School of the Vrije Universiteit Brussel (VUB), Brussels, Belgium*
WOLFGANG WALTHER • *Max-Delbrück-Center for Molecular Medicine and Experimental Clinical Research Center, Charité Berlin, Berlin, Germany*
MARC S. WEINBERG • *Antiviral Gene Therapy Research Unit, Department of Molecular Medicine and Haematology, University of the Witwatersrand Medical School, Wits, South Africa*
ZEYU XIAO • *Laboratory of Nanomedicine and Biomaterials, Department of Anesthesia, Brigham and Women's Hospital, Harvard Medical School, Boston, MA, USA*
XIANGLING XIONG • *Department of Chemistry and Department of Physiology and Functional Genomics, Shands Cancer Center and Center for Research at the Bio/nano Interface, UF Genetics Institute and McKnight Brain Institute, University of Florida, Gainesville, FL, USA*
WEI YAN • *Department of Physiology and Cell Biology, Anderson Biomedical Science, University of Nevada School of Medicine, Reno, NV, USA*
SHAO-YAO YING • *Department of Cell and Neurobiology, Keck School of Medicine, University of Southern California, Los Angeles, CA, USA*
YI-TAO YU • *Department of Biochemistry and Biophysics, University of Rochester Medical Center, Rochester, NY, USA*
XIAO-FENG ZHAO • *Department of Molecular Biology, University of Bergen, Bergen, Norway*

Chapter 1

RNA Modifications: A Mechanism that Modulates Gene Expression

John Karijolich, Athena Kantartzis, and Yi-Tao Yu

Abstract

Accuracy in the flow of genetic information from DNA to protein, or gene expression, is essential to the viability of an organisms. Pre-mRNA splicing and protein translation are two major steps in eukaryotic gene expression that necessitate the production of accurate gene products. Both processes occur in large complexes, consisting of both proteins and noncoding RNAs. Interestingly, the RNA components contain a large number of posttranscriptional modifications, including 2′-O-methylation and pseudouridylation, which are functionally important. In this chapter, we highlight the functional aspects of the modifications of spliceosomal snRNA and rRNA and provide a framework for understanding how posttranscriptional modifications are capable of influencing gene expression.

Key words: Pre-mRNA splicing, RNA modifications, 2′-O-methylation, pseudouridylation.

1. Introduction

Inherent to the expression of the majority of higher eukaryotic genes is the processing of pre-messenger RNA (pre-mRNA) and synthesis of proteins (translation). Pre-mRNA splicing is an RNA-processing reaction in which two successive transesterification reactions result in the removal of noncoding sequences, introns, and ligation of coding sequences, exons (including the 5′ and 3′ untranslated sequences), ensuing in the production of a mature mRNA. The mature mRNA can then be exported to the cytoplasm where it directs the synthesis of protein.

Both pre-mRNA splicing and protein synthesis are multistep processes coordinated by large and highly dynamic ribonucleoprotein (RNP) complexes (1). In the case of the spliceosome, the

M. Sioud (ed.), *RNA Therapeutics*, Methods in Molecular Biology 629,
DOI 10.1007/978-1-60761-657-3_1, © Springer Science+Business Media, LLC 2010

machinery responsible for intron removal, five uridyl-rich small nuclear RNAs (snRNA), namely U1, U2, U4, U5, and U6, as well as a number of proteins, orchestrate the splicing of pre-mRNAs (2–5). On the other hand, the ribosome, the machinery that directs protein synthesis, is composed of four ribosomal RNAs (rRNAs), namely 5S, 5.8S, 18S, 28S (in high eukaryotic organisms), and numerous protein factors. Over the years, the mechanism of both processes has been extensively studied, and it is now clear that in each case, the RNA components play indispensable roles in orchestrating their respective processes.

Interestingly, both spliceosomal snRNAs and rRNAs contain a variety of posttranscriptionally modified nucleotides. By far, the two most abundant modifications are 2′-O-methylation and pseudouridylation. While 2′-O-methylation is an RNA backbone-targeting reaction that results in the introduction of a methyl group at the 2′-O position of the sugar ring, pseudouridylation is a uridine-specific modification, which results in the formation of the 5-ribosyl isomer of uridine, pseudouridine (Ψ) (**Fig. 1.1**). Over the last few decades, considerable effort has been directed at elucidating the mechanism by which these modifications are introduced as well as their molecular functions. To date, progress has been most significant in the area regarding the introduction of modifications. In contrast, knowledge regarding the function of posttranscriptionally

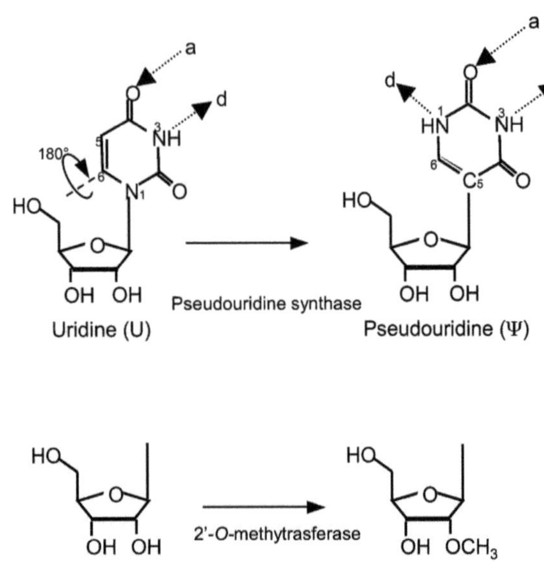

Fig. 1.1. Schematic depiction of the two most abundant modified nucleotides in snRNA and rRNA. (*Top*) Pseudouridine is a rotational isomer of uridine, in which the C–C glycosidic bond is broken to form an N–C bond, resulting in the presence of an extra hydrogen bond donor (d) while the number of hydrogen bond acceptors (a) is unchanged. (*Bottom*) Schematic representation of a 2′-O-methylated ribose.

modified nucleotides has lagged behind. This has been in part due to the lack of proper assays and experimental systems. Recently, especially over the past 10 years, there has been considerable growth in the field of RNA modification function, resulting in the accumulation of a large amount of data. This review discusses the functional aspects of the modifications of spliceosomal snRNA and rRNA and provides a framework for understanding how posttranscriptional modifications are capable of influencing gene expression.

2. Spliceosomal snRNA Modification

Prior to 1951, RNA was believed to consist of the four classical ribonucleosides (i.e., adenosine, guanosine, uridine, and cytidine). The field of RNA modification has its birth in 1951 when Cohn and Volkin reported the identification of an unknown nucleoside, subsequently identified as 5-ribosyluracil (pseudouridine), present in the RNA hydrolysates of calf liver (6). Shortly thereafter, additional modified ribonucleosides were discovered including various 2′-*O*-methylribose derivatives.

Spliceosomal snRNAs were initially identified as U-rich small nuclear RNAs (7–11). Surprisingly, however, when the exact nucleotide composition was finally determined, it became clear that these RNAs contained a number of pseudouridines and 2′-O-methylated residuals (**Fig. 1.2**). Whether these modified nucleotides are important for function in pre-mRNA splicing became a central question.

2.1. Experimental Clues that Spliceosomal snRNA Modifications Might Affect Pre-mRNA Splicing

Although the spliceosomal U snRNAs had been known to be extensively posttranscriptionally modified since their initial identification in the 1960s and 1970s, the exploration for a functional role of these modifications was not fruitful until the early 1990s. However, as early as 1988 Reddy and Busch did make note of the fact that the modified nucleotides were distributed nonrandomly throughout the snRNAs (11) (**Fig. 1.2**). In fact, modifications are particularly clustered in regions of functional significance, such as regions engaged in RNA–RNA interactions (4). Furthermore, analysis of the distribution of modified nucleotides in spliceosomal snRNAs from various organisms demonstrated conservation in the location of modifications throughout evolution.

In the early 1990s, a number of functional reconstitution systems were developed, permitting a direct assessment of the function of spliceosomal snRNAs and their modifications in pre-mRNA splicing. In particular, Patton developed an in vitro U

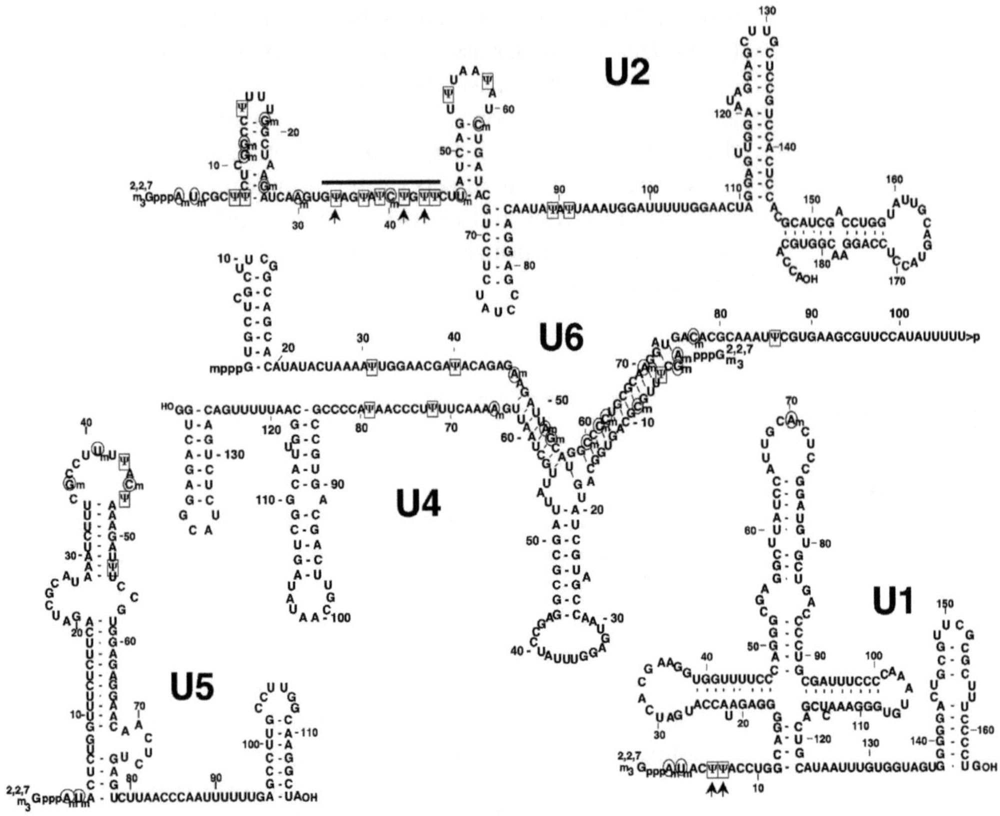

Fig. 1.2. Primary sequences and secondary structures of the vertebrate major spliceosomal snRNAs. Positions of pseudouridine residues are marked in *boxes* and 2′-O-methylated nucleotides are *circled*. The 5′ caps (2,2,7-trimethylated guanosine cap for U1, U2, U4, and U6, and a γ-methylated guanosine for U6) are also shown. The pseudouridine residues in U2 and U1 snRNAs that are conserved in yeast are marked with *arrows*. The *thick line* indicates the U2 branch site recognition region.

snRNA assembly-modification system from HeLa cell S100 and nuclear extracts that was capable of both pseudouridylating U snRNAs and assembling functional U snRNPs (12). Interested in addressing the effect of 5-fluorouridine (5FU) incorporation on snRNP assembly, he demonstrated that the presence of 5FU in U2 snRNA prevented the formation of salt-resistant complexes when analyzed by cesium-sulfate buoyant density gradient centrifugation. This data, coupled with previous data demonstrating the inhibition of pseudouridylation on 5FU-containing snRNA, was suggestive of a role for pseudouridylation in U2 snRNP biogenesis (13).

Reconstitution systems were also developed that involved the specific depletion of one of the endogenous spliceosomal snRNAs followed by the supplementation of that respective snRNA synthesized in vitro (14–28). As the in vitro synthesized snRNAs are void of modifications, the ability (or lack thereof) of the RNA to reconstitute pre-mRNA splicing would indicate whether the modifications were required for pre-mRNA splicing. Surprisingly, in vitro synthesized U1 (15), U4 (16), and U6 (21–24) were shown to be effective in reconstituting pre-mRNA splicing in mammalian extracts depleted of the respective endogenous snRNA. Furthermore, in vitro synthesized U6 snRNA was successful in rescuing pre-mRNA splicing in U6-depleted cell-free extracts prepared from *Ascaris* embryos (25, 26). In addition, in vitro synthesized U2 and U6 were capable of restoring pre-mRNA splicing in yeast cell extracts (14, 18–20, 28). These initial data created rather a conundrum, as a lack of U snRNP assembly would have been expected to negatively impact pre-mRNA splicing. However, it is important to note that none of the abovementioned reconstitution experiments monitored modification of the in vitro transcribed snRNA. In fact, based on the results of others (D. S. McPheeters, personal communication) and our own (unpublished data), in vitro transcribed U2 snRNA is readily modified when added to yeast splicing extracts. Thus, whether modified nucleotides contributed to pre-mRNA splicing was still an open question.

In 1995, the Luhrmann group studied the function of spliceosomal snRNAs in a HeLa reconstitution system (17). Interestingly, while in vitro synthesized U2 snRNA completely failed to reconstitute pre-mRNA splicing in U2 snRNA-depleted extracts, cellularly derived U2 snRNA successfully restored splicing, suggesting that the modified nucleotides in U2 snRNA might indeed play a role in splicing. Further analysis confirmed that U2 snRNA was not pseudouridylated in the extracts.

2.2. Definitive Evidence That Spliceosomal snRNA Modifications Are Required for snRNP Assembly and Pre-mRNA Splicing

In an effort to clarify the effect of modifications in pre-mRNA splicing, Yu et al. examined U2 snRNP biogenesis and spliceosome assembly in *Xenopus* oocytes (29). First, through the injection of an antisense U2 DNA oligonucleotide, endogenous U2 snRNA was depleted from *Xenopus* oocytes. The U2-depleted oocytes were then supplemented with in vitro synthesized U2 snRNA and incubated for a short period. Under these conditions, U2 modifications largely did not occur. Radiolabeled adenovirus pre-mRNA was then injected, and pre-mRNA splicing and spliceosome assembly were assayed using denaturing gel electrophoresis and native gel electrophoresis, respectively (29). Interestingly, unmodified or hypomodified U2 snRNA was unable to support splicing. Detailed analyses showed that such U2 snRNA failed to participate in the assembly of any of the higher-order splicing complexes (i.e., complexes A, B, and C). Anti-snRNP immunoprecipitation and glycerol gradient sedimentation were used to further examine U2 snRNP assembly. Experi mental data indicated that U2 snRNA lacking modifications could only form a 12S snRNP; no significant 17S snRNP, which is a mature form of U2 snRNP, was observed. These results indicated that U2 snRNA modification is required for the transition from the 12S snRNP to a functional 17S snRNP. Interestingly, however, it was observed that upon longer reconstitution periods, in vitro synthesized U2 snRNA was assembled into primarily a 17S particle, thus regaining its ability to support splicing. Analysis of the modification status of U2 snRNA over time indicated that in vitro synthesized U2 snRNA was efficiently modified in the *Xenopus* oocyte reconstitution system following prolonged incubations. Thus, a nice correlation between the modification status of U2 snRNA and its ability to participate in pre-mRNA splicing was established. In addition, through the construction of chimeric U2 snRNAs in which various regions of U2 snRNA contained no modifications, Yu et al. were able to demonstrate that modifications within the 275′-most nucleotides were required for pre-mRNA splicing (29). The functional importance of these modified nucleotides has been later confirmed by the Luhrmann group (30). More recently, using 5FU-containing U2 snRNA, a U2-specific pseudouridylation inhibitor, Zhao and Yu further dissected the importance of U2 snRNA modi fications. The experimental results revealed a requirement for pseudouridylation within the branch site recognition region for efficient pre-mRNA splicing and spliceosome assembly in *Xenopus* oocytes (31). The earlier studies failed to identify these pseudouridines (downstream from the 275′-most nucleotides) as important due to the fact that in the *Xenopus* reconstitution system these nucleotides are modified extremely fast, making it difficult to examine their importance under the particular conditions without specific pseudouridylation inhibitors.

2.3. Genetic Dissection of Spliceosomal snRNA Modification in Saccharomyces cerevisiae

The pre-mRNA splicing machinery of *S. cerevisiae* is one of the most extensively studied systems with regard to spliceosome assembly and pre-mRNA splicing catalysis. This fact coupled with the awesome power of yeast genetics makes *S. cerevisiae* an ideal organism for dissecting the function of posttranscriptionally modi fied nucleotides present in the spliceosomal snRNAs. Surprisingly, to date no 2′-O-methylations have been observed in the spliceosomal snRNAs of *S. cerevisiae* (our unpublished data). However, in 1999 Massenet et al. identified six pseudouridines, five of which are conserved in mammals (32). There are two within U1, three in the U2 branch site recognition region, and one within U5. The genes which encode the enzymes responsible for U2 snRNA pseudouridylation have all been identified (Pus1p, Ψ44; Pus7p, Ψ35; Ψ42 is catalyzed by the box H/ACA RNP snR81) (32–34).

To investigate the role of Ψ35 in U2 snRNA, Ma et al. constructed a deletion of *pus7*. Interestingly, while viable, the *pus7-Δ* displayed reduced fitness when grown in media containing high concentrations of NaCl, or when in competition with a wild-type strain (34). To further clarify the role of Ψ35, Yang et al. screened the *pus7-Δ* strain against a collection of U2 snRNA mutants (35). Semi-quantitative PCR indicated an accumulation of pre-mRNA in the *pus7-Δ* strain when in combination with U40G or U40Δ mutations in U2 snRNA. Pre-mRNA accumulation varied depending on the transcript analyzed, with fold increases ranging from 0 to >10.

A detailed analysis of the other yeast spliceosomal snRNA pseudouridylations has yet to be described. However, the fact that they are conserved throughout evolution is suggestive of function. It will be interesting to see whether the pseudouridylations present within yeast spliceosomal snRNAs function synergistically.

3. Ribosomal RNA Modifications

Ribosomal RNAs (rRNAs), integral components of the ribosome, have evolutionary conserved structures important for the process of translation despite some sequence variation among organisms. A frequent trait of rRNAs is the presence of modified nucleotides (**Fig. 1.3**). Similar to the spliceosomal snRNAs, the most common modifications in rRNAs are pseudouridylation and 2′-O-methylation.

Several lines of evidence suggest that rRNA modifications contribute to ribosome function. For instance, while rRNAs do not exhibit a significant difference in size or number of ribosomal proteins from prokaryotic to lower and higher eukaryotic organisms, the number of modifications increases along with the evolutionary complexity of the organism (36). Furthermore, three-dimensional (3D) maps deduced for *Escherichia coli* and cytoplasmic *S. cerevisiae* ribosomes indicate the clustering of modifications in functionally important regions of the ribosome, such as the peptidyl transferase center (PTC), the A, P, and E sites of tRNA and mRNA binding, the polypeptide exit tunnel, and sites of subunit–subunit interaction (37). Together, these data strongly suggest that rRNA posttranscriptional modifications influence ribosome function.

3.1. Global Analysis of rRNA Modifications

It has long been known that archaeal and eukaryotic rRNA modifications, including 2′-O-methylation and pseudouridylation, are catalyzed by box C/D RNPs and box H/ACA RNPs, respectively. Specifically, the box C/D and H/ACA snoRNAs (**Fig. 1.4**) are each associated with four evolutionary conserved core proteins of which only one is considered to have the catalytic activity required for the modification reaction. In yeast, Nop1p

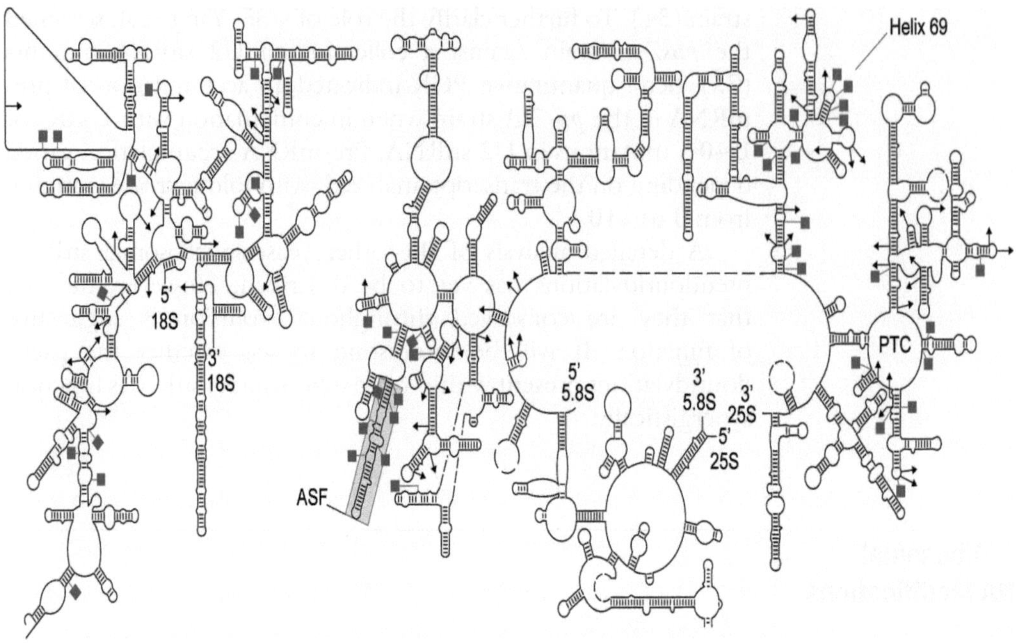

Fig. 1.3. Secondary structure depiction of *S. cerevisiae* cytoplasmic ribosome (modified from (37)). The 18S rRNA is depicted on the *left*, while the 25S–5.8S is shown on the *right*. The PTC, ASF, and Helix 69 are apparent in the structure. Modification sites in the rRNA species are illustrated as follows: pseudouridylation sites (Ψ) – *gray squares* and 2′-O-methylation sites (Nm) – arrow heads. Adopted with modification from Decatur and Fournier (2002).

(fibrillarin in human) is the 2′-*O*-methyltransferase associated with box C/D RNPs, while Cbf5p (dyskerin in humans, NAP57 in rat, and Nop60B in *Drosophila*) is the pseudouridine synthase associated with box H/ACA RNPs. Using this information, it became possible to specifically disrupt the catalytic center of the enzyme, thus permitting an examination of the effect of a global depletion of either 2′-O-methylations (Nm) or pseudouridines on ribosome biogenesis and function.

To assess the function of Nop1p in ribosome biogenesis, Tollervey et al. constructed a series of mutations within the coding sequence of NOP1 (38). *S. cerevisiae* expressing Nop1p carrying the double mutation A175V and P219S were defective both in growth and in rRNA methylation at high temperature (36°C). Surprisingly, although rRNA/pre-rRNA methylation was severely inhibited, there was only a slight impairment of 18S and 25S synthesis. However, the production of mature 60S ribosomal subunits was delayed when monitored by sucrose gradient centrifugation. Consistently, a similar mild processing defect was observed when yeast cells were exposed to the methylation inhibitor ethionine. These results suggested that rRNA methylation played only a minor role in the processing of pre-rRNA, but might contribute significantly to the assembly of ribosome subunits (or the combination of the two led to temperature-sensitive phenotype).

In 1999, point mutations within the pseudouridine synthase Cbf5p permitted an assessment of the role of rRNA pseudouridylation in ribosome biogenesis (39). Mutation of the universally conserved catalytic aspartate residue (D95 in *S. cerevisiae*) resulted in no detectable pseudouridine formation in 18S and 25S rRNAs. *S. cerevisiae* strains expressing catalytically deficient Cbf5p exhi bited severe growth defects, and contained a reduction in the level of both 18S and 25S rRNAs. In addition, sucrose gradient analysis of ribosome assembly indicated a strong defect in both 40S and 60S subunit biogenesis. These data are strongly indicative of a role for rRNA pseudouridylation in rRNA processing and subunit assembly.

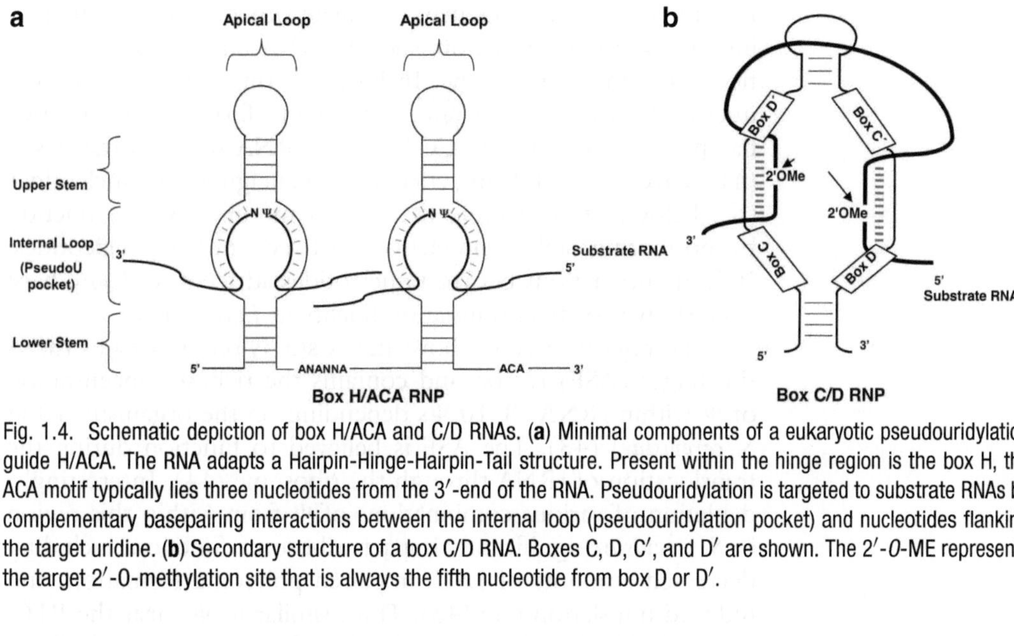

Fig. 1.4. Schematic depiction of box H/ACA and C/D RNAs. (**a**) Minimal components of a eukaryotic pseudouridylation guide H/ACA. The RNA adapts a Hairpin-Hinge-Hairpin-Tail structure. Present within the hinge region is the box H, the ACA motif typically lies three nucleotides from the 3′-end of the RNA. Pseudouridylation is targeted to substrate RNAs by complementary basepairing interactions between the internal loop (pseudouridylation pocket) and nucleotides flanking the target uridine. (**b**) Secondary structure of a box C/D RNA. Boxes C, D, C′, and D′ are shown. The 2′-*O*-ME represents the target 2′-O-methylation site that is always the fifth nucleotide from box D or D′.

3.2. Functional Analysis of rRNA Modifications by Deleting Individual snoRNAs

The generation of 3D maps for the yeast ribosome made possible the realization that modifications cluster within 3D space (37). This information, coupled with the fact that the box H/ACA RNAs responsible for directing the pseudouridylation of virtually all rRNA pseudouridine residues have been identified, permitted the use of genetics in addressing the functional role pseudouridylation plays in specific regions of the ribosome.

In 2003, the Fournier group systematically analyzed the importance of pseudouridylation within the peptidyl transferase center (PTC) by deleting the snoRNA genes responsible for guiding pseudouridylation within the PTC (40). Located within the PTC are six Ψs which are guided by five box H/ACA snoRNAs (*snR34* for Ψ_{2822}. and Ψ_{2876}; *snR46* for Ψ_{2861}; *snR10* for Ψ_{2919}; *snR37* for Ψ_{2940}; and *snR42* for Ψ_{2971}). Interestingly, deletion of any snoRNA alone (with the exception of snR10) resulted in only minor if any phenotype; however, deletion of several or all six simultaneously resulted in a significant (20–45%) decrease in translation rate when measured by in vivo [^{35}S] methionine incorporation. It should be noted that snR10 not only directs formation of Ψ_{2919} within the A loop close to tRNA binding, but it is also involved in the processing of pre-rRNA. Surprisingly, expression of a mutant, in which a point mutation in snR10 disrupts its ability to guide modification while maintaining its ability to process pre-rRNA, displayed a 30% lower rate of translation as well as altered polysome profiles (40).

As none of the modifications are within 13 Å of the proposed site of peptide bond formation, it is unlikely that the chemistry of protein synthesis is affected. However, it is reasonable to expect that these modifications may modulate other critical events taking place during translation, such as tRNA binding or translocation through the ribosome. In line with this notion, G_{2918}, the nucleotide adjacent to Ψ_{2919}, is known to form a Watson–Crick basepair with C_{75} in the CCA end of tRNA when in the A site (41). Given the stabilizing effect of Ψ on neighboring nucleotides (see below), it is possible that Ψ_{2919} modifies A site function by promoting stable interactions between nearby nucleotides. Indeed, interactions between nucleotides adjacent to Ψ_{2919} have been shown to affect translation fidelity in *E. coli* (42).

The region directly above the A site is referred to as the A site finger (ASF) region and contains the richest concentration of Ψ within rRNA (8–10 Ψs depending on the organism; 10 in *S. cerevisiae*) (43). The ASF is believed to function during the translocation of tRNA through the ribosome (44). Interestingly, depletion of various combinations of Ψ from within this region in *S. cerevisiae* significantly impacts ribosome function, including decreased amounts of LSU, impaired polysome formation, and reduced translation rate (43). Thus, similar to Ψs near the PTC, it is probable that Ψs within the ASF function synergistically to stabilize interactions between neighboring nucleotides as well as to optimize rRNA structure.

The large and small subunits of the ribosome are joined by a series of bridges, which are conserved throughout evolution. In addition to functioning as "molecular glue," holding the 40S and 60S subunits together during protein synthesis, these bridges have been proposed to have additional functions including the transmission of signals from functional centers of the subunits to the modulation of ribosome–tRNA interactions during tRNA translocation (45–47). As with other regions of the ribosome which have well-characterized functions, the intersubunit bridges possess a wealth of posttranscriptionally modified nucleotides (37).

One such structure in yeast, helix 69 (H69), contains a rich cluster of pseudouridines, which are evolutionarily conserved from *E. coli* to human (three Ψ in *E. coli*, four Ψ and one Nm in yeast, and five Ψ and one Nm in human). H69 interacts with both the A and P site tRNAs, both ribosomal subunits, and H44 (in the small subunit) which is part of the decoding center (48). Two of these modifications Ψ_{2258} and Ψ_{2260} are the most conserved pseudouridines from *E. coli* to human (37, 49). These modifications are catalyzed by RluD in *E. coli* (50, 51), hU19 in humans (52), and snr191 in yeast (53). Interestingly, a single deletion of *snR191* led to a slight growth disadvantage in yeast (53). Liang et al. systemically addressed the function of modifications within H69 of *S. cerevisiae* by constructing deletion strains lacking snoRNAs that target this region for modification. As has been observed with modifications in other regions, loss of any single or two modifications resulted in no apparent growth phenotype or translational impairment under standard conditions. However, loss of two pseudouridines at positions Ψ_{2264} and Ψ_{2266} results in increased susceptibility to the aminoglycoside neomycin. Neomycin resistance was further decreased by loss of the Am_{2256} modification. These data clearly suggest that these modifications influence ribosome structure. Furthermore, cells lacking three or more modifications from this region display numerous translation-related phenotypes including reduced translation rates (ranging from 25 to 60% depending on modifications disrupted), reduced translational fidelity, and reduced amounts of rRNA due to faster turnover (54). These data clearly demonstrate that posttranscriptional modifications can affect both ribosome synthesis and function, and add to the notion that posttranscriptionally modified nucleotides function in concert to manipulate ribosome performance.

While most research regarding rRNA modification has been focused on pseudouridylation, attempts have been made to address the role of individual rRNA 2′-O-methylations. However, single deletions of box C/D snoRNAs targeting upward of 51 2′-O-methylations have been made in yeast and have not shown an effect on either ribosome function or cell growth (55). More recently, Esguerra et al. screened 20 individual box C/D snoRNA deletion strains for phenotypic consequences in response to various environmental challenges (56). Surprisingly, all 20 individual deletion strains displayed phenotypes, primarily during the phase of adaptation to the environmental stress. Although this data is indicative of a function for individual modifications, further detailed molecular and biochemical dissection is required to understand the nature of these phenotypes. Overall, deletions of individual snoRNAs targeting rRNA within a specific 3D region of the ribosome, an approach that has been quite successful in eliciting the function of pseudouridines, has yet to be attempted for 2′-O-methylation.

3.3. rRNA Modification and Translational Recoding

During translation elongation, the mRNA is translated on the ribosome following strict rules of decoding. Although this process is thought to be very accurate, there has been an accumulation of data over the past 15 years, which indicate that the frequency of unconventional decoding, or "recoding," can reach as high as 40% (57, 58). Such phenomena may include events such as readthrough of stop codons, frameshifting, and ribosomal hopping (59–61)

Recently, several assays have been developed to study the role of rRNA modifications in translational fidelity. For instance, in vivo bicistronic dual-luciferase reporters have been used to quantitatively measure changes in programmed −1 and programmed +1 ribosomal frameshifting (PRF), suppression of nonsense codons, and fidelity of aminoacylated-tRNA (aa-tRNA) selection (62, 63). Interestingly, yeast deficient in both the box C/D snoRNA snR52 and the 2′-*O*-methyltransferase Sbp1p displayed significant defects in −1 PRF, while all snoRNA deletion strains examined thus far have failed to allow the detection of any defects in +1 PRF. Remarkably, however, several snoRNA deletion strains were hyperaccurate in their ability to recognize translation termination codons. Specifically *snR37Δ*, *snR10Δ*, and a double mutant *spb1DA/snR52Δ* strains were all hyperaccurate at mediating translation termination (63).

More recently, we have characterized translation fidelity and recoding in the *cbf5D95A* strain where ribosomal RNAs are void of pseudouridine (see above) (Karijolich, Kantartzis, and Yu, unpublished data). Interestingly, the efficiency of programmed −1 and +1 frameshifting, as well as cap-independent translation is severely impaired. Rather surprisingly, yeasts lacking detectable pseudouridylation were several fold more efficient at translation termination than their wild-type counterpart. Along the same line, it has been demonstrated that cap-independent translation is severely impaired in hypomorphic dyskerin mice, which have a reduced level of rRNA pseudouridylation (64, 65).

Although a fairly large amount of functional data has been gathered regarding rRNA modifications and translation fidelity, a clear and concise picture of how these modifications function is still lacking. Nonetheless, it is clear that the majority of these modifications do not affect cell viability or ribosome performance when deleted alone, at least under "normal" laboratory conditions. Importantly, the data accumulated thus far suggests that these modifications function synergistically to modulate the structure and function of the ribosome.

4. Cytotoxicity Associated with 5FU Is a Result of Inhibition of Pseudouridylation

5FU is a chemotherapeutic agent that is prescribed for a variety of malignancies including bowel, breast, stomach, and esophagus. Although 5FU has been in use for nearly half a century, the mechanism behind its therapeutic effect has yet to be established (66–68). Initially it was hypothesized that 5FU affects DNA metabolism. 5FU, when converted into 5-fluorodeoxyuridine monophosphate (5FdUMP), inhibits the activity of thymidylate synthase, which is required for the conversion of deoxyuridine monophosphate (dUMP) to deoxythymidine monophosphate (dTMP). Lack of synthesis of dTMP, in turn, results in an inhibitory effect on the production of its downstream product dTTP and hence affecting DNA synthesis. Consequently, dUMP accumulates, resulting in elevated synthesis of its downstream product dUTP and the incorporation of dUTP into DNA, thus resulting in DNA damage. Paradoxically, however, when 5FU-exposed cells are treated with thymidine, which can be converted into dTMP through the action of thymidine kinase (a pathway that is independent of the thymidylate synthase pathway), 5FU-mediated cytotoxic and apoptotic effects remain, suggesting that DNA metabolism is not the primary target of 5FU.

Given that 5FU can be readily converted into 5-fluorouridine triphosphate (5FUTP), a ribonucleotide analog that can be incorporated into RNAs, it has been proposed that 5FU may directly affect RNA metabolism. Indeed, past research has shown that 5FU is readily incorporated into the long single pre-rRNA transcript that is processed to become the 18S, 5.8S, and 28S rRNAs. In this regard, it has been reported that 5FU-treated cells may have defects in rRNA processing (67, 69–72). However, whether this defect is a direct result from 5FU incorporation remains largely unclear.

More recently, data from other labs, as well as our own, now suggest that the cytotoxic effect of 5FU is mainly RNA based and is a result of the inhibition of pseudouridylation. For instance, 5FU-treated HeLa cells show a dramatic accumulation of pre-mRNA, which is accompanied by a significant reduction in the amount of pseudouridine present in U2 snRNA (73). Furthermore, in contrast to the fact that U2 snRNA isolated from uracil-treated HeLa cells efficiently reconstitute pre-mRNA splicing in *Xenopus* oocytes, U2 snRNA purified from 5FU-treated HeLa cells failed to reconstitute pre-mRNA splicing (73). Further demonstrating a role for the process of pseudouridylation in the cytotoxic effects of 5FU, Hoskins and Butler showed that a mutation in the catalytic subunit of Cbf5p, the pseudouridine synthase-associated box H/ACA RNPs (see above), suppresses the 5FU hypersensitivity of an yeast *rrp6-Δ* strain (74). Given that ribosomal RNAs are modified almost exclusively by box H/ACA RNPs in yeast, it would be of interest to see whether rRNA pseudouridylation is affected in the 5FU-treated *rrp6-Δ* strain. Moreover, the effects of 5FU on translation fidelity and recoding are unknown and warrant investigation.

5. Biophysical Contribution of Posttranscriptionally Modified Nucleotides

To date the data are clear, posttranscriptional modifications within the spliceosomal snRNAs and rRNA are important for pre-mRNA splicing and protein synthesis, respectively. However, the mechanism by which these modifications exert the effect are only beginning to be unraveled. It is well established that both 2′-O-methylation and pseudouridylation have distinct chemical properties from their unmodified counterparts. These distinct chemical properties have the potential to impact numerous aspects of the modified RNA, including structure, thermal stability, and biochemical interactions (75). In each case, the structural, thermodynamic, and biochemical contributions depend on the structural context and can extend beyond the site of modification. In this regard, 1H NMR, UV, and CD spectroscopy have demonstrated that short RNA fragments containing pseudouridine are significantly more stable than if the same RNA contained uridine (76). Conformation stabilization appears to be an intrinsic property of pseudouridine at the nucleotide level, and is mediated by both an increase in base stacking and the ability to coordinate a water molecule through the extra hydrogen bond present (76, 77). Similarly, data also suggest that 2′-O-methylation is capable of leading to increased stability in RNA conformations. For instance, 2′-O-methylation alters the hydration sphere around the oxygen resulting in the blockage of sugar edge interactions (78–80). In addition, the methyl group alters the ability of the ribose to engage in hydrogen bonding. Furthermore, it also plays a role in protecting the RNA from hydrolysis by alkali and nucleases. In addition, there is a direct correlation between an organism's optimal growth temperature and the number of 2′-O-methylations present in rRNA (81), as was reported for tRNA (75, 82, 83), suggesting that 2′-O-methylation may indeed result in increased conformational stability.

In recent years, significant efforts have been made to understand the molecular mechanism by which U2 snRNA pseudouridylation contributes to pre-mRNA splicing. In this regard, the crystal structure of a self-complementary RNA designed to mimic the *S. cerevisiae* U2 snRNA–branchpoint interaction was determined in the absence of pseudouridine (84). Interestingly, the adenosines 5′ of the expected adenosine were bulged out. Subsequently, the Greebaum group determined solution structures of the U2 snRNA–branchpoint interaction either in the presence or absence of pseudouridine (85, 86). NMR data coupled with 2-aminopurine fluorescence titration data indicated that the presence of the pseudouridine was required for the extrahelical positioning of the expected branchpoint adenosine. This has recently been revalidated by Lin and Kielkopf who determined the crystal structure of the U2 snRNA–branchpoint interaction in the presence of pseudouridine (87). In support of the biophysical data, Valadkhan and Manley have shown that the change of a single uridine in the branch site recognition region of U2 snRNA to pseudouridine greatly enhances the production of X-RNA, a product generated by a splicing-related branching reaction in a cell- and protein-free system (88, 89).

6. Concluding Remarks

In the past decade or so, research into the functionality of posttranscriptional modifications, in particular 2′-O-methylation and pseudouridylation, has quickened. While the data indicate that modified nucleotides are important for fully functional spliceosomes and ribosomes, exactly how the modifications exert their effects is unclear. Furthermore, many fundamental questions remain unexplored. For instance, are posttranscriptional modifications reversible? Do modifications have a role during the catalytic phase of pre-mRNA splicing and/or protein synthesis? To answer this question, a detailed structural mechanistic analysis seems to be necessary. Furthermore, there are a number of modi fied residues within the spliceosomal U snRNAs and rRNAs that have never been tested for a function. With so many unanswered questions, the field of RNA modification is sure to see some exciting results in the near future.

Acknowledgments

We would like to thank the members of the Yu lab for insightful discussions. Our work was supported by grant GM62937 (to Yi-Tao Yu) from the National Institute of Health. John Karijolich was supported by a NIH Institutional Ruth L. Kirschstein National Research Service Award GM068411.

References

1. Staley, J.P. and Woolford, J.L., Jr. (2009) Assembly of ribosomes and spliceosomes: complex ribonucleoprotein machines. *Curr Opin Cell Biol*, **21**, 109–118.
2. Jurica, M.S. and Moore, M.J. (2003) Pre-mRNA splicing: awash in a sea of proteins. *Mol Cell*, **12**, 5–14.
3. Staley, J.P. and Guthrie, C. (1998) Mechanical devices of the spliceosome: motors, clocks, springs, and things. *Cell*, **92**, 315–326.
4. Yu, Y.T., Scharl, E.C., Smith, C.M., and Steitz, J.A. (1999) *The RNA World*, 2nd edn. Cold Spring Harbor Laboratory Press, Cold Spring Harbor, NY, pp. 487–524.
5. Valadkhan, S. (2005) snRNAs as the catalysts of pre-mRNA splicing. *Curr Opin Chem Biol*, **9**, 603–608.
6. Cohn, W.E. and Volkin, E. (1951) Nucleoside-5′-phosphates from ribonucleic acid. *Nature*, **167**, 483–484.
7. Hodnett, J.L. and Busch, H. (1968) Isolation and characterization of uridylic acid-rich 7 S ribonucleic acid of rat liver nuclei. *J Biol Chem*, **243**, 6334–6342.
8. Weinberg, R.A. and Penman, S. (1968) Small molecular weight monodisperse nuclear RNA. *J Mol Biol*, **38**, 289–304.
9. Muramatsu, M. and Busch, H. (1965) Studies on the nuclear and nucleolar ribonucleic acid of regenerating rat liver. *J Biol Chem*, **240**, 3960–3966.
10. Lerner, M.R. and Steitz, J.A. (1979) Antibodies to small nuclear RNAs complexed with proteins are produced by patients with systemic lupus erythematosus. *Proc Natl Acad Sci USA*, **76**, 5495–5499.
11. Reddy, R. and Busch, H. (1988) *Structure and Function of Major and Minor Small Nuclear Ribonucleoprotein Particles.* Springer-Verlag Press, Heidelberg, pp. 1–37.
12. Patton, J.R. (1993) Ribonucleoprotein particle assembly and modification of U2 small nuclear RNA containing 5-fluorouridine. *Biochemistry*, **32**, 8939–8944.
13. Patton, J.R. (1993) Multiple pseudouridine synthase activities for small nuclear RNAs. *Biochem J*, **290 (Pt 2)**, 595–600.
14. McPheeters, D.S., Fabrizio, P., and Abelson, J. (1989) In vitro reconstitution of functional yeast U2 snRNPs. *Genes Dev*, **3**, 2124–2136.
15. Will, C.L., Rumpler, S., Klein Gunnewiek, J., van Venrooij, W.J., and Luhrmann, R. (1996) In vitro reconstitution of mammalian U1 snRNPs active in splicing: the U1-C protein enhances the formation of early (E) spliceosomal complexes. *Nucl Acids Res*, **24**, 4614–4623.
16. Wersig, C. and Bindereif, A. (1992) Reconstitution of functional mammalian U4 small nuclear ribonucleoprotein: Sm protein binding is not essential for splicing in vitro. *Mol Cell Biol*, **12**, 1460–1468.
17. Segault, V., Will, C.L., Sproat, B.S., and Luhrmann, R. (1995) In vitro reconstitution of mammalian U2 and U5 snRNPs active in splicing: Sm proteins are functionally interchangeable and are essential for the formation of functional U2 and U5 snRNPs. *EMBO J*, **14**, 4010–4021.
18. Fabrizio, P., McPheeters, D.S., and Abelson, J. (1989) In vitro assembly of yeast U6 snRNP: a functional assay. *Genes Dev*, **3**, 2137–2150.
19. Fabrizio, P. and Abelson, J. (1990) Two domains of yeast U6 small nuclear RNA required for both steps of nuclear precursor messenger RNA splicing. *Science*, **250**, 404–409.
20. Fabrizio, P. and Abelson, J. (1992) Thiophosphates in yeast U6 snRNA specifically affect pre-mRNA splicing in vitro. *Nucl Acids Res*, **20**, 3659–3664.
21. Wolff, T. and Bindereif, A. (1992) Reconstituted mammalian U4/U6 snRNP complements splicing: a mutational analysis. *EMBO J*, **11**, 345–359.
22. Wolff, T., Menssen, R., Hammel, J., and Bindereif, A. (1994) Splicing function of mammalian U6 small nuclear RNA: conserved positions in central domain and helix I are essential during the first and second step of pre-mRNA splicing. *Proc Natl Acad Sci USA*, **91**, 903–907.
23. Wolff, T. and Bindereif, A. (1995) Mutational analysis of human U6 RNA: stabilizing the intramolecular helix blocks the spliceosomal assembly pathway. *Biochim Biophys Acta*, **1263**, 39–44.
24. Wolff, T. and Bindereif, A. (1993) Conformational changes of U6 RNA during the spliceosome cycle: an intramolecular helix is essential both for initiating the U4–U6 interaction and for the first step of slicing. *Genes Dev*, **7**, 1377–1389.
25. Yu, Y.T., Maroney, P.A., and Nilsen, T.W. (1993) Functional reconstitution of U6 snRNA in nematode *cis*- and *trans*-splicing: U6 can serve as both a branch acceptor and a 5′ exon. *Cell*, **75**, 1049–1059.
26. Yu, Y.T., Maroney, P.A., Darzynkiwicz, E., and Nilsen, T.W. (1995) U6 snRNA function in nuclear pre-mRNA splicing: a

phosphorothioate interference analysis of the U6 phosphate backbone. *RNA*, **1**, 46–54.
27. Pan, Z.Q. and Prives, C. (1989) U2 snRNA sequences that bind U2-specific proteins are dispensable for the function of U2 snRNP in splicing. *Genes Dev*, **3**, 1887–1898.
28. McPheeters, D.S. and Abelson, J. (1992) Mutational analysis of the yeast U2 snRNA suggests a structural similarity to the catalytic core of group I introns. *Cell*, **71**, 819–831.
29. Yu, Y.T., Shu, M.D., and Steitz, J.A. (1998) Modifications of U2 snRNA are required for snRNP assembly and pre-mRNA splicing. *EMBO J*, **17**, 5783–5795.
30. Donmez, G., Hartmuth, K., and Luhrmann, R. (2004) Modified nucleotides at the 5′ end of human U2 snRNA are required for spliceosomal E-complex formation. *RNA*, **10**, 1925–1933.
31. Zhao, X. and Yu, Y.T. (2004) Pseudouridines in and near the branch site recognition region of U2 snRNA are required for snRNP biogenesis and pre-mRNA splicing in Xenopus oocytes. *RNA*, **10**, 681–690.
32. Massenet, S. and Branlant, C. (1999) A limited number of pseudouridine residues in the human atac spliceosomal UsnRNAs as compared to human major spliceosomal UsnRNAs. *RNA*, **5**, 1495–1503.
33. Ma, X., Yang, C., Alexandrov, A., Grayhack, E.J., Behm-Ansmant, I., and Yu, Y.T. (2005) Pseudouridylation of yeast U2 snRNA is catalyzed by either an RNA-guided or RNA-independent mechanism. *EMBO J*, **24**, 2403–2413.
34. Ma, X., Zhao, X., and Yu, Y.T. (2003) Pseudouridylation (Psi) of U2 snRNA in *S. cerevisiae* is catalyzed by an RNA-independent mechanism. *EMBO J*, **22**, 1889–1897.
35. Yang, C., McPheeters, D.S., and Yu, Y.T. (2005) Psi35 in the branch site recognition region of U2 small nuclear RNA is important for pre-mRNA splicing in *Saccharomyces cerevisiae*. *J Biol Chem*, **280**, 6655–6662.
36. Lapeyre, B. (2005) *Conserved Ribosomal RNA Modification and Their Putative Roles in Ribosome Biogenesis and Translation*. Springer, Heidelberg/Berlin.
37. Decatur, W.A. and Fournier, M.J. (2002) rRNA modifications and ribosome function. *Trends Biochem Sci*, **27**, 344–351.
38. Tollervey, D., Lehtonen, H., Jansen, R., Kern, H., and Hurt, E.C. (1993) Temperature-sensitive mutations demonstrate roles for yeast fibrillarin in pre-rRNA processing, pre-rRNA methylation, and ribosome assembly. *Cell*, **72**, 443–457.
39. Zebarjadian, Y., King, T., Fournier, M.J., Clarke, L., and Carbon, J. (1999) Point mutations in yeast CBF5 can abolish in vivo pseudouridylation of rRNA. *Mol Cell Biol*, **19**, 7461–7472.
40. King, T.H., Liu, B., McCully, R.R., and Fournier, M.J. (2003) Ribosome structure and activity are altered in cells lacking snoRNPs that form pseudouridines in the peptidyl transferase center. *Mol Cell*, **11**, 425–435.
41. Kim, D.F. and Green, R. (1999) Base-pairing between 23S rRNA and tRNA in the ribosomal A site. *Mol Cell*, **4**, 859–864.
42. O'Connor, M. and Dahlberg, A.E. (1993) Mutations at U2555, a tRNA-protected base in 23S rRNA, affect translational fidelity. *Proc Natl Acad Sci USA*, **90**, 9214–9218.
43. Piekna-Przybylska, D., Przybylski, P., Baudin-Baillieu, A., Rousset, J.P., and Fournier, M.J. (2008) Ribosome performance is enhanced by a rich cluster of pseudouridines in the A-site finger region of the large subunit. *J Biol Chem*, **283**, 26026–26036.
44. Komoda, T., Sato, N.S., Phelps, S.S., Namba, N., Joseph, S., and Suzuki, T. (2006) The A-site finger in 23 S rRNA acts as a functional attenuator for translocation. *J Biol Chem*, **281**, 32303–32309.
45. Bashan, A., Agmon, I., Zarivach, R., Schluenzen, F., Harms, J., Berisio, R., Bartels, H., Franceschi, F., Auerbach, T., Hansen, H.A. et al. (2003) Structural basis of the ribosomal machinery for peptide bond formation, translocation, and nascent chain progression. *Mol Cell*, **11**, 91–102.
46. Cochella, L. and Green, R. (2005) An active role for tRNA in decoding beyond codon:anticodon pairing. *Science*, **308**, 1178–1180.
47. Rodnina, M.V., Daviter, T., Gromadski, K., and Wintermeyer, W. (2002) Structural dynamics of ribosomal RNA during decoding on the ribosome. *Biochimie*, **84**, 745–754.
48. Yusupov, M.M., Yusupova, G.Z., Baucom, A., Lieberman, K., Earnest, T.N., Cate, J.H., and Noller, H.F. (2001) Crystal structure of the ribosome at 5.5 Å resolution. *Science*, **292**, 883–896.
49. Ofengand, J. (2002) Ribosomal RNA pseudouridines and pseudouridine synthases. *FEBS Lett*, **514**, 17–25.
50. Raychaudhuri, S., Conrad, J., Hall, B.G., and Ofengand, J. (1998) A pseudouridine synthase required for the formation of two universally conserved pseudouridines in ribosomal RNA is essential for normal growth of *Escherichia coli*. *RNA*, **4**, 1407–1417.
51. Huang, L., Ku, J., Pookanjanatavip, M., Gu, X., Wang, D., Greene, P.J., and Santi, D.V.

(1998) Identification of two *Escherichia coli* pseudouridine synthases that show multisite specificity for 23S RNA. *Biochemistry*, **37**, 15951–15957.
52. Bortolin, M.L. and Kiss, T. (1998) Human U19 intron-encoded snoRNA is processed from a long primary transcript that possesses little potential for protein coding. *RNA*, **4**, 445–454.
53. Badis, G., Fromont-Racine, M., and Jacquier, A. (2003) A snoRNA that guides the two most conserved pseudouridine modifications within rRNA confers a growth advantage in yeast. *RNA*, **9**, 771–779.
54. Liang, X.-H., Liu, Q., and Fournier, M.J. (2007) rRNA modifications in an intersubunit bridge of the ribosome strongly affect both ribosome biogenesis and activity. *Mol Cell*, **28**, 965–977.
55. Lowe, T.M. and Eddy, S.R. (1999) A computational screen for methylation guide snoRNAs in yeast. *Science*, **283**, 1168–1171.
56. Esguerra, J., Warringer, J., and Blomberg, A. (2008) Functional importance of individual rRNA 2′-O-ribose methylations revealed by high-resolution phenotyping. *RNA*, **14**, 649–656.
57. Grentzmann, G., Ingram, J.A., Kelly, P.J., Gesteland, R.F., and Atkins, J.F. (1998) A dual-luciferase reporter system for studying recoding signals. *RNA*, **4**, 479–486.
58. Gesteland, R.F., Weiss, R.B., and Atkins, J.F. (1992) Recoding: reprogrammed genetic decoding. *Science*, **257**, 1640–1641.
59. Baranov, P.V., Gurvich, O.L., Hammer, A.W., Gesteland, R.F., and Atkins, J.F. (2003) Recode 2003. *Nucl Acids Res*, **31**, 87–89.
60. Baranov, P.V., Gesteland, R.F., and Atkins, J.F. (2002) Recoding: translational bifurcations in gene expression. *Gene*, **286**, 187–201.
61. Namy, O., Rousset, J.P., Napthine, S., and Brierley, I. (2004) Reprogrammed genetic decoding in cellular gene expression. *Mol Cell*, **13**, 157–168.
62. Harger, J.W. and Dinman, J.D. (2003) An in vivo dual-luciferase assay system for studying translational recoding in the yeast *Saccharomyces cerevisiae*. *RNA*, **9**, 1019–1024.
63. Baxter-Roshek, J.L.P., Alexey N., and Dinman, J.D. (2006) Optimization of ribosome structure and function by rRNA base modification. *PLoS ONE*, **2**, 174.
64. Ruggero, D., Grisendi, S., Piazza, F., Rego, E., Mari, F., Rao, P.H., Cordon-Cardo, C., and Pandolfi, P.P. (2003) Dyskeratosis congenita and cancer in mice deficient in ribosomal RNA modification. *Science*, **299**, 259–262.
65. Yoon, A., Peng, G., Brandenburger, Y., Zollo, O., Xu, W., Rego, E., and Ruggero, D. (2006) Impaired control of IRES-mediated translation in X-linked dyskeratosis congenita. *Science*, **312**, 902–906.
66. Longley, D.B., Harkin, D.P., and Johnston, P.G. (2003) 5-Fluorouracil: mechanisms of action and clinical strategies. *Nat Rev Cancer*, **3**, 330–338.
67. Ghoshal, K. and Jacob, S.T. (1997) An alternative molecular mechanism of action of 5-fluorouracil, a potent anticancer drug. *Biochem Pharmacol*, **53**, 1569–1575.
68. Parker, W.B. and Cheng, Y.C. (1990) Metabolism and mechanism of action of 5-fluorouracil. *Pharmacol Ther*, **48**, 381–395.
69. Greenhalgh, D.A. and Parish, J.H. (1989) Effects of 5-fluorouracil on cytotoxicity and RNA metabolism in human colonic carcinoma cells. *Cancer Chemother Pharmacol*, **25**, 37–44.
70. Herrick, D. and Kufe, D.W. (1984) Lethality associated with incorporation of 5-fluorouracil into preribosomal RNA. *Mol Pharmacol*, **26**, 135–140.
71. Cory, J.G., Breland, J.C., and Carter, G.L. (1979) Effect of 5-fluorouracil on RNA metabolism in Novikoff hepatoma cells. *Cancer Res*, **39**, 4905–4913.
72. Greenhalgh, D.A. and Parish, J.H. (1990) Effect of 5-fluorouracil combination therapy on RNA processing in human colonic carcinoma cells. *Br J Cancer*, **61**, 415–419.
73. Zhao, X. and Yu, Y.T. (2007) Incorporation of 5-fluorouracil into U2 snRNA blocks pseudouridylation and pre-mRNA splicing in vivo. *Nucl Acids Res*, **35**, 550–558.
74. Hoskins, J. and Butler, J.S. (2008) RNA-based 5-fluorouracil toxicity requires the pseudouridylation activity of Cbf5p. *Genetics*, **179**, 323–330.
75. Agris, P.F. (1996) The importance of being modified: roles of modified nucleosides and Mg^{2+} in RNA structure and function. *Prog Nucl Acid Res Mol Biol*, **53**, 79–129.
76. Davis, D.R. (1995) Stabilization of RNA stacking by pseudouridine. *Nucl Acids Res*, **23**, 5020–5026.
77. Arnez, J.G. and Steitz, T.A. (1994) Crystal structure of unmodified tRNA(Gln) complexed with glutaminyl-tRNA synthetase and ATP suggests a possible role for pseudouridines in stabilization of RNA structure. *Biochemistry*, **33**, 7560–7567.
78. Auffinger, P. and Westhof, E. (1997) Rules governing the orientation of the

2′-hydroxyl group in RNA. *J Mol Biol*, **274**, 54–63.
79. Auffinger, P. and Westhof, E. (1998) Hydration of RNA base pairs. *J Biomol Struct Dyn*, **16**, 693–707.
80. Helm, M. (2006) Post-transcriptional nucleotide modification and alternative folding of RNA. *Nucl Acids Res*, **34**, 721–733.
81. Noon, K.R., Bruenger, E., and McCloskey, J.A. (1998) Posttranscriptional modifications in 16S and 23S rRNAs of the archaeal hyperthermophile *Sulfolobus solfataricus*. *J Bacteriol*, **180**, 2883–2888.
82. Agris, P.F., Koh, H., and Soll, D. (1973) The effect of growth temperatures on the in vivo ribose methylation of *Bacillus stearothermophilus* transfer RNA. *Arch Biochem Biophys*, **154**, 277–282.
83. Kowalak, J.A., Dalluge, J.J., McCloskey, J.A., and Stetter, K.O. (1994) The role of posttranscriptional modification in stabilization of transfer RNA from hyperthermophiles. *Biochemistry*, **33**, 7869–7876.
84. Berglund, J.A., Rosbash, M., and Schultz, S.C. (2001) Crystal structure of a model branchpoint-U2 snRNA duplex containing bulged adenosines. *RNA*, 7, 682–691.
85. Newby, M.I. and Greenbaum, N.L. (2001) A conserved pseudouridine modification in eukaryotic U2 snRNA induces a change in branch-site architecture. *RNA*, 7, 833–845.
86. Newby, M.I. and Greenbaum, N.L. (2002) Sculpting of the spliceosomal branch site recognition motif by a conserved pseudouridine. *Nat Struct Biol*, **9**, 958–965.
87. Lin, Y. and Kielkopf, C.L. (2008) X-ray structures of U2 snRNA-branchpoint duplexes containing conserved pseudouridines. *Biochemistry*, **47**, 5503–5514.
88. Valadkhan, S. and Manley, J.L. (2001) Splicing-related catalysis by protein-free snRNAs. *Nature*, **413**, 701–707.
89. Valadkhan, S. and Manley, J.L. (2003) Characterization of the catalytic activity of U2 and U6 snRNAs. *RNA*, **9**, 892–904.

Chapter 2

Quantitative Analysis of RNA Modifications

John Karijolich, Athena Kantartzis, and Yi-Tao Yu

Abstract

RNA modifications impact numerous cellular processes such as pre-mRNA splicing and protein synthesis. The elucidation of the mechanisms by which these modifications impact cellular processes necessitates the ability to both detect and quantify the presence of these modifications within RNA molecules. Here, we present a detailed procedure that allows the detection and quantification of RNA base modifications. This procedure involves a number of techniques, including oligonucleotide-affinity selection, site-specific cleavage and radiolabeling, nuclease digestion, and thin layer chromatography.

Key words: RNA modifications, pseudouridine, 2′-*O*-methylation, RNase H, U2 snRNA, site-specific radiolabeling.

1. Introduction

Posttranscriptionally modified ribonucleotides were first identified in the hydrolysates of RNA more than a half a century ago (1, 2). It is now accepted that virtually all species of RNA contain posttranscriptional modifications. However, despite the fact that over 50 years have passed since the first identification of noncanonical ribonucleotides, the function of many remains undefined (*see* **Chapter 1**).

A prerequisite to elucidating the functions of posttranscriptional modifications is the knowledge of their existence in exact location (detection) as well as the amount present (quantitation). Prior to the early 1990s, detection of RNA base modifications was both time consuming and laborious, requiring a combination of techniques including in vivo radiolabeling, nuclease digestion, and chromatography (or fingerprinting) (3–7). The advent of

M. Sioud (ed.), *RNA Therapeutics*, Methods in Molecular Biology 629,
DOI 10.1007/978-1-60761-657-3_2,

primer extension/reverse transcription-based approaches greatly facilitated research regarding posttranscriptional modifications. These methods are primarily based on the fact that certain modified nucleotides will result in a premature stop during a primer-extension reaction. For instance, the chemical derivatization of pseudouridine with *N*-cyclohexyl-*N′*-(2-morpholinoethyl)-carbodiimid-methop-toluolsulfonate (CMC) blocks reverse transcription one nucleotide prior to the CMC-modified pseudouridine (8, 9). Similarly, the presence of a 2′-*O*-methylated nucleotide results in a premature stop when primer extension is carried out at low dNTP concentrations (10).

Though these experimental techniques ease the burden of detecting specific posttranscriptional modifications, there are several caveats to these approaches. For instance, with the exception of 2′-*O*-methylation (the sugar ring modification), most base modifications require chemical derivatization to induce premature stops during primer extension. As a variety of base modifications exist within RNA, identification of the specific chemical modifier and reaction conditions for all known modifications is a formidable task (11). Furthermore, while these primer extension-based methods are suitable for detection, they are not quantitative. Other disadvantages of these approaches include the dependency on visual observation of premature stops by gel electrophoresis, which is not very sensitive and may lead to errors in the identification of posttranscriptional modifications.

More advanced techniques based on mass spectroscopy have also been developed (12). However, these techniques are not practical to utilize in the common laboratory as they require expensive and specialized equipments (i.e., mass spectrometers and high-pressure liquid chromatography systems). In addition, they are not amenable for all modifications. Recently, a promising ligation-based approach has been described which takes advantage of T4 DNA ligase's ability to discriminate modified nucleotides (13, 14). While this technique has the potential to be utilized in high-throughput screening for modified ribonucleotides, it is currently not optimized for all modifications.

Here we describe an approach that when coupled with the primer extension-based approaches provides an extremely effective way of detecting and quantifying modified nucleotides in a variety of RNAs. Based on the fact that RNase H cleavage occurs only at sites where the 2′-OH of RNA is not modified, cleavage can be directed to a specific nucleotide of interest through the use of 2′-*O*-methyl RNA–DNA chimeric oligonucleotides (15–18) or even DNAzymes (11, 19). Following radiolabeling of the cleaved RNA, the RNA can be digested to single nucleotides by ribonucleases. The digested nucleotides can then be separated by thin layer chromatography (TLC), and the nucleotide of interest can be visualized by autoradiography (**Fig. 2.1**). We have

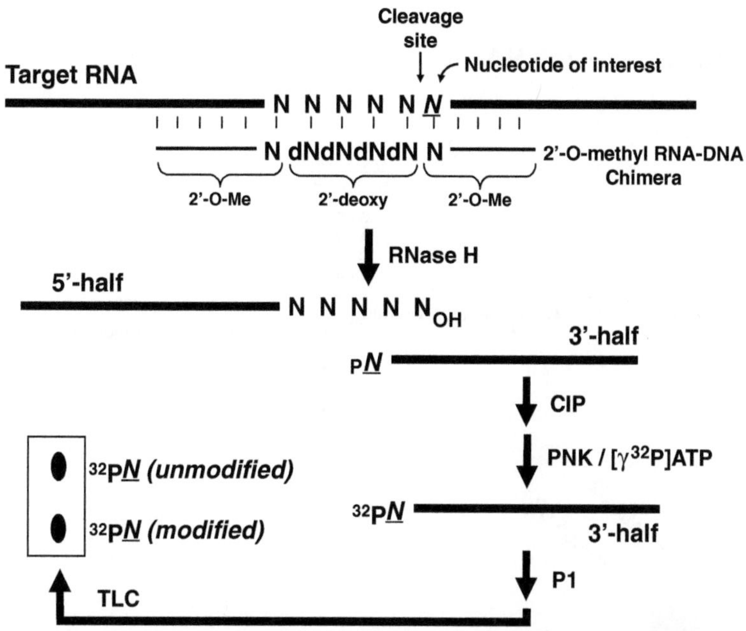

Fig. 2.1. The method for detecting and quantifying base modifications is schematized. The *thick lines* and *thin lines* represent target RNA and 2′-*O*-methyl RNA–DNA chimera, respectively. See text for detailed description.

successfully implemented this approach in the detection and quantitation of numerous base modifications including pseudouridylation and 5-fluorouridylation (11, 18, 20). As an example of application, below we apply the method to quantify the pseudouridylation at position 34 within mouse brain U2 snRNA. Our analysis indicates that the uridine at this position is nearly 100% converted to pseudouridine (**Fig. 2.2**).

2. Materials

2.1. Purification of U2 snRNA from Mouse Brain

1. Trizol reagent.
2. Dounce tissue grinder.
3. Biotinylated antisense U2 2′-*O*-methyl oligonucleotide complimentary to nucleotides 158–177 of mouse U2 snRNA [UmCmCmUmGmGmAmGmGmUmAmCmUmGmCm AmAmUmAmCmBBB, where B stands for biotin–TEG (triethylene glycol)].
4. NET-2-$MgCl_2$ buffer: 50 mM Tris–HCl (pH 7.5), 150 mM NaCl, 0.05% (v/v) NP-40, and 2 mM $MgCl_2$.

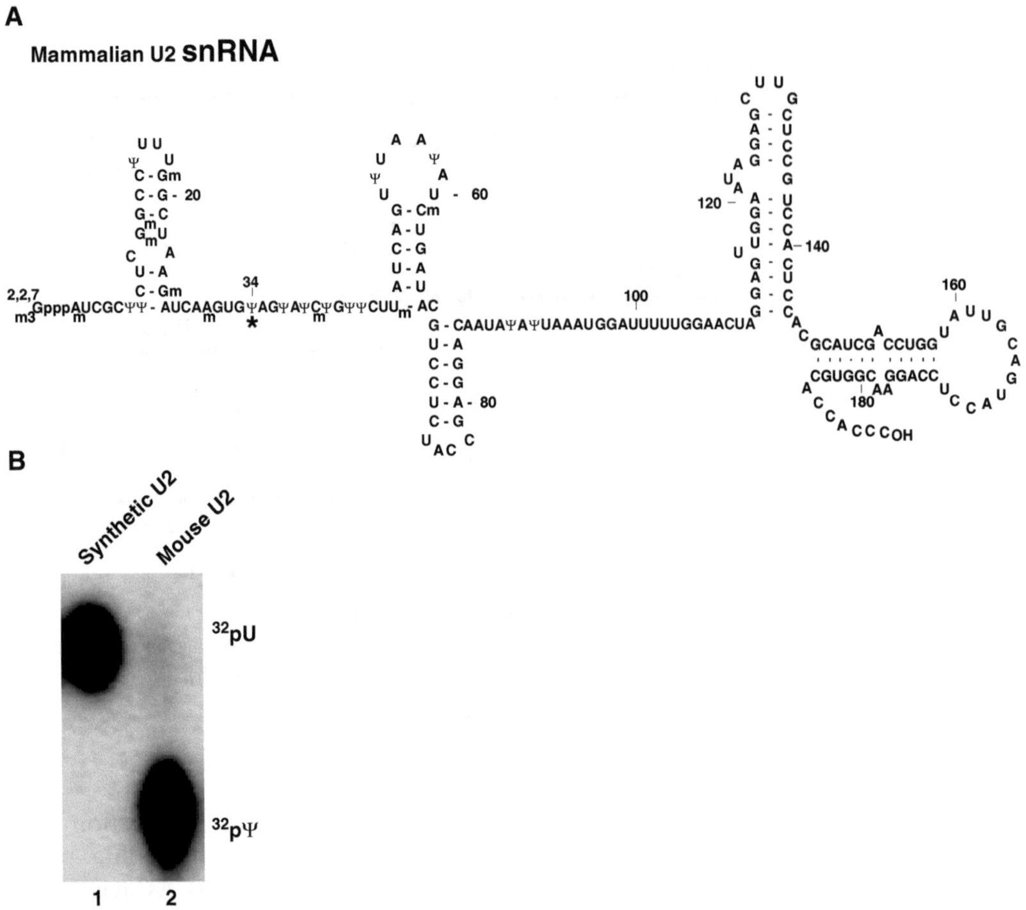

Fig. 2.2. (**a**) The primary sequence and the secondary structure of mammalian U2 snRNA are shown. The modified nucleotides, including 2′-*O*-methylated residuals (Nm) and pseudouridines (Ψ), are indicated. The *asterisk* indicates the pseudouridine at position 34, which is quantified in (**b**). (**b**) Using RNase H cleavage directed by a 2′-*O*-methyl RNA–DNA chimera targeting the phosphodiester bond between positions 33 and 34, both synthetic mammalian U2 and mouse brain U2 were cleaved. The 3′ RNA fragments were gel purified, and the 5′ phosphate of both fragments was further replaced with ^{32}P through dephosphorylation and consequent rephosphorylation. Both fragments were then treated with nuclease P1 to completion, and the resulting mononucleotides were subjected to TLC analysis. Lane 1, in vitro synthesized mammalian U2; lane 2, mouse brain U2. The spots corresponding to uridylate and pseudouridylate are indicated.

5. Preblocking mix: 100 μg/mL glycogen and 100 μg/mL tRNA in 50 mM WB50.
6. WB50: 20 mM Tris–HCl (pH 7.6), 0.01% NP-40, 50 mM NaCl, 1.5% NaN_3.
7. WB250: 20 mM Tris–HCl (pH 7.6), 0.05% NP-40, 250 mM NaCl, 0.1% NaN_3.
8. Streptavidin agarose beads.
9. Dissociation buffer: 10 mM Tris–HCl (pH 7.5), 0.1% sodium dodecyl sulfate (SDS), and 0.5 mM ethylenediaminetetraacetic acid (EDTA).

10. Formamide loading buffer: 95% formamide, 10 mM EDTA, 0.1% xylene cyanol FF, 0.1% bromophenol blue.
11. PCA: Tris–HCl (pH 7.5)—buffered phenol/chloroform/isoamyl alcohol (50:49:1).
12. G50 buffer: 20 mM Tris–HCl (pH 7.5), 300 mM sodium acetate, 2 mM EDTA, 0.25% SDS.
13. Glycogen: 10 mg/mL.
14. Polyacrylamide gel solution (40%).
15. Urea.
16. TBE (10×; Omnipur, EMD chemicals).
17. Ammonium persulfate.
18. TEMED.
19. Chloroform.
20. Isopropanol.
21. Ethanol.
22. Autoclaved distilled water.
23. Electrophoresis apparatus.

2.2. RNase H Site-Specific Cleavage Directed by 2′-O-Methyl RNA–DNA Chimera

1. 2′-*O*-Methyl RNA–DNA chimera: For the purposes of this protocol the sequence of the chimera used is specific for cleavage between positions 33 and 34 of mouse U2 snRNA: UmAmdCdAdCdTUmGmAm UmCmUmUmAm GmCmCm.
2. RNase H buffer (2×): 40 mM Tris–HCl (pH 7.5), 20 mM $MgCl_2$, 200 mM KCl, 50 mM DTT, 10% sucrose.
3. RNasin (20 units/μL).
4. RNase H.
5. Formamide loading buffer: 95% formamide, 10 mM EDTA, 0.1% xylene cyanol FF, 0.1% bromophenol blue.
6. PCA: Tris–HCl (pH 7.5)—buffered phenol/chloroform/isoamyl alcohol (50:49:1).
7. G50 buffer: 20 mM Tris–HCl (pH 7.5), 300 mM sodium acetate, 2 mM EDTA, 0.25% SDS.
8. Glycogen: 10 mg/mL.
9. Polyacrylamide gel solution (40%).
10. Urea.
11. TBE (10×).
12. Ammonium persulfate.
13. TEMED.
14. Chloroform.

15. Isopropanol.
16. Ethanol.
17. Autoclaved distilled water.
18. Electrophoresis apparatus.

2.3. Radiolabeling of the Cleaved U2 snRNA (3′ Half)

1. Dephosphorylation buffer (10×).
2. Calf intestinal alkaline phosphatase (CIAP).
3. Sodium acetate solution: 3 M NaAc, pH 5.2.
4. G50 buffer: 20 mM Tris–HCl (pH 7.5), 300 mM sodium acetate, 2 mM EDTA, 0.25% SDS.
5. PCA: Tris–HCl (pH 7.5)—buffered phenol/ chloroform/isoamyl alcohol (50:49:1).
6. Ethanol.
7. Glycogen: 10 mg/mL.
8. Polynucleotide kinase (PNK).
9. PNK buffer (10×).
10. [γ-^{32}P]ATP (6,000 Ci/mmol; DuPont NEN).
11. Polyacrylamide gel solution (40%).
12. Formamide loading buffer: 95% formamide, 10 mM EDTA, 0.1% xylene cyanol FF, 0.1% bromophenol blue.
13. Urea.
14. TBE (10×).
15. Ammonium persulfate (APS).
16. TEMED.
17. Electrophoresis apparatus.

2.4. Detection and Quantification of U2 snRNA Pseudouridylation by TLC

1. Nuclease P1.
2. TLC PEI membrane.
3. TLC buffer: 70% isopropanol, 15% HCl, 15% dH_2O.
4. Sodium acetate.
5. PhosphorImager.

3. Methods

3.1. Purification of U2 snRNA from Mouse Brain

1. In a 15-mL tube, mince 200 mg of mouse brain into small pieces using scissors and resuspend the tissue with 4 mL Trizol reagent.
2. Using a pre-chilled Dounce tissue grinder, homogenize the tissue by passing them through the grinder for about

30 times (*see* **Note 1**). The entire homogenization procedure should be carried out on ice.

3. Extract the homogenized sample by adding 800 μL of chloroform to the sample and mix by vortexing. Centrifuge at 13,000×*g* for 5 min at 4°C.
4. Collect the aqueous phase in a new 15-mL tube and add 2 mL of isopropanol. Store the tube at –80°C.
5. In a 1.5-mL microfuge tube, add 200 μL of streptavidin agarose beads (100 μL total bed volume) and microfuge the tube for 10 s at 5,000×*g*. Add 250 μL of preblocking mix and rotate the tube on a rotator for 20 min at 4°C.
6. Wash the beads by adding 1 mL of cold WB50 and microfuge for 1 min at 5,000×*g*. Remove the supernatant and resuspend the beads with 1 mL of WB50. Repeat this step two more times without resuspending the beads in WB50 the last time.
7. Resuspend the beads in 1 mL WB250 and microfuge for 1 min at 5,000×*g*. Repeat this step one more time.
8. Resuspend the beads in 400 μL of WB250 and aliquot 50 μL (~10 μL total bed volume) in a 1.5-mL microfuge tube.
9. Obtain the total mouse RNA tube from step 4 and microfuge at 13,000×*g* for 15 min at 4°C and discard the supernatant.
10. Resuspend the pellet with 50 μL of NET-2-$MgCl_2$ buffer and transfer to a 1.5-mL microfuge tube.
11. Add 200 pmol of the biotinylated antisense U2 oligonucleotide to the tube and vortex briefly. In parallel, set up an identical binding reaction with an irrelevant biotinylated oligonucleotide as a control.
12. Heat the mixture for 2 min at 95°C and then incubate for 10 min at 65°C and 30 min at 30°C.
13. Obtain the resuspended beads from step 8 and microfuge the tube for 10 s at 5,000×*g*. Remove the supernatant (bead volume is ~10 μL).
14. Resuspend the beads with 50 μL of the RNA/oligo mix and bring up the volume to 300 μL with NET-2-$MgCl_2$ buffer. Gently nutate the tube at 4°C for 1.5 h.
15. Microfuge the tube briefly at 4°C for 30 s at 5,000×*g* and discard the supernatant.
16. Wash the beads five times at 4°C, each time with 1 mL of WB250 (microfuge for 1 min at 5,000×*g* each time).
17. Add 250 μL of dissociation buffer to the pelleted beads and incubate at 85°C for 15 min.

18. Microfuge the tube for 30 s at 5,000×*g* and collect the supernatant in a new 1.5-mL microfuge tube.
19. Add 250 μL of G50 buffer to the supernatant (up to 500 μL) and then add 500 μL of PCA. Vortex vigorously for 30 s and microfuge for 5 min at 13,000×*g*. Collect the aqueous phase in a new 1.5-mL microfuge tube.
20. Add 1 μL of glycogen and 1 mL of 100% ethanol to the aqueous phase. Vortex briefly and then place the tube on dry ice for 10 min. Microfuge at 13,000×*g* for 15 min to precipitate the RNA. Remove and discard the supernatant promptly.
21. Resuspend the RNA pellet in 2 μL of autoclaved distilled water, mix well with 4 μL of formamide loading buffer, heat at 95°C for 3 min, chill on ice immediately.
22. Meanwhile, make an 8% polyacrylamide–7 M urea gel by mixing 23.75 g of urea in 5 mL of 10× TBE, 10 mL polyacrylamide gel solution. Bring the volume to 50 mL with autoclaved distilled water and stir on a hot plate until the urea has completely dissolved. Add 0.5 mL of 10% APS and 30 μL TEMED, mix briefly, and pour into pre-taped gel plates (~25 cm wide and ~40 cm height).
23. Load the RNA sample (from step 21) on the 8% polyacrylamide–7 M urea gel. Electrophorese for ~1 h at 30 W.
24. Place the gel on an intensifying screen and visualize the RNA under 254 nm UV light and excise the RNA with a razor. Place the gel slice in a 1.5-mL microfuge tube.
25. Add 450 μL of G50 buffer to the tube and place on dry ice for 5 min. Transfer to room temperature for elution overnight (16 h).
26. Microfuge the tube containing the gel slice at 13,000×*g* for 5 min. Transfer the supernatant to a new 1.5-mL microfuge tube. Extract the supernatant with 500 μL of PCA and precipitate the gel-purified RNA with 1 mL of 100% ethanol (using 1 μL of glycogen as carrier) as previously described.
27. Resuspend the pellet with 10 μL of autoclaved distilled water and quantify the concentration using UV/VIS spectroscopy.

3.2. RNase H Site-Specific Cleavage Directed by 2′-O-Methyl RNA–DNA Chimeras

1. In a 1.5-mL microfuge tube, mix 1 μL (~1.2 pmol) of U2 snRNA with 5 pmol of the 2-*O*-methyl RNA–DNA chimera in 4 μL of water.
2. Heat the mixture at 95°C for 3 min and slowly cool down to room temperature. Microfuge the tube briefly.

3. Meanwhile, place 1 μL (20 units) of RNasin, 1 μL (2 units) of RNase H, and 7 μL of 2× RNase H buffer in a new 1.5-mL microfuge tube (*see* **Notes 2** and **3**). Keep the tube on ice.
4. Transfer the RNase H mixture to the tube containing the hybridized U2 snRNA. Mix gently by pipetting.
5. Incubate the resulting mixture at 37°C for 1 h.
6. Recover the two cleaved RNA fragments (5′ and 3′ halves) via PCA extraction and ethanol precipitation as in steps 19 and 20.
7. Resuspend the recovered RNA pellet in 2 μL of autoclaved distilled water, mix well with 4 μL of formamide loading buffer, heat at 95°C for 3 min, chill on ice immediately.
8. Meanwhile, prepare an 8% polyacrylamide–7 M urea gel as in step 22.
9. Load the sample on the 8% polyacrylamide–7 M urea gel. Electrophorese for ~1 h at 30 W.
10. Place the gel on an intensifying screen and visualize the two 5′- and 3′-RNA halves under 254 nm UV light (*see* **Note 4**) and excise the band corresponding to the 3′ half with a razor. Place the gel slice in a 1.5-mL microfuge tubes.
11. Add 450 μL of G50 buffer to the tube and place on dry ice for 5 min. Transfer to room temperature for elution overnight (16 h).
12. Microfuge the tube containing the gel slice at 13,000×*g* for 5 min. Transfer the supernatant to a new 1.5-mL microfuge tube. Extract the supernatant with 500 μL of PCA and precipitate with 1 mL of 100% ethanol (using 1 μL of glycogen as carrier) as previously described.
13. Resuspend the pellet with 10 μL of autoclaved distilled water and quantify the concentration using UV/VIS spectroscopy.

3.3. Dephosphorylation and Rephosphorylation of the 3′ Half of U2 snRNA

1. Dephosphorylate 0.15 pmol of the 3′ half of U2 snRNA in a 10 μL reaction containing 1× dephosphorylation buffer, and 1 unit of CIAP for 45 min at 50°C (*see* **Note 5**).
2. Following the reaction, add 250 μL autoclaved dH_2O and extract the sample with 300 μL of PCA. Add 25 μL of sodium acetate solution and precipitate with 1 mL of 100% ethanol (using 1 μL of glycogen) as previously described.
3. Rephosphorylate the recovered RNA at its 5′ terminus for 30 min at 37°C in a 10 μL reaction containing 1× phosphorylation buffer, 0.15 pmol of 3′ half U2 snRNA, 150 μCi of [γ-^{32}P]ATP, and 10 units of T4 PNK.

4. Add 250 μL of G50 buffer, PCA extract once, and then ethanol precipitate the RNA as previously described.
5. Resuspend the 5′-radiolabeled 3′ half of U2 snRNA in 2 μL autoclaved distilled water. Add 4 μL of formamide loading buffer, heat at 95°C for 3 min, chill on ice immediately.
6. Meanwhile, prepare an 8% polyacrylamide–7 M urea gel as described in **Section 3.1**, step 22.
7. Load the sample on a 8% polyacrylamide–7 M urea gel. Electrophorese for ~1 h at 30 W.
8. Locate the band by autoradiography and excise the band. Place gel slice in a 1.5-mL microfuge tube.
9. Add 450 μL of G50 buffer to the tube and place on dry ice for 5 min. Transfer to room temperature for elution overnight (16 h).
10. Microfuge the tube containing the gel slice at 13,000×*g* for 5 min. Transfer the supernatant to a new 1.5-mL microfuge tube. Extract the supernatant with 500 μL of PCA and precipitate with 1 mL of 100% ethanol (using 1 μL of glycogen) as previously described. (*Do not resuspend pellet here.*)

3.4. Detection and Quantification of U2 snRNA Pseudouridylation

1. Resuspend the 5′-radiolabeled 3′ half of U2 snRNA with nuclease P1 (200 μg/mL) in 3 μL of 20 mM sodium acetate (pH 5.2) for 1 h at 37°C.
2. Dot 1,000 cpm of the digested nucleotide mixture on cellulose TLC PEI membrane approximately 1 cm from one of the edges.
3. Place the edge of the TLC PEI membrane closest to the mixture in TLC buffer and wait until the front of the buffer is three-fourth the length of the membrane.
4. Visualize labeled uridylate and pseudouridylate by autoradiography (*see* **Fig. 2.1**) and the ratio of uridylate to pseudouridylate can be determined using a PhosphorImager (*see* **Note 6**).

The used chimeric oligonucleotide should be specific for cleavage at the desired positions. In our case, the chimera is designed to guide the cleavage between positions 33 and 34 (pseudouridylation at position 34 is targeted). RNA-cleaving ribozymes or RNA-cleaving DNAzymes can also be used to direct site-specific cleavage (11, 19, 21). The use of a DNAzyme would conveniently leave a 5′-OH on the downstream target fragment, thus making dephosphorylation unnecessary (11, 19).

4. Notes

1. Number of passes could vary in order to reach complete homogenization depending on the tissue and the cell type.
2. Units of RNase H added may vary according to supplier.
3. RNase H from different suppliers may have different cleavage specificities (22). RNase H purchased through Amersham cleaves RNA specifically at the site 3′ to the nucleotide that base pairs with the 5′-most deoxynucleotide of the chimera (22).
4. Based on the fact that RNase H cleaves RNA only at sites where the 2′ position of the sugar is not modified (2′-OH), cleavage at 2′-*O*-methylated residues is completely blocked. Thus, the degree of resistance to RNase H cleavage quantitatively reflects the level of 2′-*O*-methylation (18). Accordingly, this method can also be used to quantify 2′-*O*-methylation. To this end, end labeling (for example, 3′-end labeling with ^{32}pCp and RNA ligase) of RNA is desirable, because this will allow an accurate measurement/quantification of the level of cleavage.
5. Shrimp alkaline phosphatase (SAP), which is sensitive to heat (65°C), may be a better choice for the dephosphorylation reaction. Heat inactivation of SAP after the dephosphorylation reaction could allow the omission of PCA extraction and ethanol precipitation, which would otherwise be needed to remove active phosphatase (see CIAP-catalyzed dephosphorylation reaction above).
6. More than 70 modifications have Rf values mapped on cellulose plates using three solvent systems (19, 23–27). Thus, an extremely comprehensive reference guide is available for detecting and quantifying a variety of modifications.

Acknowledgments

We would like to thank the members of the Yu laboratory for discussion and inspiration. Our work was supported by grant GM62937 (to Yi-Tao Yu) from the National Institute of Health. John Karijolich was supported by a NIH Institutional Ruth L. Kirschstein National Research Service Award GM068411.

References

1. Cohn, W.E. and Volkin, E. (1951) Nucleoside-5′-phosphates from ribonucleic acid. *Nature*, **167**, 483–484.
2. Grosjean, H. (2005) *Modification and Editing of RNA: Historical Overview and Important Facts to Remember*. Springer-Verlag, Berlin Heidelberg.
3. Gupta, R.C. and Randerath, K. (1979) Rapid print–readout technique for sequencing of RNA's containing modified nucleotides. *Nucl Acids Res*, **6**, 3443–3458.
4. Maden, B.E. (1986) Identification of the locations of the methyl groups in 18 S ribosomal RNA from *Xenopus laevis* and man. *J Mol Biol*, **189**, 681–699.
5. Maden, B.E. (1988) Locations of methyl groups in 28 S rRNA of Xenopus laevis and man. Clustering in the conserved core of molecule. *J Mol Biol*, **201**, 289–314.
6. Kuchino, Y., Mizushima, H., and Nishimura, S. (1990) *Nucleotide Sequence Analysis and Identification of Modified Nucleotides of tRNA*. CRC Press, Boca Raton.
7. Reddy, R., Henning, D., Epstein, P., and Busch, H. (1981) Primary and secondary structure of U2 snRNA. *Nucl Acids Res*, **9**, 5645–5658.
8. Bakin, A. and Ofengand, J. (1993) Four newly located pseudouridylate residues in *Escherichia coli* 23S ribosomal RNA are all at the peptidyl transferase center: analysis by the application of a new sequencing technique. *Biochemistry*, **32**, 9754–9762.
9. Bakin, A.V. and Ofengand, J. (1998) Mapping of pseudouridine residues in RNA to nucleotide resolution. *Methods Mol Biol*, **77**, 297–309.
10. Maden, B.E., Corbett, M.E., Heeney, P.A., Pugh, K., and Ajuh, P.M. (1995) Classical and novel approaches to the detection and localization of the numerous modified nucleotides in eukaryotic ribosomal RNA. *Biochimie*, **77**, 22–29.
11. Zhao, X. and Yu, Y.T. (2004) Detection and quantitation of RNA base modifications. *RNA*, **10**, 996–1002.
12. Crain, P.F. (1998) *Detection and Structure Analysis of Modified Nucleosides in RNA by Mass Spectrometry*. American Society of Microbiology Press, Washington, DC.
13. Saikia, M., Dai, Q., Decatur, W.A., Fournier, M.J., Piccirilli, J.A., and Pan, T. (2006) A systematic, ligation-based approach to study RNA modifications. *RNA*, **12**, 2025–2033.
14. Dai, Q., Fong, R., Saikia, M., Stephenson, D., Yu, Y.T., Pan, T., and Piccirilli, J.A. (2007) Identification of recognition residues for ligation-based detection and quantitation of pseudouridine and N6-methyladenosine. *Nucl Acids Res*, **35**, 6322–6329.
15. Yu, Y.T. (1999) Construction of 4-thiouridine site-specifically substituted RNAs for cross-linking studies. *Methods*, **18**, 13–21.
16. Yu, Y.T. (2000) Site-specific 4-thiouridine incorporation into RNA molecules. *Methods Enzymol*, **318**, 71–88.
17. Lapham, J. and Crothers, D.M. (2000) Site-specific cleavage of transcript RNA. *Methods Enzymol*, **317**, 132–139.
18. Yu, Y.T., Shu, M.D., and Steitz, J.A. (1997) A new method for detecting sites of 2′-O-methylation in RNA molecules. *RNA*, **3**, 324–331.
19. Hengesbach, M., Meusburger, M., Lyko, F., and Helm, M. (2008) Use of DNAzymes for site-specific analysis of ribonucleotide modifications. *RNA*, **14**, 180–187.
20. Yu, Y.T. and Steitz, J.A. (1997) A new strategy for introducing photoactivatable 4-thiouridine ((4S)U) into specific positions in a long RNA molecule. *RNA*, **3**, 807–810.
22. Lapham, J., Yu, Y.T., Shu, M.D., Steitz, J.A., and Crothers, D.M. (1997) The position of site-directed cleavage of RNA using RNase H and 2′-O-methyl oligonucleotides is dependent on the enzyme source. *RNA*, **3**, 950–951.
21. Santoro, S.W. and Joyce, G.F. (1997) A general purpose RNA-cleaving DNA enzyme. *Proc Natl Acad Sci USA*, **94**, 4262–4266.
23. Silberklang, M., Prochiantz, A., Haenni, A.L., and Rajbhandary, U.L. (1977) Studies on the sequence of the 3′-terminal region of turnip-yellow-mosaic-virus RNA. *Eur J Biochem*, **72**, 465–478.
24. Kuchino, Y., Hanyu, N., and Nishimura, S. (1987) Analysis of modified nucleosides and nucleotide sequence of tRNA. *Methods Enzymol*, **155**, 379–396.
25. Keith, G. (1995) Mobilities of modified ribonucleotides on two-dimensional cellulose thin-layer chromatography. *Biochimie*, **77**, 142–144.
26. Grosjean, H., Keith, G., and Droogmans, L. (2004) Detection and quantification of modified nucleotides in RNA using thin-layer chromatography. *Methods Mol Biol*, **265**, 357–391.
27. Grosjean, H., Motorin, Y., and Morin, A. (1998) *RNA-Modifying and RNA-Editing Enzymes: Methods for Their Identification*. American Society of Microbiology Press, Washington, DC.

Chapter 3

Advances in RNA Sensing by the Immune System: Separation of siRNA Unwanted Effects from RNA Interference

Mouldy Sioud

Abstract

Small interfering RNAs (siRNAs) are routinely used as a genetic tool and hold promise for a range of therapeutic applications. However, one of the hurdles of making these agents a real therapeutic modality includes the activation of innate immunity and off-target effects. Therefore, the use of siRNAs in functional genomics and therapies depends on the development of new strategies to overcome these unwanted effects. It appears that the major innate immune response to chemically synthesized siRNAs is mediated by TLR7 and/or TLR8 in immune cells. Importantly, it has also been shown that the replacement of uridines with their 2′-modified counterparts can prevent immune activation. Similarly, 2′-modifications, particularly at the seed sequence reduced the number of unwanted off-target genes without interfering with siRNA silencing potency of the anticipated target gene. This chapter describes how to separate gene silencing from immunostimulation. Also, it discusses the impact of these findings on the design of effective cancer vaccines.

Key words: RNA interference, small interfering RNAs, innate immunity, Toll-like receptors, 2′-ribose modifications, off-target effects, cancer vaccines.

1. Introduction

RNA interference (RNAi) is a recently described gene silencing pathway by which long double-stranded RNA can silence genes by degrading mRNA in sequence-specific manner. During this process, the long dsRNA is processed into 21–23 nucleotide-active intermediates, known as small interfering RNAs (siRNAs). Although gene silencing was first described by Fire and colleagues

M. Sioud (ed.), *RNA Therapeutics*, Methods in Molecular Biology 629,
DOI 10.1007/978-1-60761-657-3_3, © Springer Science+Business Media, LLC 2010

in *Caenorhabditis elegans* (1), a similar phenomenon termed post-transcriptional gene silencing (PTGS) had been described years ago in plants, and now it is believed to function as a surveillance system for blocking the function of harmful RNAs such as viral RNAs (2).

RNAi was first used extensively in *C. elegans*, *Drosophila melanogaster*, and plants as a gene knockdown technology. However, its use as a reverse genetic tool in mammalian cells remained because vertebrates react to long dsRNA by activating the interferon pathway (3–6). Long double-stranded RNAs are synthesized during the replication of many viruses and it is a potent activator of innate immunity. However, Elbashir and colleagues demonstrated that, in contrast to long dsRNA, the active intermediates of RNAi siRNAs can be used to inhibit gene expression in mammalian cells without triggering the IFN response (7). Thereafter, the use of chemically synthesized siRNAs to study gene functions in mammalian systems became a standard laboratory technique (8, 9). Also, several studies have demonstrated the efficacy of siRNAs in animal models of human diseases upon local or systemic administration (8, 9). Theoretically, the mRNA encoding any protein that is associated with a disease can be cleaved selectively by siRNAs.

Naturally occurring siRNAs are derived from the cleavage of long dsRNA by the endonuclease Dicer. Similar to all RNase III enzymes, Dicer leaves two nucleotide (nt) 3′ overhangs and 5′ phosphate groups. These siRNA duplexes are then incorporated into a multiprotein complex, the RNA-induced silencing complex (RISC). In contrast to long dsRNAs, synthetic siRNAs enter directly into the RNAi pathway (**Fig. 3.1**). Subsequent to unwinding of the duplex, the antisense strand (guide strand) guides the RISC to recognize and cleave target mRNA sequences (7). The catalytic activity of RISC is mediated by Argonaute 2 (AGO2) protein, (10, 11), a highly basic protein that contains two common domains, PAZ and PIWI domains. The PIWI domain is essential for interaction with Dicer and contains the nuclease activity that cleaves target mRNAs. AGO2 is also responsible for cleavage of the siRNA passenger strand, thus facilitating the formation of functional RISC complexes (12, 13, *see* **Fig. 3.1**). Analysis of the crystal structures of a siRNA guide strand associated with AGO2 PIWI domain identified a seed sequence (nucleotides 2–8) that directs target mRNA recognition by RISC (14).

Although the discovery of RNAi has provided a powerful tool to investigate gene function and drug target validation (*see* **Chapters 9, 10, 11, 12, 13,** and **14**), several recent studies underscore the off-target and activation of innate immunity by siRNAs (15–18). Therefore, the development of strategies that block these unwanted effects is important for siRNA therapeutic

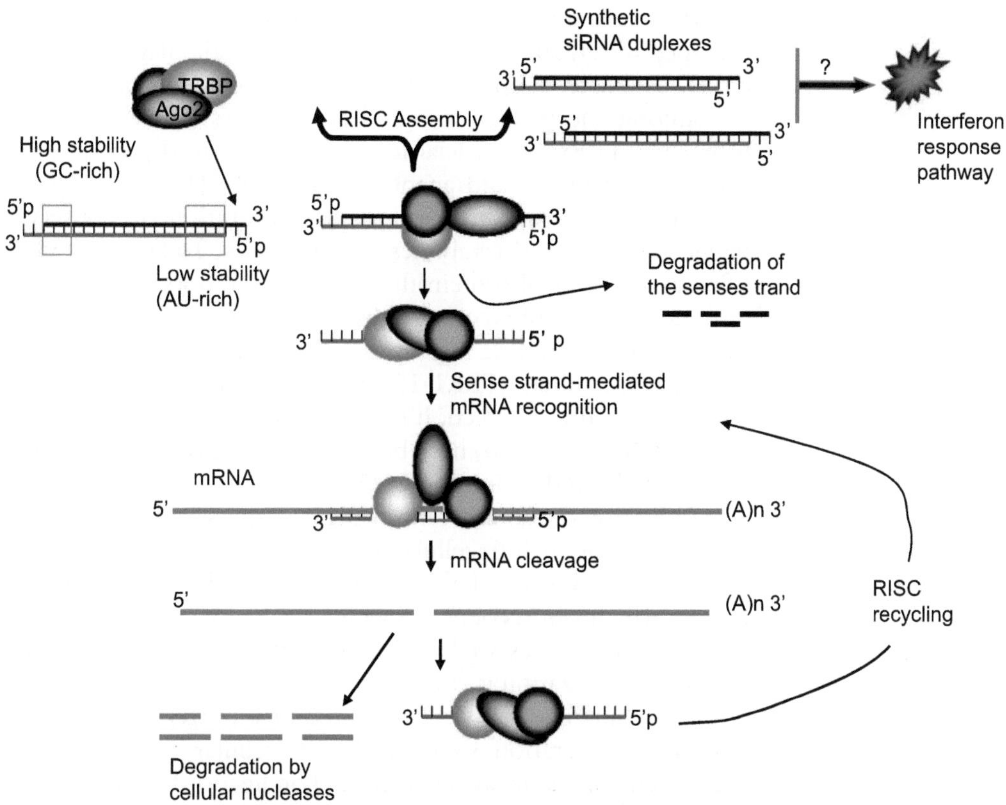

Fig. 3.1. Schematic representation of gene silencing by siRNAs. In contrast to long double-stranded RNAs, siRNAs are directly loaded into a multiprotein complex termed RNA-induced silencing complex (RISC) where the sense strand with high 5′-stability is cleaved by the nuclease AGO2 resulting in strand separation. Subsequently, the RISC containing the antisense strand (guide strand) seeks out and binds to complementary mRNA sequences. Bound mRNA molecules are then cleaved by AGO2 and cleaved mRNA fragments are rapidly degraded by cellular nucleases. Following dissociation, the active RISC is able to recycle and cleave additional mRNA molecules.

applications. However, if we view immune activation as beneficial for certain diseases, immunostimulatory siRNAs may contribute to the activation of innate and adaptive immunity against cancer and viral-infected cells (*see* **Section 8**).

2. Sensing of Pathogens by TLRs

Notably, the innate immune system has developed germline pattern recognition receptors (PPR) that promote rapid responses to microbial pathogens during the invading phase. These receptors recognize conserved pathogen-associated molecular patterns (PAMPs), which are not present in the host and are usually important for pathogenicity and/or survival of the pathogens (19, 20).

PAMPs are unique to microorganisms such as lipopolysaccharide, peptidoglycan, capsular structures, bacterial flagellin, bacterial DNA, bacterial lipids, viral RNAs, and viral glycoproteins.

Among the pattern-recognition receptors (PRRs), Toll-like receptors (TLRs) are crucial for pathogen-derived products and activation of innate and adaptive immunity (21, 22). They are type I transmembrane proteins which are evolutionarily conserved between insects and vertebrates. In *Drosophila*, Toll was first identified as an essential protein that controls the dorsal–ventral patterning of the embryo and subsequently as a key protein for the antifungal immune response in the adult (23, 24). In vertebrates, 13 members (TLR1-13) have been reported so far which are essential in the detection of pathogens. In humans, 10 functional TLRs (TLR1-10) have been identified and shown to detect pathogen-derived compounds (25).

TLRs are transmembrane proteins expressed either on the cell surface or in intracellular endocytic vesicles or organelles. The ectodomains are characterized by a number of leucine-rich repeats which can recognize a wide range of viral, bacterial, and fungal structures, as well as interact with other TLRs (hetero- or homodimerization is essential for activity). All the intracellular domains contain a Toll-interleukin receptor (TIR) domain which links the recognition signal with intracellular pathways. Two major signaling pathways are activated on ligand-driven TLR activation. One pathway requires the adaptor molecule MyD88 and results in the production of type I interferon or NF-κB-dependent proinflammatory cytokines including TNF-α and interleukin 12. The second pathway requires the adaptor molecule TRIF and primarily induces the production of type I IFN (25). In immune cells, TLRs that recognize nucleic acids are expressed exclusively in endosomes. These include TLR3, TLR7/8, and TLR9, which sense double-stranded dsRNAs, single-stranded ss RNA, and ss DNA, respectively. Other TLRs, in contrast, reside on the cell membrane. These include TLR2, TLR4, TLR5, and TLR11, which recognize lipopeptides, lipopolysaccharides, flagellin, and propelling, respectively.

3. Recognition of Nucleic Acids by Cytosolic PKR and RIG-1 Receptors

As part of the innate defense mechanism against invading pathogens, the mammalian immune system is activated by microbial RNA and DNA leading to the production of type I interferon and proinflammatory cytokines. The first sensor of dsRNA identified was the dsRNA-dependent protein kinase (PKR) that phosphorylates serine and threonine residues of target proteins (26).

Most human cells constitutively express a low level of PKR that remain inactive. However, upon binding to dsRNA, PKR forms a homodimer resulting in its autophosphorylation and activation. Activated PKR phosphorylates a large number of substrates, particularly the translation initiation factor elF-2α leading to translation arrest and induction of apoptosis, an essential step in antiviral resistance (26). PKR can also activate the NF-κB signaling pathway via the phosphorylation of IKKβ. It should be noted that recognition of dsRNAs by PKR is sequence independent and the presence of interferon upregulates its expression.

A second protein that is stimulated by long dsRNA is 2′-5′ oligoadenylate synthetase (OAS), which is expressed constitutively and also upregulated by IFN-α and IFN-β during antiviral responses (27). This interferon-induced enzyme catalyzes the formation of 2′-5′-linked oligoadenylates from ATP that activate a latent ribonuclease, called RNase L that degrades both cellular and viral RNAs. Although both OAS and PKR are implicated in antiviral immunity, PKR and RNase L are mainly IFN effectors and not absolutely required for the initial phase of IFN production. Indeed, PKR gene targeting in mice showed that it is not essential for interferon responses to viral

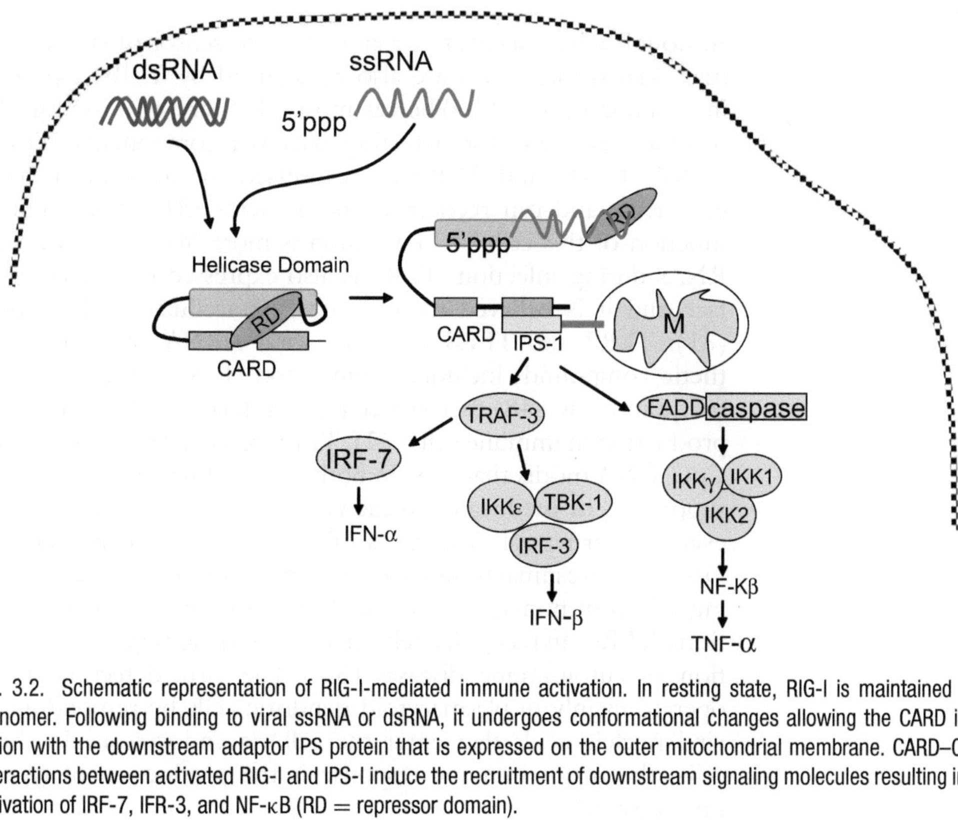

Fig. 3.2. Schematic representation of RIG-I-mediated immune activation. In resting state, RIG-I is maintained as a monomer. Following binding to viral ssRNA or dsRNA, it undergoes conformational changes allowing the CARD interaction with the downstream adaptor IPS protein that is expressed on the outer mitochondrial membrane. CARD–CARD interactions between activated RIG-I and IPS-I induce the recruitment of downstream signaling molecules resulting in the activation of IRF-7, IFR-3, and NF-κB (RD = repressor domain).

infection. Therefore, other cytoplasmic receptors may be involved. More recently, two additional intracellular cytosolic DExD/H box RNA-helicases retinoic-acid-inducible gene I (RIG-I) and melanoma differentiation-associated gene 5 (MDA-5) were identified as a main cytoplasmic sensor of viral RNA (28). Mice deficient for RIG-1 were found to be highly susceptible to viral infection (29). Both helicases are widely expressed in inactive form, and like other antiviral proteins they are also upregulated by IFN-α/-β. RIG-I encodes a caspase recruitment domain (CARD) at the N terminus, in addition to an RNA helicase domain. The RNA helicase domain requires ATPase activity and is responsible for viral dsRNA recognition and induction of conformational changes leading to the interaction of the RIG-I CARD domain with another CARD-containing adaptor protein, known as IPS-1, MAVS, Cardif, or VISA (30). IPS-1 is an outer mitochondrial membrane-binding protein. IPS-1 activates IRF3 and IFR-7 through TBK1/IKKi, resulting in the production of IFN-β production (30, **Fig. 3.2**). Mitochondrial retention of IPS-1 is essential for IRF3, IRF7, and NF-κB activation by RIG-1 (30).

4. Recognition of Nucleic Acids by TLRs

Although RIG-I seems to be an important sensor of viral RNAs, microbial nucleic acids are also recognized by TLRs, especially in immune cells (22). Whereas most TLRs are expressed in the plasma membrane for detecting bacterial components, TLR3, TLR7, TLR8, and TLR9 are expressed in intracellular compartments and can recognize nucleic acids (21). The immune function of this cellular localization is more likely to sense viral RNAs during infection. TLR3 is also expressed on the cell surface and it is believed to recognize extracellular viral dsRNAs (31). TLR7 and TLR8 recognize viral ssRNA and small synthetic compounds including imiquimod (R-837, S-26308) and resiquimod (R-848) that induce type I interferon IFN and IL-12 production in immune cells (32). TLR9 recognizes unmethylated CpG-DNA motifs that exist in both viral and bacterial DNA, but suppressed or methylated in the vertebrate genomes (33). However, the structural differences of eukaryotic versus prokaryotic DNA are presumably not the only mechanism for distinguishing self from non-self nucleic acids because under certain conditions, TLRs can recognize self-nucleic acids resulting in the induction of autoimmune diseases (34). The virus-detecting TLRs operate mainly in plasmacytoid dendritic cells by responding to viral nucleic acids that enter the cell via endocytosis. In these cells, the major immune response is the production of type 1 interferon (35).

5. Molecular Basis for Immune Sensing of Single- and Double-Stranded RNAs

Although the recognition of RNA by TLRs is well documented, it remains unclear how these receptors discriminate viral from host nucleic acids. Because they all recognize RNAs in endocytic compartments, one hypothesis is that discrimination is achieved through endosomal exclusion of self nucleic acids. The discovery of RNAi in mammalian cells has excited the study of immune tolerance of siRNAs. Initially, siRNAs appeared to evade immune recognition and the shutdown of cellular proteins expression that can occur following the interaction of long double-stranded RNA (>30 nucleotides) with intracellular RNA receptors, particularly PKR (7). However, we and others have shown that they can activate immune responses in mammalian cells through TLR-dependent and TLR-independent mechanisms (36–38). In immune cells, the response is mainly mediated through TLR7 in mice and TLR7/8 in humans. Indeed, TLR7 knockout mice did not mount immune activation in response to siRNAs (37). With the use of microarrays, we have demonstrated that in peripheral blood mononuclear cells over 400 were affected by the activation of TLR by either double-stranded siRNAs or single-stranded siRNAs (39). Genes encoding for proinflammatory cytokines, interferons, and Mx proteins are among the genes that are significantly induced. Mx proteins are IFN-induced GTPases that form complexes with dynamin disrupting trafficking or activity of viral polymerases, thereby interfering with viral replication. Induction of innate immune response by RNA can have significant unspecific effects, primary via the actions of IFNs and inflammatory cytokines. Although there is a need for analyzing the immunostimulatory potential of siRNAs prior to clinical applications, the adjuvant effects of siRNA may enhance cellular immunity (*see* **Chapters 26** and **29**).

In addition to cellular compartmentalization, structural and sequence modifications are thought to contribute to the recognition of siRNAs by endosomal TLR7 and TLR8. Initial experiments indicate that some types of secondary structures and/or specific nucleotides are responsible for the activation of NF-κB signaling pathway by siRNAs in human monocytes (15). Monocytes are circulating peripheral blood cells that can be differentiated by cytokines into macrophages of different phenotypes as well as into dendritic cells. Although a defined and universal sequence motif recognized by TLR7 and/ TLR8 as described for TLR9 (the CpG motif) has not been identified yet, two reports suggested that specific nucleotides need to be present within the siRNA in order to activate innate immunity. In the first report, the induction of interferon was attributed to the activation of TLR7 by a GU-based "UGUGU" immunostimulatory sequence (40). In the second report, however, the stimulation was attributed to a RNA motif (5′-GUCCUUCAA-3′) that is recognized by TLR7 in the context of siRNA duplexes and the activity does not depend on GU content (40). Previously, we have shown that the induction of proinflammatory cytokines double- and single-stranded siRNAs cannot be easily suppressed by selecting siRNA sequences without the GU dinucleotides. Indeed, several siRNA sequences without GU bases induced TNF-α production in human blood cells. Although the precise nature of the RNA motifs responsible for innate immune activation is not known, we have shown that TLR8/7-dependent cytokine production is triggered by the presence of uracil in RNA and that the potency of TLR7/8 agonists correlates with the number of total uridine moieties (41). Replacement of uridines with adenosines abrogated immune activation (41).

6. Overcoming siRNA Immune Activation

Much of the current interest in the mechanisms involved in RNA sensing by the immune system was generated by the observation that siRNA can activate innate immunity (15, 18). Considering the high frequency of uridines and/or GU dinucleotides in messenger RNAs, it is more likely that a high proportion of self and non-self chemically made siRNA sequences will activate innate immunity. Therefore, it would be desirable to develop strategies that evade immune activation. At least two distinct strategies to alleviate immune activation by siRNAs can be applied. First, the use of delivery agents that avoid the delivery of siRNA into the endosomes. Second, the use of chemically modified siRNAs. A wide variety of chemical modifications that confer nuclease resistance have been successfully used in RNA oligonucleotides and ribozymes, many of these modifications can be imported directly into siRNAs. In this respect, Morrissey and colleagues showed that the incorporation of various 2′-modified nucleotides that included DNA bases, 2′O-methyl purines, 2′F-pyrimidines, terminal inverted-dt bases, and PS linkage modification at selected positions can abrogate siRNA immune activation (42). However, the chemical modifications that block immune activation must be chosen carefully so as not to inhibit siRNA silencing activity. Fortunately, we have shown that replacement of only uridine bases with their 2′-fluoro, 2′-deoxy, or 2′-*O*-methyl-modified counterparts can block immune recognition of siRNAs by TLRs without reducing siRNA silencing potency (41, **Fig. 3.3a, b**). Also, 2′-uridine-modified single-stranded RNAs did not activate innate

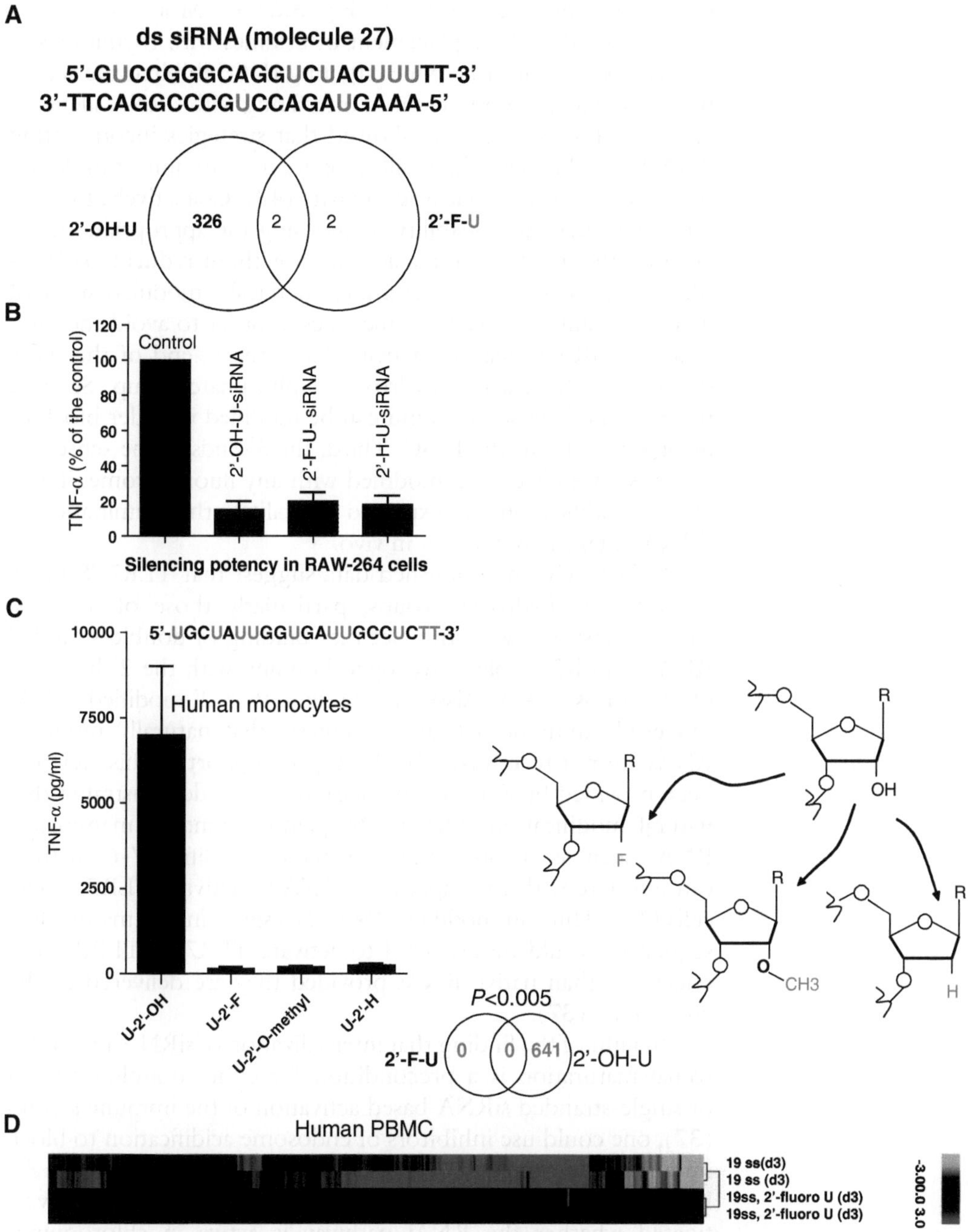

Fig. 3.3. Modified RNAs evade TLR recognition. (**a**) Double-stranded siRNAs targeting mouse TNF-α activate innate immunity in human PBMC, whereas their 2′-uridine-modified counterparts did not. As shown unmodified siRNA altered the expression of 328 genes, whereas only 4 genes were affected by its 2′-fluoro-modified version. (**b**) Inhibition of TNF-α gene expression by the modified siRNA in RAW 364 mouse macrophages. (**c**) Analysis of TNF-α gene expression in human monocytes in response to unmodified and 2′-uridine-modified single-stranded (ss) siRNA targeting mouse basigin. (**d**) Global gene expression in human PBMC in response to unmodified and 2′-fluoro uridine-modified basigin ss siRNA (38, 43).

immunity in human blood cells (**Fig. 3.3c, d**). More recently, we have shown that the replacement of uridines with thymidines in the sense strand of a ribozyme targeting IDO inhibited the activation of innate immunity without interfering with siRNA silencing activity. Other studies also showed that strategies incorporating 2′-OMe-modified residues will generate non-immunostimulatory siRNA with sustained cleavage activity (43). Collectively, the current data offer the possibility of choosing the appropriate modifications that evade immune activation without reducing siRNA-silencing potency. When designing chemically modified siRNAs, it is important to consider some rules in order to avoid the inhibition of siRNA cleavage activity. First, the 5′-end of the sense strand must have a free hydroxyl or phosphate group. Second, the 5′-end of the sense strand can be modified in order block its incorporation into the RISC. Third, the 3′-ends of the sense and antisense strands can be modified with any fluorochrome. These 3′-end modifications are expected to facilitate the examination of siRNA uptake and imaging in vivo.

Collectively, the published data suggest that TLR7/8 interact with the hydroxyl groups, particularly those of uridines. This is consistent with the fact that binding of double-stranded RNA to TLR3 involves hydrogen bonding with the 2′-hydroxyl of the ribose (44). Also, the finding that 2′-modified RNAs can evade immune activation suggests that naturally modified RNAs are not recognized by TLR7/8. Support of this view has been provided by Karikó and colleagues, who demonstrated that natural modifications that are frequently found in mammalian RNA (such as pseudouridine, 5′-methylcytidine, 2′-*O*-methyl) can interfere with the capacity of RNA to activate TLR7 in DC cells (45). Thus, unmodified RNA corresponding to mammalian sequences would be expected to activate TLR7 or TLR8 more effectively than native RNAs provided they are delivered to the endosomes (37).

In light of the finding that internalization of siRNA and endosomal maturation is a precondition for either double-stranded or single-stranded siRNA-based activation of the immune system (37), one could use inhibitors of endosome acidification to block immune activation (37). Because chloroquine and bafilomycin A1 blocked the immunological activity of siRNAs, we then investigated whether the RNAi pathway is active in chloroquine-treated human cells. The data indicate that the silencing activity of siRNAs was not inhibited, even enhanced, by chloroquine, which is known to induce the neutralization and the swelling of endosome vesicles. Similar results were obtained with bafilomycin A1.

7. Suppressive 2′-Modified RNAs

The finding that unmodified, but not 2′-modified RNA, are potent triggers of innate immunity raised the question about the difference in their structures that might be relevant for binding to TLR7/8. Therefore, which step is affected by 2′-modifications, and why cannot 2′-modified RNAs trigger immune activation. One way to address the first question is to assess whether 2′-modified RNA could antagonize with immunostimulatory RNAs to trigger TLR7/8 signaling. Studies of transfected human monocytes show that 2′-*O*-methyl-modified RNAs abrogate the activation of TLR7 by immunostimulatory RNAs (46). Of considerable interest is that 2′-*O*-methyl-modified RNAs suppressed immune activation at very low concentrations (47, **Fig. 3.4**). In addition, we have shown that they can effectively inhibit immune activation by a variety of immunostimulatory sequences including bacterial and mitochondrial RNAs (47). Also, chemically modified RNA can antagonize with immunostimulatory RNA to activate IDO, an immunosuppressive enzyme

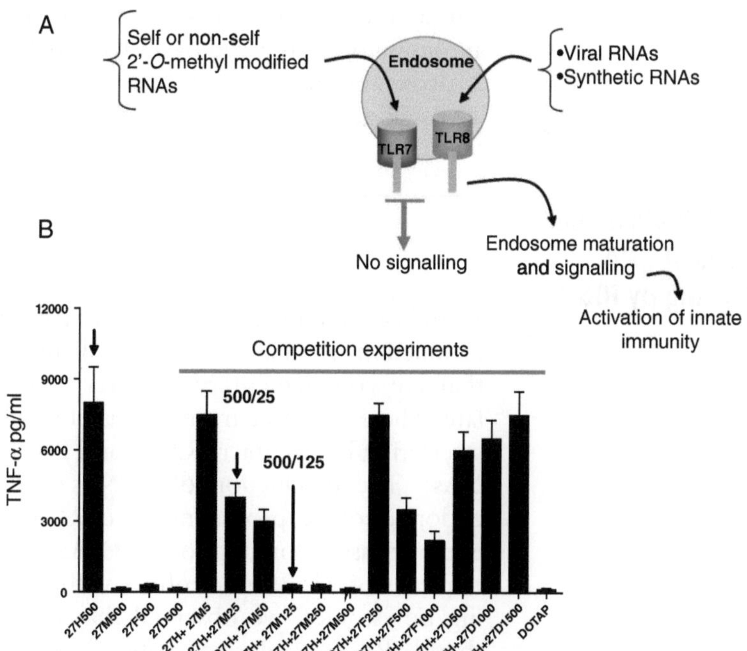

Fig. 3.4. 2′-*O*-methyl-modified RNAs alter TLR signaling. (**a**) Schematic representation of TLR7/8 recognition of RNAs. (**b**) TNF-α production in human monocytes in response to unmodified ss siRNA 27 (27H) either alone or with combination with various concentrations of 2′-*O*-methyl (27 M), 2′-fluoro (27F), or 2′-deoxy (27D). The numbers indicate the used RNA concentrations in ng/mL (47).

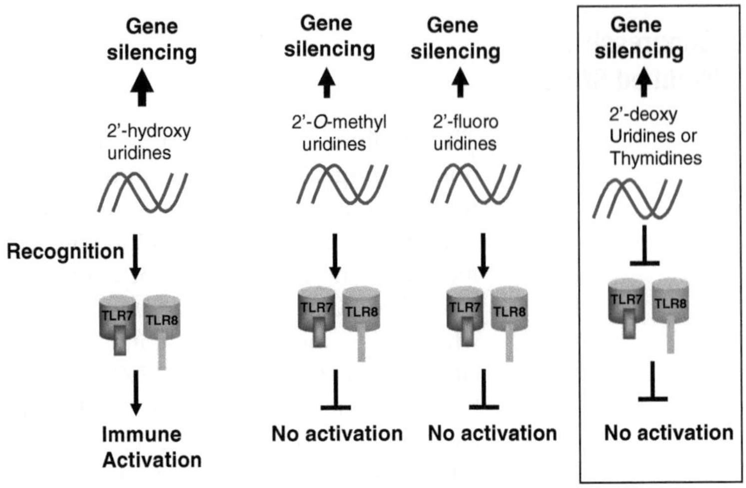

Fig. 3.5. Separation of immunostimulation from RNA interference. For details, see the text.

(48). In accordance with our data, Robbins and colleagues have reported that 2′-modified immunostimulatory RNAs can function as TLR7/8 antagonist by inhibiting TLR7 signaling induced by immunostimulatory RNAs or loxoribine in both murine and human cells (49). Suppressive 2′-modified RNAs should represent a new class of agents that may be useful in the treatment of autoimmunity triggered by TLR7 and TLR8 signaling, given that 2′-deoxyuridines or thymidine-modified siRNAs did not bind to TLRs (**Fig. 3.5**).

8. The Molecular Basis of RNA Sensing by RIG-I

As indicated above, gene silencing pathways are initiated by the production of small RNAs (20–30 nucleotides) with sequences that are complementary to parts of the transcripts that regulate. There are three main classes of small regulatory RNA: small interfering RNAs, microRNAs, and PIWI-interacting RNAs (piRNAs). The production of siRNAs and miRNAs relies on the endonucleolytic processing of dsRNA precursors by Dicer, an end nuclease belonging to the RNAase III family, yielding siRNA duplexes of 21–25 nucleotides with 2 bases 3′ overhangs. During our studies, we observed that cytoplasmic delivery of synthetic siRNAs by electroporation into peripheral blood mononuclear cells such as monocytes did not induce the production of inflammatory cytokines and type I interferon, whereas the same sequences when delivered by lipid did (37). Thus, synthetic siRNAs are not detected by RIG-1, a new identified viral RNA sensor. However, our data do not explain why ds siRNA are not sensed by RIG-I (37, **Fig. 3.2**). Recently, structural studies indicated that innate immunity can be triggered by synthetic siRNA duplexes harboring 2-base 3′ overhangs, but not by siRNAs with blunt ends (50). The authors showed that RIG-I can bind siRNAs with or without 2-base 3′ overhangs, however, only those with blunt ends could trigger its ATPase activity. This observation implies that endogenous shRNAs or miRNA harboring the Dicer signature, 2-base 3′ overhang, are not perfect stimulators of RIG-I. Thus, the structures of the 5′-ends between shRNAs (substrate for Dicer) and non-self dsRNAs such as viral RNAs are critical for self-discrimination and non-self discrimination (51, 52).

The predominant form of naturally occurring dsRNAs in mammalian cells is derived from endogenously expressed miRNAs that constitute a large class of noncoding small RNAs involved in gene regulation in a variety of organisms ranging from plants to mammalians (53, *see* **Chapters 18** and **19**). Presently, more than 1,000 potential human miRNAs have been identified and numerous have been experimentally validated. Usually, miRNAs are transcribed from endogenous genes by RNA polymerase II as long RNA precursor called primary miRNA (pri-miRNA), containing one or more distinct miRNAs. In the nucleus, the RNA precursors are processed by Drosha to 60–80 nt RNA hairpin intermediate, bearing 2 base 3′ overhang, called a pre-miRNA. The Drosha cleavage site was shown to be 11 base pairs from the stem single-stranded RNA junction (54). Processed pre-miRNAs are then transported from the nucleus to the cytoplasm by exportin-5, where its 2 base 3′-overhang is recognized by Dicer leading to the generation of mature miRNAs that can evade RIG-I activation.

During our studies, we have also found that innate immunity is not activated by synthetic single-stranded sequences (21 nt) when delivered to the cytoplasm via electroporation (37). These RNAs do not contain 2 base 3′ overhang because they are single stranded. To examine the contribution of RIG-I in sensing exogenous RNAs, adherent PBMCs were transfected with either T7-transcribed siRNAs or chemically synthesized siRNAs. The inhibition of endosome maturation by chloroquine abrogated the immunostimulatory activity of chemically made siRNAs, but not the T7-made siRNAs (55). In addition, the immunostimulatory effect of the T7-made siRNAs was not inhibited with 2-aminopurine, a specific inhibitor of PKR (55). So, which are the cytoplasmic factors that sense in vitro transcribed RNA but not chemically made siRNAs? Additional studies from other investigators showed that RIG-I senses single-stranded RNA-bearing 5′-triphosphate, a specific signature of viral and in vitro transcribed RNAs (56). Most recently, the recognition of triphosphorylated RNA was found to be mediated by the C-terminal regulatory domain of RIG-1 (57). In mammals, cellular RNAs undergo processing in order to mask or to remove the 5′-triphosphate moiety. However, the data do not explain why certain endogenous RNA with 5′-triphosphates escape RIG-I recognition. Naturally occurring 2′-modifications might protect cellular RNA-bearing 5′-triphosphate from being detected by RIG-I. It should be noted that MDA5, the most closely related protein of RIG-I, is also an IFN-inducible protein. Interestingly, a recent study indicates that MDA-5 is preferentially activated by dsRNA larger than 2 kb, whereas RIG-1 is activated by dsRNA shorter than 1 kb, and without a 5′-triphosphate end (58). Thus, it seems that RIG-1 and MDA-5 complement each other in preferentially recognizing dsRNAs of different length. And the recognition of dsRNA by RIG-1 does not depend on 5′-triphosphate ends.

9. Effect of Chemical Modifications on siRNA Off-Target Effects

In mammalians, miRNAs target transcripts through imperfect base pairing to multiple sites in 3′ untranslated regions (UTRs). Watson–Crick base pairing to the 5′-end of miRNAs, especially the so-called seed sequence that comprises nucleotides 2–7, is crucial for targeting. In addition, recent data indicated that imperfect miRNA–mRNA hybrids with central bulges (nucleotides 9–12) enable translational inhibition or exonucleolytic mRNA decay. Furthermore, miRNA can guide DNA methylation (59). The diversity of mechanisms used by microRNAs to silence gene expression highlights the need to revisit the specificity of siRNAs. Accordingly, microarray studies showed that siRNAs can have off-target effects that must be thoroughly investigated prior to use (60). The most commonly used strategy to ensure siRNA target specificity is the basic local alignment search tool BLAST. However, short sequence stretches may not be detected by BLAST program. In addition, the identification of such sequences does not necessarily indicate the occurrence of off-target effects. Similarly, the absence of short homologies will not rule out off-target effects. The best way to deal with this problem is to analyze global gene expression, specifically when siRNAs are going to be used in functional genomics or in therapy. Previously, we have found that 2′-uridine modifications of siRNAs not only evade the activation of TLR7/8 but they can block most of the TLR-independent effects including off-target effects (39). Although the evading mechanisms remain to be investigated, it is possible that the interaction of siRNA sequences with unintended cellular mRNAs is affected by chemical modifications. Modifications of RNA might particularly be disruptive for siRNA binding to mismatched sequences. In accordance with our data, Jackson and colleagues found that chemical modification of siRNA such as

2′-*O*-methyl substitutions can minimize siRNA off-target effects. Interestingly, a single 2′-*O*-methyl group at the second position of the guide strand inhibited off-target gene silencing effects without affecting siRNA silencing of the intended target gene (60). Collectively, the data offer a simple strategy for reducing siRNA off-target effects.

10. Potential Beneficial Effects of Immunostimulatory siRNAs

Among the innate immune sensors that link innate and adaptive immunity, dendritic cells (DCs) play an important role in immunity and are the only cell type capable of initiating adaptive immune responses by activating naive T cells (61, 62). These cells are generated from either myeloid or lymphoid bone marrow progenitors that home to sites of potential antigens entry, where they differentiate locally into immature DCs (**Fig. 3.6**).

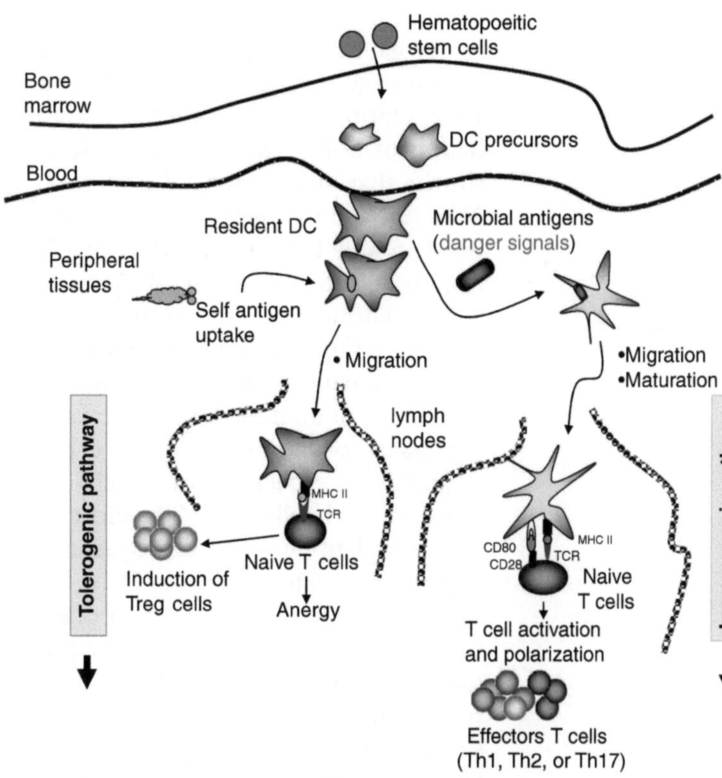

Fig. 3.6. Activation of T cells by DCs. Under physiological conditions, DCs can capture antigens derived from self and non-self compounds with the same efficiency. However, only antigens captured in the presence of "danger signals" such as microbes and inflammatory cytokines can induce DC maturation resulting in the expression of CD80/86 molecules. The interaction between these molecules expressed on DCs and CD28 expressed on T cells is crucial for the activation of naive T cells.

Recent studies demonstrated that both immunity and tolerance are controlled by DCs. In the absence of "danger signals" immature DCs mediate peripheral tolerance, leading to T-cell anergy and/or deletion due to the absence of appropriate costimulation by CD80/CD86 molecules (61). Following antigen capture in the presence of maturation signals, DCs undergo a complex maturation that entails upregulation of major histocompatibility complex class I and class II molecules, costimulatory molecules such as CD40, CD80, and CD86, and the production of IL12. Among the signaling pathways that activate DC maturation, TLR signaling pathway is the most effective. Unlike infectious pathogens, tumors do not induce an effective inflammatory response leading to DC activation. In addition, tumor microenvironment can protect tumor cells from immune destruction. In this respect, soluble immunosuppressive factors, and membrane-bound molecules including transforming growth factor-β, interleukin IL-10, prostaglandin E2, and CTLA-4 represent a barrier for antitumor immunity (63). Interfering with the expression of these factors might potentiate antitumor T-cell effector function in vivo.

Notably, pathogen-mediated maturation of DCs is mediated mainly through the TLRs that are expressed on immature DCs. Optimal DC maturation might therefore require a combination of both cytokines and TLR ligands. Previously, we have shown that stimulation of DCs with immunostimulatory siRNAs can induce their maturation to secrete cytokines, including IL6 and IL12 (37). While IL12 is required for Th-1 type response, IL6 may render $CD4^+$ effector T-cells refractory to T reg cell-mediated suppression. To extend these findings, we have tested the possibility of designing bifunctional siRNAs capable of inducing TLR signaling and simultaneously blocking the expression of immunosuppressive factors (55). The inappropriate expression of immunosuppressive cytokines and other negative regulators is expected to hamper immunity against tumors and virus-infected cells (64). Therefore, the development of agents that stimulate DCs and subsequently suppress the expression of negative regulators, such as IL10, TGF-β, and SOCS proteins would facilitate the development of effective cancer vaccines. In a recent study, we have assessed the possibility of combining gene silencing and immunostimulation in one siRNA molecule. Immature monocyte-derived dendritic cells incubated with anti-IL10 siRNAs produced cytokines (e.g., IL6, TNF-α, IL12), upregulated the expression of the costimulatory molecules (e.g., CD80, CD86), MHC class II molecules, and the chemokine receptors CCR7. Also, IL10 siRNAs enhanced the ability of DCs to activate T cells in MLR assays (55). Thus, the possibility of triggering endogenous IL12 and IL6 production through deliberate activation of TLR7 or TLR8 pathway via IL-10 siRNAs is of considerable interest. In addition to IL10, bifunctional siRNAs against other key factors involved in immune suppression were developed. These include SOCS1, STAT-3, TGF-β and indoleamine 2,3-dioxygenase (IDO), a rate-limiting enzyme in tryptophan catabolism that is known to be involved in fetus tolerance during pregnancy (65). The SOCS proteins have been identified as inhibitors of cytokines signaling and shown to function in a classical feedback loop (66). They regulate signaling via the Jak/Stat pathway and modulate DC function by switching off interferon-γ and/or IL12 signaling during immune responses. By targeting SOCS1 with conventional siRNAs in DCs, recently Chen and colleagues demonstrated that antigen-specific antitumor immunity can be enhanced (67).

With respect to IDO, we have previously shown that monocytes expressing IDO acquire an immunosuppressive phenotype capable of suppressing T-cell proliferation (48). More recently, we designed a very effective anti-IDO siRNA that can effectively block IDO gene expression in human primary cells such as monocytes and DCs (68). Depending on the design and chemical modifications, we also showed that it is possible to design either monofunctional siRNA devoid of immunostimulation or bifunctional siRNA with gene silencing and immunostimulatory activities. Interestingly, immature monocytes-derived DCs that had been transfected with siRNA-bearing 5′-triphosphate-activated T cells, indicating that, even in the absence of added external stimuli such as tumor necrosis factor-α, those DCs were sufficiently mature to initiate T-cell activation (*see* **Chapter 26**). Collectively, the data indicate that it is possible to modulate T-cell maturation and function with RNAi technology. We are confident that clinical researchers will implant RNAi technology in DC-based vaccination to interfere with the negative regulatory mechanisms used by DCs to induce peripheral tolerance (*see* **Chapters 5** and **29**).

References

1. Jorgensen, R. (1990) Altered gene expression in plants due to trans interactions between homologous genes. *Trends Biotechnol*, **8**, 340–344.
2. Fire, A., Xu, S., Montgomery, M.K., Kostas, S.A., Driver, S.E., and Mello, C.C. (1998) Potent and specific genetic interference by double-stranded RNA in *Caenorhabditis elegans*. *Nature*, **391**, 806–811.
3. Svoboda, P., Stein, P., Hayashi, H., and Schultz, R.M. (2000) Selective reduction of dormant maternal mRNAs in mouse oocytes by RNA interference. *Development*, **127**, 4147–4156.
4. Sen, G.C. (2001) Viruses and interferons. *Annu Rev Microbiol*, **55**, 255–281.
5. Zamore, P.D., Tuschl, T., Sharp, P.A., and Bartel, D.P. (2000) RNAi: double-stranded RNA directs the ATP-dependent cleavage of mRNA at 21 to 23 nucleotide intervals. *Cell*, **101**, 25–33.
6. Elbashir, S.M., Lendeckel, W., and Tuschl, T. (2001) RNA interference is mediated by 21- and 22-nucleotide RNAs. *Genes Dev*, **15**, 188–200.
7. Elbashir, S.M., Harborth, J., Lendeckel, W., Yalcin, A., Weber, K., and Tuschl, T. (2001) Duplexes of 21-nucleotide RNAs mediate

RNA interference in cultured mammalian cells. *Nature*, **411**, 494–498.
8. Sioud, M. (2004) Therapeutic siRNAs. *Trends Pharmacol Sci*, **25**, 22–28.
9. Hannon, G.J. and Rossi, J.J. (2004) Unlocking the potential of the human genome with RNA interference. *Nature*, **431**, 371–378.
10. Liu, J., Carmell, M.A., Rivas, F.V., Marsden, C.G., Thomson, J.M., Song, J.J., Hammond, S.M., Joshua-Tor, L., and Hannon, G.J. (2004) Argonaute2 is the catalytic engine of mammalian RNAi. *Science*, **305**, 1437–1441.
11. Song, J.J., Smith, S.K., Hannon, G.J., and Joshua-Tor, L. (2004) Crystal structure of Argonaute and its implications for RISC slicer activity. *Science*, **305**, 1434–1447.
12. Matranga, C., Tomari, Y., Shin, C., Bartel, D.P., and Zamore, P.D. (2005) Passenger-strand cleavage facilitates assembly of siRNA into Ago2-containing RNAi enzyme complexes. *Cell*, **123**, 607–620.
13. Rand, T.A., Petersen, S., Du, F. et al. (2005) Wang: Argonaute2 cleaves the anti-guide strand of siRNA during RISC activation. *Cell*, **123**, 621–629.
14. Ma, J.B., Yuan, Y.R., Meister, G., Pei, Y., Tuschl., T., and Partel, D.J. (2005) Structural basis for 5′-end-specific recognition of guide RNA by the *A. fulgidus* Piwi protein. *Nature*, **434**, 666–670.
15. Sioud, M. and Sorensen, D.R. (2003) Cationic liposome-mediated delivery of siRNAs in adult mice. *Biochem Biophys Res Commun*, **312**, 1220–1225.
16. Jackson, A.L., Bartz, S.R., Schelter, J., Kobayashi, S.V., Burchard, J., Mao, M., Li, B., Cavet, G., and Linsley, P.S. (2003) Expression profiling reveals off-target gene regulation by RNAi. *Nat Biotechnol*, **21**, 635–637.
17. Semizarov, D., Frost, L., Sarthy, A., Kroeger, P., Halbert, D.N., and Fesik, S.W. (2003) Specificity of short interfering RNA determined through gene expression signatures. *Proc Natl Acad Sci USA*, **100**, 6347–6352.
18. Sledz, C.A., Holko, M., de Veer, M.J., Silverman, R.H., and Williams, B.R. (2003) Activation of the interferon system by short-interfering RNAs. *Nat Cell Biol*, **5**, 834–839.
19. Parkin, J. and Cohen, B. (2001) An overview of the immune system. *Lancet*, **357**, 1777–1789.
20. Janeway, C.A. and Medzhitov, R., Jr. (2002) Innate immune recognition. *Annu Rev Immunol*, **20**, 197–216.
21. Takeda, K. and Akira, S. (2005) Toll-like receptors in innate immunity. *Int Immunol*, **17**, 1–14.
22. Sioud, M. (2006) Innate sensing of self and non-self RNAs by Toll-like receptors. *Trends Mol Med*, **12**, 167–716.
23. Brennan, C.A. and Anderson, K.V. (2004) Drosophila: the genetics of innate immune recognition and response. *Annu Rev Immunol*, **22**, 457–483.
24. Lemaitre, B., Nicolas, E., Michaut, L., Reichhart, J.M., and Hoffmann, J.A. (1996) The dorsoventral regulatory gene cassette spatzle/Toll/cactus controls the potent antifungal response in Drosophila adults. *Cell*, **86**, 973–983.
25. Akira, S. and Takeda, K. (2004) Toll-like receptor signaling. *Nat Rev Immunol*, **4**, 499–511.
26. Li, S., Peters, G.A., Ding, K., Zhang, X., Qin, J., and Sen, G.C. (2006) Molecular basis for PKR activation by PACT or dsRNA. *Proc Natl Acad Sci USA*, **103**, 10005–10010.
27. Samuel, C.E. (2001) Antiviral actions of interferons. *Clin Microbiol Rev*, **14**, 778–809.
28. Kato, H., Takeuchi, O., Sato, S., Yoneyama, M., Yamamoto, M., Matsui, K., Uematsu, S., Jung, A., Kawai, T., Ishii, K.J., Yamaguchi, O., Otsu, K., Tsujimura, T., Koh, C.S., Reis e Sousa, C., Matsuura, Y., Fujita, T., and Akira, S. (2006) Differential roles of MDA5 and RIG-I helicases in the recognition of RNA viruses. *Nature*, **441**, 101–105.
29. Kato, H., Sato, S., Yoneyama, M., Yamamoto, M., Uematsu, S., Matsui, K., Tsujimura, T., Takeda, K., Fujita, T., Takeuchi, O., and Akira, S. (2005) Cell type-specific involvement of RIG-I in antiviral response. *Immunity*, **23**, 19–28.
30. Meylan, E., Tschopp, J., and Karin, M. (2006) Intracellular pattern recognition receptors in the host response. *Nature*, **442**, 39–44.
31. Alexopoulou, L., Holt, A.C., Medzhitov, R., and Flavell, R.A. (2001) Recognition of double-stranded RNA and activation of NF-kappaB by Toll-like receptor 3. *Nature*, **413**, 732–738.
32. Heil, F., Hemmi, H., Hochrein, H., Ampenberger, F., Kirschning, C., Akira, S., Lipford, G., Wagner, H., and Bauer, S. (2004) Species-specific recognition of single-stranded RNA via Toll-like receptor 7 and 8. *Science*, **303**, 1526–1529.
33. Krieg, A.M. (2002) CpG motifs in bacterial DNA and their immune effects. *Annu Rev Immunol*, **20**, 709–760.
34. Brentano, F., Kyburz, D., Schorr, O., Gay, R., and Gay, S. (2005) The role of Toll-like receptor signaling in the pathogenesis of arthritis. *Cell Immunol*, **233**, 90–96.

35. Cao, W. and Liu, Y.J. (2007) Innate immune functions of plasmacytoid dendritic cells. *Curr Opin Immunol*, **19**, 24–30.
36. Kariko, K., Bhuyan, P., Capodici, J., and Weissman, D. (2004) Small interfering RNAs mediate sequence-independent gene suppression and induce immune activation by signaling through Toll-like receptor 3. *J Immunol*, **172**, 6545–6549.
37. Sioud, M. (2005) Induction of inflammatory cytokines and interferon responses by double-stranded and single-stranded siRNAs is sequence-dependent and requires endosomal localization. *J Mol Biol*, **348**, 1079–1090.
38. Hornung, V., Guenthner-Biller, M., Bourquin, C., Ablasser, A., Schlee, M., Uematsu, S., Noronha, A., Manoharan, M., Akira, S., de Fougerolles, A., Endres, S., and Hartmann, G. (2005) Sequence-specific potent induction of IFN-alpha by short interfering RNA in plasmacytoid dendritic cells through TLR7. *Nat Med*, **11**, 263–270.
39. Cekaite, L., Furset, G., Hovig, E., and Sioud, M. (2007) Gene expression analysis in blood cells in response to unmodified and 2′-modified siRNAs reveals TLR-dependent and independent effects. *J Mol Biol*, **365**, 90–108.
40. Judge, A.D., Sood, V., Shaw, J.R., Fang, D., McClintock, K., and MacLachlan, I. (2005) Sequence-dependent stimulation of the mammalian innate immune response by synthetic siRNA. *Nat Biotechnol*, **23**, 457–462.
41. Sioud, M. (2006) Single-stranded small interfering RNA are more immunostimulatory than their double-stranded counterparts: a central role for 2′-hydroxyl uridines in immune responses. *Eur J Immunol*, **36**, 1222–1230.
42. Morrissey, D.V., Lockridge, J.A., Shaw, L., Blanchard, K., Jensen, K., Breen, W., Hartsough, K., Machemer, L., Radka, S., Jadhav, V., Vaish, N., Zinnen, S., Vargeese, C., Bowman, K., Shaffer, C.S., Jeffs, L.B., Judge, A., MacLachlan, I., and Polisky, B. (2005) Potent and persistent in vivo anti-HBV activity of chemically modified siRNAs. *Nat Biotechnol*, **23**, 1002–1007.
43. Judge, A.D., Bola, G., Lee, A.C., and MacLachlan, I. (2006) Design of noninflammatory synthetic siRNA mediating potent gene silencing in vivo. *Mol Ther*, **13**, 494–505.
44. Bell, J.K., Askins, J., Hall, P.R., Davies, D.R., and Segal, D.M. (2006) The dsRNA binding site of human Toll-like receptor 3. *Proc Natl Acad Sci USA*, **103**, 8792–8797.
45. Kariko, K., Buckstein, M., Ni, H., and Weissman, D. (2005) Suppression of RNA recognition by Toll-like receptors: the impact of nucleoside modification and the evolutionary origin of RNA. *Immunity*, **23**, 165–175.
46. Sioud, M. (2007) RNA interference and innate immunity. *Adv Drug Deliv Rev*, **59**, 153–163.
47. Sioud, M., Furset, G., and Cekaite, L. (2007) Suppression of immunostimulatory siRNA-driven innate immune activation by 2′-modified RNAs. *Biochem Biophys Res Commun*, **361**, 122–126.
48. Furset, G., Floisand, Y., and Sioud, M. (2008) Impaired expression of indoleamine 2, 3-dioxygenase in monocyte-derived dendritic cells in response to Toll-like receptor-7/8 ligands. *Immunology*, **123**, 263–271.
49. Robbins, M., Judge, A., Liang, L., McClintock, K., Yaworski, E., and MacLachlan, I. (2007) 2′-*O*-methyl-modified RNAs act as TLR7 antagonists. *Mol Ther*, **15**, 1663–1669.
50. Marques, J.T., Devosse, T., Wang, D., Zamanian-Daryoush, M., Serbinowski, P., Hartmann, R., Fujita, T., Behlke, M.A., and Williams, B.R. (2006) A structural basis for discriminating between self and nonself double-stranded RNAs in mammalian cells. *Nat Biotechnol*, **24**, 559–565.
51. Robbins, M.A., Li, M., Leung, I., Li, H., Boyer, D.V., Song, Y., Behlke, M.A., and Rossi, J.J. (2006) Stable expression of shRNAs in human $CD34^+$ progenitor cells can avoid induction of interferon responses to siRNAs in vitro. *Nat Biotechnol*, **24**, 566–571.
52. Sioud, M. (2006) RNA interference below the immune radar. *Nat Biotechnol*, **24**, 521–522.
53. Bartel, D.P. (2004) MicroRNAs: genomics, biogenesis, mechanism, and function. *Cell*, **116**, 281–297.
54. Han, J., Lee, Y., Yeom, K.H., Nam, J.W., Heo, I., Rhee, J.K., Sohn, S.Y., Cho, Y., Zhang, B.T., and Kim, V.N. (2006) Molecular basis for the recognition of primary microRNAs by the Drosha-DGCR8 complex. *Cell*, **125**, 887–901.
55. Furset, G. and Sioud, M. (2007) Design of bifunctional siRNAs: combining immunostimulation and gene-silencing in one single siRNA molecule. *Biochem Biophys Res Commun*, **352**, 642–649.
56. Hornung, V., Ellegast, J., Kim, S., Brzozka, K., Jung, A., Kato, H., Poeck, H., Akira, S., Conzelmann, K.K., Schlee, M., Endres, S., and Hartmann, G. (2006) 5′-Triphosphate

RNA is the ligand for RIG-I. *Science*, **314**, 994–997.
57. Cui, S., Eisenächer, K., Kirchhofer, A., Brzózka, K., Lammens, A., Lammens, K., Fujita, T., Conzelmann, K.K., Krug, A., and Hopfner, K.P. (2008) The C-terminal regulatory domain is the RNA 5′-triphosphate sensor of RIG-I. *Mol Cell*, **29**, 169–179.
58. Kato, H., Takeuchi, O., Mikamo-Satoh, E., Hirai, R., Kawai, T., Matsushita, K., Hiiragi, A., Dermody, T.S., Fujita, T., and Akira, S. (2008) Length-dependent recognition of double-stranded ribonucleic acids by retinoic acid-inducible gene-I and melanoma differentiation-associated gene 5. *J Exp Med*, **205**, 1601–1610.
59. Bao, N., Lye, K.W., and Barton, M.K. (2004) MicroRNA binding sites in Arabidopsis class III HD-ZIP mRNAs are required for methylation of the template chromosome. *Dev Cell*, 7, 653–662.
60. Jackson, A.L., Burchard, J., Leake, D., Reynolds, A., Schelter, J., Guo, J., Johnson, J.M., Lim, L., Karpilow, J., Nichols, K., Marshall, W., Khvorova, A., and Linsley, P.S. (2006) Position-specific chemical modification of siRNAs reduces "off-target" transcript silencing. *RNA*, **12**, 1197–1205.
61. Banchereau, J. and Steinman, R.M. (1998) Dendritic cells and the control of immunity. *Nature*, **392**, 245–252.
62. Rossi, M. and Young, J.W. (2005) Human dendritic cells: potent antigen-presenting cells at the crossroads of innate and adaptive immunity. *J Immunol*, **175**, 1373–1381.
63. Pardoll, D. (2003) Does the immune system see tumors as foreign or self? *Annu Rev Immunol*, **21**, 807–839.
64. Grutz, G. (2005) New insights into the molecular mechanism of interleukin-10-mediated immunosuppression. *J Leukoc Biol*, **77**, 3–15.
65. Munn, D.H., Zhou, M., Attwood, J.T., Bondarev, I., Conway, S.J., Marshall, B., Brown, C., and Mellor, A.L. (1998) Prevention of allogeneic fetal rejection by tryptophan catabolism. *Science*, **281**, 1191–1193.
66. Alexander, W.S. and Hilton, D.J. (2004) The role of suppressors of cytokine signaling (SOCS) proteins in regulation of the immune response. *Annu Rev Immunol*, **22**, 503–529.
67. Shen, L., Evel-Kabler, K., Strube, R., and Chen, S.Y. (2004) Silencing of SOCS1 enhances antigen presentation by dendritic cells and antigen-specific anti-tumor immunity. *Nat Biotechnol*, **22**, 1546–1553.
68. Flatekval, G.F., and Sioud, M. (2009) Modulation of dendritic cell function and maturation with mono- and bifunctional small interfering RNAs targeting indoleamine 2,3 dioxygenase. *Immunology*, **128**, e837–e848. DOI: 10.1111/j.1365-2567.2009.03093.x.

Chapter 4

Progress in siRNA Delivery Using Multifunctional Nanoparticles

Weiwei Gao, Zeyu Xiao, Alex Radovic-Moreno, Jinjun Shi, Robert Langer, and Omid C. Farokhzad

Abstract

Nanoparticles made from synthetic polymers have been developed to deliver small interfering RNA (siRNA). For successful siRNA delivery, these nanoparticles need to efficiently encapsulate siRNA, actively target sites of interest, and release siRNA intracellularly. This chapter reviews recent progress using a multifunctional approach to design and engineer polymeric nanoparticles for siRNA delivery.

Key words: Polymer, nanoparticles, siRNA, delivery, RNAi.

1. Introduction

1.1. Emerging of Synthetic Vehicles for siRNA Delivery

In 2001, synthetic double-stranded small interfering RNAs (siRNA) were found to induce sequence-specific, post-transcriptional gene silencing through the RNA interference (RNAi) pathway (1). Following this discovery, the therapeutic potential of the siRNAs was immediately realized (2–4). Synthetic siRNAs have emerged as potential drug candidates that can, when delivered properly, harness the RNAi pathway in a consistent and predictable way (5). In addition, synthetic siRNAs are amenable to chemical modification and scaled-up production (6). With these advantages, the initial in vivo therapeutic effects of siRNAs were established using free (un-encapsulated) siRNAs in for various disease indications (7, 8).

M. Sioud (ed.), *RNA Therapeutics*, Methods in Molecular Biology 629,
DOI 10.1007/978-1-60761-657-3_4, © Springer Science+Business Media, LLC 2010

However, free siRNAs are not ideal to produce efficient and predictable therapeutic effects. Within minutes after intravenous administration, the majority of the injected siRNA dose is removed from circulation by hepatic and renal clearance. Hence, only a very small percentage of the administered dose remains available for the target cells or tissues. This small percentage of siRNA is additionally subject to nuclease degradation by non-specific nucleases in the blood. Moreover, the efficiency of siRNA passively entering target cells is usually very low (9). Although local administration can avoid some difficulties encountered after intravenous administration, in many cases the targeted cells or tissues cannot be readily reached, or the extent of disease covers an area that is too large for a local treatment. To overcome these difficulties, there is a distinct need to develop efficient siRNA delivery systems (*see* **Chapters** 7, **8**, and **9**).

Earlier siRNA delivery efforts were focused on viral delivery systems such as those developed for gene delivery. Over the millions of years, viruses have evolved sophisticated replication machinery for the delivery of their genetic payloads to host cells. Many viral delivery systems have been explored for siRNA delivery (10–12). Despite their high transduction efficiency, viral delivery approaches face serious challenges. Several gene therapy trials based on viral delivery systems have produced adverse effects (13). In addition, high costs for producing large amounts of viral stocks for clinic use and limited quantities of nucleic acids that can be packaged for therapy also limit the applications of viral delivery systems. Although viral delivery systems are further engineered to overcome these limitations (14), it is highly desirable to develop non-viral alternatives.

1.2. Multifunctional Polymeric Nanoparticles for siRNA Delivery

Synthetic polymer-based nanoparticles have been developed to deliver a variety of therapeutic agents (15–17). They have recently emerged as alternatives to viral vehicles for the delivery of nucleic acids, including siRNAs (18). Nanoparticles made from these polymers need to encapsulate siRNAs and protect them from clearance and degradation (**Fig. 4.1**). Ideally they need to be functionalized with targeting ligands, which may potentially improve the delivery specificity and efficiency. Moreover, since siRNA molecules are hydrophilic and negatively charged, after the nanoparticles enter the cells through ligand-mediated endocytosis, the released siRNAs cannot spontaneously diffuse across the plasma membrane. Instead, they may be trapped in endosomal/lysosome compartments and be subject to lysosomal degradation. Therefore, for efficient siRNA delivery, nanoparticles should incorporate certain mechanisms for siRNA to escape from the endosomal compartment and enter cytoplasm. Furthermore, once in the cytoplasm, the siRNA needs to be bioactive for efficient gene silencing. In order to meet the criteria for effective

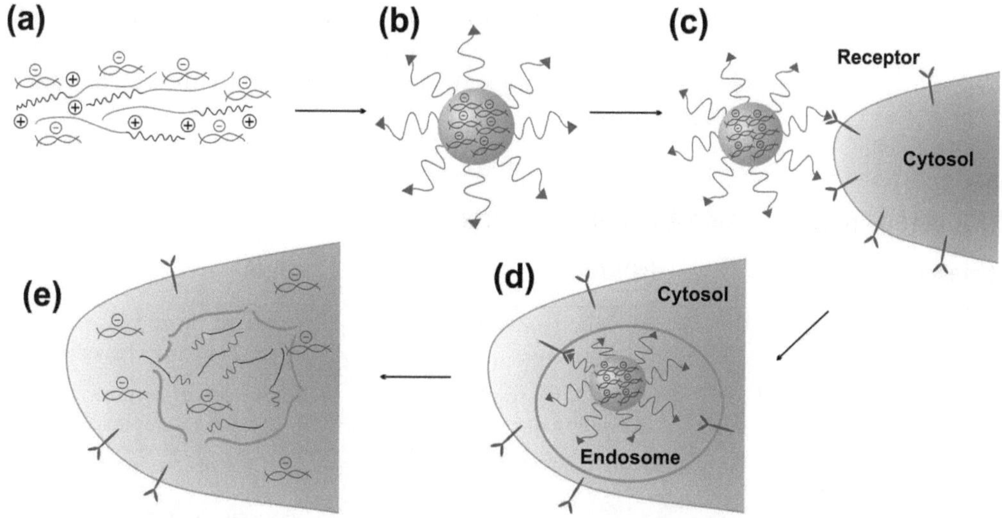

Fig. 4.1. Schematic of polymeric nanoparticles for siRNA delivery; (**a**) cationic polymers encapsulate anionic siRNA through self-assembly; (**b**) nanoparticles with encapsulated siRNA and target ligand; (**c**) nanoparticle targeting cell surface receptors; (**d**) nanoparticle uptake through receptor-mediated endocytosis; (**e**) endosome disruption and siRNA intracellular release.

delivery of siRNAs, multiple functions may need to be integrated into a single nanoparticle.

Polymers are attractive in developing multifunctional nanoparticles as they allow for advanced synthetic and conjugation chemistry. In addition, precise engineering of nanoparticles can be realized through blending of copolymers. Polymers with both hydrophobic and hydrophilic segments can self-assemble in aqueous solutions and spontaneously form stable particles on the nanometer scale. For siRNA delivery, these polymeric particles may include cationic segments for efficient encapsulation of siRNA into the particle core. Poly(ethylene glycol) (PEG) is often the hydrophilic block of choice to stabilize nanoparticles and prevent immune responses. Surface functionalization with targeting ligands can be performed both before and after nanoparticle formulation through conjugation chemistry. Besides encapsulation, siRNA can be also directly conjugated with polymers (polymer conjugates) for subsequent delivery. Despite the different structures of polymeric nanoparticles and polymer conjugates, they are both designed and engineered to integrate key functions that are essential for the delivery of siRNA into the cytoplasm.

1.3. Gene Delivery Versus siRNA Delivery

Significant knowledge has been gained through earlier gene delivery and therapy studies, especially in areas of DNA encapsulation and cytoplasmic delivery. Although siRNA and double-stranded DNA (dsDNA) share many common properties, studies have shown distinct differences between them in delivery (19). These differences are summarized in **Table 4.1**. Due to these differences, the knowledge gained from gene delivery studies cannot always be directly applied to siRNA delivery without experimental verification.

Table 4.1
siRNA versus dsDNA: a side-by-side comparison (19)

Differences	dsDNA	siRNA
Stability	Stable because of the sugar deoxyribose	Prone to degradation due to the 2′-hydroxyl group
Molecular topography and complex size	Condensable resulting smaller complex size	Rigid rod often resulting excessive complex size
Cellular location of mechanistic activation	Host cell nucleus	Cytoplasm
Duration of therapeutic effect	Vary, can be transient, long-term, or permanent transgene expression	Typically 3–7 days

Notably, the differences in molecular structures between dsDNA and siRNA not only affect how they interact with delivery vehicles during the encapsulation but also lead to changes in the efficacy of a particular transfection vector. For example, dsDNAs are larger molecules but can condense into particles of the nanometer scale. SiRNAs, though displaying much smaller molecule size compared with dsDNAs, behave as stiff rods. It has been hypothesized that molecular weight of siRNA is too low for efficient interaction with polymers. Thus they tend to form disordered and loose complexes with the polymers, resulting in less stable particle formulation and reduced delivery efficacy. To increase siRNA delivery efficiency, Bolcato-Bellemin et al. reported that by adding short complementary A5–8/T5–8 overhangs, the siRNA molecules exhibited increased binding and formed large "gene-like" structures with enhanced stability, improved RNase protection, and better gene silencing (20). Similarly, Li et al. showed another example to illustrate that dsDNA can facilitate siRNA encapsulation. They incorporated high molecular weight carrier DNA (e.g., calf thymus DNA ~50 kbp) into the nanoparticle (21). The formulation with the carrier DNA showed a 10–30% decrease in particle size and a 20–80% increase in delivery efficiency compared with the formulation lacking carrier DNA.

1.4. Chapter Outline

As discussed above, siRNA packaging, specific targeting, and intracellular release are essential requirements when designing nanoparticles for siRNA delivery. This chapter highlights the recent progress that has been made to meet these requirements using synthetic polymeric nanoparticles for siRNA delivery.

2. siRNA Packaging

Efficient siRNA encapsulation and subsequent release are fundamental requirements when using polymeric nanoparticles for siRNA delivery. The nanoparticles need not only to carry siRNA but also to protect siRNA from degradation.

2.1. siRNA Encapsulation Using Conventional Cationic Polymers

Several traditional cationic polymers, such as *poly*(ethylenimine) (PEI) and *poly*-L-lysine (PLL), have been extensively studied for siRNA delivery because they can condense the negatively charged siRNA through electrostatic interaction into small particles referred to as "polyplexes". An additional advantage of these cationic polymers is that they can bind to the negatively charged mammalian cell membranes and facilitate uptake of the associated nucleic acid into cells. However, the cationic segments are not favored for in vivo applications due to the material toxicity.

Many studies are still focused on modifying traditional cationic polymers to overcome the above difficulties. One strategy is to PEGylate the polymer. The PEG shield can provide a barrier to reduce non-specific binding caused by the positive charge (22). Another strategy is to cross-link PEI with reversible disulfide bonds. The nanoparticles are therefore prone to be degraded into small fragments, which have less toxicity (23). Zintchenko et al. reported simple amine modifications such as acetylation of primary amines and introducing negatively charged propionic acid or succinic acid groups to the PEI polymer structure (24). The resulting siRNA delivery system maintained high efficiency in gene knockdown. In particular, succinylation of branched PEI resulted in up to tenfold lower polymer toxicity in comparison to unmodified PEI. Shim et al. adopted another strategy to reduce polymer toxicity by ketalizing branched PEI (25). In this study, siRNA efficiently dissociated from the polyplexes made of ketalized PEI upon hydrolysis. In vitro studies further demonstrated that ketalization enhanced transfection efficiency of the polyplexes and reduced cellular toxicity.

2.2. Cationic Polymers Rationally Designed for siRNA Encapsulation

To improve the efficiency and minimize the toxicity of traditional cationic polymers, efforts have been made toward the rational design and synthesis of new polymers for siRNA delivery. A number of these new polymer systems have shown promising results in siRNA delivery.

A novel amphiphilic and cationic tri-block copolymer consisting of monomethoxy PEG, *poly*(3-caprolactone), and *poly*(2-aminoethyl ethylene phosphate) denoted as mPEG45-b-PCL100-b-PPEEA12 was designed and synthesized for siRNA delivery (26). Nanoparticles formulated using this polymer allowed for siRNA loading after nanoparticle formation through charge interactions without changing nanoparticle spherical morphology. The siRNA-loaded nanoparticles showed effective silencing of GFP expression on HEK293 cells without noticeable cytotoxicity. Tietze et al. reported the development of a new polymer based on β-propionamide-cross-linked oligoethylenimine. Using transferrin as targeting ligand, the polyplex effectively knocked down nuclear protein Ran expression without obvious cytotoxicity (27). Jeong et al. recently developed a reducible *poly*(amido ethylenimine) synthesized by addition copolymerization of triethylenetetramine and cystamine bisacrylamide (28). In studies using the new polymer and linear PEI to form polyplexes for siRNA delivery, substantial dissociation and intracellular distribution of siRNA were observed after 5 h of transfection, but not with linear PEI formulations. The new polymer also showed better suppression of VEGF in human PC3 prostate cancer cell line with lower corresponding cytotoxicity.

Poly (β-amino ester) (PbAE) is another class of pH-sensitive and biodegradable cationic polymers that may be utilized for intracellular pH-dependent drug release (29). PbAEs are synthesized through Michael addition of amine- and acrylate-terminated monomers. In an acidic microenvironment (pH ~6.5), PbAE undergoes rapid dissolution and releases its content. PbAE nanoparticles have been utilized to deliver small molecule drugs (30), DNA (31), and siRNA (32, 33). Recently, Jere et al. developed PbAE/PEI/PEG hybrid nanoparticles to deliver siRNA to lung cancer cells (34). The new system successfully silenced enhanced green fluorescence protein (EGFP) and oncoprotein Akt1 expression and was found to be much safer than standard PEI25K. Tseng et al. incorporated tertiary amines and PEG into *poly*(ester urethane) to synthesize a soluble *poly*(amino ester glycol urethane) (PaEGU) to formulate PaEGU/siRNA polyplexes for gene silencing (35). After optimization, the polyplexes showed lower cytotoxicity compared with Lipofectamine. Significant (up to 92%) transient knockdown of constitutive luciferase expression was achieved in the recombinant human HT-1080 fibroblast model via anti-luciferase siRNA delivery.

"Lipidoids" are another novel class of biomaterials that have been developed for siRNA delivery by Anderson et al. (36). They are lipid-like molecules containing tertiary amines and can form siRNA–lipidoid complexes by mixing. Recently, synthetic methods were developed to allow for rapid synthesis of a large library of over 1,200 structurally diverse lipidoids. From this library, lipidoids that facilitate high levels of specific silencing of endogenous gene transcripts were identified when formulated with both double-stranded siRNA and single-stranded antisense RNA targeting micro-RNA. The safety and efficacy of lipidoids have been evaluated in rodents and non-human primates, suggesting their potential for both local and systemic delivery of RNA therapies.

2.3. Using Non-cationic Polymers for siRNA Delivery

To avoid the toxicity associated with cationic polymers, non-cationic polymers have also been explored for siRNA delivery. For example, Toub et al. reported a nanoparticle formulation using biodegradable, non-cationic *poly*(isobutyl cyanoacrylate) to form liposomes for siRNA delivery. In vivo delivery of siRNA targeting the EWSYFli1 transcription resulted in effective tumor growth inhibition (37). As another example of non-cationic polymer, *poly*(lactic-co-glycolic acid) (PLGA) has been widely used in drug delivery, including siRNA delivery (38–40). PLGA is biodegradable, biocompatible, and allows for sustained release of siRNA at the site of interest. Since PLGA has been approved by FDA for clinical use, nanoparticle formulations using PLGA hold potential for rapid clinical translation. Recently, Imamura et al. explored siRNA delivery using PLGA encapsulation. In this study, tumor growth in the Eker (*Tsc2* mutant) rat model of hereditary renal cancer was successfully inhibited with the silencing of the *Erc/*mesothelin gene (41). Woodrow et al. used PLGA nanoparticles to load high amounts of siRNA (42). In this study, a single dose of siRNA-loaded nanoparticles to the mouse female reproductive tract caused efficient and sustained gene silencing. Knockdown of gene expression was observed proximal (in the vaginal lumen) and distal (in the uterine horns) to the site of topical delivery. In addition, nanoparticles were found to penetrate deep into the epithelial tissue.

2.4. Polymer–siRNA Conjugates

In the above nanoparticle formulations, siRNA is encapsulated through either charge interactions with polymers or physical entrapment into polymer matrix. In addition, siRNA can be directly conjugated to polymer chains for delivery. siRNAs have been directly conjugated to small targeting ligands including peptides (43), cholesterols (44, 45), and aptamers (46, 47). Recently, reversible cross-linkers were applied to conjugate siRNA to small polymer chains for delivery (48, 49). Heredia et al. recently reported a straightforward synthetic method to prepare pyridyl disulfide end-functionalized *poly*(PEG acrylate) by RAFT polymerization for siRNA conjugation and delivery (50). To improve siRNA delivery efficiency, Oishi et al. conjugated siRNA to pH-sensitive polymers (51). In this design, one end of the polymer is lactosylated for targeting. The other end was conjugated to siRNA via an acid-labile linkage of β-thiopropionate, which was readily cleaved in low pH endosomal compartments.

Recently, Mirus Bio Corporation developed a rational approach to siRNA delivery, which they termed "Dynamic

PolyConjugates" (52). An amphipathic *poly*(vinyl ether) termed PBAVE composed of butyl and amino vinyl ethers was designed as the backbone of the delivery vehicle. The siRNA is grafted to PBAVE through a reversible disulfide linkage, and this covalent linkage prevented displacement of the siRNA from the polymer en route to the target cell. Likewise, other functional components, including PEG, targeting ligands, and endosomolytic components, were also reversibly conjugated to the polymer backbone to form the siRNA delivery vehicle. Using this delivery technology, effective knockdown of two endogenous liver genes, apolipoprotein B and peroxisome proliferator-activated receptor alpha, was reported.

3. Targeting in siRNA Delivery

Nanoparticles can target diseased cells or tissues after systemic administration through passive targeting (e.g., the drug carrier accumulates in the interstitial compartment of the diseased tissue by enhanced permeability and retention) or active targeting (e.g., the drug carrier penetrates into cellular compartment through the targeting ligand—cell receptor interaction). Among these mechanisms, active targeting has attracted most attention because it may enhance delivery efficacy as well as reduce adverse effects on normal tissues. A variety of targeting ligands have been engineered and applied for siRNA delivery.

3.1. Monoclonal Antibodies and Antibody Fragments

Monoclonal antibodies (mAbs) are a class of molecules commonly used for targeting purposes. Although highly specific targeting can be achieved with mAbs, their large and complex molecular structure and high production cost often hinder applications in drug delivery. Antibody fragments retain the high antigen-binding specificity of antibodies, while their smaller molecular size helps to enhance tissue penetration and ease of engineering into multifunctional delivery systems. Thus, antibody fragments have become an attractive alternative for targeting. Common targeting antibody fragments include $F(ab)_2$, Fab, or even smaller variants such as Fv and single-chain variable fragments.

Both mAbs and antibody fragments have been conjugated to cationic polymers as targeting ligands for gene delivery. However, only a few studies of polymer–antibody conjugates were reported for siRNA delivery. Song et al. reported the development of antibody–protamine fusion proteins as vehicles for receptor-directed siRNA delivery (53, 54). In this study, the fusion protein (F105-P) was designed with the protamine coding sequence linked to the C terminus of the heavy chain Fab fragment of an HIV-1 envelope antibody and further bound to siRNAs. The fusion successfully silenced HIV-1 capsid gene (gag) and inhibited HIV replication in hard-to-transfect, HIV-infected primary T cells. Recently, mAbs mono-biotinylated with siRNA were engineered to cross the blood–brain barrier, therefore acting as a molecular "Trojan horse" (55). Peer et al. conjugated b7 integrin antibodies to polymeric nanoparticles and targeted specific leukocyte subsets involved in gut inflammation. Nanoparticles were loaded with cyclin D1 siRNA and they reversed experimentally induced colitis in mice by suppressing leukocyte proliferation and T-helper 1 (Th1) cytokine expression (56).

3.2. Small Peptide-Based Targeting Molecules

Short peptides have also been successfully used to bind to targets with high specificity and affinity. They are attractive as targeting molecules due to their small size, low immunogenicity, high stability, and ease of manufacturing. Recently, the development of peptide phage libraries, bacterial peptide display libraries, plasmid peptide libraries, and new screening technologies have made their selection much easier, contributing to their popularity as targeting ligands. Short peptide chains can be directly conjugated to siRNAs. Kumar et al. recently used a short peptide derived from rabies virus glycoprotein (RVG) as a targeting ligand to cross the blood–brain barrier (57). This 29-amino acid peptide was found to specifically bind to the acetylcholine receptor expressed in neuronal cells. In this study, the chimeric peptide was synthesized by adding nonamer arginine residues at the C terminus of RVG for siRNA conjugation. This RVG-9R peptide was able to bind and transduce siRNA to neuronal cells in vitro and shows specific gene silencing within the brain in vivo.

Transferrin (Tf) is a glycoprotein that delivers iron to cells. Since the transferrin receptor (TfR) is overexpressed on tumor cells, Tf has been used as ligand for tumor-targeted delivery and has been tested in human clinical trials (58, 59). Notably, human Tf can bind to the TfR in rodent and non-human primate species and therefore could be taken from pre-clinical studies into human patients without lengthy reformulation steps (60). By using cyclodextrin-containing polycations, Heidel et al. reported the first targeted delivery of siRNA in non-human primates (60). Recently, Tietze et al. conjugated transferrin with a β-propionamide-cross-linked oligoethylenimine-based polyplex for targeting and successfully silenced the Ran protein in murine neuroblastoma cells (27).

Protein transduction domains, also called cell penetrating peptides, are a class of small cationic peptides of approximately 10–30 amino acids in length. They can interact with cell surfaces and induce cellular internalization through a number of mechanisms (61). Peptides with an exposed arginine–glycine–aspartate (RGD) sequence can target integrin receptors that mediate cell adhesion to the extracellular matrix. Schiffelers et al. used PEI nanoparticles PEGylated with RGD peptide at the distal end to deliver siRNA (62). The nanoparticles showed effective targeting to tumor neovasculature-expressing integrins and successfully inhibited tumor angiogenesis by silencing the VEGF-2 gene.

3.3. Aptamers

Aptamers are single-stranded nucleic acids (~15 kDa), which fold into unique conformations that can bind to targets with high affinity and specificity. Aptamers are selected through an in vitro process of systemic evolution of ligands by exponential enrichment (SELEX) against specific targets (63). With the development of SELEX, more than 200 aptamers have now been isolated and applied for various uses.

Using nanoparticles made from PLGA–PEG copolymers, aptamers have been used as targeting ligands to deliver chemotherapy drugs in vivo (39, 40). Aptamer-mediated siRNA delivery so far has been mainly focused on using direct aptamer–siRNA conjugates (47). For example, siRNA targeting PLK1 and BCL2 in prostate cancer cells was directly conjugated to aptamers that specifically bind to prostate-specific membrane antigen (PSMA) (64). siRNA–aptamer conjugates were internalized by PSMA-positive cells, which resulted in target protein knockdown and cell death. In contrast, conjugates were not internalized by PSMA-negative prostate cancer cells. Inhibition of tumor growth was further demonstrated in a xenograft model of prostate cancer. A similar strategy was also used to deliver the anti-HIV siRNAs through fusion to an anti-gp120 aptamer, which resulted in the successful inhibition of HIV replication (65).

3.4. Small Molecules

Small molecules have gained attention as a class of targeting molecules due to their unique advantages. They are cost-effective, robust, and can endure harsh chemical synthetic processes.

Folic acid is a well-studied small molecule widely used as targeting ligands against folate receptors which are overexpressed in human tumors (66). In one study, folic acid-conjugated PEI differentially targeted tumors over healthy tissue (67). Lactose has also been used as a targeting ligand, and cells expressing asialoglycoprotein receptors can recognize compounds bearing terminal galactose moieties. Oishi et al. conjugated siRNA to lactosylated PEG through an acid-labile linkage of β-thiopropionate, followed by complexation with poly(L-lysine) (51). Effective gene silencing was observed from targeting human hepatoma cells, HuH-7.

Another small molecule, *N*-acetylgalactosamine (NAG), was used in the PolyConjugates (52). This small ligand was directly conjugated to the PBAVE backbone and used to target hepatocytes. It has been shown that NAG facilitates the preferential accumulation of polyconjugates in hepatocytes. Only minimal association with non-parenchymal cells in liver sinusoids was observed.

4. siRNA Cytoplasm Release

4.1. Proton Sponge Effect

Nanoparticles functionalized with targeting ligands enter cells through either passive or receptor-mediated endocytosis. However, siRNA is negatively charged and hydrophilic. Therefore, after being trapped inside the endosomal compartment, the siRNA molecules cannot spontaneously cross the plasma membrane to reach the cytoplasm and are degraded by lysosomal enzymes upon endosomal to lysosomal sorting. Endosomal escape thus becomes a key step for successful siRNA delivery. One typical method to enhance endosomal escape is to use cationic polymers with a pKa slightly below physiological pH. Due to the acidic nature of endosome, cationic polymers absorb protons and increase the osmotic pressure inside the endosomal compartment. This "proton sponge effect" leads to the disruption of plasma membrane and subsequent release of siRNA into cytoplasm (68).

It has been shown that histidine residues with a *pKa* of 6.4 can enable endosomal buffering and escape. siRNA delivery using reducible polycations containing histidine monomers efficiently reduced hepatitis B viral surface antigen expression and corresponding viral titer from Alexander cells (69). In another study using histidine–lysine-rich polymers to deliver Raf-1 siRNA, effective knockdown of Raf-1 mRNA was observed and resulted in apoptosis in several cell lines (70). Besides endosomal escape, histidine has been found to facilitate nucleic acid condensation and membrane fusion, which may be beneficial for siRNA delivery (71).

There are other strategies that may not directly cause endosome disruption but may aid nanoparticle escape from the endosome. Takea et al. developed PEG-detachable polyplex micelles, where the PEG layer surrounding the particle core was conjugated through reversible disulfide bonds which could be detached in the reducing endosomal compartment (72). PEG detachment is assumed to enhance endosomal escape by increasing the interaction between the exposed nanoparticle cationic core and the endosomal membrane. PEG detachment also favors the release of siRNA from the nanoparticle due to the removal of the surrounding PEG layer.

4.2. Protein Transduction Domains (PTD)

As mentioned above, PTD peptides can facilitate siRNA packaging and cell targeting. More importantly, some of these peptides are especially capable of escaping the endosome and gain access to the cytoplasm. Although the mechanism used by PTD peptides to penetrate membranes is not well understood, their effect on siRNA cytoplasm delivery has been established from numerous studies. It remains challenging to integrate these peptides as components into complex nanoparticle structures while preserving their membrane penetrating properties. Recently, Holowka et al. prepared vesicles composed of arginine-rich PTDs (73). The particles can entrap water-soluble species and be processed to different sizes and scaled up. In this nanoparticle design, the PTD segments carry out dual functions of directing structure for vesicle formation and providing functionality for efficient intracellular delivery of the vesicles. This is a good example of the unique synergy between nanoscale self-assembly and inherent peptide functionality in designing multifunctional nanoparticles for siRNA delivery.

5. Final Remarks

Novel approaches to multifunctional nanoparticle design and engineering have resulted in significant improvements in the performance of nanoparticles in pre-clinical and clinical studies. This chapter has reviewed recent progress made in the application of nanoparticle technology for siRNA delivery. With continued research and development it is expected that RNAi approaches will result in a new class of impactful drugs for medical applications.

Acknowledgment

We thank Juliana Chan for critical review of the chapter. Authors' research was supported by National Institutes of Health Grants CA119349 and EB003647 and Koch–Prostate Cancer Foundation Award in Nanotherapeutics.

References

1. Elbashir, S.M., Harborth, J., Lendeckel, W. et al. (2001) Duplexes of 21-nucleotide RNAs mediate RNA interference in cultured mammalian cells. *Nature*, **411**, 494–498.
2. Milhavet, O., Gary, D.S., and Mattson, M.P. (2003) RNA interference in biology and medicine. *Pharmacol Rev*, **55**, 629–648.
3. Cheng, J.C., Moore, T.B., and Sakamoto, K.M. (2003) RNA interference and human disease. *Mol Genet Metab*, **80**, 121–128.
4. Leung, R.K.M. and Whittaker, R.A. (2005) RNA interference: from gene silencing to gene specific therapeutics. *Pharmacol Ther*, **107**, 222–239.
5. Drude, I., Dombos, V., Vauleon, S. et al. (2007) Drugs made of RNA: development and applicaiton of engineered RNAs for gene therapy. *Mini Rev Med Chem*, **7**, 912–931.
6. Paula, D.D., Bentley, M.V.L.B., and Mahato, R.I. (2007) Hydrophobization and

bioconjugation for enhanced siRNA delivery and targeting. *RNA*, **13**, 431–456.
7. Fougerolles, A.R. de, Vornlocher, H.-P., Maraganore, J. et al. (2007) Interfering with disease: a progress report on siRNA-based therapeutics. *Nat Rev Drug Discovery*, **6**, 443–453.
8. Fougerolles, A.R. de. (2008) Delivery vehicles for small interfering RNA in vivo. *Human Gene Ther*, **19**, 125–132.
9. Feinberg, E.H. and Hunter, C.P. (2003) Transport of dsRNA into cells by the transmembrane protein SID-1. *Science*, **301**, 1545–1547.
10. Barton, G.M. and Medzhitov, R. (2002) Retroviral delivery of small interfering RNA into primary cells. *Proc Natl Acad Sci*, **99**, 14943–14945.
11. Morris, K.V. and Rossi, J.J. (2006) Lentiviral-mediated delivery of siRNA for antiviral therapy. *Gene Ther*, **13**, 553–558.
12. Devroe, E. and Silver, P.A. (2004) Therapeutic potential of retroviral RNAi vectors. *Expert Opin Biol Ther*, **4**, 319–327.
13. Somia, N. and Verma, I.M. (2000) Gene therapy: trials and tribulations. *Nat Rev Genet*, **1**, 91–99.
14. Schaffer, D.V., Koerber, J.T., and Lim, K. (2008) Molecular engineering of viral gene delivery vehicles. *Annu Rev Biomed Eng*, **10**, 169–194.
15. Duncan, R. (2003) The dawning era of polymer therapeutics. *Nat Rev Drug Discovery*, **2**, 347–360.
16. Peer, D., Karp, J.M., Hong, S. et al. (2007) Nanocarriers as an emerging platform for cancer therapy. *Nat Nanotechnol*, **2**, 751–760.
17. Davis, M.E., Chen, Z., and Shin, D.M. (2008) Nanoparticle therapeutics: an emerging treatment modality for cancer. *Nat Rev Drug Discovery*, **7**, 771–782.
18. Putnam, D. (2006) Polymers for gene delivery across length scales. *Nat Mater*, **5**, 439–451.
19. Gary, D.J., Puri, N., and Won, Y.-Y. (2007) Polymer-based siRNA delivery: perspectives on the fundamental and phenomenological distinctions from polymer-based DNA delivery. *J Control Release*, **121**, 64–73.
20. Bolcato-Bellemin, A.-L., Bonnet, M.-E., Creusat, G. et al. (2007) Sticky overhangs enhance siRNA-mediated gene silencing. *Proc Natl Acad Sci*, **104**, 16050–16055.
21. Li, S.-D., Chen, Y.-C., Hackett, M.J. et al. (2007) Tumor-targeted delivery of siRNA by self-assembled nanoparticles. *Mol Ther*, **16**, 163–169.
22. Malek, A., Czubzayk, F., and Aigner, A. (2008) PEG grafting of polyethylenimine (PEI) exerts different effects on DNA transfection and siRNA-induced gene targeting efficacy. *J Drug Target*, **16**, 124–139.
23. Breunig, M., Hozsa, C., Lungwitz, U. et al. (2008) Mechanistic investigation of poly(ethylene imine)-based siRNA delivery: disulfide bonds boost intracellular release of the cargo. *J Control Release*, **130**, 57–63.
24. Zintchenko, A., Philipp, A., Dehshahri, A. et al. (2008) Simple modifications of branched PEI lead to highly efficient siRNA carriers with low toxicity. *Bioconjugate Chem*, **19**, 1448–1455.
25. Shim, M.S. and Kwon, Y.J. (2008) Controlled delivery of plasmid DNA and siRNA to intracellular targets using ketalized polyethylenimine. *Biomacromolecules*, **9**, 444–455.
26. Sun, T.-M., Du, J.-Z., Yan, L.-F. et al. (2008) Self-assembled biodegradable micellar nanoparticles of amphiphilic and cationic block copolymer for siRNA delivery. *Biomater*, **29**, 4348–4355.
27. Tietze, N., Pelisek, J., Philipp, A. et al. (2008) Induction of apoptosis in murine neuroblastoma by systemic delivery of transferrin-shielded siRNA polyplexes for downregulation of Ran. *Oligonucleotides*, **18**, 161–174.
28. Jeong, J.H., Christensen, L.V., Yockman, J.W. et al. (2007) Reducible poly(amido ethylenimine) directed to enhance RNA interference. *Biomaterials*, **28**, 1912–1917.
29. Littlea, S.R. and Kohane, D.S. (2008) Polymers for intracellular delivery of nucleic acids. *J Mater Chem*, **18**, 832–841.
30. Koa, J., Park, K., Kim, Y.-S. et al. (2007) Tumoral acidic extracellular pH targeting of pH-responsive MPEG-poly (β-amino ester) block copolymer micelles for cancer therapy. *J Control Release*, **123**, 109–115.
31. Little, S.R., Lynn, D.M., Ge, Q. et al. (2004) Poly-β-amino ester-containing microparticles enhance the activity of nonviral genetic vaccines. *Proc Natl Acad Sci*, **101**, 9534–9539.
32. Lynn, D.M., Anderson, D.G., Putnam, D. et al. (2001) Accelerated discovery of synthetic transfection vectors: parallel synthesis and screening of a degradable polymer library. *J Am Chem Soc*, **123**, 8155–8156.
33. Anderson, D.G., Lynn, D.M., and Langer, R. (2003) Semi-automated synthesis and screening of a large library of degradable cationic polymers for gene delivery. *Angew Chem Int Ed*, **42**, 3153–3158.

34. Jere, D., Xu, C.-X., Arote, R. et al. (2008) Poly(β-amino ester) as a carrier for si/shRNA delivery in lung cancer cells. *Biomaterials*, **29**, 2535–2547.
35. Tseng, S.-J. and Tang, S.-C. (2007) Development of poly(amino ester glycol urethane)/siRNA polyplexes for gene silencing. *Bioconjuate Chem*, **18**, 1383–1390.
36. Akin, A., Andreas, Z., Michael, G. et al. (2008) A combinatorial library of lipid-like materials for delivery of RNAi therapeutics. *Nat Biotechnol*, **26**, 561–569.
37. Toub, N., Bertrand, J.-R., Tamaddon, A. et al. (2006) Efficacy of siRNA nanocapsules targeted against the EWS-Fli1 oncogene in Ewing Sarcoma. *Pharm Res*, **23**, 892–900.
38. Cheng, J., Teply, B.A., Sherifi, I. et al. (2007) Formulation of functionalized PLGA-PEG nanoparticles for in vivo targeted drug delivery. *Biomaterials*, **28**, 869–876.
39. Farokhzad, O.C., Cheng, J., Teply, B.A. et al. (2006) Targeted nanoparticle-aptamer bioconjugates for cancer chemotherapy in vivo. *Proc Natl Acad Sci*, **103**, 6315–6320.
40. Gu, F., Zhang, L., Teply, B.A. et al. (2007) Precise engineering of targeted nanoparticles by using self-assembled biointegrated block copolymers. *Proc Natl Acad Sci*, **105**, 2586–2591.
41. Imamura, O., Okada, H., Takashima, Y. et al. (2008) siRNA-mediated Erc gene silencing suppresses tumor growth in Tsc2 mutant renal carcinoma model. *Cancer Lett*, **268**, 278–285.
42. Woodrow, K.A., Cu, Y., Booth, C.J. et al. (2009) Intravaginal gene silencing using biodegradable polymer nanoparticles densely loaded with small-interfering RNA. *Nat Mater*, **8**, 526–533.
43. Emerich, D.F. and Thanos, C.G. (2008) Multifunctional peptide-based nanosystems for improving delivery and molecular imaging. *Curr Opin Mol Ther*, **10**, 132–139.
44. Kubo, T., Zhelev, Z., Ohba, H. et al. (2007) Modified 27-nt dsRNAs with dramatically enhanced stability in serum and long-term RNAi activity. *Oligonucleotides*, **17**, 445–464.
45. Wolfrum, C., Shi, S., Jayaprakash, K.N. et al. (2007) Mechanisms and optimization of in vivo delivery of lipophilic siRNAs. *Nat Biotechnol*, **25**, 1149–1157.
46. Nicolas, J., Mantovani, G., and Haddleton, D.M. (2007) Living radical polymerization as a tool for the synthesis of polymer-protein/peptide bioconjugates. *Macromol Rapid Commun*, **28**, 1083–1111.
47. Chu, T.C., Twu, K.Y., Ellington, A.D. et al. (2006) Aptamer mediated siRNA delivery. *Nucl Acids Res*, **34**, e73.
48. Moad, G., Rizzardo, E., and Thang, S.H. (2008) Toward living radical polymerization. *Acc Chem Res*, **41**, 1133–1142.
49. Stenzel, M.H. (2008) RAFT polymerization: an avenue to functional polymeric micelles for drug delivery. *Chem Commun*, **30**, 3486–3503.
50. Heredia, K.L., Nguyen, T.H., Chang, C.-W. et al. (2008) Reversible siRNA-polymer conjugates by RAFT polymerization. *Chem Commun*, **28**, 3245–3247.
51. Oishi, M., Nagasaki, Y., Itaka, K. et al. (2005) Lactosylated poly(ethylene glycol)-siRNA conjugate through acid-labile b-thiopropionate linkage to construct pH-sensitive polyion complex micelles achieving enhanced gene silencing in hepatoma cells. *J Am Chem Soc*, **127**, 1624–1625.
52. Rozema, D.B., Lewis, D.L., Wakefield, D.H. et al. (2007) Dynamic polyconjugates for targeted in vivo delivery of siRNA to hepatocytes. *Proc Natl Acad Sci*, **104**, 12982–12987.
53. Song, E., Lee, S.-K., Wang, J. et al. (2003) RNA interference targeting Fas protects mice from fulminant hepatitis. *Nat Med*, **9**, 347–351.
54. Vornlocher, H.-P. (2006) Antibody-directed cell-type-specific delivery of siRNA. *Trends Mol Med*, **12**, 1–3.
55. Peer, D., Park, E.J., Morishita, Y. et al. (2008) Systemic leukocyte-directed siRNA delivery revealing cyclin D1 as an anti-inflammatory target. *Science*, **319**, 627–630.
56. Song, E., Zhu, P., Lee, S.-K. et al. (2005) Antibody mediated in vivo delivery of small interfering RNAs via cell-surface receptors. *Nat Biotechnol*, **23**, 709–717.
57. Kumar, P., Wu, H., McBride, J.L. et al. (2007) Transvascular delivery of small interfering RNA to the central nervous system. *Nature*, **448**, 39–43.
58. Head, J.F., Wang, F., and Elliott, R.L. (1997) Antineoplastic drugs that interfere with iron metabolism in cancer cells. *Advan Enzyme Regul*, **37**, 147–169.
59. Rainov, N.G. and Soling, A. (2005) Technology evaluation: TransMID, KS Biomedix/Nycomed/Sosei/PharmaEngine. *Curr Opin Mol Ther*, **7**, 483–492.
60. Heidel, J.D., Yu, Z., Liu, J.Y.-C. et al. (2007) Administration in non-human primates of escalating intravenous doses of targeted nanoparticles containing ribonucleotide reductase subunit M2 siRNA. *Proc Natl Acad Sci*, **104**, 5715–5721.

61. Lundberg, M., Wikstrom, S., and Johansson, M. (2003) Cell surface adherence and endocytosis of protein transduction domains. *Mol Ther*, **8**, 143–150.
62. Schiffelers, R.M., Ansari, A., Xu, J. et al. (2004) Cancer siRNA therapy by tumor selective delivery with ligand-targeted sterically stabilized nanoparticle. *Nucl Acids Res*, **32**, e149.
63. Tuerk, C. and Gold, L. (1990) Systematic evolution of ligands by exponential enrichment: RNA ligands to bacteriophage T4 DNA polymerase. *Science*, **249**, 505–510.
64. McNamara, J.O., Andrechek, E.R., Wang, Y. et al. (2006) Cell type–specific delivery of siRNAs with aptamer-siRNA chimeras. *Nat Biotechnol*, **24**, 1005–1015.
65. Zhou, J., Li, H., Li, S. et al. (2008) Novel dual inhibitory function aptamer–siRNA delivery system for HIV-1 therapy. *Mol Ther*, **16**, 1481–1489.
66. Zhao, X.B., Li, H., and Lee, R.J. (2008) Targeted drug delivery via folate receptors. *Expert Opin Drug Deliv*, **5**, 309–319.
67. Kima, S.H., Jeonga, J.H., Choa, K.C. et al. (2005) Target-specific gene silencing by siRNA plasmid DNA complexed with folate-modified poly(ethylenimine). *J Control Release*, **104**, 223–232.
68. Behr, J.-P. (1997) The proton sponge: a trick to enter cells the viruses did not exploit. *CHIMIA*, **51**, 34–36.
69. Stevenson, M., Ramos-Perez, V., Singh, S. et al. (2008) Delivery of siRNA mediated by histidine-containing reducible polycations. *J Control Release*, **130**, 46–56.
70. Leng, Q., Scaria, P., Lu, P. et al. (2008) Systemic delivery of HK Raf-1 siRNA polyplexes inhibits MDA-MB-435 xenografts. *Cancer Gene Ther*, **15**, 485–495.
71. Kumar, V.V., Pichon, C., Refregiers, M. et al. (2003) Single histidine residue in head-group region is sufficient to impart remarkable gene transfection properties to cationic lipids: evidence for histidine-mediated membrane fusion at acidic pH. *Gene Ther*, **10**, 1206–1215.
72. Takae, S., Miyata, K., Oba, M. et al. (2008) PEG-detachable polyplex micelles based on disulfide-linked block catiomers as bioresponsive nonviral gene vectors. *J Am Chem Soc*, **130**, 6001–6009.
73. Holowka, E.P., Sun, V.Z., Kamei, D.T. et al. (2007) Polyarginine segments in block copolypeptides drive both vesicular assembly and intracellular delivery. *Nat Mater*, **6**, 52–57.

Chapter 5

Optimized Protocols for siRNA Delivery into Monocytes and Dendritic Cells

Anne Mobergslien and Mouldy Sioud

Abstract

Over the past decade, immunotherapy has emerged as a promising alternative form of cancer treatment with the potential to eradicate tumour metastasis. Unfortunately, its curative potential is in general limited by immunosuppressive proteins that negatively regulate dendritic (DC) and/or T-cell function. The recent discovery of RNA interference (RNAi) has facilitated the study of gene function in immune cells and recent data indicate that DC maturation, function, and survival can be modulated by small interfering RNAs (siRNAs) targeting genes involved in immune suppression. This chapter describes detailed protocols for introducing siRNAs into human monocytes and dendritic cells using standard electroporation techniques.

Key words: RNA interference, siRNAs, dendritic cells, electroporation.

1. Introduction

One main reason for the failure of cancer vaccines is believed to be the presence of several immunosuppressive mechanisms that hamper DC activation. Therefore, the development of agents that stimulate DC function and/or suppress the expression of immunosuppressive factors, such as interleukin (IL)10 and indoleamine 2,3-dioxygenase (IDO) will facilitate the design of effective cancer vaccines. IL10 has been associated with the development of tolerogenic DCs, and IDO is an enzyme involved in inhibiting of the activation of alloreactive T cells, leading to tumour evasion of the host immune system (1–4). IDO-positive DCs are expected to convert naïve T cells into regulatory T cells, thus inducing tolerance instead of T-cell activation. Also,

M. Sioud (ed.), *RNA Therapeutics*, Methods in Molecular Biology 629,
DOI 10.1007/978-1-60761-657-3_5, © Springer Science+Business Media, LLC 2010

IDO-positive DCs produce various tryptophan metabolites such as *N*-formylkynurenine with a general immunosuppressive effect on T-cell activation (2) Thus, blocking IDO expression with siRNAs might enhance the efficacy of cancer vaccines.

To interfere with the negative regulatory mechanisms in DCs such as the expression of IL-10, we have used RNA interference (RNAi), a sequence-specific gene silencing mechanism triggered by double-stranded RNA (5, 6). It has been shown to be a powerful tool to study gene function in a broad range of organisms. Small interfering RNAs (siRNAs), the effector of RNAi, hold great promise as a therapeutic tool (7, 8). They can be used alone or in combination with other traditional therapies. We and others have shown that chemically made siRNAs can activate innate immunity, when delivered to the endosomes via cationic liposomes (9–11). Both single- and double-stranded siRNAs induce cytokines and interferon expression in a sequence-specific manner and the response is mainly mediated by TLR7/8 in humans. Activation of the interferon pathway by siRNAs is thought to hamper the therapeutic use of siRNAs. However, a combinatorial strategy that blocks the expression of immunosuppressive factors and simultaneously activate innate immunity could lead to more effective cancer therapies. In this respect, we have shown that targeting IL10 expression by bifunctional siRNA can block IL10 expression and activate TLR7/8 signalling in human monocytes and dendritic cells (12). In the same report, we also showed that cytoplasmic delivery of in vitro transcribed siRNA bearing 5′-triphosphate-inhibited gene expression and induced cytokine production through more likely the activation of the retinoid acid-inducible gene-I (RIG-1) that recognizes a variety of RNA viruses (13). Specifically, RIG-1 senses RNA containing a terminal 5′-triphosphate (*see* **Chapter 26**).

With the aim of breaking self-tolerance to tumour antigens, we also developed a new strategy that is based on the observation that DCs might have an additional level of self–nonself discrimination that functions at the level of the phagosome maturation (14). Under physiological conditions, self-antigens are not processed and presented by DCs to T cells. Apparently, the activation of endosomal toll-like receptors is required for both nonself and self-antigen processing and presentation by DCs (14, 15). Thus, the development of agents that stimulate DC maturation through endosomal TLRs and simultaneously inhibit the expression of immunosuppressive factors and/or regulatory T-cell function might break self-tolerance and activate high-affinity autoreactive T cells. Based on this reasoning, we have applied a novel DC-based vaccine (DC loaded with leukemia antigens that have been transfected with an IL-10 siRNA capable of coordinately activating DC via TLR7/8) in a rat model of acute myeloid leukaemia (AML). The vaccine promoted specific T-cell responses in association with reduced tumour load after i.p. challenge with AML cells (16).

As indicated above, DCs are critical for bridging the innate and the adaptive immune response. Also, they play a critical role in tumour immunity. In addition to the commercially available nucleofection kit, we have evaluated BTX electroporation on siRNA delivery to DCs (17). Electroporation is a useful tool to transfect siRNA in vitro into human cell lines that are refractory to lipid-based transfection strategies, and it was employed to test the recognition of siRNA by cytosolic RNA-sensing receptors in human monocytes and DCs (9). Also, the technique has been used by other investigators to load DC with mRNA encoding tumour antigens (18, 19) (*see* **Chapter 27**). Using the BTX instrument and 0.5 μg (37.5 pmol) of siRNA4 targeting IDO (**Table 5.1**), we almost abolish IDO gene expression in DC (**Fig. 5.3a, c**). In this chapter, we describe a detailed protocol for siRNA delivery into human monocytes and DC via BTX electroporation. Notably, the first generation of the resulting siRNA therapeutics are now being evaluated in clinical trials, raising significant interest in this emerging area of medical research.

Table 5.1
siRNA sequences

siRNA	Sequence
IDO siRNA-1	S: 5′-GCCUCCUAUUUUGGUUUAUTT-3′ A: 3′-TTCGGAGGAUAAAACCAAAUA-5′
IDO siRNA-2	S: 5′-GCAGCGUCUUUCAGUGCUUTT-3′ A: 3′-TTCGUCGCAGAAAGUUACGAA-5′
IDO siRNA-3	S: 5′-CCCUUCAAGUGUUUCACCATT-3′ A: 3′-TTGGGAAGUUCACAAAGUGGU-5′
IDO siRNA-4	S: 5′-CCGUGAGUUUGUCCUUUCATT-3′ A: 3′-TTGGCACUCAAACAGGAAAGU-5′
Scrambled IDO siRNA-4	S: 5′-GAUUUCGUACUCUGUGCUCTT-3′ A: 3′TTCUAAAGCAUGAGGCACGAG-5′
Control siRNA β-Galactosidase	S: 5′-UUGAUGUGUUUAGUCGCUATT-3′ A: 3′-TTAACUACACAAAUCAGCGAU-5′
IL10 siRNA	S: 5′-AGGAUCAGCUGGACAACUUTT-3′ A: 3′-TTUCCUAGUCGACCUGUUGAA-5′

S = Sense strand (target site)
A = Antisense strand

2. Materials

2.1. Isolation of Blood Monocytes

1. Buffy coat from healthy adult donors.
2. RMPI 1640 tissue culture medium.

3. RMPI 1640 supplemented with 10% fetal calf serum (FCS), penicillin (100 μg/mL), and streptomycin (100 μg/mL). Complete medium.
4. Lymphoprep™.
5. Pasteur pipettes.
6. Z1 Coulter Particle Counter.
7. Cell scraper.
8. Granulocyte macrophage colony stimulating factor (GM-CSF).
9. Interleukin-4.
10. Tumour necrosis factor-α (TNF-α).

2.2. siRNA Electroporation into Monocytes and Dendritic Cells

1. Electroporation apparatus BTX ECM 830.
2. 4 mm Gap cuvettes.
3. Chemically synthesized siRNAs.
4. 6-well cell culture plates.
5. RMPI 1640 supplemented with 10% FCS and antibiotics.
6. Human monocytes and dendritic cells.

2.3. Nucleofection of siRNA into DCs

1. Nucleofactor kit (Amaxa, Biosciences).
2. Transfection solutions supplemented with Amaxa transfection kit.
3. Transfection cuvettes supplemented with Amaxa transfection kit.
4. RMPI 1640 supplemented with 10% FCS and antibiotics.
5. Human monocytes and dendritic cells.
6. 6-well cell culture plates.

2.4. Preparation of Protein Extracts

1. Phosphate-buffered saline (PBS): Dissolve 8 g NaCl, 1.16 g Na_2HPO_4, 0.2 g KH_2PO_4, and 0.2 g KCl in 1 L distilled water. Adjust the pH to 7.0.
2. Lysis buffer: 10% NP40 in PBS, and 10 μg/mL each of aprotinin, pepstatin, and leupeptin.
3. Sample buffer (2×): 125 mM Tris-HCl, pH 6.8, 4% SDS, 20% glycerol, 0.1% bromophenol blue, 5% β 2-mercaptoethanol.
4. Eppendorf centrifuge.
5. γ-Globulin (10 mg/mL).
6. Protein assay kit (BioRad laboratories).
7. Disposable cuvettes.

2.5. Protein Separation in Polyacrylamide Gels

1. Vertical electrophoresis plate assembly with comb.
2. Electrophoresis power supply-EPS 600.
3. 30% Acrylamide mix, 29% (w/v) acrylamide, 1% bisacrylamide, prepared in distilled water.
4. 10% SDS, 10% ammonium sulphate (APS), and TEMED.
5. Stacking gel buffer: 1 M Tris-HCl, pH 6.8.
6. Resolving gel buffer: 1 M Tris-HCl, pH 8.8.
7. Electrophoresis buffer 10×: Dissolve 144 g glycine, 30 g Tris-base, and 10 g SDS in 1 L distilled water.
8. Coloured protein markers (Bio-rad).

2.6. Western Blotting

1. BioRad Trans-Blot.
2. Transfer buffer 10×: Dissolve 144 g glycine, and 30 g Tris-base into 1 L distilled water.
3. Transfer buffer 1×: 100 mL 10× transfer buffer, 200 mL methanol, and 700 mL distilled water.
4. Nitrocellulose transfer membrane.
5. Chromatography Whatman paper.
6. Phosphate-buffered saline with 0.1% Tween (PBS-T).
7. Blocking buffer: 5% milk in PBS-T.
8. Antibody dilution buffer: 1% milk in PBS-T.
9. Anti-indoleamine 2,3 dioxygenase (IDO) antibody.
10. Anti-mouse immunoglobulin-HRP conjugated.
11. Anti-human β-actin polyclonal antibodies.
12. Enhanced chemiluminescent (ECL) reagent.
13. High-performance chemiluminescence's film.
14. Stripping buffer: 1×: Tris 7.57 g/L, SDS 20 g/L, pH 6.8. Autoclave.
15. β-Mercaptoethanol.

3. Methods

3.1. Isolation of Mononuclear Cells

1. Transfer the buffy coat into a 75 cm^3 cell culture flask.
2. Add 50 mL RPMI medium and mix gently.
3. Carefully layer 20 mL of the diluted blood over 10 mL Lymphoprep™ in a 50 mL falcon tube. Alternatively, Lymphoprep™ can be underlayered.
4. Centrifuge at 800×*g* for 30 min in swing-out rotor.

5. Subsequent to centrifugation, the mononuclear cells form a distinct band at the sample/medium interface.
6. Remove the cells from the interface using a Pasteur pipette.
7. Dilute the harvested cells with medium and pellet the cells by centrifugation for 5 min at 800×*g*.
8. Wash the cells twice with serum-free RPMI 1640 medium or PBS.
9. Subsequent to washing, resuspend the cells in 50–100 mL complete RPMI 1640 medium.
10. Dispense 50 × 10^6 cells per 25 mL complete medium into 75 cm^3 cell culture flasks and then incubate at 37°C for 1–2 h.
11. Subsequently, carefully aspirate the medium and wash the adherent cells with 10 mL warmed RPMI 1640 complete medium.
12. Subsequent to washing, add 10–15 mL complete medium and carefully harvest the cells by scraping (*see* **Note 1**).
13. Transfer the cells to 50 mL falcon tubes, and then centrifuge at 800×*g* for 5 min.
14. Remove the supernatant, resuspend the cells in complete RPMI 1640 medium, and then determine the cell number.
15. Disperse the appropriate cell numbers into 25- or 75-cm^3 tissue culture flasks.

3.2. Generation of Immature and Mature Dendritic Cells

Immature monocyte-derived dendritic cells (immoDCs) were generated by culturing monocytes in the presence of GM-CSF (25 ng/mL) and IL-4 (50 ng/mL) for 4–5 days. To generate mature monocyte-derived DCs (mmoDCs), the cells were further stimulated with TNF-α (50 ng/mL) for two following days.

3.3. Electroporation of siRNAs into Human Monocytes and Dendritic Cells

Prior to electroporation, wash the cells twice with serum free medium and then resuspend at 5–10 × 10^6 cells/0.5 mL serum free medium.

1. Dispense RMPI complete medium into 6-well tissue culture plate (2.5 mL/well) and incubate at 37°C.
2. Prepare Eppendorf tubes with the desired siRNA concentrations.
3. Wash the electroporation cuvettes with serum free medium.
4. Mix the cells (500 μL) with the desired siRNA concentration, transfer the mixtures into the electroporation cuvette, and then incubate at 4°C for 2–5 min.
5. During the incubation time, set up the electroporation parameters:

- Volt: 500
- Pulse length: 2 ms
- Pulse: 01
- Interval: 01.0 s
- Polarity: Unipolar

6. Subsequently, place the metal walls in the cuvette against the metal walls in the electroporator, and then pulse (*see* **Note 2**).
7. Transfer the electroporated cells immediately to 2.5 mL pre-warmed RMPI complete medium and incubate the cells at 37°C in humidified 5% CO_2 for at least 1 h.
8. To induce IDO gene expression, add INF-γ (500 ng/mL).
9. Incubate at 37°C in humidified 5% CO_2 for 15–20 h prior analysing siRNA effects.

3.4. siRNA Delivery to Monocytes Using the Nucleofection Method

1. Before transfection, dispense RMPI complete medium into 6-well tissue culture plate (2.5 mL/well) and incubate at 37°C.
2. Count the cells, calculate the number of cells needed, and spin down. We usually use 5–10^6 cells per transfection.
3. In Eppendorf tubes, dispense the siRNA concentrations needed. Usually, we use 0.1–3 μg/transfection.
4. Select program Y-001 on the nucleofector apparatus.
5. Resuspend the cell pellet in 100 μL transfection solution provided in the kit (*see* **Note 3**).
6. Transfer 100 μL of the cell solution to an Eppendorf tube containing the siRNA, mix gently, and then transfer the solution to the transfection cuvette.
7. Insert the cuvette in the nucleofector instrument, the metal walls facing you, and press the parameter X.
8. When the transfection is completed, it is essential to add 0.5 mL of complete RPMI medium to the cells as soon as possible.
9. Transfer the cells from the cuvette (≈0.6 mL) to the 6-well plate using the white plastic pipette provided with the electroporation cuvettes.
10. The machine has to be reset between each transfection by pressing the parameter X.

3.5. Preparation of Protein Extracts

1. Subsequent to overnight transfection, harvest the cells by centrifugation for 5 min at 12,000×*g*.
2. Wash the cells twice with PBS.

3. Add to the cell pellet lysing buffer (≈30 μL/10[6] cells), mix well, and then place the cell on ice for 30 min. Vortex at least once during the incubation.
4. Subsequently, centrifuge the cells at 12,000×*g* for 10 min.
5. Save the supernatant which contains the cytoplasmic proteins, and then measure the protein concentration using, for example, the Bio-rad protein assay kit.

3.6. SDS-Polyacrylamide Gel

Polyacrylamide gel electrophoresis (PAGE) is a powerful tool for separating and identifying mixtures of proteins. Several systems exist for performing PAGE and consideration should be given to which system best suits a given sample prior to running the samples.

1. Assemble plates according to the manufacturer's instructions. Prepare 10% PAGE according to **Table 5.2**.
2. Pour the resolving gel, leaving space for the stacking gel.
3. Gently overlay the acrylamide with water-saturated isobutanol.
4. Allow the gel to polymerize for 30–60 min. A very sharp liquid–gel interface will be visible when the gel has polymerized.
5. Pour off the overlay and rinse the top of the gel several times with water.
6. Blot any remaining water with a paper towel.
7. Prepare a 5% stacking gel according to **Table 5.1**.
8. Pour the stacking gel directly onto the resolving gel.
9. Insert the comb immediately.
10. Allow the gel to polymerize for at least 30 min.

Table 5.2
Examples of SDS polyacrylamide gels

Separating gel	5.0%	7.5%	10%	12%	15%	20%	Stacking gel	
Tris-HCl pH 8.8 1 M (mL)	3.75	3.75	3.75	3.75	3.75	3.75	Tris-HCl pH 8.8 1 M (mL)	630
10% SDS (μL)	100	100	100	100	100	100	10% SDS (μL)	50
Bis-acrylamide 40% (mL)	1.25	1.88	2.5	3.13	3.75	5	Bis-acrylamide 40% (μL)	600
H_2O (mL)	4.8	4.2	3.6	2.97	2.35	1.1	H_2O (mL)	3.7
Ammonium-persulphate (μL)	50	50	50	50	50	50	Ammonium-persulphate (μL)	25
TEMED (μL)	5	5	5	5	5	5	TEMED (μL)	5

11. Prepare samples while the gel is polymerizing.
12. Carefully remove the comb, and rinse each well with water or 1× electrophoresis buffer.
13. Place the gel into the electrophoresis chamber.
14. Add 1× electrophoresis buffer to cover the wells.

3.7. Sample Preparations

1. To prepare the protein samples, mix protein extracts (up to 60 μg) 10 μL, and 10 μL of 2× sample buffer. Mix well.
2. Heat samples at 95°C for 5 min.
3. Place on ice until ready to use.
4. Centrifuge the samples at 300×*g* for 20 s, and then gently load the same sample volume beneath the buffer in each well. Loading the sample too fast will lead to diffusion of the sample in the well. Also, if a well is left empty, adjacent samples may spread.
5. Electrophorese at 100 V for 5 min, and then at 120 V until the blue dye has migrated to the bottom of the gel.

3.8. Western Blots

Protein transfer efficiency in blotting applications is dependent upon multiple factors, including the gel percentage, gel thickness, protein size, transfer conditions, and type and quality of the membrane. Therefore, transfer conditions must be adjusted to address these varying factors. The protocol given below is adjusted for the siRNA experiments done with human monocytes and DC cells. All washes are done at room temperature using TBS-T buffer, unless otherwise indicated.

1. Equilibrate the nitrocellulose membrane for 2–5 min in 10 mL 1× transfer buffer. While equilibrating, assemble filter pad and filter paper in the cathode side of the transfer module and soak in 1× transfer buffer.
2. Place gel large plate down, break open, and remove the small plate, being careful to leave the gel on large plate. Cut off and remove wells (stacking gel).
3. Place wet filter paper on the gel, remove bubbles, and flip over.
4. Assemble the transfer cassette in the following order, from the bottom (cathode unit) side-1 filter pad, Whatman paper, polyacrylamide gel, nitrocellulose membrane, Whatman paper, filter pad (anode unit).
5. Transfer the assembled cassette to the inner electrophoresis chamber.
6. Place an ice block and a magnetic stir-bar in the outer chamber and then fill the chamber with 1× transfer buffer to just cover the top of the assembled cassette.

7. Transfer for 1 h at 100 V, 400 mA at 4°C.
8. When the transfer is complete, disassemble the cassette, and remove the membrane from the gel. The coloured molecular weight marker should be visible on the membrane (*see* **Note 5**).
9. Block overnight in 50 mL of 5% milk PBS-T.
10. The next day, remove the membrane from the blocking buffer, and wash with PBS-T for 10 min.
11. Subsequent to washing, incubate the membrane with 5–10 mL diluted primary antibody (for IDO, 1/1,000) for 1 h at room temperature with gentle mixing.
12. Remove the primary antibody and wash the membrane three times with PBS-T for 5 min each (*see* **Note 5**).
13. Add dilute HRP-conjugated secondary antibody and incubate for 1 h at room temperature with gentle mixing.
14. Discard the secondary antibody and wash the membrane three times with PBS-T for 5 min each (*see* **Note 5**). During the final wash, get plastic wrap cut and spread out on bench top. Also, prepare the ECL detection solution by mixing A and B solutions. For a small blot, use 2 mL from each.
15. Drain membrane well and place it on plastic wrap.
16. Add ECL detection solution and activate for 1 min.
17. Drain membrane with cell paper and place in plastic wrap. Make sure that there are no air bubbles between the membrane and the plastic wrap. Place the membrane in an X-ray film cassette and expose to film for 30 s to 30 min.

3.9. Membrane Stripping and Actin Detection

When a satisfactory exposure has been obtained for IDO, strip the membrane in order to analyse actin protein levels (internal control).

1. Soak the membrane in 45 mL stripping buffer supplemented with 350 μL β-mercaptoethanol. Use a 50 mL falcon tube.
2. Incubate at 55°C in a water bath for 30 min with gentle mixing.
3. Subsequently, wash the membrane with PBS-T for 10 min.
4. Block the membrane for 1 h in 20 mL of 5% milk in PBS-T at room temperature with gentle mixing.
5. Fill the pipette with 12 mL lymphoprep and add it from the bottom of the tube carefully, such that the lymphoprep lies as a separate layer at the bottom.
6. Add the dilute primary anti-actin antibody and then process the membrane as indicated above (**Section 3.8**, Steps 12–17).

Under our experimental conditions, the transfection efficiency was up to 95%, consistent with the high percentage of IDO knock down in transfected cells, 95–99% (**Figs. 5.1** and **5.2**). Also, transfected cells showed that 75–85% of total cells are viable cells as assessed by trypan blue exclusion test. Using Gene-Pulser II (pulsed conditions: 300 V, 150 μF, 100 Ω) and 5 μg siRNA (375 pmol), Prechtel and colleagues achieved only 60% reduction of lamin A/C expression in DC (20). On the basis of these results, we think that the electroporation of siRNA into DCs using BTX instrument should be adapted in the clinic. Any gene involved in immunosuppression such as IDO, IL10, TGF-β, and SOCS-1 can be targeted by siRNAs. **Figure 5.3a** shows that siRNA maintained its silencing activity after freezing and thawing of DCS as outlined in **Fig. 5.3b**.

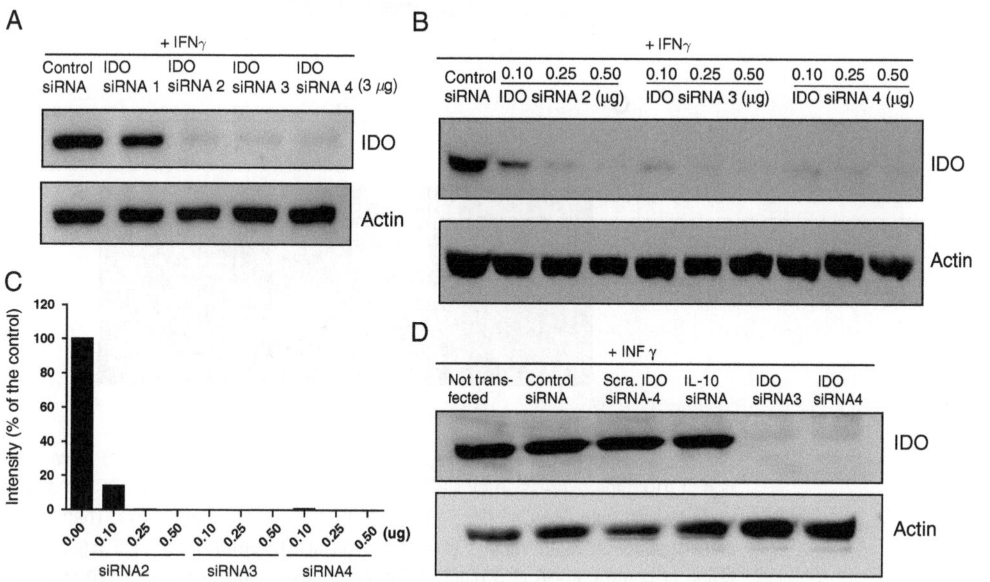

Fig. 5.1. Silencing potency of IDO siRNAs in human monocytes. (**a**) Downregulation of IDO gene expression by siRNAs. Freshly isolated monocytes were transfected with anti-IDO siRNAs (3 μg/225 pmol each) or control siRNA targeting β-galactosidase (3 μg/225 pmol) using nucleofection. Subsequently, IDO expression was induced with IFN-γ (500 U/mL) for 18 h and IDO protein levels were determined by Western blots. (**b**) Concentration-dependent downregulation of IDO gene expression by siRNAs. Freshly isolated monocytes were transfected with various concentrations of IDO siRNAs and then processed as in (**a**). (**c**) Quantification of the IDO signals shown in (**b**). The control intensity is defined as 100%. (**d**) Anti-IDO siRNAs are specific. Monocytes were transfected with the indicated siRNA duplexes (1 μg/75 pmol) and processed as in (**a**). Sca = a scrambled IDO siRNA 4. Not transfected but IFNγ-stimulated cells were included in these experiments.

4. Notes

1. Differentiation factors such as GM-CSF and IL-4 can be added directly to adherent monocytes. Therefore, there is no need for harvesting the cells by scraping.

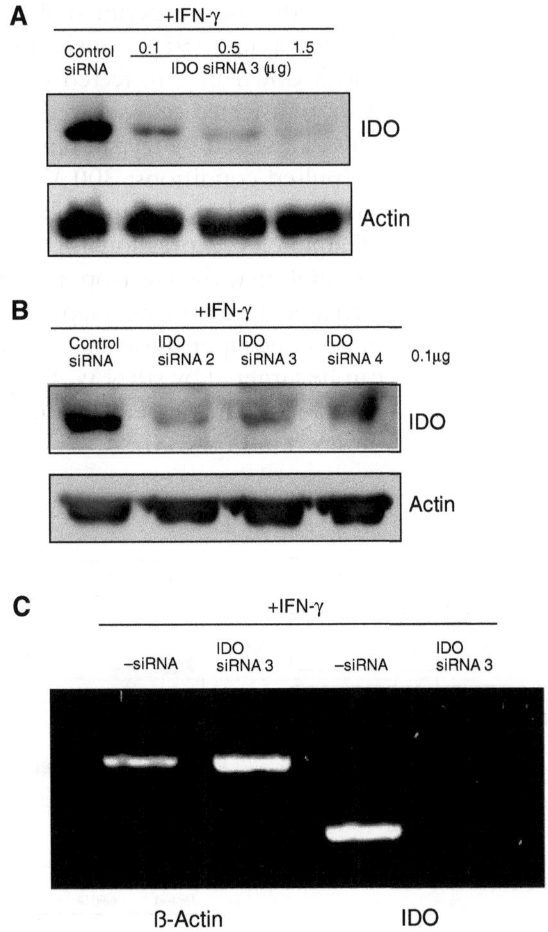

Fig. 5.2. Silencing potency of IDO siRNA in human dendritic cells. (**a**) Downregulation of IDO gene expression by siRNAs in immoDCs. To generate immoDCs, monocytes were treated with GM-CSF (25 ng/mL) and IL-4 (50 ng/mL) for 5 days. Then the cells were transfected with various concentrations of IDO siRNA3 (0.1–1.5 μg/7.5–112.5 pmol) or control siRNA (1.5 μg/112.5 pmol). Subsequently, IDO gene expression was induced by IFN-γ (500 U/mL) for 18 h and IDO protein levels were determined by Western blots. (**b**) Downregulation of IDO gene expression by siRNAs in mmoDCs. To generate mmoDCs, immoDCs were stimulated with TNF-α (50 ng/mL) for 2 days, and then they were washed, transfected with various IDO siRNAs (0.1 μg/7.5 pmol each), and processed as in (**a**). (**c**) RT-PCR analysis of IDO mRNA levels. ImmoDCs were transfected with siRNA-3 (0.4 μg/30 pmol) or control siRNA (0.4 μg/30 pmol) and then IDO expression was induced with IFN-γ for 18 h. Subsequently, total RNA was prepared and IDO and β-actin mRNA levels were analysed by RT-PCR as described in **Sections 2 and 3**.

2. If the electroporation was successful you should see some air bubbles immediately after pulsing.
3. The cells should not stay in the transfection solution for more than 15 min, so when having many transfections it is better to resuspend the cells in groups.

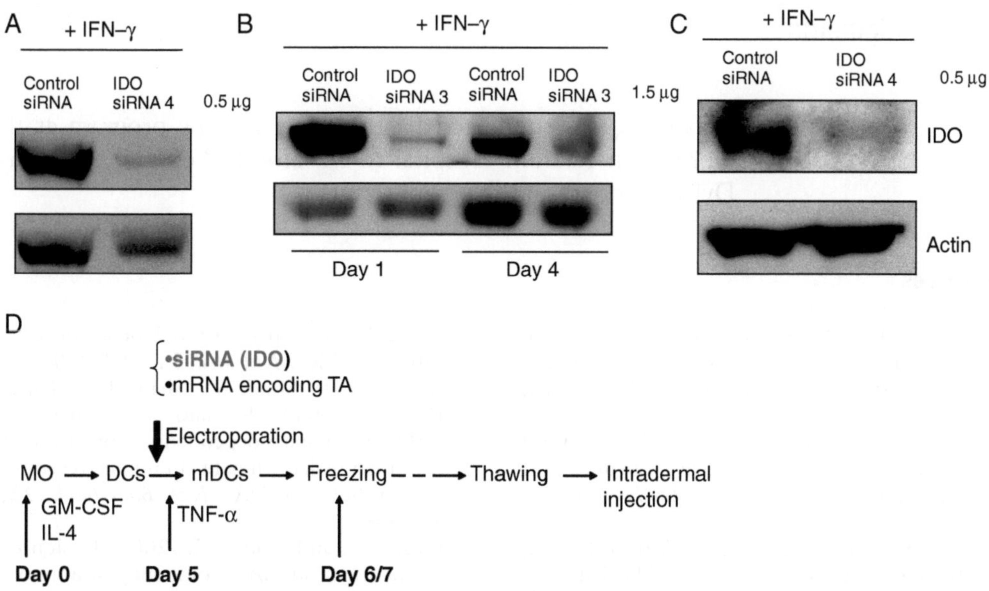

Fig. 5.3. Efficient siRNA gene silencing in DCs using the BTX electroporation method. (**a**) Transfection of IDO siRNAs with BTX electroporation in immoDCs. Around 5×10^6 cells were transfected with IDO siRNA4 (0.5 µg/37.5 pmol) or control siRNA (0.5 µg/37.5 pmol) and then IDO expression was induced with IFN-γ (500 U/mL) for 18 h. Subsequently, IDO and actin protein levels were determined by Western blots. (**b**) IDO siRNA remains functional during DC maturation. ImmoDCs were transfected with IDO siRNA-3 (1.5 µg/112.5 pmol) or control siRNA (1.5 µg/112.5 pmol) and then the cells were cultured in the presence of IL-4, GM-CSF, and TNF-α (50 ng/mL) to induce maturation. The expression of IDO following 18 h stimulation with IFN-γ was analysed by Western blots on day 1 or day 4 subsequent to transfection (day 0). (**c**) IDO siRNA remains functional after DC freezing and thawing. ImmoDCs were transfected with IDO siRNA4 (0.5 µg/37.5 pmol) or control siRNA (0.5 µg/37.5 pmol) and then they were cultured in the presence of IL-4, GM-CSF, and TNF-α (50 ng/mL) for 2 days to induce maturation. Subsequently, the cells were frozen in liquid nitrogen. On day 2 after freezing, the cells were thawed, cultured in complete medium supplemented with IFN-γ (500 U/mL) for 18 h, and then IDO protein levels were determined by Western blots. (**d**) Schematic representation of DC vaccine preparation. PBMC monocytes (MO) are differentiated into immature DC (imDC) by incubation with GM-CSF and IL-4 for 5 days and electroporated with mRNA encoding tumour antigens and siRNA targeting IDO siRNA. Subsequently, the cells are stimulated with TNF-α for 1 or more days to induce maturation. Transfected DCs are cryopreserved and frozen until use. This DC vaccine can be processed and administered to patients as described in **Chapter 27**.

4. The diluted antibody solution can be used up to four times. Therefore, store at –20°C.
5. Subsequent to transfer, it is also possible to colour the membrane with Ponceau stain before blocking in order to check transfer efficiency. Wash the Ponceau stain off with PBS-T prior to blocking.
6. To increase transfer efficiency use the lowest concentration gel that will resolve the protein of interest, avoid using gels ≥1 mm thick, decrease the concentration of methanol to optimize transfer efficiency of protein >150 kDa, longer transfer time may be used to ensure complete transfer of larger proteins (>60 kDa).

Acknowledgments

This work was supported by the Gene therapy program at the Norwegian Radium Hospital to M. Sioud. We thank Dr. Anne Dybwad for editing the manuscript.

References

1. Grütz, G. (2005) New insights into the molecular mechanism of interleukin-10-mediated immunosuppression. *J Leukoc Biol*, **77**, 3–15.
2. Munn, D.H. and Mellor, A.L. (2004) IDO expression by dendritic cells: tolerance and tryptophan catabolism. *Nat Rev Immunol*, **4**, 762–774.
3. Munn, D.H., Zhou, M., Attwood, J.T., Bondarev, I., Conway, S.J., Marshall, B., Brown, C., and Mellor, A.L. (1998) Prevention of allogeneic fetal rejection by tryptophan catabolism. *Science*, **281**, 1191–1193.
4. Munn, D.H., Sharma, M.D., Hou, D., Baban, B., Lee, J.R., Antonia, S.J., Messina, J.L., Chandler, P., Koni, P.A., and Mellor, A.L. (2004) Expression of indoleamine 2,3-dioxygenase by plasmacytoid dendritic cells in tumor-draining lymph nodes. *J Clin Invest*, **114**, 280–290.
5. Fire, A., Xu, S., Montgomery, M.K., Kostas, S.A., Driver, S.E., and Mello, C.C. (1998) Potent and specific genetic interference by double-stranded RNA in *Caenorhabditis elegans*. *Nature*, **391**, 806–811.
6. Elbashir, S.M., Harborth, J., Lendeckel, W., Yalcin, A., Weber, K., and Tuschl, T. (2001) Duplexes of 21-nucleotides RNAs mediate RNA interference in cultured mammalian cells. *Nature*, **411**, 494–498.
7. Sioud, M. (2004) Therapeutic siRNAs. *Trends Pharmacol Sci*, **25**, 22–28.
8. Behlke, M.A. (2006) Progress towards in vivo use of siRNAs. *Mol Ther*, **13**, 644–670.
9. Sioud, M. (2005) Induction of inflammatory cytokines and interferon responses by double-stranded and single-stranded siRNAs is sequence-dependent and requires endosomal localization. *J Mol Biol*, **348**, 1079–1090.
10. Hornung, V., Guenthner-Biller, M., Bourquin, C., Ablasser, A., Schlee, M., Uematsu, S., Noronha, A., Manoharan, M., Akira, S., de Fougerolles, A., Endres, S., and Hartmann, G. (2005) Sequence-specific potent induction of IFN-alpha by short interfering RNA in plasmacytoid dendritic cells through TLR7. *Nat Med*, **11**, 263–270.
11. Judge, A.D., Sood, V., Shaw, J.R., Fang, D., McClintock, K., and MacLachlan, I. (2005) Sequence-dependent stimulation of the mammalian innate immune response by synthetic siRNA. *Nat Biotechnol*, **23**, 457–462.
12. Furset, G. and Sioud, M. (2007) Design of bifunctional siRNAs: combining immunostimulation and gene-silencing in one single siRNA molecule. *Biochem Biophys Res Commun*, **352**, 642–649.
13. Takeuchi, O. and Akira, S. (2007) Recognition of viruses by innate immunity. *Immunol Rev*, **220**, 214–224.
14. Watts, C. (2004) The bell tolls for phagosome maturation. *Science*, **304**, 976–977.
15. Sioud, M. (2008) Does understanding of immune activation by RNA predict the design of safe siRNAs. *Front Biosci*, **13**, 4379–4392.
16. Iversen, P.O., Semaeva, E., Sørensen, D.R., Wiig, H., and Sioud, M. (2009) Dendritic cells loaded with tumor antigens and a dual immunostimulatory and anti-interleukin-10-specific small interference RNA prime T lymphocytes against leukemic cells. *Transl Oncol*, **2**, 242–246.
17. Flatekval, G.F. and Sioud, M. (2009) Modulation of dendritic cell maturation and function with mono- and bifunctional small interfering RNAs targeting indoleamine 2,3-dioxygenase. *Immunology*, **128**, e837–e848. DOI: 10.1111/j.1365-2567.2009.03093.x.
18. Van Tendeloo, V.F., Ponsaerts, P., Lardon, F., Nijs, G., Lenjou, M., Van Broeckhoven, C., Van Bockstaele, D.R., and Berneman, Z.N. (2001) Highly efficient gene delivery by mRNA electroporation in human hematopoietic cells: superiority to lipofection and passive pulsing of mRNA and to electroporation of plasmid cDNA for tumor antigen loading of dendritic cells. *Blood*, **98**, 49–56.
19. Michiels, A., Tuyaerts, S., Bonehill, A., Corthals, J., Breckpot, K., Heirman, C.,

Van Meirvenne, S., Dullaers, M., Allard, S., Brasseur, F., van der Bruggen, P., and Thielemans, K. (2005) Electroporation of immature and mature dendritic cells: implications for dendritic cell-based vaccines. *Gene Ther*, **12**, 772–782.

20. Prechtel, A.T., Turza, N.M., Theodoridis, A.A., Kummer, M., and Steinkasserer, A. (2006) Small interfering RNA (siRNA) delivery into monocyte-derived dendritic cells by electroporation. *J Immunol Methods*, **311**, 139–152.

Chapter 6

Systemic Delivery of Synthetic siRNAs

Dag R. Sørensen and Mouldy Sioud

Abstract

RNA interference is a biological process for gene silencing that can be harnessed for the development of new drugs. However, a major obstacle to the use of small interfering RNAs (siRNAs) as therapeutics is their delivery across the plasma membrane of cells in vivo. A range of solutions for this challenge have been described, including cationic lipids, high-pressure injection, viral vectors, and chemical modifications of the siRNAs. This chapter describes cationic lipid delivery of siRNAs to adult mice.

Key words: RNA interference, siRNA, cationic lipids.

1. Introduction

Novel tools for evaluating gene function in vivo such as ribozymes and RNA interference (RNAi) are emerging as the most highly effective strategies (1, 2). RNAi is a sequence-specific posttranscriptional gene silencing, which is triggered by double-stranded RNA (dsRNA). This process in which dsRNA mediated the degradation of homologous transcript was first described in the nematode worm *Caenorhabditis elegans* (3). Long dsRNAs are cleaved into small interfering RNAs with two-nucleotide 3′ overhangs and 5′-phosphate termini by Dicer. Thereafter, the processed siRNAs are incorporated into a multicomponent nuclease complex, known as RISC, containing Ago2 and other accessory proteins such as Tar-RNA-binding protein. Ago2 is the enzyme responsible for siRNA strand selection and cleavage of the target mRNA in the middle of the siRNA mRNA recognition site (10 nt from the 5′-end of the guide siRNA) (2, 4, 5).

M. Sioud (ed.), *RNA Therapeutics*, Methods in Molecular Biology 629,
DOI 10.1007/978-1-60761-657-3_6, © Springer Science+Business Media, LLC 2010

Although virtually any gene can be efficiently silenced by chemically made siRNAs, the therapeutic potential of RNAi requires the development of a delivery agent that can be administered efficiently, safely, and repeatedly (*see* **Chapters 4** and **9**). Cationic lipids represent one of the few examples that can meet these requirements (6). In this respect, several in vivo studies have shown that cationic liposomes such as DOTAP can enhance the uptake of siRNAs (2, 7). More sophisticated lipid mixtures and peptide structures are being developed for in vivo delivery of siRNAs (2, 8). In this chapter, we describe liposomal delivery of siRNA into adult mice.

2. Materials

1. Chemically made siRNAs.
2. Cationic liposomes DOTAP.
3. Transfection buffer 5X: 100 mM HEPES, 750 mM NaCl, pH 7.4.
4. RPMI culture medium supplemented with 10% foetal bovine serum and antibiotics.
5. Sterile syringes 1, 5, and 10 mL.
6. Glass tubes (5 and 10 mL).
7. Polystyrene and eppendorf tubes.
8. Tissue culture dishes (25 and 75 cm^3).
9. 96-well cell culture plates.
10. Pasteur pipettes.

3. Methods

3.1. Isolation of Peritoneum Macrophages for In Vitro Testing

In this study we have used BALB/C mice, 6–8 weeks of age and weighting 20–25 g.

To increase the number of residual peritoneal cells, 1 day before cell isolation, inject the mice (i.p.) with 10 μg of DOTAP in 200 μL transfection buffer (*see* **Note 1**).

1. The next day, kill the mice by cervical dislocation and place the animals into 95% ethanol.
2. Restrain the mice in the supine position and make an incision in the abdominal wall and gently lift the abdominal wall.
3. With a syringe, inject 5–6 mL of saline to the abdominal cavity; massage the abdomen for 1 min.

4. Make a small hole to admit Pasteur pipette by cutting through the skin and the muscle (*see* **Note 2**).
5. Aspirate the fluid by Pasteur pipette, while holding the muscle up by a forceps. Place the fluid into a 5 mL glass tube.
6. Spin down and wash the cells with RPMI medium.

3.2. Isolation of Adherent Peritoneal Cells

1. Plate peritoneal cells at 10×10^6/20 mL complete RPMI medium in a 75 cm^3 culture dish and incubate at 37°C for 1–2 h.
2. Gently aspirate non-adherent cells and then gently wash the adherent cells (macrophages) with 20 mL warm medium.
3. Add 10 mL complete RPMI medium and harvest the adherent cells by gentle scrapping.
4. Determine the cell number.

3.3. Transfection of Adherent Peritoneal Cells

There are several techniques available to introduce siRNA into primary cells, e.g., cationic liposomes and electroporation. It should be noted that electroporation routinely worked well for the delivery of siRNAs into primary cells (*see* **Chapter 10**). The experimental conditions described in **Chapter 10** are also suitable for transfecting mouse cells, including macrophages, monocytes, and dendritic cells.

3.4. In Vivo Delivery of siRNAs

The major obstacle for the effective application of RNAi in vivo is the delivery of the siRNA to the target cells. Although more work is needed, several lipid-based transfection reagents have been successfully used for local in vivo application (2, 9). We have been exploring the use of liposomes for testing synthetic siRNAs in animals. Whatever the type of cationic liposomes used, the formulated complexes should have a net positive charge. Therefore, it is desirable to test a range of cationic liposome concentrations and siRNA ratio.

3.5. Intraperitoneal Delivery (i.p.)

1. In a sterile microcentrifuge tube mix 100 μg of siRNA (up to 20 μL) and 80 μL of 1X transfection buffer.
2. In a separate polystyrene tube mix 100–200 μL (1 μg/μL) of DOTAP and 300–600 μL of transfection buffer. Ratio of siRNAs and DOTAP is 1:1 and 2:1, respectively.
3. Transfer the siRNA mixture to the polystyrene tube containing the DOTAP.
4. Mix gently by pipetting several times and incubate at room temperature for 30 min. The final volume should be around 500 or 800 μL.
5. Complete to 1 mL with 1X transfection buffer, mix gently, and then inject i.p. into mice.

6. After the desired time of postinjection (12–48 h), investigate the gene silencing potency of the designed siRNAs (*see* **Notes 3–6**).

3.6. Intravenous Delivery (i.v.)

1. In a sterile microcentrifuge tube mix 50–100 μg siRNA (up to 20 μL) and 40 μL of transfection buffer.
2. In a separate polystyrene tube mix 50–100 μL (1 μg/μL) of DOTAP and 90 μL of transfection buffer. Charge ratio of siRNAs and DOTAP is 1:1 (*see* **Note 7**).
3. Transfer the siRNA mixture to the polystyrene tube containing the DOTAP.
4. Mix gently by pipetting several times and incubate at room temperature for 30 min. The final volume should be around 200 μL.
5. Inject the mixture into the tail vein.

3.7. In Vivo Uptake of siRNA by Peritoneal Cells

The most important aspect of in vivo delivery of siRNAs is the careful optimization of the delivery conditions. In the case of siRNA delivery, some cationic lipids have many advantages in terms of enhancing the binding to cells and blood circulation time. For reach reagent, the analysis of siRNA uptake should give some information about the factors that govern the distribution of siRNA/lipid complexes in vivo. For therapeutic purpose, it is therefore important to identify the organs where the complexes siRNA are taken easily. Using the i.p. delivery protocol described in **Section 3.5**, a FITC-labeled siRNA (100 μg) was delivered to peritoneal cells. Subsequent to overnight treatment, peritoneal cells were isolated and then analyzed by flow cytometry. As shown in **Fig. 6.1**, around 30% of peritoneal cells were in vivo transfected.

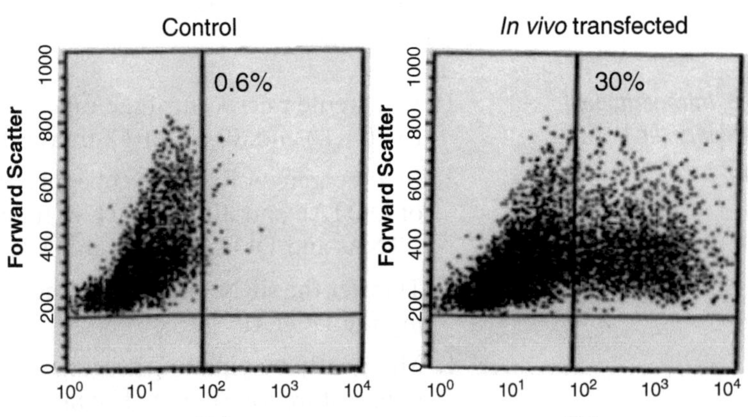

Fig. 6.1. In vivo uptake of siRNAs by peritoneal cells. Subsequent to i.p. delivery of FITC-labeled siRNA molecules, cells within the peritoneal lavage were washed with RPMI medium and then examined by flow cytometry. Nearly 30% of the cells showed siRNA uptake.

4. Notes

1. Under physiological conditions, the peritoneum contains some residual macrophages. However, they are not sufficient for in vitro studies. Liposomes such as DOTAP can recruit macrophages into the peritoneal cavity.
2. The peritoneal fluid can be aspirated by a syringe with 25 G needle. In this case, there is no need of making a small hole within the abdominal cavity.
3. Before analyzing siRNA activity in peritoneal cells or other organs, transfection efficiency should be analyzed. For this purpose it will be desirable to use 3′-FITC-labeled siRNAs or include a small amount of FITC-labeled DNA or RNA oligonucleotide.
4. The maximum period available for observation of the siRNA effect on gene expression after a single treatment depends on the stability of the protein after elimination of mRNA. Therefore, it is important to determine the half-life of the target protein(s) after a single injection.
5. Changes in gene expression can be directly monitored from protein extracts by Western and/or Northern blots. If desired, cytokine contents can be measured in the peritoneal lavage fluids using commercially available ELISA kits.
6. The condensation that occurs during complex formation is progressive and within minutes may result in the precipitation of large aggregates that are not suitable for i.v. delivery. Small particles are desirable. In this respect, cationic polyspermine could be a good carrier system for i.v. delivery.

References

1. Sioud, M. (2001) Nucleic acid enzymes as a novel generation of anti-gene agents. *Curr Mol Med*, **1**, 575–588.
2. Behlke, M.A. (2006) Progress towards in vivo use of siRNAs. *Mol Ther*, **13**, 644–666.
3. Fire, A., Xu, S., Montgomery, M.K., Kostas, S.A., Driver, S.E., and Mello, C.C. (1998) Potent and specific genetic interference by double-stranded RNA in *Caenorhabditis elegans*. *Nature*, **391**, 806–811.
4. Elbashir, S.M., Harborth, J., Lendeckel, W., Yalcin, A., Weber, K., and Tuschl, T. (2001) Duplexes of 21-nucleotide RNAs mediate RNA interference in cultured mammalian cells. *Nature*, **411**, 494–498.
5. Sioud, M. (2004) Therapeutic siRNAs. *Trends Pharmacol Sci*, **25**, 22–28.
6. Templeton, S.N. (2002) Liposomal delivery of nucleic acids in vivo. *DNA Cell Biol*, **21**, 859–867.
7. Sørensen, D.R., Leirdal, M., and Sioud, M. (2003) Gene silencing by systemic delivery of synthetic siRNAs in adult mice. *J Mol Biol*, **327**, 761–766.
8. Morrissey, D.V., Lockridge, J.A., Shaw, L., Blanchard, K., Jensen, K., Breen, W., Hartsough, K., Machemer, L., Radka, S., Jadhav, V., Vaish, N., Zinnen, S., Vargeese, C., Bowman, K., Shaffer, C.S., Jeffs, L.B., Judge, A., MacLachlan, I., and Polisky, B. (2005) Potent and persistent in vivo anti-HBV activity of chemically modified siRNAs. *Nat Biotechnol*, **23**, 1002–1100.
9. Sioud, M. (2005) On the delivery of small interfering RNAs into mammalian cells. *Expert Opin Drug Deliv*, **2**, 639–651.

Chapter 7

What Are the Key Targeted Delivery Technologies of siRNA Now?

Mouldy Sioud

Abstract

Whilst significant advances have been made in the delivery of nucleic acids to mammalian cells, most of the used strategies do not distinguish between normal and cancer cells. The same challenge is also facing radioactive- and chemo-therapies which are highly toxic and poorly tolerated due to limited tumor specificity. Regardless of the nature of the drug, there is a need for developing a technology platform which targets drugs only to tumors cells, leaving normal cells undamaged. Among the targeting strategies, receptor-targeted delivery provides an innovative strategy to selectively direct therapeutics to cancer cells. Receptor-binding ligands (e.g., peptides, antibodies, aptamers) can be incorporated into gene delivery vesicles or directly conjugated to siRNA in the hope in promoting their localization in target cell expressing the cognate receptors. The present chapter discusses the current progress made in the specific delivery of siRNAs.

Key words: RNAi, siRNA, random peptide libraries, hormone peptides, peptide analogues, endocytose, cell surface receptors.

1. Introduction

The main goal of any therapy is to eradicate pathogenic cells such as malignant cells, while sparing normal cells. The advantages of being able to avoid delivering bioactive agents (cytotoxic drugs, radionuclides, and cytokines) to healthy tissues/cells are numerous. For example, specific delivery of cytotoxic drugs to cancer cells could alleviate the problem of side effects because high concentrations of the drug within tumors could be attained without affecting normal tissues. Several experimental approaches have been used in order to achieve a preferential accumulation

M. Sioud (ed.), *RNA Therapeutics*, Methods in Molecular Biology 629,
DOI 10.1007/978-1-60761-657-3_7, © Springer Science+Business Media, LLC 2010

of therapeutics in target cells (e.g., antibodies, liposomes displaying an enhanced retention at tumor sites). In principle, ligands that recognize cell receptors that are preferentially or specifically expressed by the target cells can be conjugated to delivery agents to promote specific cellular uptake via receptor-mediated endocytosis (1). To achieve specific delivery to target cells, different strategies have been pursued, including the use of random peptides and antibody libraries to identify ligands (peptides or antibodies) that bind to membrane receptors expressed on target cells (2). Furthermore, changes in the tumor environment such as increased expression of surface proteases and angiogenesis can be taken advantages of to improve the therapeutic index of existing and new therapeutic approaches such as siRNAs (3).

In vivo, siRNA delivery can be achieved by a number of strategies including lipid-based formulations (3), atelocollagen (4), nanoparticles (5), and magnetofection (6). As the most used delivery agents can enter all cell types, specificity must be built into the delivery agents or expressed shRNAs. Although some studies reported on the specific delivery of siRNAs to target cells, the identification of effective targeting strategies remains the most challenging technical step toward the transformation of RNA technology into modern medicine (7).

2. Targeting via Antibodies

As indicated in the introduction, cell surface molecules that exhibit high specificity for tumor cells represent good therapeutic targets. Antibodies directed against these targets can be used as drugs and/or delivery agents (8). At present, there are several examples of effective antibodies to treat various cancers. Two of such examples are Herceptin and Rituxan (9). Herceptin targets Her-2 antigen expressed by a subset of breast cancer patients. Rituxan is an anti-CD20 antibody effective against non-Hodgkin's lymphoma. Although some unmodified mAbs exhibited some therapeutic potency, their effects tend to be varying and ultimately not curative when not used in combination with classical chemotherapy. Furthermore, these antibodies need to be given to patients in massive doses to show a therapeutic effect, thus leading to serious side effects. As a result, investigators have developed new technologies to arm antibodies with drugs, toxins, or radionuclides. Unfortunately, the use of toxins and radionuclides in combination with these antibodies complicates the preclinical and clinical testing necessary to gain approval.

One approach to circumvent the delivery of siRNA to target cells is to conjugate the siRNA to antibodies. Alternatively, liposomes-formulated siRNAs can be coated with antibodies against tumor markers (10, 11). Previous studies have shown that protamine-conjugated antibodies specific for ErbB2 or LFA-1 receptors can deliver siRNAs to cell expressing these receptors (11). Similarly, a second study described a method to specifically deliver siRNA to tumor cells using transferrin as a targeting ligand (12). Anti-CD40 siRNA-containing immunoliposomes that were coated with dendritic cell-specific DEC-205 monoclonal antibody resulted in DC-specific cell targeting in vitro and in vivo (13). Using avidin–biotin technology, Xia and colleagues demonstrated that siRNA can be targeted to antibody directed to the insulin receptor (14). To target T cells, Kumar et al. have used an anti-CD7-specific single-chain antibody conjugated to a 9-mer arginine peptide and showed that the construct can deliver anti-viral siRNAs to naïve T cells in mice (15). Collectively, these examples illustrated the feasibility of conjugating siRNAs to antibody recognizing cancer cells.

3. Targeting via Peptides

Compared to antibodies, peptides have the advantage of increased tumor penetration and low immunogenicity (2, 16). The most intensely studied CPPs include those derived from the transduction domains of the HIV transcription factor TAT (HIV-TAT peptide) or the *Drosophila melanogaster* homeobox protein Antennapedia (penetratin), together with synthetic oligomers of arginines (R4-R16). They often include a relatively large proportion of positive amino acids, lysine and arginine. The charge peptides allow them to interact strongly with negative charges on the plasma membrane. Beyond interacting with cells, a number of CPPs cross cellular membranes alone or attached to a much larger cargo ranging from small molecule drugs to antisense oligonucleotides, siRNAs, plasmids, peptides, and proteins. Although CPPs are capable of mediating the internalization of therapeutics, their cell selectivity is limited. Therefore, there is a growing interest in developing peptides that recognize specific receptors expressed by, for example, cancer cells. Such targeted delivery would increase effectiveness by lowering drug concentrations required throughout the body while raising the effective drug concentration in the tumor area.

Traditionally, receptor- or cell-binding peptides are identified rationally via structure–activity studies that involve the synthesis of a large number of peptides for in vitro and in vivo testing (17). However, the advances in combinatorial and biological peptide libraries have made it possible to select specific binding peptides for membrane receptors expressed by tumor cells (2, 18). It should be noted that random peptide libraries include virtually all possible sequences of small peptides that can mimic conformational structures of both continuous and discontinuous epitopes (19, 20). Biopanning of these libraries on purified proteins or on whole cells has led to the selection of a large number of targeting peptides with high binding specificity (2). Some of the selected peptides are listed in **Table 7.1**. In principle, these cell-binding peptides can be linked to any therapeutic agents to increase their accumulation in the cells that express the cognate peptide receptors.

In the case of cancer cells, targeting peptides are often selected for their binding to known cell surface receptors involved in biological functions such as cell proliferation, motility, evasion, and metastasis (21, 22). Among the attractive receptors, fibroblast growth factor (FGFR) and epidermal growth factor receptors (EGFR) families are often overexpressed in human tumors (23). The epidermal growth factor receptor type 2, known as ErbB-2/HER-2, is a carcinoma-associated tyrosine kinase receptor whose activation is responsible for cancer cell survival, proliferation, and metastases. Thus, ErbB2 receptor-binding peptides should be important for drug- or siRNA-targeted strategies. In this respect, several ErbB2-binding peptides have been selected from random peptide phage libraries and one the selected peptides (KCCYSL) bound to the extracellular domain of the human purified ErbB-2 receptor (24, 25).

In addition to cancer cells, differentially expressed cell surface markers on endothelial cells in angiogenic vessels of tumors should be excellent targets for site-specific targeting. The accessibility of endothelial cell markers by the blood stream makes them more attractive targets than their counterparts expressed on tumor cells (26). To identify endothelial cell-binding peptides, random peptide phage libraries have been screened for binding to the vasculature of various organs by injecting the phage libraries intravenously into live animals (27–29). This in vivo selection strategy has the advantage over in vitro selection strategies in that one can select in whole animal peptides that bind to tumors. Indeed, Ruoslahti and colleagues have demonstrated that this innovative in vivo screening method is feasible leading to selection of $\alpha_v\beta_3$ and $\alpha_v\beta_5$ integrin receptor-binding peptides (27, 28). A phage displaying a peptide containing the Arg-Gly-Asp (RGD) motif homed to tumors when injected intravenously into tumor-bearing mice. Also, phages displaying a cyclic CDCRGDCFC peptide (RGD-4C) exhibited 10–20 times higher tumor-homing ability than the negative control phages (29, 30). Interestingly, coupling of doxorubicin to RGD-4C-targeting peptide led to the design of effective and less toxic agents than the doxorubicin alone (30, 31). Furthermore, the RGD-4C-targeting peptide and

Table 7.1
Target-specific peptides selected from random peptide phage libraries

Peptide	Cellular targets
Vasculature of various tumors	
CD**RGD**CFC, ACDC**RGD**CFCG	$\alpha_v\beta_3$, $\alpha_v\beta_5$ integrins
C**NGR**CVSGCAGRC, CVC**NGR**MEC, **NGR**AHA	Aminopeptidase N
CPGPEGAGC	Aminopeptidase P
TAASGVRSMH, LTLRWVGLMS	NG2 proteoglycan
C**GSL**VRC, C**GLS**DSC,	Tumor vasculature
LRIKRKRRKRKKTRK, NRSTHI	IC-12 rat trachea
SMSIARL, VSFLEYR	Mice prostate
ATWLPPR, RRKRRR, ASSSYPLIHWRPWAR	VEGF
CTTHWGFTLC	Gelatinase
Cell surface of various tumors	
KNGPWYAYTGRO, NWAVWXKR, YXXEDLRRR, XXPVDHGL	Surface idiotype of SUP-88 human B-cell lymphoma
LVRSTGQFV, LVSPSGSWT, ALRPSGEWL, AIMASGQWL, QILASGRWL, RRPSHAMAR, DNNRPANSM, LQDRLRFAT, PLSGDKSST	Surface idiotype of human chronic lymphocytic lymphoma (CLL)
IELLQAR	HL 60 human lymphoma and B-16 mouse melanoma
CVFXXXYXXC, CXFXXXYXYLMC, CVXYCXXXXCYVC, CVXYCXXXXCWXC	Prostate-specific antigen (PSA)
DPRATPGS	LNCaP prostate cancer
HLQLQPWYPQIS	WAC-2 human neuroblastoma
VPWMEPAYQRFL	MDA-MB435 breast cancer
TSPLNIHNGQKL	Head and neck cancer cell lines
SPLW/F,R/K,N/H,**S**, V/H,**L**	ECV304 endothelial cell line
RLTGGKGVG	HEp-2 human laryngeal carcinoma
LTVX**PW**X	Human breast tumor
KCCYSL	ErbB2 tyrosine kinase type 1 receptor

other peptides were conjugated to several therapeutic agents such as apoptotic peptides (32) and radioactive agents for tumor imaging and therapy (33). Because of these encouraging results, it would be attractive to conjugate siRNAs to RGD-4C peptide for tumor targeting. In this respect, systemic administration of anti-HIF-1 siRNA with RGD peptide-targeted delivery systems resulted in significant tumor growth inhibition than a nontargeted delivery system (34). Also, a self-assembling nanoparticle with siRNA and PEI-PEGylated with an RGD peptide ligand attached to the distal end of the PEG as a means to target tumor neovasculature-expressing intergrins was designed. Intravenous administration into tumor-bearing mice resulted in selective tumor uptake, siRNA sequence-specific inhibition of protein expression within the tumor, and inhibition of both tumor angiogenesis and growth rate (35).

Toward the identification of cancer cell-targeting peptides for nucleic acid delivery, we have described the selection of breast cancer cell-specific peptides from random peptide phage libraries (36). In these experiments, the breast cancer cell line SKBR3 was used as affinity matrix for selection. Subsequent to biopanning, internalized phages were selected, amplified, and the sequences of their displayed peptides were deduced from the DNA sequences. One of the selected peptide was able to deliver antisense oligonucleotides to breast cell line SKBR3. More recently, Wang and colleagues have conjugated the vitamin E to LTVSPWY peptide and demonstrated that the conjugates were preferentially taken up by ErbB2-expressing cancer cells (37), thus further validating the potential use of the LTVSPWY peptide as targeting agent. To test the feasibility of this delivery system, siRNA-targeting ErbB2 was incubated with the protamine-LTVSPWY peptide for 1 h at RT and then added to SKBR3 cells. Subsequent to 48 h transfection time, protein extracts were prepared and the expression of ErbB2 was analyzed by a Western blot (**Fig. 7.1**). A significant reduction in survivin expression was evident in cells treated with the siRNA peptide formulations (37). It should be noted

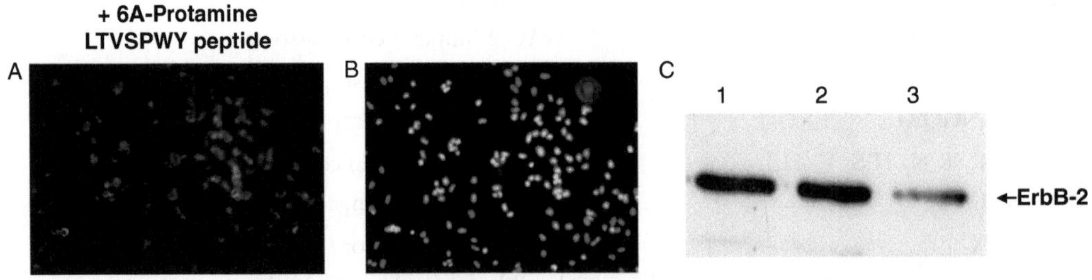

Fig. 7.1. Cancer cell targeting using the LTVSPWY peptide. (**a**) Binding of the protamine LTVSPWY peptide to breast cancer cell line SKBR3. Cells were incubated with 6A-conjugated peptides for 1 h at 37°C and, subsequently, were analyzed with an epifluorescence microscope. The protamine LTVSPWY peptide sequence is 6A-**C**RSQSRSRYYRQRQRSRRRRRRSA**LTVSPWY**. (**b**) Cell nuclei were counterstained with Hoechst staining. (**c**) Downregulation of ErbB2 expression in SKBR3 cells. The siRNA molecules targeting the ErbB2 mRNA were incubated with the protamine LTVSPWY peptide for 1 h at room temperature and then added to SKBR3 cells growing in ex vivo 15 medium (lane 3). As a control, free siRNA molecules were added to the cells (lane 2). Following 48 h transfection time, cytoplasmic proteins were prepared and the expression of the ErbB2 gene was analyzed with Western blotting. Lane 1, untreated cells.

that the ErbB2 receptor is overexpressed in 30–50% of primary breast cancers and its expression is low in most normal adult tissues, thus making an attractive target for peptide-mediated delivery into ErbB2-positive tumor cells.

4. Peptide Hormone Conjugates

As indicated above, cancer targeting is usually achieved by adding to the drug delivery vesicles a ligand moiety specifically directed to receptor expressed on cancer cells. Recent studies indicated that several hormone receptors are not only expressed by normal human tissues but also overexpressed in a large variety of human cancers, thus permitting an in vivo targeting of tumors for diagnostic and therapeutic purposes (38–40). For example, the receptors for luteinizing hormone-releasing hormone (LHRH) are overexpressed in breast, ovarian, and prostate cancers (41). The NPY receptors are mainly expressed in specific endocrine tumors and epithelial malignancies, particularly breast cancer cells (38). The human gastrin-releasing peptide receptor (GRPr) is overexpressed on a variety of human cancers, including prostate, breast, lung, and pancreatic cancers (40). Somatostatin receptors are overexpressed in many neuroendocrine tumors, including small-cell lung cancer, carcinoid tumor, insulinoma, gastrinoma, and medullary thyroid cancer (39). The epidermal growth factor receptor is highly expressed by glioblastomas when compared to surrounding non-tumor cells such as astrocytes (42). The examples described above clearly show that different hormone and growth factor receptors are overexpressed on the cell membrane of cancer cells. Thus, the coupling of their peptide ligands to siRNAs is anticipated to target systemically siRNAs to specific cell populations (37).

It should be noted that the concept of using radiolabeled peptide hormones or derivatives for molecular imaging and therapy has been established by several groups. For instance, the coupling of cytotoxic drugs to LHRH peptide enhanced their accumulation in tumor cells expressing the LHRH receptor (43). More recently a LHRH peptide analogue was coupled to the distal end of the PEG–siRNA conjugate for ovarian cancer cell targeting (44). Peptide analogues of human gastrin-releasing peptides exhibited an excellent targeting potency of cancer cells. Similarly, somatostatin analogues were developed and used for diagnostic imaging of somatostatin receptor-positive tumors. It should be noted that the somatostatin receptor is the first peptide hormone receptor identified for receptor-targeted diagnosis and therapy of tumors. Other peptides targeting epidermal growth factor

(EGF), glutathione, laminin receptor, estrogen receptors, EGF receptor were also developed (45). Because the clinical data with some of the hormone peptides described above are encouraging, we thought to conjugate siRNAs to these peptides in order to target siRNAs to specific cell types such as cancer cells (3). In these experiments, the siRNA was designed with a thiol group in the 5′-end of the sense strand and the peptide with a cysteine residue at the N-terminus (**Fig. 7.2**). Western blots analysis was used to investigate the relative knock down of survivin expression in MCF-7 that expresses the GRP receptor. As shown in **Fig. 7.2**, the siRNA/GRP conjugates are able to block survivin gene expression in cancer breast cell line MCF-7. These results are encouraging and should set the stage for in vivo testing.

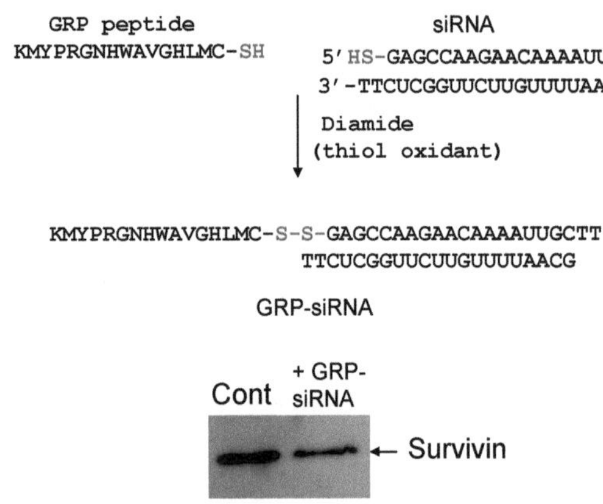

Fig. 7.2. GRP peptide-mediated delivery of siRNAs. SiRNAs were directly conjugated to the GRP peptide as outlined in (**a**) and then added to MCF7 cells growing in ex vivo medium. Subsequent to 48 h incubation time, cytoplasmic protein extracts were prepared and then the expression of the survivin was analyzed by Western blotting. Cont, untreated cells.

5. Chemical Modifications to Enhance Targeted Delivery

A second strategy to enhance the binding of siRNAs to specific cells and enhance their biodistribution is to use chemical modifications (46). During the last years, a number of chemical modifications have been developed to improve the in vivo properties of nucleic acids. Each modification offers a different potential for tailing the properties of siRNAs. Although more investigations are needed, certain terminal conjugates have been reported to improve or direct cellular uptake. For example, siRNAs conjugated with cholesterol improve in vitro and in vivo cell uptake by liver cells (47). The cholesterol was coupled to one of the RNA strands to create cholesterol siRNA conjugates. Attachment of the RNA duplex to the cholesterol increased binding to human serum albumin and increase serum biodistribution to various organs, particularly the liver. The coupling of the folic acid to a siRNA against protein kinase α also led to selective delivery to tumor cells (7). Similarly, a folate-linked nanoparticles were evaluated as potential targeting agents of siRNA to human nasopharyngeal KB cells, which overexpressed folate receptor (48).

In addition of directing the siRNAs to target cells, previous studies have shown that RNAi can be controlled temporally. Photocaging is one specific class of modifications for siRNAs that inhibit their bioactivity until exposure to near-ultraviolet light (49, 50). These light-controllable strategies can be employed to achieve control over delivery of siRNA molecules to intended target cells.

6. Targeting via RNA Aptamers

Aside from serving as reservoir of information and catalysis, RNA can function as molecular recognition. RNA molecules binding to their targets with high affinity were first developed for proteins that are known to naturally bind RNA using a technique known as systematic evolution of ligands by exponential enrichment (51). Subsequently, the technique was extended for isolation of high-affinity binders against growth factors and other functional molecules. Indeed, a number of aptamers have now been generated that can target specific cell surface antigens and cell types, including tumor biomarkers and tumor cells (*see* **Chapters 23** and **24**). In addition of exhibiting high affinity and exquisite specificity to cognate ligands, aptamers have the advantage that they can be chemically modified with relative ease to improve their stability and bioavailability. In addition to their therapeutic use, aptamers have also the potential to deliver other therapeutics, including nanoparticles-encapsulated drugs, siRNAs, and toxins in a cell type-specific manner. By conjugating an aptamer that is specific for prostate-specific membrane antigen (PSMA), a cell surface glycoprotein found in abundance on prostate cancer cells, it has been shown that specific delivery to prostate cancer cells can be achieved (52, 53). Similarly, Rossi and colleagues demonstrated cell type-specific delivery of anti-human immunodeficiency virus siRNAs through fusion of an anti-gp120 aptamer (54). Wullner and colleagues used an aptamer that bind to PSMA linked to siRNA specific for eukaryotic elongation factor 2 mRNA (55). Although further studies are required, the examples indicated above show that RNA aptamers could be used to deliver siRNA into human cells.

7. Transcriptional Targeting of siRNAs

Short synthetic siRNA duplexes can be introduced into cells by transfection and have emerged as a valuable tool for analysis of gene function (56). However, the silencing effect is transient. To circumvent this problem, several groups have developed DNA expression vectors for RNAi in human cells that express short hairpin siRNA (shRNAs) under the control of a RNA polymerase III or a pol II promoter (56). To date, pol III promoters are used most frequently because it is possible to express small RNAs that carry the structural feature of siRNAs (**Fig. 7.3a, b**).

Although shRNA vectors are widely used as strategy for the analysis of gene function, currently siRNA vectors have some limitations: (1) siRNA expression cannot be controlled in a time- or tissue-specific manner because pol III promoters are constitutively active in all mammalian cells and regulated expression from

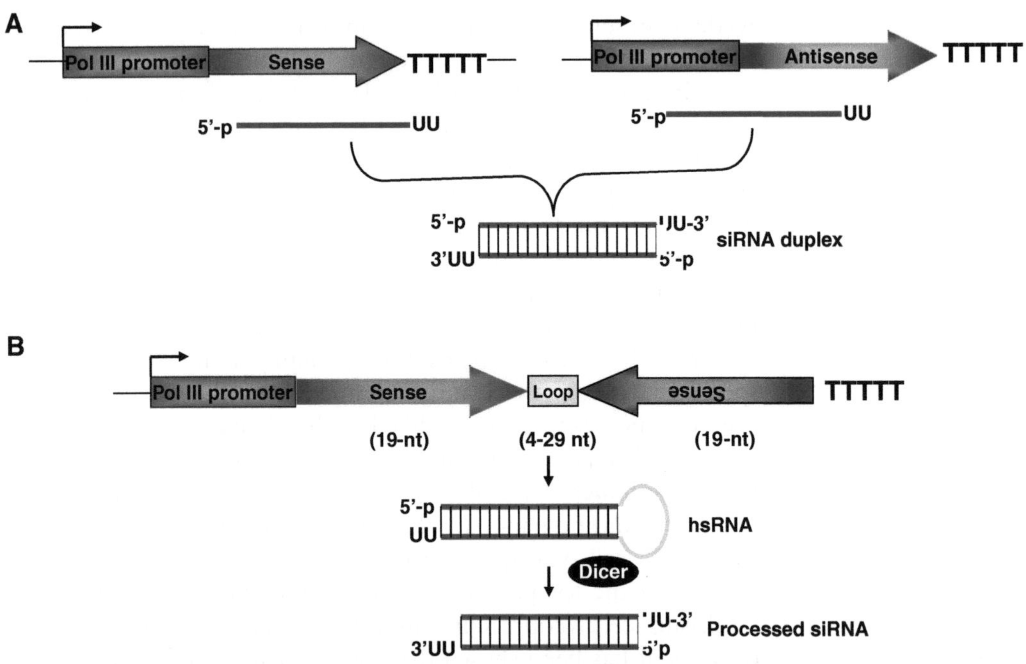

Fig. 7.3. Pol III expression strategy. (**a**) Cellular expression of small interfering RNAs (siRNAs). Tandem-type expression strategy. Both the sense and the antisense strands are expressed from different RNA polymerase (pol) III promoters (H1 or U6). The transcriptional termination of five thymidines is added to the 3′ of the sense and antisense sequences. After transcription, sense and antisense RNAs hybridize and form a duplex siRNA in the transfected cells. (**b**) Hairpin-type expression strategy. Sense sequence and its inverted sequence are expressed from H1 or U6 promoters. Both sequences can be separated by a loop sequence of 4–29 nucleotides (nt). The transcriptional termination of five thymidines is added to the 3′ end of the sense-inverted sequence. In the case of the H1 promoter, the transcript ends with two uridine 3′ overhangs. The transcribed RNA forms a hairpin with a loop structure that is processed by Dicer into a functional siRNA.

pol III promoters is more difficult in comparison with pol II promoters. (2) Only one siRNA sequence is expressed from each promoter. (3) Constitutive expression of siRNAs might abrogate the cytoplasmic transport of micro-RNAs, a new class of small noncoding RNAs with essential biological functions (*see* **Chapter 17**). In this respect, a recent study has reported fatality in mice owing to competition between shRNAs and miRNAs for limiting cellular factors such as exportin 5 (57).

In addition to sequence-specific inhibition of the target gene, siRNAs also have off-target effects that could be detrimental for healthy cells (58, 59). Indeed, transcriptional profiling studies have shown that siRNA duplexes can potentially silence multiple genes in addition to the intended target. Therefore, there is an urgent need for controlled expression systems of shRNAs. To overcome this potential problem, regulated H1 and U6 expression vectors have been developed. For example, a doxycycline-regulated form of the H1 promoter to drive the expression of siRNAs against β catenin in colorectal cancer was designed (60). Although these modified promoters have provided significant insights into the regulation of siRNA expression, they do not restrict the expression of the shRNAs to pathogenic cells and therefore their use in gene therapy is limited (61).

One way to target pathogenic cells and avoid healthy cells is to place the shRNA under the control of a promoter that is transcriptionally active only in pathogenic cells such as cancer cells, but not normal cells. Several studies have shown that a number of genes are expressed exclusively or predominantly in cancer cells when compared to normal counterparts. Moreover, various promoters have already been evaluated for transcriptional targeting in cancer gene therapy including the prostate-specific antigen promoter for prostate cancer and the tyrosinase gene promoter for melanoma (62). Among the genes that are preferentially expressed in tumors, survivin is a perfect candidate because it is expressed in many human cancer types but not normal adult tissues (63).

To test whether the survivin promoter could express siRNA in cancer cells, we first PCR amplified the promoter sequence and then modified the sequence to allow the cloning of shRNAs (64). **Figure 7.4** illustrates the cloning strategy. When an siRNA-targeting GFP was expressed under the survivin promoter, an inhibition of the target gene was seen only in cancer cells but not in normal cells (64), thus confirming the specificity of the designed promoter (**Fig. 7.4b, c**). More, recently Song and colleagues described a vector for RNAi in which a synthetic siRNA is expressed under the PSA promoter (65). Interestingly, reduced gene expression was achieved in a tissue-specific and hormone-dependent manner. Both approaches offer a number of important potential advantages when contrasted with modified U1 and U6

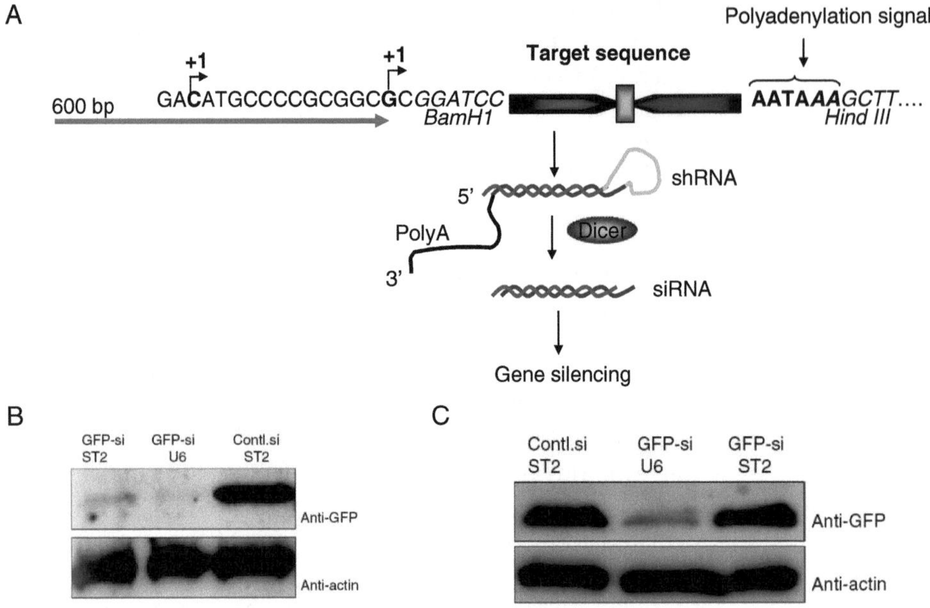

Fig. 7.4. Expression of siRNA under the survivin promoter. (**a**) A partial sequence of the survivin promoter showing the *Bam*HI and *Hin*dIII cloning sites. The natural transcription starts are kept and a polyA stop signal is added at the end of the construct. The survivin promoter (ST2)-mediated siRNA expression inhibited the expression of GFP in survivin-positive cancer cell lines (**b**) but not in survivin negative normal cells. (**c**) In these experiments, cells were cotransfected with GFP-encoding plasmid in combination with anti-GFP shRNA under the control of the survivin promoter (64).

vectors. These include the following: (1) siRNA can be expressed only or preferentially in cancer cells. (2) It is possible to express a long single transcript resembling the miRNA primary transcripts, allowing the generation of several siRNAs from each survivin promoter. Notably, optimized tissue-specific promoters for shRNA expression are expected to facilitate the therapeutic applications of siRNAs.

As discussed in **Chapter 2**, long double-stranded RNAs (dsRNAs) frequently expressed in virus-infected cells activate several anti-viral cellular pathways leading to cell death. Therefore, these non-specific effects of long dsRNAs as death inducer can be used to selectively kill cancer cells by expressing long dsRNAs under the control of the survivin promoter. In this respect, our preliminary data indicate that the expression of 40 nt long dsRNA targeting the survivin mRNA can block the expression of the survivin gene and trigger the interferon pathway resulting in enhanced cell. This method which explores the specificity of the survivin promoter should be applicable to any gene.

In addition to peptides and antibodies, targeting can be achieved through local administration (e.g., intraocular injection and inhalation, intratumoral injection). In this respect, intraocular injection of a VEGF-specific siRNA is being assessed as a possible treatment of macular degeneration and inhalation of siRNA is being assessed as a possible treatment for infection with respiratory syncytial virus. Other encouraging examples of local targeting were also described.

Collectively, the studies discussed in this chapter indicate that the identification of targeting ligand to cancer cells and the development of ligand-targeted therapy will help us to improve therapeutic efficacy and reduce side effects of siRNAs.

References

1. Garanger, E., Boturyn, D., and Dumy, P. (2007) Tumor targeting with RGD peptide ligands-design of new molecular conjugates for imaging and therapy. *Anticancer Agents Med Chem*, 7, 552–558.
2. Shadidi, M. and Sioud, M. (2003) Selective targeting of cancer cells using synthetic peptides. *Drug Resist Updat*, **6**, 363–371.
3. Sioud, M. (2009) Targeted delivery of antisense oligonucleotides and siRNAs into mammalian cells. *Method Mol Biol*, **487**, 61–82.
4. Takeshita, F., Hokaiwado, N., Honma, K., Banas, A., and Ochiya, T. (2009) Local and systemic delivery of siRNAs for oligonucleotide therapy. *Methods Mol Biol*, **487**, 83–92.
5. Moore, A. and Medarova, Z. (2009) Imaging of siRNA delivery and silencing. *Methods Mol Biol*, **487**, 93–110.
6. Mykhaylyk, O., Zelphati, O., Hammerschmid, E., Anton, M., Rosenecker, J., and Plank, C. (2009) Recent advances in magnetofection and its potential to deliver siRNAs in vitro. *Methods Mol Biol*, **487**, 111–146.
7. Sioud, M. (2005) On the delivery of small interfering RNAs into mammalian cells. *Expet Opin Drug Deliv*, **2**, 639–651.
8. van Dijk, M.A. and van de Winkel, J.G. (2001) Human antibodies as next generation therapeutics. *Curr Opin Chem Biol*, **5**, 368–374.
9. Ross, J., Gray, K., Schenkein, D., Greene, B., Gray, G.S., Shulok, J., Worland, P.J., Celniker, A., and Rolfe, M. (2003) Antibody-based therapeutics in oncology. *Expert Rev Anticancer Ther*, **3**, 107–121.
10. Song, E., Zhu, P., Lee, S.K., Chowdhury, D., Kussman, S., Dykxhoorn, D.M., Feng, Y., Palliser, D., Weiner, D.B., Shankar, P., Marasco, W.A., and Lieberman, J. (2005) Antibody mediated in vivo delivery of small interfering RNAs via cell-surface receptors. *Nat Biotechnol*, **23**, 709–717.
11. Peer, D., Zhu, P., Carman, C.V., Lieberman, J., and Shimaoka., M. (2007) Selective gene silencing in activated leukocytes by targeting siRNAs to the integrin lymphocyte function-associated antigen-1. *Proc Natl Acad Sci USA*, **104**, 4095–4100.
12. Lee, J.H., Engler, J.A., Collawn, J.F., and Moore, B.A. (2001) Receptor mediated uptake of peptides that bind the human transferrin receptor. *Eur J Biochem*, **268**, 2004–. 2012.
13. Zheng, X., Vladau, C., Zhang, X., Suzuki, M., Ichim, T.E., Zhang, Z.X., Li, M., Carrier, E., Garcia, B., Jevnikar, A.M., and Min, W.P. (2009) A novel in vivo siRNA delivery system specifically targeting dendritic cells and silencing CD40 genes for immunomodulation. *Blood*, **113**, 2646–2654.
14. Xia, C.F., Boado, R.J., and Pardridge, W.M. (2008) Antibody-mediated targeting of siRNA via the human insulin receptor using avidin-biotin technology. *Mol Pharm*, **6**, 747–751.
15. Kumar, P., Ban, H.S., Kim, S.S., Wu, H., Pearson, T., Greiner, D.L., Laouar, A., Yao, J., Haridas, V., Habiro, K., Yang, Y.G., Jeong, J.H., Lee, K.Y., Kim, Y.H., Kim, S.W., Peipp, M., Fey, G.H., Manjunath, N., Shultz, L.D., Lee, S.K., and Shankar, P. (2008) T cell-specific siRNA delivery suppresses HIV-1 infection in humanized mice. *Cell*, **134**, 577–586.
16. Aina, O.H., Sroka, T.C., Chen, M.L., and Lam, K.S. (2002) Therapeutic cancer targeting peptides. *Biopolymers*, **66**, 184–199.
17. Patel, D.S., Dessalew, N., Iqbal, P., and Bharatam, P.V. (2007) Structure-based approaches in the design of GSK-3 selective inhibitors. *Curr Protein Pept Sci*, **8**, 352–364.
18. Romanov, V.I. (2003) Phage display selection and evaluation of cancer drug targets. *Curr Cancer Drug Targets*, **3**, 119–129.
19. Falciani, C., Lozzi, L., Pini, A., and Bracci, L. (2005) Bioactive peptides from libraries. *Chem Biol*, **12**, 417–426.
20. Houghten, R.A., Pinilla, C., Blondelle, S.E., Appel, J.R., Dooley, C.T., and Cuervo, J.H.

(1991) Generation and use of synthetic peptide combinatorial libraries for basic research and drug discovery. *Nature*, **354**, 84–86.
21. Fukuda, M.N., Ohyama, C., Lowitz, K., Matsuo, O., Pasqualini, R., Ruoslahti, E., and Fukuda, M. (2000) A peptide mimic of E-selectin ligand inhibits sialyl Lewis X-dependent lung colonization of tumor cells. *Cancer Res*, **60**, 450–456.
22. Campa, M.J., Serlin, S.B., and Patz, E.F. (2002) Development of novel tumor imaging agents with phage-display combinatorial peptide libraries. *Acad Radiol*, **9**, 927–932.
23. Alaoui-Jamali, M.A. and Qiang, H. (2003) The interface between ErbB and non-ErbB receptors in tumor invasion: clinical implications and opportunities for target discovery. *Drug Resist Updat*, **6**, 95–107.
24. Urbanelli, L., Ronchini, C., Fontana, L., Menard, S., Orlandi, R., and Monaci, P. (2001) Targeted gene transduction of mammalian cells expressing the HER2/neu receptor by filamentous phage. *J Mol Biol*, **313**, 965–976.
25. Karasseva, N.G., Glinsky, V.V., Chen, N.X., Komatireddy, R., and Quinn, T.P. (2002) Identification and characterization of peptides that bind human ErbB-2 selected from a bacteriophage display library. *J Protein Chem*, **21**, 287–296.
26. Folkman, J. (2002) Role of angiogenesis in tumor growth and metastasis. *Semin Oncol*, **29**, 15–18.
27. Pasqualini, R. and Ruoslahti, E. (1996) Organ targeting in vivo using phage display peptide libraries. *Nature*, **380**, 364–366.
28. Pasqualini, R., Koivunen, E., and Ruoslahti, E. (1997) Alpha v integrins as receptors for tumor targeting by circulating ligands. *Nat Biotechnol*, **15**, 542–546.
29. Pasqualini, R., Koivunen, E., Kain, R., Lahdenranta, J., Sakamoto, M., Stryhn, A., Ashmun, R.A., Shapiro, L.H., Arap, W., and Ruoslahti, E. (2000) Aminopeptidase N is a receptor for tumor-homing peptides and a target for inhibiting angiogenesis. *Cancer Res*, **60**, 272–722.
30. Arap, W., Pasqualini, R., and Ruoslahti, E. (1998) Cancer treatment by targeted drug delivery to tumor vasculature in a mouse model. *Science*, **279**, 377–380.
31. de Groot, F.M., Broxterman, H.J., Adams, H.P., van Vliet, A., Tesser, G.I., Elderkamp, Y.W., Schraa, A.J., Kok, R.J., Molema, G., Pinedo, H.M., and Scheeren, H.W. (2002) Design, synthesis, and biological evaluation of a dual tumor-specific motive containing integrin-targeted plasmin-cleavable doxorubicin prodrug. *Mol Cancer Ther*, **1**, 901–911.
32. Cheng, J.Q., Jiang, X., Fraser, M., Li, M., Dan, H.C., Sun, M., and Tsang, B.K. (2002) Role of X-linked inhibitor of apoptosis proteins in chemoresistance in ovarian cancer: possible involvement of the phosphoinositide-3 kinase/Akt pathway. *Drug Resist Updat*, **5**, 131–146.
33. Su, Z.F., Liu, G., Gupta, S., Zhu, Z., Rusckowski, M., and Hnatowich, D.J. (2002) In vitro and in vivo evaluation of a Technetium-99m-labeled cyclic RGD peptide as a specific marker of alpha(v)beta(3) integrin for tumor imaging. *Bioconjug Chem*, **13**, 561–570.
34. Wang, X.L., Xu, R., Wu, X., Gillespie, D., Jensen, R., and Lu, Z.R. (2009) Targeted systemic delivery of a therapeutic siRNA with a multifunctional carrier controls tumor proliferation in mice. *Mol Pharm*, **6**, 738–746.
35. Schiffelers, R.M., Ansari, A., Xu, J., Zhou, Q., Tang, Q., Storm, G., Molema, G., Lu, P.Y., Scaria, P.V., and Woodle, M.C. (2004) Cancer siRNA therapy by tumor selective delivery with ligand-targeted sterically stabilized nanoparticle. *Nucl Acids Res*, **32**, e149.
36. Shadidi, M. and Sioud, M. (2003) Identification of novel carrier peptides for the specific delivery of therapeutics into cancer cells. *FASEB J*, **17**, 256–258.
37. Wang, X.F., Birringer, M., Dong, L.F. et al. (2007) A peptide conjugate of vitamin E succinate targets breast cancer cells with high ErbB2 expression. *Cancer Res*, **67**, 3337–3344.
38. Körner, M. and Reubi, J.C. (2007) NPY receptors in human cancer: a review of current knowledge. *Peptides*, **28**, 419–425.
39. Riccabona, G. and Decristoforo, C. (2003) Peptide targeted imaging of cancer. *Cancer Biother Radiopharm*, **18**, 675–687.
40. Hoffman, T.J., Quinn, T.P., and Volkert, W.A. (2001) Radiometallated receptor-avid peptide conjugates for specific in vivo targeting of cancer cells. *Nucl Med Biol*, **28**, 527–539.
41. Dharap, S.S. and Minko, T. (2003) Targeted proapoptotic LHRH-BH3 peptide. *Pharm Res*, **20**, 889–896.
42. Shir, A. and Levitzki, A. (2001) Gene therapy for glioblastoma: future perspective for delivery systems and molecular targets. *Cell Mol Neurobiol*, **21**, 645–656.
43. Dharap, S.S., Wang, Y., Chandna, P., Khandare, J.J., Qiu, B., Gunaseelan, S., Sinko, P.J., Stein, S., Farmanfarmaian, A., and Minko, T. (2005) Tumor-specific targeting of an anticancer drug delivery system by LHRH peptide. *PNAS*, **102**, 12962–12967.

44. Kim, S.H., Jeong, J.H., Lee, S.H., Kim, S.W., and Park, T.G. (2008) LHRH receptor-mediated delivery of siRNA using polyelectrolyte complex micelles self-assembled from siRNA-PEG-LHRH conjugate and PEI. *Bioconjug Chem*, **19**, 2156–2162.
45. Huang, P.S. and Oliff, A. (2001) Drug-targeting strategies in cancer therapy. *Curr Opin Genet Develop*, **11**, 104–110.
46. Behlke, M.A. (2009) Chemical modification of siRNAs for in vivo use. *Oligonucleotides*, **18**, 305–319.
47. Soutschek, J., Akinc, A., Bramlage, B., Charisse, K., Constien, R., Donoghue, M., Elbashir, S., Geick, A., Hadwiger, P., Harborth, J., John, M., Kesavan, V., Lavine, G., Pandey, R.K., Racie, T., Rajeev, K.G., Röhl, I., Toudjarska, I., Wang, G., Wuschko, S., Bumcrot, D., Koteliansky, V., Limmer, S., Manoharan, M., and Vornlocher, H.P. (2004) Therapeutic silencing of an endogenous gene by systemic administration of modified siRNAs. *Nature*, **432**, 173–178.
48. Yoshizawa, T., Hattori, Y., Hakoshima, M., Koga, K., and Maitani, Y. (2008) Folate-linked lipid-based nanoparticles for synthetic siRNA delivery in KB tumor xenografts. *Eur J Pharm Biopharm*, **70**, 718–725.
49. Mikat, V. and Heckel, A. (2007) Light-dependent RNA interference with nucleobase-caged siRNAs. *RNA*, **13**, 2341–2347.
50. Casey, J.P., Blidner, R.A., and Monroe, W.T. (2009) Caged siRNAs for spatiotemporal control of gene silencing. *Mol Pharm*, **6**, 669–685.
51. Tuerk, C. and Gold, L. (1990) Systematic evolution of ligands by exponential enrichment: RNA ligands to bacteriophage T4 DNA polymerase. *Science*, **249**, 505–510.
52. McNamara, J.O., Andrechek, E.R., Wang, Y., Viles, K.D., Rempel, E.R., Gilboa, E., Sullenger, B.A., and Giangrande, P.H. (2006) Cell type-specific delivery of siRNAs with aptamer-siRNA chimeras. *Nat Biotechnol*, **24**, 1005–1015.
53. Chu, T.C., Twu, K.Y., Ellington, A.D., and Levy, M. (2006) Aptamer mediated siRNA delivery. *Nucl Acids Res*, **34**, e73.
54. Zhou, J., Li, H., Li, S., Zaia, J., and Rossi, J.J. (2008) Novel dual inhibitory function aptamer-siRNA delivery system for HIV-1 therapy. *Mol Ther*, **16**, 1481–1489.
55. Wullner, U., Neef, I., Eller, A., Kleines, M., Tur, M.K., and Barth, S. (2008) Cell-specific induction of apoptosis by rationally designed bivalent aptamer-siRNA transcripts silencing eukaryotic elongation factor 2. *Curr Cancer Drug Targets*, **8**, 554–565.
56. Hannon, G.J. and Rossi, J.J. (2004) Unlocking the potential of the human genome with RNA interference. *Nature*, **431**, 371–378.
57. Grimm, D., Streetz, K.L., Jopling, C.L., Storm, T.A., Pandey, K., Davis, C.R., Marion, P., Salazar, F., and Kay., M.A. (2006) Fatality in mice due to oversaturation of cellular microRNA/short hairpin RNA pathways. *Nature*, **441**, 537–541.
58. Jackson, A.L., Bartz, S.R., Schelter, J., Kobayashi, S.V., Burchard, J., Mao, M., Li, B., Cavet, G., and Linsley., P.S. (2003) Expression profiling reveals off-target gene regulation by RNAi. *Nat Biotechnol*, **21**, 635–637.
59. Lin, X., Ruan, X., Anderson, M.G., McDowell, J.A., Kroeger, P.E., Fesik, S.W., and Shen., Y. (2005) siRNA-mediated off-target gene silencing triggered by a 7 nt complementation. *Nucl Acids Res*, **33**, 4527–4535.
60. Van de Wetering, M. et al. (2003) Specific inhibition of gene expression using a stably integrated, inducible small-interfering-RNA vector. *EMBO Rep*, **4**, 609–615.
61. McIntyre, G.J. and Fanning., G.C. (2006) Design and cloning strategies for constructing shRNA expression vectors. *BMC Biotechnol*, **6**, 1.
62. Saukkonen, K. and Hemminki, A. (2004) Tissue-specific promoters for cancer gene therapy. *Expert Opin Biol Ther*, **4**, 683–696.
63. Altieri, D.C. (2003) Validating survivin as a cancer therapeutic target. *Nat Rev Cancer*, **3**, 46–54.
64. Huynh, T., Wälchli, S., and Sioud., M. (2006) Transcriptional targeting of small interfering RNAs into cancer cells. *Biochem Biophys Res Commun*, **350**, 854–859.
65. Song, J., Pang, S., Lu, Y., Yokoyama, K.K., Zheng, J.Y., and Chiu, R. (2004) Gene silencing in androgen-responsive prostate cancer cells from the tissue-specific prostate-specific antigen promoter. *Cancer Res*, **64**, 7661–7663.

Chapter 8

Using OligoWalk to Identify Efficient siRNA Sequences

David H. Mathews

Abstract

RNA interference (RNAi) has emerged as an important tool in science and in medicine. Small-interfering RNAs (siRNAs) can be used to knockdown gene expression of specific mRNAs. In practice, a number of factors influence whether an siRNA sequence will elicit RNAi and knockdown target gene expression. One factor that significantly influences the efficiency of an siRNA is the effect of RNA secondary structure. Self-structure in either the siRNA sequence or the target mRNA at the binding site may prevent gene silencing. This chapter provides protocols for using the OligoWalk software package to design efficient siRNAs. OligoWalk considers the effect of target and guide strand self-structures and also local sequence features in siRNA design. OligoWalk can be run either locally by compiling the software or through a convenient web interface. OligoWalk is freely available at http://rna.urmc.rochester.edu/cgi-bin/server_exe/oligowalk/oligowalk_form.cgi.

Key words: RNA interference, small inhibitory RNAs, RNA secondary structure, RNA thermodynamics, RNA partition function.

1. Introduction

The discovery of RNA interference (RNAi) fundamentally changed our understanding of gene regulation (1). Messenger RNA (mRNA) expression can be regulated by short, complementary RNAs. If these RNAs are perfectly complementary, such as with a small inhibitory RNA (siRNA), expression of the mRNA is reduced, generally because of decay of the mRNA (2). If the RNAs are only partially complementary, i.e., they contain non-canonical pairs, such as with a typical microRNA (miRNA), the expression is reduced by either mRNA decay or other mechanisms (3). RNAi is involved in a number of different cellular

M. Sioud (ed.), *RNA Therapeutics*, Methods in Molecular Biology 629,
DOI 10.1007/978-1-60761-657-3_8, © Springer Science+Business Media, LLC 2010

processes, including development and immunity. RNAi is also used extensively as a tool to knock out gene expression to infer gene function and their significant efforts underway to use RNAi for human therapies (4, 5). Excellent reviews that provide more details about both mechanism and use of RNAi are available (2, 6, 8).

In principle, controlling gene expression with an siRNA is simple. A duplex with two 21 nucleotide strands and a central 19-bp helix is introduced into the cell. The two unpaired nucleotides in each strand are at the 3′ end. One of the strands is exactly complementary to the target mRNA and is called the guide strand. The guide strand will be loaded into the RNA-induced silencing complex (RISC) and the complementarity to the target results in the RISC identifying and then cleaving the target, resulting in the decay of the mRNA.

In practice, it was quickly realized that not all sequences complementary to a given target will work equally well at gene silencing. One of the first breakthroughs in choosing an siRNA sequence that efficiently silences the target was the discovery that there needs to be asymmetry in the stability of the duplex (9, 10). The 5′ end of the guide strand needs to form less stable base pairs than does the 3′ end of that sequence. Therefore, the asymmetry in helical stability identifies which of the two strands is loaded onto the RISC. Subsequently, a number of other empirical sequence factors were discovered to influence efficiency (11).

Another factor that influences the silencing efficiency of an siRNA is RNA structure. Self-structure of either the siRNA or the target mRNA can interfere with the formation of the required siRNA–mRNA duplex (12, 13). For example, if the mRNA is inaccessible to hybridization at the site to which the siRNA is complementary because of pre-existing structure, the siRNA will be unable to hybridize. Several groups demonstrated the importance of avoiding self-structure in the target using computational means (14, 17).

In this chapter, protocols are outlined for the use of the program OligoWalk to choose siRNAs efficient for expression knockdown of a specific mRNA (14). The software can be run either through a public web server (protocol A) (18) or locally by compiling source code (protocol B) on a computer with a Unix (Unix, Linux, or Macintosh OS X) operating system. The web server requires no initial setup, but local compilation provides more user control over the exact details of the calculation.

1.1. OligoWalk Calculation

OligoWalk considers the equilibrium of the siRNA guide strand hybridization with the target mRNA as shown in **Fig. 8.1** (19). The desired product is the siRNA–mRNA duplex, which will elicit cleavage of the mRNA by the RISC. To form this duplex, the siRNA duplex, as delivered to the cell, needs to unwind to

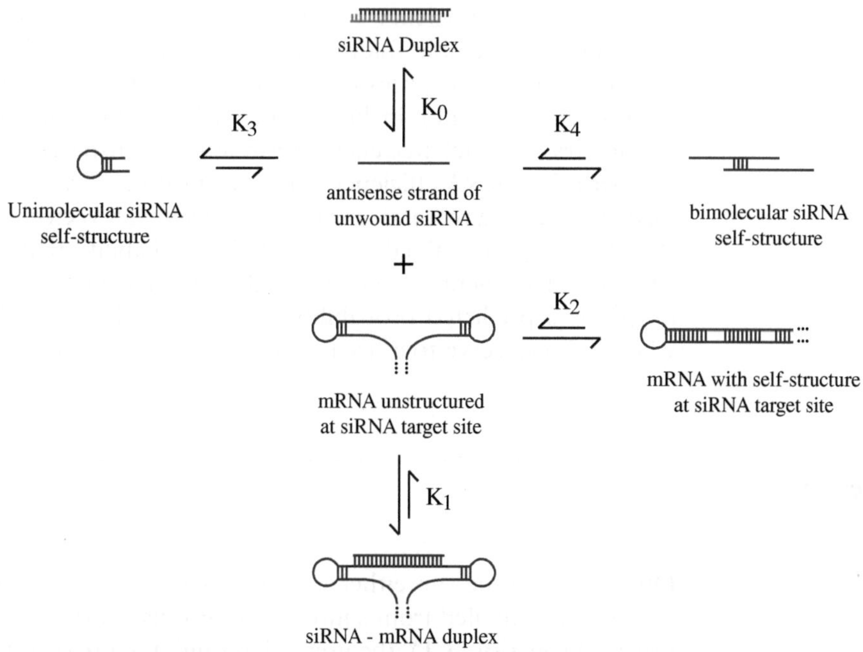

Fig. 8.1. The equilibria considered by OligoWalk for siRNA–mRNA hybridization. For each equilibrium, the longer arrow demonstrates the favored direction, i.e., the direction in which the Gibbs free energy change is negative. Calculations are performed with the latest nearest neighbor parameters for predicting free energy changes (20, 21). K_0 is the unwinding of the siRNA duplex. K_1 is the association of the unstructured antisense strand with the mRNA, unstructured at the target site. K_2 measures the extent of local structure in the mRNA that needs to be broken for siRNA to be able to hybridize. K_3 and K_4 measure the extent antisense siRNA self-structure for unimolecular and bimolecular structures, respectively. Efficient siRNAs tend to have less hybridization strength in the siRNA duplex, K_0, and with the target, K_1 (14). Furthermore, efficient siRNAs have reduced self-structure stabilities, K_3 and K_4, and hybridize to mRNA regions with reduced self-structure, K_2 (14).

make the guide strand available for hybridization to the mRNA. This guide strand can form unimolecular or bimolecular self-structures that interfere with the desired product. Furthermore, self-structure in the mRNA at the hybridization site also interferes with the desired product.

OligoWalk calculates the extent to which self-structures interfere with the siRNA–mRNA duplex by calculating Gibbs free energy changes for each equilibrium in **Fig. 8.1** (20, 21). The lower a free energy change, i.e., the more negative the value, the stronger the self-structure. The free energy changes are calculated by considering the entire ensemble of possible secondary structures, where each structure is weighted according to a probability of occurring at equilibrium. This is accomplished by calculating partition functions (22). The partition function approach is significantly more accurate than considering a set of predicted low free energy structures because those structures are more prone to errors than ensemble averages (14, 22, 23).

Finally, the efficiency of each possible siRNA is predicted utilizing the ensemble folding free energy changes and a set of 21 local sequence features that were identified to correlate with siRNA efficiency (11), including the siRNA duplex stability and asymmetry (9). Each free energy term was demonstrated to correlate with the siRNA efficiency with statistical significance (14). A machine learning method, a classification support vector machine (SVM) (24), was trained to predict the probability that a given siRNA will efficiently knockdown the targeted gene. The final output is a predicted probability for each possible siRNA as to whether it can serve to efficiently knockdown message.

2. Materials

OligoWalk can be run either via a web server or locally from an executable compiled from source code. For use of the web server protocol (**Section 3.1**), the user will require the use of a standard web browser, such as Apple Safari, Microsoft Internet Explorer, or Mozilla Firefox. For local compilation of OligoWalk (protocol B), a C++ compiler will be required, such as GNU gcc or the Intel C compiler. A standard Unix Makefile is provided to direct the package build. To select siRNAs, LIBSVM (24) and Perl (www.perl.org) must be installed.

3. Methods

3.1. Web Protocol for OligoWalk Use

This protocol outlines the steps to use the OligoWalk web server, located on the Mathews lab website. Start by pointing a web browser to http://rna.urmc.rochester.edu/cgi-bin/server_exe/oligowalk/oligowalk_form.cgi. The calculation is run using an example sequence, firefly luciferase, GenBank U47298.

3.1.1. Main Page Form

Figure 8.2 shows the main page for entering the sequence. Type a sequence name in the field labeled "Sequence name." This will help identify the sequence if more than one calculation is performed. Next, one of two methods can be used to specify the mRNA sequence. One is to paste the raw mRNA sequence into the "Primary sequence" field. **Figure 8.2** shows the firefly luciferase sequence pasted in place. Note that this sequence is treated as RNA so that any Ts that appear in the sequence are automatically converted to U for the calculation. The alternative

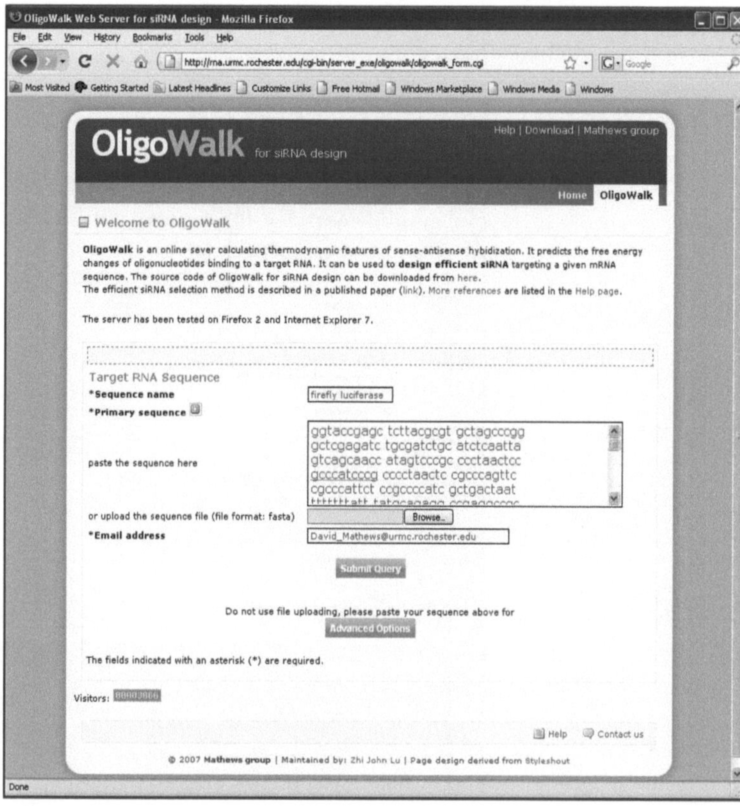

Fig. 8.2. OligoWalk web server input form.

method is to upload a sequence in FASTA format. To do this, click the "Browse..." button to choose the sequence from your local hard drive. Finally, provide your e-mail address. This is used so that the server can provide notice via e-mail when the calculation is complete.

At this time, either the calculation can be started, by clicking "Submit Query," or advanced options can be chosen to customize the calculation, by clicking "Advanced Options." Note that, as the server indicates, if advanced options are to be used, the sequence must be pasted into the "Primary sequence" field, rather than by upload.

3.1.2. Advanced Server Options

The OligoWalk web server calculation is optimized for prediction of efficient siRNA to a specified mRNA sequence (18). There is no need to change the default settings for siRNA design, but advanced options are made available to allow experimentation. **Figure 8.3** shows the input form for modifying advanced options. The screenshot shows the page defaults, which are those used for a default calculation. For most options, the green arrow to the right of the field can be clicked to receive online help.

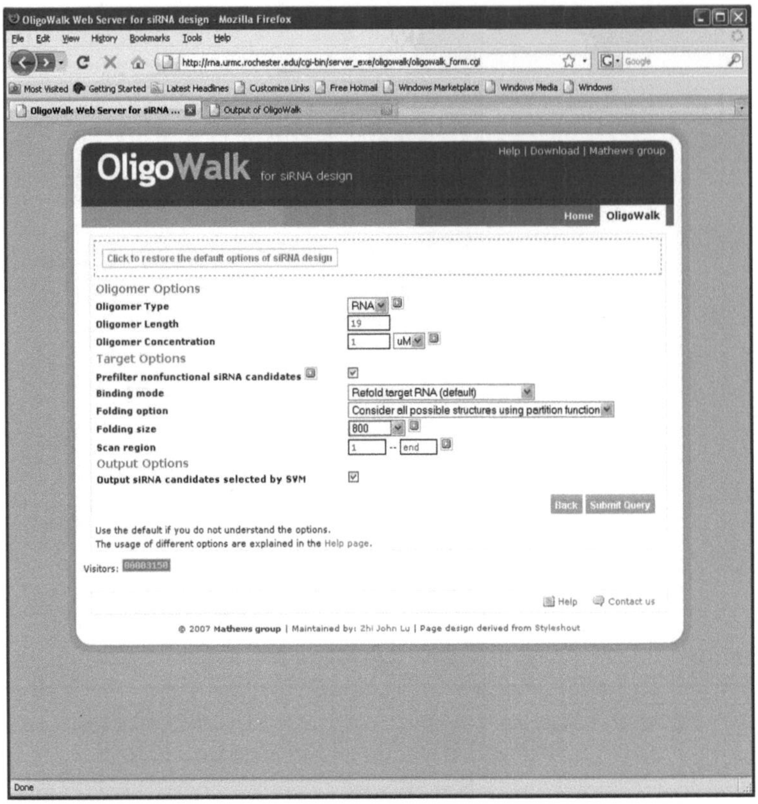

Fig. 8.3. OligoWalk web server advanced options form.

Furthermore, a link to the "Help page" is available to provide detailed descriptions of the options.

Two options can be changed to affect the treatment of the siRNA strands, listed under "Oligomer Options." The web server allows the use of only RNA oligonucleotides, but downloading and compiling OligoWalk locally (**Section 3.2**) allows the use of DNA oligonucleotides. This can be helpful in the design of antisense deoxyoligonucleotides (19, 25, 26). Both the length of the oligonucleotides and the intracellular oligonucleotide concentration can be changed by changing the values in the respective text boxes. The oligomer concentration does not affect the SVM estimation of siRNA efficiency but does affect the calculation of the total binding free energy change as reported in the free energy tables (*see* **Section 3.1.3**).

A number of options can be manually changed for handling of the target sequence. First, by default, a prefilter is used to narrow down the set of candidate siRNAs for which hybridization thermodynamics are calculated. The filter is based on empirical rules determined by Reynolds et al. (11) and it is used to save significant computation time. On the web server, the prefilter can be disabled if target secondary structure is not considered in the calculation. If calculations are needed for each possible siRNA while also considering target structure, the program must be run locally (**Section 3.2**).

A number of different types of calculations are available for handling the target mRNA secondary structure. "Binding mode" and "Folding option" change the type of calculation. For binding mode, three options are available that specify the way in which the target structure is considered. "Refold target RNA (default)" is an equilibrium calculation where the target mRNA is allowed to refold in response to the breaking of base pairs when the guide strand hybridizes. "Break local structure" is a calculation where only base pairs at the guide strand hybridization site are broken, but no new base pairs are allowed to form. Finally, "Not consider target structure" neglects the cost of opening mRNA secondary structure as though the mRNA had no self-structure. When target secondary structure is considered, there are three computational methods available for calculating the cost of opening target secondary structure, which are set with the "Folding option." "Consider all possible structures using partition function" is the default method and it is the most rigorous method because it considers all possible secondary structures at equilibrium both before and after binding (14). It is compatible only with the "Refold target RNA (default)" method for "Binding mode" because it is applicable only at equilibrium. "Consider suboptimal structures" predicts a set of low free energy structures (21, 27) and considers the cost of opening target secondary structure for the structures in the set. The contribution that each suboptimal structure makes in determining the overall cost is weighted according to a Boltzmann probability. For "Only consider optimal structure," only the lowest free energy structure is predicted (21) and the cost of opening pairs in only the lowest free energy structure is determined.

Folding size restricts the region in the target sequence with which the target site nucleotides can pair. Global folding ("all") provides the greatest correlation between prediction and experiment but is time consuming. It was empirically determined that a folding size of 800 nucleotides provides similar correlation to global folding and is much faster (14). For long sequences (greater than 800 nucleotides), "Folding Size" defaults to 800 nucleotides and other lengths can be chosen up to 1,000 nucleotides. If longer folding regions are desired, OligoWalk must be run locally (**Section 3.2**). For sequences shorter than 800 nucleotides, "Folding size" defaults to "all," i.e., global structure prediction.

Finally, the regions of interest along a long target mRNA can be specified to reduce the calculation time by adjusting the "Scan region." By default, the entire length of the target is considered viable for siRNA hybridization. Considerable calculation time can be saved by specifying a region of interest, where an index specifies the nucleotide in the mRNA to which the 3′ end of the guide strand would hybridize.

After specifying advanced options, the calculation is started by clicking the "Submit Query" button. Another option, "Back," is provided to revise the settings on the main submission form (**Fig. 8.2**).

3.1.3. Server Output

After submitting the query, the user is taken to the output page as illustrated in **Fig. 8.4**. This output page contains links to tables of candidate siRNA sequences and the folding free energy changes. For most calculations, these data will take time to generate. **Table 8.1** provides example calculation times (18). The results will be available at this page when the calculation is complete. The server will also send e-mail notifying the user when the calculation is complete. The e-mail contains a unique URL (web address) at which the prediction results can be obtained.

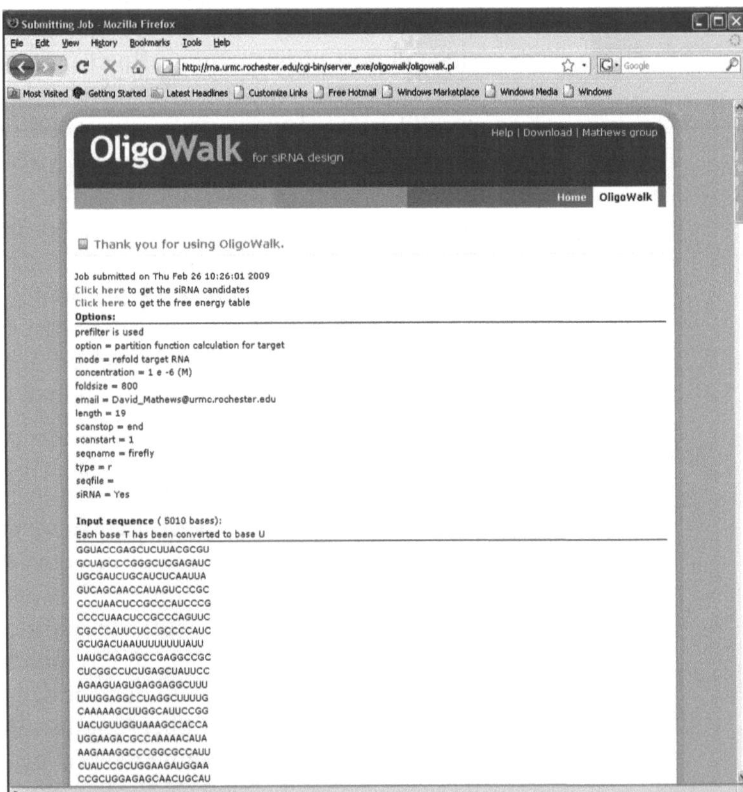

Fig. 8.4. OligoWalk server output form.

The candidate siRNA sequence table provides the probability that candidate siRNAs will elicit gene knockdown. The candidates are sorted in descending order of probability that the siRNA will be efficient. Note that a prefilter step, using local sequence rules identified by Reynolds et al. (11), is used to narrow the candidate

Table 8.1
Example OligoWalk calculation times (18)

Target mRNA (GenBank ID)	Sequence length (nucleotide)	Time (h:min:sec)	Memory (MB)
NM_020548	730	0:57:17	93
M60857	851	3:55:53	110
NM_002870	1211	6:09:25	112
NM_002467	2189	6:36:42	113
AJ272212	3460	6:53:05	117

The benchmarks were performed with the default options: The siRNA was a 19-base RNA; the folding size of the target was 800 nucleotides centering on the binding site (full length if the whole target has less than 800 nucleotides); the partition function calculation was conducted; the entire mRNA was scanned; and the prefilter was on. The time cost is similar for long sequence because the prefilter is on and the number of candidates being folded is limited to about the same number for each sequence. The calculations were submitted and benchmarked on the OligoWalk web server, which is a cluster with five compute nodes, managed by Sun Grid Engine. Each node has a 3.2- or 3.4-GHz Pentium 4 processor running Fedora Linux.

list to those that meet local sequence criteria before the structure prediction calculations are performed, so not all possible guide strand sequences appear in the table.

The folding free energy change table provides the self-structure folding free energy changes for candidate siRNAs. The folding free energy change for each of the individual folding equilibria as shown in **Fig. 8.1** is reported. Additionally, an overall association folding free energy change is provided for the association of the single-stranded guide strand with the target mRNA (19).

For the example target used here, firefly luciferase (5,010 nucleotides), there are 140 candidate siRNAs that are predicted to have a greater than 90% chance of eliciting efficient gene knockdown. In a prior experiment, a 198-nucleotide region of this gene was targeted at every second position by siRNAs (9). In this calculation, the top hit predicted to be an efficient siRNA target in the 198-nucleotide region is to index 1873 (UCUCUCUGAUUUUUCUUGC), with a predicted probability of eliciting knockdown of 94.9%. Experimentally, this siRNA was found to induce more than 90% gene silencing (9).

3.2. Protocol for Local Compilation and Use of OligoWalk

This alternative protocol outlines the use of OligoWalk on a Unix system. The first steps are to download and compile both OligoWalk and LIBSVM (24). This will require the use of a C++ compiler, such as the GNU gcc compiler. After the installation, efficient siRNAs can be predicted.

3.2.1. Download and Install OligoWalk

OligoWalk can be downloaded from the Mathews lab website at http://rna.urmc.rochester.edu/cgi-bin/server_exe/oligowalk/download_form.cgi. The code is provided using the GNU public license, but registration is requested. The registration information is used to track the number of independent downloads of the software and to issue critical bug notices. Fill the fields for Name, Institution, Department, and E-mail and then click "Download Now."

OligoWalk is downloaded as a gzipped tar file. Extract the software using the command "`tar -xvzf oligowalk_src.tar.gz`". This will create a directory called "oligowalk_src," which contains several subdirectories, including "exe, NN_data, src, and svm." The "oligowalk_src" directory contains a file named "README" that provides help in compiling and using OligoWalk.

Next, build OligoWalk by issuing the command, "`make OligoWalk`" in the "src" directory. The enclosed Makefile provides instructions for building OligoWalk. By default, it assumes that the gcc compiler will be used with the g++ command. If other compilers will be used, manually edit the Makefile. Example compiler commands and flags are provided for SGI, SUN, Intel, and IBM compilers.

3.2.2. Download and Install LIBSVM

LIBSVM is available for download at http://www.csie.ntu.edu.tw/~cjlin/libsvm/ and it is accompanied by fairly extensive documentation (24). Download the.tar.gz file and extract it using the command "`tar -xvzf libsvm-2.88.tar.gz`". In the directory "libsvm-2.88," use the command "`make`" to build the executables. This directory also contains a "README" file that provides instructions for building and using LIBSVM. Copy the executables "svm-scale" and "svm-predict" to the directory "oligowalk_src/svm" using the cp command.

3.2.3. Running OligoWalk

To design siRNAs, the Perl script siRNAWalk.pl in the root directory of OligoWalk should be used. It wraps the calls to both the OligoWalk program and to LIBSVM. An example sequence (example.seq) and example option file (example.options) are provided, so the program can be run using the command "`siRNAwalk.pl example.seq example.options`". By default, this will create output files called example.energy and example.siRNA, which contain the same information as the html tables from the OligoWalk web server, the equilibria free energy changes and a ranked list of candidate siRNAs, respectively. These files are tab-delimited plain text files.

Sequences are input using the.seq file format, explained in **Fig. 8.5**. The calculation options can be changed by modifying the example.options file. In the options file, comments start with the "#" character. Each of the calculation options can be changed and the details of these options are explained in **Section 3.2.2** above and in the online OligoWalk help at http://rna.urmc.rochester.edu/servers/oligowalk/help.html.

```
;
RL4000
GCUAUUUUG
GUGGAAUUGG
UAGACACGAU
ACUCUUAAGA
UGUAUUACUU
UACAGUAUGA
AGGUUCAAGU
CCUUUAAAUA
GCACCA1
```

Fig. 8.5. Sequence file format. OligoWalk uses the.seq format for sequence input. The file starts with comment lines, denoted by ";". At least one comment is required. The subsequent line contains a title, which is "RL4000" in this case. The following lines contain the sequence written 5′ to 3′, where white space is ignored. The sequence terminates with a "1." Sequences can be written conveniently using the RNAstructure program for Microsoft Windows, available from the Mathews lab at http://RNA.urmc.rochester.edu/RNAstructure.html

Note that example.seq is a short sequence. For sequences longer than 800 nucleotides, it is recommended that the folding option "foldsize" be changed to 800. As explained in **Section 3.1.2**, limiting the folding region in the target to 800 nucleotides is an optimal tradeoff between accuracy and computation time.

4. Notes

A number of considerations determine the knockdown efficiency of a given siRNA. For a given candidate siRNA sequence, OligoWalk predicts the knockdown efficiency by considering the equilibrium binding affinity of the guide strand to the target, the asymmetry in the helical stability, and a set of empirical local sequence features (14). The equilibrium binding affinity is reduced by self-structure in the guide strand or self-structure in the mRNA at the target binding site. A machine learning method, using an SVM, was trained to make predict the probability that a candidate will be efficient.

On average, the positive predictive value of siRNA design is 78.6%, meaning that about four of five siRNAs designed by OligoWalk will elicit gene knockdown, defined here as at least 50% reduction in protein expression. Interestingly, the sensitivity is 22.9%, meaning that less than a quarter of efficient siR-

NAs are predicted as efficient siRNAs by this method. In other words, OligoWalk is stringent in selection and tends to not find all siRNAs that will elicit knockdown. Overall, this is the best approach to siRNA design because most users are interested in having a small set of putative siRNA sequences of which most will elicit knockdown. It is rare that a user would want to know all sequences that can function in gene knockdown for a specific message.

The OligoWalk siRNA efficiency prediction was trained on a dataset of siRNA experiments performed with homogeneous conditions (28). The performance was then tested using a database of experiments performed with diverse experimental conditions (29). The reported performance should therefore apply realistically to different types of experiments, performed under a variety of conditions.

OligoWalk does not incorporate any method for checking off-target effects. Users will want to use a search, such as BLAST, to check for hybridization of putative siRNA guide strands to off-target mRNAs, which might result in non-specific effects. NCBI provides a convenient web interface for performing BLAST searches at http://blast.ncbi.nlm.nih.gov/Blast.cgi.

Acknowledgments

The author thanks Zhi John Lu for helpful comments. This contribution was supported by the National Institutes of Health grant R01GM076485 to the author.

References

1. Fire, A., Xu, S., Montgomery, M.K., Kostas, S.A., Driver, S.E., and Mello, C.C. (1998) Potent and specific genetic interference by double-stranded RNA in *Caenorhabditis elegans. Nature*, **391**, 806–811.
2. Wu, L. and Belasco, J.G. (2008) Let me count the ways: mechanisms of gene regulation by miRNAs and siRNAs. *Mol Cell*, **29**, 1–7.
3. Liu, J. (2008) Control of protein synthesis and mRNA degradation by microRNAs. *Curr Opin Cell Biol*, **20**, 214–221.
4. Castanotto, D. and Rossi, J.J. (2009) The promises and pitfalls of RNA-interference-based therapeutics. *Nature*, **457**, 426–433.
5. Sioud, M. (2004) Therapeutic siRNAs. *Trends Pharmacol Sci*, **25**, 22–28.
6. Siomi, H. and Siomi, M.C. (2009) On the road to reading the RNA-interference code. *Nature*, **457**, 396–404.
7. Moazed, D. (2009) Small RNAs in transcriptional gene silencing and genome defence. *Nature*, **457**, 413–420.
8. Zamore, P.D. and Haley, B. (2005) Ribo-gnome: the big world of small RNAs. *Science*, **309**, 1519–1524.
9. Khvorova, A., Reynolds, A., and Jayasena, S.D. (2003) Functional siRNAs and miRNAs exhibit strand bias. *Cell*, **115**, 209–216.
10. Schwarz, D.S., Hutvagner, G., Du, T., Xu, Z., Aronin, N., and Zamore, P.D. (2003) Asymmetry in the assembly of the RNAi enzyme complex. *Cell*, **115**, 199–208.
11. Reynolds, A., Leake, D., Boese, Q., Scaringe, S., Marshall, W.S., and Khvorova, A. (2004)

Rational siRNA design for RNA interference. *Nat Biotechnol*, **22**, 326–330.
12. Ameres, S.L., Martinez, J., and Schroeder, R. (2007) Molecular basis for target RNA recognition and cleavage by human RISC. *Cell*, **130**, 101–112.
13. Far, R.K. and Sczakiel, G. (2003) The activity of siRNA in mammalian cells is related to structural target accessibility: a comparison with antisense oligonucleotides. *Nucl Acids Res*, **31**, 4417–4424.
14. Lu, Z.J. and Mathews, D.H. (2007) Efficient siRNA selection using hybridization thermodynamics. *Nucl Acids Res*, **36**, 640–647.
15. Heale, B.S., Soifer, H.S., Bowers, C., and Rossi, J.J. (2005) siRNA target site secondary structure predictions using local stable substructures. *Nucl Acids Res*, **33**, e30.
16. Shao, Y., Chan, C.Y., Maliyekkel, A., Lawrence, C.E., Roninson, I.B., and Ding, Y. (2007) Effect of target secondary structure on RNAi efficiency. *RNA*, **13**, 1631–1640.
17. Tafer, H., Ameres, S.L., Obernosterer, G., Gebeshuber, C.A., Schroeder, R., Martinez, J., and Hofacker, I.L. (2008) The impact of target site accessibility on the design of effective siRNAs. *Nat Biotechnol*, **26**, 578–583.
18. Lu, Z.J. and Mathews, D.H. (2008) OligoWalk: an online siRNA design tool utilizing hybridization thermodynamics. *Nucl Acids Res*, **36**, W104–W108.
19. Mathews, D.H., Burkard, M.E., Freier, S.M., Wyatt, J.R., and Turner, D.H. (1999) Predicting oligonucleotide affinity to nucleic acid targets. *RNA*, **5**, 1458–1469.
20. Xia, T., SantaLucia, J., Jr., Burkard, M.E., Kierzek, R., Schroeder, S.J., Jiao, X., Cox, C., and Turner, D.H. (1998) Thermodynamic parameters for an expanded nearest-neighbor model for formation of RNA duplexes with Watson–Crick pairs. *Biochemistry*, **37**, 14719–14735.
21. Mathews, D.H., Disney, M.D., Childs, J.L., Schroeder, S.J., Zuker, M., and Turner, D.H. (2004) Incorporating chemical modification constraints into a dynamic programming algorithm for prediction of RNA secondary structure. *Proc Natl Acad Sci USA*, **101**, 7287–7292.
22. Mathews, D.H. (2004) Using an RNA secondary structure partition function to determine confidence in base pairs predicted by free energy minimization. *RNA*, **10**, 1178–1190.
23. Layton, D.M. and Bundschuh, R. (2005) A statistical analysis of RNA folding algorithms through thermodynamic parameter perturbation. *Nucl Acids Res*, **33**, 519–524.
24. Chang, C.C., Hsu, C.W., and Lin, C.J. (2000) The analysis of decomposition methods for support vector machines. *IEEE Trans Neural Netw*, **11**(4), 1003–1008.
25. Lu, Z.J. and Mathews, D.H. (2008) Fundamental differences in the equilibrium considerations for siRNA and antisense oligodeoxynucleotide design. *Nucl Acids Res*, **36**, 3738–3745.
26. Matveeva, O.V., Mathews, D.H., Tsodikov, A.D., Shabalina, S.A., Gesteland, R.F., Atkins, J.F., and Freier, S.M. (2003) Thermodynamic criteria for high hit rate antisense oligonucleotide design. *Nucl Acids Res*, **31**, 4989–4994.
27. Zuker, M. (1989) On finding all suboptimal foldings of an RNA molecule. *Science*, **244**, 48–52.
28. Huesken, D., Lange, J., Mickanin, C., Weiler, J., Asselbergs, F., Warner, J., Meloon, B., Engel, S., Rosenberg, A., Cohen, D. et al. (2005) Design of a genome-wide siRNA library using an artificial neural network. *Nat Biotechnol*, **23**, 995–1001.
29. Shabalina, S.A., Spiridonov, A.N., and Ogurtsov, A.Y. (2006) Computational models with thermodynamic and composition features improve siRNA design. *BMC Bioinform*, 7, 65.

Chapter 9

Jet-Injection of Short Hairpin RNA-Encoding Vectors into Tumor Cells

Wolfgang Walther, Ulrike Stein, and Hermann Lage

Abstract

The use of the RNA interference (RNAi) through the expression of small hairpin RNA (shRNA) is a promising approach for efficient gene silencing for therapeutic applications. In this chapter, we describe the in vivo reversal of the classical MDR1/P-glycoprotein (MDR1/P-gp)-mediated multidrug resistance (MDR) phenotype by shRNA. For local intratumoral delivery of naked shRNA-encoding vector constructs, the nonviral jet-injection was used. This jet-injector system uses compressed air to inject small volumes (5–10 μL) of naked nucleic acid solutions into tumor tissues. Furthermore, the design of the jet-injector allows multiple injections. Under our experimental design, the delivery of plasmid DNA encoding anti-MDR shRNA by jet-injection into human MDR1/P-gp overexpressing MaTu/ADR breast cancer xenografts resulted in a decrease of MDR1 mRNA expression level to more than 90%. Accordingly, the corresponding MDR1/P-gp protein is no longer detectable in the tumors after anti-MDR1 shRNA vector injection. Furthermore, combination of two intratumoral jet-injections of anti-MDR1 shRNA vectors with two intravenous administrations of doxorubicin is sufficient for a complete reversal of the MDR phenotype in association with tumor growth inhibition.

Key words: Gene therapy, nonviral gene transfer, jet-injection, RNA interference, short hairpin RNA, multidrug resistance.

1. Introduction

Since the first description of the RNA interference (RNAi) mechanism in the nematode *Caenorhabditis elegans* (1), RNAi began to replace alternative gene-silencing techniques such as antisense and ribozyme technologies. Moreover, the finding that the RNAi pathway can be triggered in mammalian cells in response to double-stranded small interfering RNAs (siRNAs) of ~21 nt in

M. Sioud (ed.), *RNA Therapeutics*, Methods in Molecular Biology 629,
DOI 10.1007/978-1-60761-657-3_9, © Springer Science+Business Media, LLC 2010

length (2) has facilitated the use of such technology in many biomedical research laboratories worldwide. In consequence, the RNAi is now a widely distributed, well-established technology for high-throughput analyses as well as for functional studies. Above and beyond its potential as a powerful laboratory tool, RNAi platform also offers the possibility for clinical applications.

In complex disorders, such as cancer, various cellular pathways composed of multiple interacting factors are altered in normal function. Thus, RNAi technology provides a strategy for the development of novel targeted anticancer agents (3). Besides proteins acting in classical carcinogenesis-associated pathways, also factors involved in therapy resistance, in particular multidrug resistance (MDR), are well suited to be targeted by therapeutic RNAi effectors.

MDR is one major reason for failure of cancer chemotherapy. Elevated expression of the ATP-binding cassette (ABC) transporter MDR1/P-glycoprotein (MDR1/P-gp; ABCB1) causes the "classical" MDR phenotype (4). This phenotype is characterized by cross-resistance against structurally and functionally non-related anticancer agents, such as Vinca alkaloids, anthracyclines, or taxanes. MDR1/P-gp has been shown to confer MDR in cultured cells and has also been implicated in clinical MDR, since it is expressed in almost 50% of human neoplasms.

Numerous approaches aim at the inhibition of MDR1/P-gp function or expression to reverse the "classical" MDR phenotype for improved effectiveness of chemotherapy. Since the early 1980s, a broad spectrum of compounds has been examined for their capability to overcome MDR1/P-gp-mediated MDR. However, clinical trials with these compounds have not been successful so far (5, 6). Thus, development of new compounds or strategies that are capable of circumventing or reducing MDR1/P-gp-mediated MDR with improved clinical efficacy and applicability is required.

Although, siRNAs appear as the simplest way to trigger RNAi as therapeutic agents in cancer gene therapy, siRNAs mediate only transient therapeutic effects (3). Therefore, expression vectors encoding siRNA-like short hairpin RNAs (shRNAs) have been developed to allow long-term production of therapeutic RNAs in targeted cells. In this regard, MDR1/P-gp-mediated MDR has been successfully reversed by application of the RNAi in different tumor models (7, 8).

Plasmid vectors containing shRNA expression cassettes commonly utilize RNA-polymerase III-specific promoters, e.g., the H1-RNA promoter, or the U6-RNA promoter. These RNA-polymerase III-depending promoters have a defined start of transcription and a termination signal consisting of five consecutive thymidine residues (T5). Therewith, these promoters can be used to direct the synthesis of small RNA molecules of interest lacking a poly adenosine poly(A) tail. Cleavage of the RNA transcript at the termination site is after the second uridine. Thus, RNA-polymerase III promoters direct the synthesis of small RNAs that are similar to the ends of chemically synthesized siRNAs containing two 3′-overhanging thymidines or uridines. In shRNAs, the sequence of interest consists of a 19-nt sequence homologous to the target mRNA, linked with a 5–11-nt spacer sequence to the reverse complement of the same 19-nt target-specific sequence. The synthesized RNA transcript folds back to its complementary strand to form a 19-base pair shRNA molecule, which is then processed by the cellular endoribonuclease-III-like enzyme Dicer to a corresponding siRNA and passed into the gene-silencing RNAi pathway.

As in all gene therapy protocols, the efficient delivery of shRNA-encoding therapeutic DNA is still one of the major problems for successful clinical application. Besides the development of virus-based delivery strategies, various techniques for nonviral transfer systems have been developed and are under intense investigation (9). In this context, the transfer of naked DNA for gene therapy represents an alternative to viral and liposomal gene transfer technologies (10). Various in vitro and in vivo procedures, such as simple needle injection, particle bombardment, in vivo electroporation, or jet-injection, are employed to deliver naked DNA into the desired cells or tissues (11–14). These nonviral techniques have numerous advantages, such as avoidance of utilization of recombinant viral particles, reduced or no immunostimulatory potential, and no toxicity. The majority of these nonviral gene transfer technologies are used in the context of DNA vaccination studies. For these intradermal or intramuscular applications, use of naked DNA has proven to be the most efficient DNA vaccine against viral infections or cancer diseases in numerous animal models (15–18).

Among the various nonviral gene delivery technologies, jet-injection has gained increasing acceptance, since this technique allows gene transfer into different tissues with deeper penetration of naked DNA (19–21). The jet-injection technology is based on jets of small volumes ($\leq 10\ \mu L$), which are ejected with high velocity and generate the force to deeply penetrate the targeted tissues and to transfect the affected area. Furthermore, this jet-injection system provides the possibility of simultaneous application of more than only one DNA construct into one tissue leading to the expression of two or more different gene products. Jet-injection generates broad areas of transgene expression within the jet-injected tissue. The in vivo application of this technology does not induce tissue damage or significant inflammatory reactions at jet-injection sites (22). In previous studies, a jet-injector prototype (Swiss Injector) **(Fig. 9.1)** was tested and optimized for nonviral gene therapy protocols in different syngeneic mouse

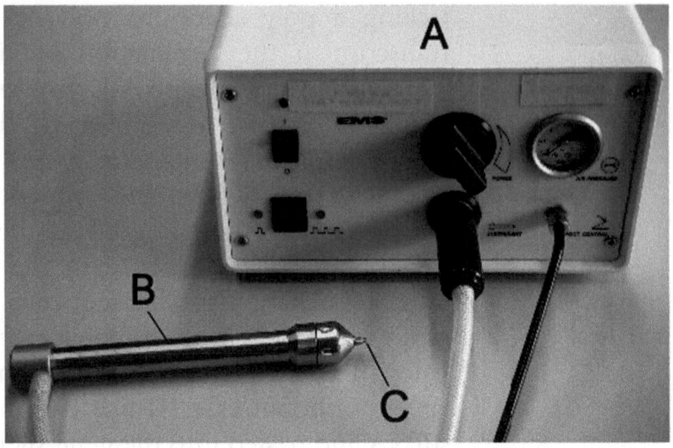

Fig. 9.1. The jet-injector is consisting of the pressure control unit (**a**) and hand-hold handpiece (**b**). The jet-injector can be filled with approximately 200 μL DNA-containing solutions. The DNA solutions are applied through the nozzle (**c**) into the target tissue for in vivo transfection. The volumes of single jet-injections are ranging from 5 to 10 μL per injection. The jet-injector uses pressurized air to produce the high-speed liquid jets. The pressure can be adjusted to 2.8–3.0 bar according to the application for the gene transfer. The in vivo studies for intratumoral gene transfer were carried out at a pressure of 3.0 bar, which produces highest transfer efficiencies in tumor models.

and xenotransplanted human tumor models using reporter gene constructs and vectors expressing therapeutic genes (22–25). A current clinical phase I gene transfer trial demonstrated the safe and efficient application of jet-injection for gene transfer in tumor lesions of cancer patients (26).

In this chapter, the jet-injection technology was used for in vivo delivery of naked therapeutic shRNA-expressing vector DNA into multidrug-resistant tumor xenografts derived from the human carcinoma cell line MaTu/ADR. A complete in vivo restoration of chemosensitivity to the clinically important anti-cancer drug and MDR1/P-gp substrate doxorubicin by the intratumorally jet-injected anti-MDR1 shRNA-encoding DNA was found. Therapeutic efficacy of the combined gene- and chemotherapy is reflected by reduced in vivo tumor growth.

2. Materials

2.1. Construction of Anti-MDR1 shRNA Expression Vectors

1. Plasmid: psiRNA-hH1zeo G2.
2. Bacteria: GT116; genotype: F^- mcrA Δ(mrr-hsdRMS-mcrBC) Φ80lacZΔM15 ΔlacX74 recA1 endA1 Δdcm ΔsbcC-sbcD.
3. Media: Fast-Media Zeo X-Gal agar and Fast-Media Zeo TB.

2.2. In Vivo Jet-Injection of shRNA-Encoding Plasmids

1. Jet-injector device consisting of the jet-injector hand piece and the control unit (**Fig. 9.1**), manufactured by the EMS Medical Systems SA, Nyon, Switzerland (*see* **Note 1**).
2. Access to pressurized air.
3. Anti-MDR1 shRNA expression cassette-containing plasmid vectors (*see* **Section 3.1**).
4. Radenarkon for i.p. injections (40 mg/kg body weight, Jansen Cilag, Neus, Germany).
5. Doxorubicin for i.v. injection (8 mg/kg body weight, Sigma, St. Louis, MO, USA).

2.3. Detection of shRNA Expression in Jet-Injected Tumor Tissues

2.3.1. Northern Blot Detection of shRNA Expression

1. TBE buffer: 90 mM Tris-base, 90 mM boric acid, 2 mM EDTA, pH 8.0.
2. γ-[^{32}P]-ATP (*see* **Note 2**).
3. T4 polynucleotide kinase.
4. Pre hybridization buffer: 6X SSC, 10X Denhardt's reagent, 0.2% SDS.
5. Hybridization buffer: 0.1X SSC, 5X Denhardt's reagent, 0.2% SDS.
6. Wash buffer: 6X SSC, 0.2% SDS.
7. 20X SSC, pH 7.0: 3 M NaCl, 0.3 M sodium citrate.
8. UV-Crosslinker UVC1000.

2.3.2. Quantitative Real-Time RT-PCR Detection of shRNA Expression

1. mirVana miRNA Isolation Kit (Ambion, Austin, TX, USA).
2. QuantiMir RT Kit (SBI, Mountain View, CA, USA).
3. LightCycler instrument.
4. LightCycler capillaries.
5. LightCycler FastStart DNA Master Mix.
6. Oligonucleotide forward primers (antisense-MDR1 siRNA or miR-16) and Universal reverse primer directed against the tagged miRNA (*see* **Note 3**).
7. RelQuant Software.

2.4. Analysis of MDR1 Downregulation by shRNAs

2.4.1. Northern Blot for Detection of MDR1 mRNA

1. BioMax MR films.
2. Exposure cassette.
3. Hybridization flasks.
4. Megaprime™ DNA Labeling System.
5. 25X MOPS buffer: 5 M 3-(*N*-morpholino)propanesulfonic acid (MOPS), 12.5 M sodium acetate, 0.25 M EDTA.
6. Nylon transfer membrane Hybond-N$^+$.

7. Phosphate-buffered saline (PBS), pH 7.4.
8. α-[^{32}P]-dCTP (*see* **Note 4**).
9. 2X RNA Loading Dye Solution (Fermentas GmbH, St. Leon-Rot, Germany).
10. RNeasy® Mini Kit.
11. 20X SSC, pH 7.0; 3 M NaCl, 0.3 M sodium citrate.
12. Trans-Blot BD Semi-Dry Transfer Cell.
13. UV-Crosslinker UVC1000.
14. Wash buffer I: 2X SSC, 0.1% SDS.
15. Wash buffer II: 0.1X SSC, 0.1% SDS.
16. Whatman paper.

2.4.2. Quantitative Real-Time RT-PCR for Detection of MDR1 mRNA Expression

1. LightCycler FastStart DNA Master HybProbe.
2. LightCycler h-G6PDH Housekeeping Gene Set.
3. LightCycler capillaries (Roche Diagnostics).
4. LightCycler instrument (Roche Diagnostics).
5. Oligonucleotide primer, Fluorescein-labeled probe, and LCRed640-labeled probe (*see* **Note 5**).
6. RelQuant Software.

2.5. Analysis of RNAi-Mediated Downregulation of MDR1/P-gp by Immunohisto-chemistry

1. Cryomicrotome.
2. Light microscope.
3. 0.9% hydrogen peroxide (H_2O_2).
4. Monoclonal MDR1/P-gp-specific antibody C219.
5. Secondary, HRP-labeled antibody.
6. Diaminobenzidine.
7. Mayer's Hemalum.
8. Glycergel mounting medium.
9. 10X PBS buffer.

3. Methods

In this section, the experimental strategy for reversal of MDR in vivo by downregulation of MDR1/P-gp by jet-injection of shRNA-encoding plasmids is described. Before starting with in vivo experiments, it is highly recommendable to evaluate the gene-silencing efficiency of the shRNA-encoding plasmids in an in vitro model. For this approach, several cancer cell lines with overexpression of MDR1/P-gp are available (*see* **Note 6**). Likewise before starting the experiments, it is important to check whether the cancer cell model of choice is growing as xenografts in nude mice. In the protocol described here, the multidrug-resistant human cancer cell line MaTu/ADR was used. This cell line was derived from the parental cell line MaTu (27) by long-term in vitro exposure to doxorubicin. The acquired MDR phenotype of MaTu/ADR cells is predominantly caused by a doxorubicin-induced MDR1/P-gp expression, but not by modulations of alternative MDR-related factors such as alternative ABC transporters (28). The MaTu cell line was originally described as derived from a human mammary carcinoma, but in the meantime it was found, by nanoplex PCR DNA profiling, that it may be a mutated derivative of HeLa cells (29). Since both cell variants, MaTu and MaTu/ADR, are growing very well as xenografts, they represent a suitable model for MDR1/P-gp-targeted overcoming approaches of classical MDR1/P-gp-mediated MDR.

Following the confirmation of the high gene-silencing efficacy of the used anti-MDR1 shRNA-encoding plasmid vectors in MaTu/ADR cells in vitro, the in vivo experiments should be initiated. The effects of the shRNA on the targeted gene can be monitored using various experimental settings such as real-time RT-PCR, Northern blots, and immunohistochemistry of cryosections.

3.1. Construction of Anti-MDR1 shRNA Expression Vectors

1. Select the target sequences for shRNAs (*see* **Note 7**) or construct an shRNA expression vector on the basis of validated chemically synthesized siRNAs (*see* **Note 8**) (**Fig. 9.2**).
2. Synthesize two suited homologous single-stranded DNA (ssDNA) molecules by commercial suppliers.
3. Anneal the ssDNA molecules by incubation of 2 μL of each complementary ssDNA oligonucleotide (25 μM) in 6 μL 0.5 M NaCl. Complete to 20 μL by water.
4. Incubate the annealing mixture at 80°C for 2 min followed by cooling to 35°C.
5. Clone the annealed dsDNA, consisting of the anti-MDR1 sense sequence, a 5-nt 3′-CCACC-5′ spacer sequence, the anti-MDR1 antisense sequence, and *Bbs*I-specific 5′-overhangs, into the *Bbs*I restriction site of the expression vector psiRNA-hH1zeo.
6. Transform the shRNA-encoding expression vector in the *Escherichia coli* strain GT116 (*see* **Note 9**, **Fig. 9.2**).
7. Confirm the correct insertion of the specific shRNA-encoding DNA molecules by restriction and sequencing.

3.2. In Vivo Jet-Injection of shRNA-Encoding Plasmids

1. Prior to jet-injection, tumor xenotransplants (*see* **Note 10**) were grown subcutaneously to an approximate tumor size of 6 × 6 mm. Significantly smaller tumors are difficult to

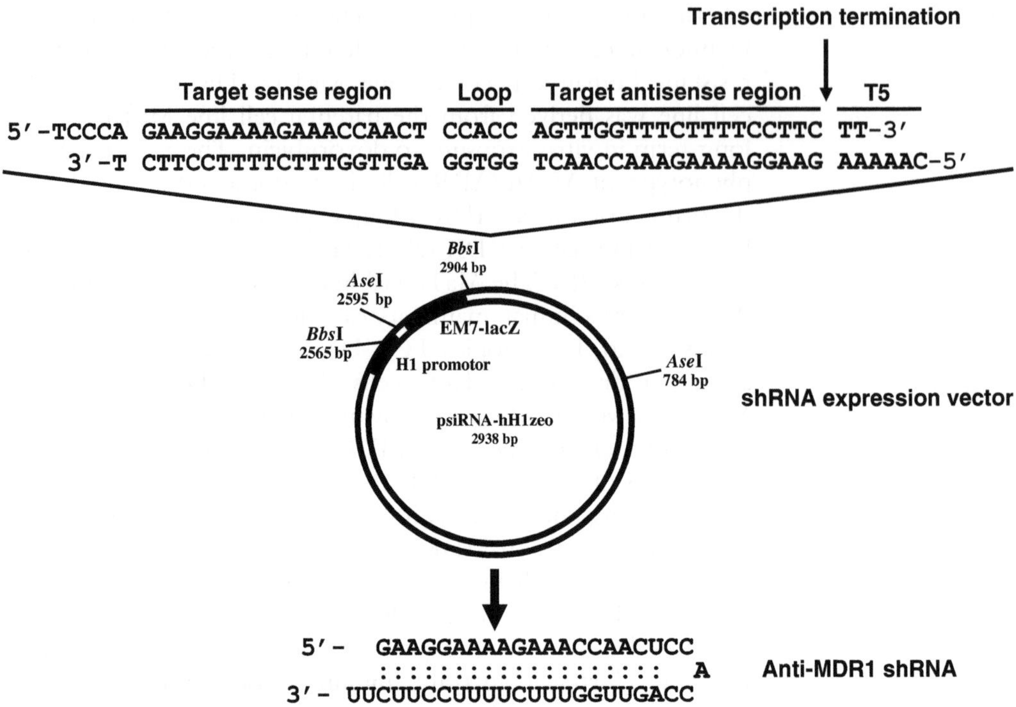

Fig. 9.2. Design of a plasmid vector containing an anti-MDR1/P-gp shRNA expression cassette. The shRNA-encoding plasmid consists of *Bbs*I-specific 5′- and 3′-asymmetric cohesive overhangs that are not compatible, of the target sense and antisense sequence separated by a loop structure, and of the termination site composed by five thymidines (5 T). Chemically synthesized shRNA-encoding DNA oligonucleotides are cloned into the shRNA expression vector psiRNA-hH1zeo. Digestion of psiRNA-hH1zeo with the restriction endonuclease *Bbs*I results in lost of a region encoding an active ß-galactosidase (LacZ) producing EM7-lacZ α-peptide. Finally, the vector drives the synthesis of an anti-MDR1P-gp shRNA which is intracellularly processed by Dicer into the corresponding siRNA triggering the specific gene-silencing RNAi pathway.

inject. Animals were anesthetized with Radenarkon. The sterile DNA solution is then filled into the chamber of the jet-injector head using needle and syringe (*see* **Note 1**).

2. For jet-injection into the tumor tissue place the injector nozzle firmly on the skin surface. To ensure that the jet can travel an optimal path through the tissue at a given setting of the apparatus, use the longest axis of the respective tumor for jet-injection. Co-lateral injection paralleling the dorsoventral line should be preferred to ensure the maximum path for jet-injection gene transfer.
3. Apply four jet-injections of 10 μL each through the skin into the tumor tissue at 3.0 bar, since in mice tumor models this application schedule showed best gene transfer results. This represents a total DNA amount of 40 μg DNA per animal at a DNA concentration of 1 μg/μL.
4. Time points of jet-injection should be empirically determined. However, in a previous experiment jet-injections

were applied at days 19 and 27 post tumor cell inoculation. Since the injected volumes are relatively small, the fluid usually retains within the injected tumor tissue without seepage. At the site of jet-injection minor bleeding might occur, which can easily be stopped with an epinephrine sponge. Jet-injection with this apparatus is well tolerated by the animals.

5. After jet-injection animals are kept for 1–3 days before sacrifice for expression analyses (this time can certainly be extended depending on the specific experimental approach). Tumors should be excised, snap frozen in liquid nitrogen, and kept at –80°C until utilization for shRNA expression detection, MDR1/P-gp mRNA, and protein detection analyses. For these analyses, complete tumor xenografts or cryosections prepared from the tumors should be utilized.
6. For therapeutic experiments, animals are jet-injected following the schedule as described in steps 1–5. Since previous in vitro and in vivo experiments indicated that the peak of MDR1/P-gp downregulation following jet-injection of anti-MDR1 siRNAs or shRNA-encoding expression vectors was achieved after 2–3 days (30–32), subsequent doxorubicin i.v. treatment of 8 mg/kg body weight should be carried out at days 22 and 30.
7. In the therapeutic experiments tumor growth should be monitored during the entire experiment with the measurements of two perpendicular tumor diameters using the spheroid equation: tumor volume = $[(\text{tumor width})^2 \times \text{tumor length}] \times 0.5$.

3.3. Detection of shRNA Expression in Jet-Injected Tumor Tissues

Assessment of shRNA expression was performed indirectly by detection of the corresponding intracellular processed siRNA molecules by Northern blot or quantitative real-time RT-PCR.

3.3.1. Northern Blot Detection of shRNA Expression

1. Isolate micro-RNA (miRNA) from a xenograft by using a suitable miRNA isolation kit.
2. Denaturate 5 μg miRNA in RNA loading buffer at 65°C for 10 min and chill on ice.
3. Separate miRNA on a 15% denaturating polyacrylamide gel in TBE buffer at 250 V.
4. Incubate the gel for 10 min in 10X SSC.
5. Transfer RNA for 24 h to a nylon membrane using reverse capillary blotting as described by Zhou et al. (33).
6. Immobilize miRNAs on nylon membrane by UV-crosslinking.
7. Label siRNA-specific oligonucleotides with $[^{32}P]$ using T4-polynucleotide kinase and γ-$[^{32}P]$-ATP according to manufacturer's instructions (*see* **Note 2**).

8. Prehybridize blot membrane with fixed miRNAs for 1 h at 65°C in prehybridization buffer.
9. Hybridize 10 pmol [^{32}P]-labeled siRNA-specific oligonucleotides to blot membrane for 16 h at the suited temperature (*see* **Note 11**) in hybridization buffer.
10. Wash blot membrane twice with wash buffer for 30 min at room temperature.
11. Expose BioMax MR film overnight at –80°C.

3.3.2. Quantitative Real-Time RT-PCR Detection of shRNA Expression

For quantitative shRNA, respectively, siRNA expression analysis, real-time RT-PCR is applicable. Different well-established real-time PCR platforms are commonly used. In this protocol, the use of a LightCycler instrument is described.

1. Isolate miRNA from frozen tumor xenografts by using a suitable miRNA isolation kit.
2. For cDNA synthesis, perform reverse transcription of 500 ng of tagged miRNA using a QuantiMir RT Kit.
3. Perform a real-time PCR with LC FastStart DNA Master Mix (Roche Diagnostics, Mannheim, Germany) containing 500 nmol/L forward primers (antisense-MDR1 siRNA or miR-16) and 500 nmol/L Universal reverse primer directed against the tagged miRNA (*see* **Note 3**).
4. Use the RelQuant Software to evaluate the results of the real-time PCR runs.

3.4. Analysis of MDR1 mRNA Downregulation by shRNAs

Downregulation of the RNAi target mRNA, i.e., here the MDR1-specific mRNA, can be measured by Northern blot or quantitative real-time RT-PCR.

3.4.1. Northern Blot for Detection of MDR1 mRNA Expression

1. Isolate total cellular RNA from a xenograft by using a suitable RNA isolation kit.
2. Separate 10 μg of total cellular RNA on a 1% agarose gel containing 6% formaldehyde.
3. Transfer RNA to a nylon membrane using reverse capillary blotting as described by Zhou et al. (33).
4. Perform a standard RT-PCR using cDNA from a cancer cell line expressing MDR1/P-gp using MDR1-fw and MDR-1 rev primers (*see* **Note 5**) to amplify the target sequence.
5. Label PCR-generated MDR1-specific cDNA probe with [^{32}P] using a Megaprime™ DNA Labeling System according to manufacturer's instructions.
6. Prehybridize blot membrane for 30 min at 58°C.
7. Hybridize [^{32}P] dCTP-labeled MDR1/P-gp-specific and [^{32}P] dCTP-labeled aldolase (loading control) cDNA fragments to blot membrane overnight at 58°C.

8. Wash blot membrane twice with wash buffer I for 30 min at 21°C and twice with wash buffer II for 30 min at 55°C.
9. Expose BioMax MR film overnight at –80°C.

3.4.2. Quantitative Real-Time RT-PCR for Detection of MDR1 mRNA Expression

For quantitative mRNA expression analysis, a real-time RT-PCR protocol is suited. There are different well-established real-time RT-PCR methodologies available. Here, the application of a LightCycler instrument is depicted detecting the PCR product by means of the hybridization format using a fluorescein-labeled probe and an LCRed640-labeled probe. The reaction mixtures were composed as described in manufacturer's instructions.

1. Isolate total cellular RNA from cryosections by using a suitable RNA Isolation Kit. For cDNA synthesis, perform reverse transcription of RNA by using the MuLV Reverse Transcriptase. Produce an appropriate amount cDNA of any mRNA to last for all following real-time PCRs (5–10 μg) as a calibrator.
2. Perform a dilution series of any cDNA containing the target gene sequence of MDR1/P-gp and the reference gene sequence of glucose-6-phosphate dehydrogenase (*see* **Note 12**).
3. Establish standard curves for the target gene and the reference gene by performing a real-time PCR. Utilize each dilution of the dilution series, the calibrator, and a non-template control as duplicates.
4. Perform a real-time PCR to quantify the cDNA of interest in the samples using target gene primers and probes for one run and reference gene primers and probes for a second run to normalize the samples (*see* **Note 5**). Perform all runs with duplicate samples at least.
5. Use the RelQuant Software to evaluate the results of the real-time PCR runs.

3.5. Analysis of RNAi-Mediated Downregulation of MDR1/P-gp by Immunohistochemistry

1. Prepare cryosections of 6–10 μm thickness from the stored tumor xenografts at –20°C using a cryomicrotome and transfer the sections to suited coated microscope slides which should be air dried (*see* **Note 13**).
2. Wash the slides once with 1X PBS and fix the tissue with acetone (1:1 with PBS) for 20 min at –20°C.
3. For blocking the endogenous peroxidase activity, incubate the air-dried slides in 0.9% H_2O_2 for 8 min.
4. Perform the immunological staining reaction by using the mouse mAb C219 directed against MDR1/P-gp in a 1:10 dilution for 2 h at room temperature.

5. Perform in parallel control reactions by the substitution of specific antibody with a primary mouse-negative control antibody.
6. Subsequently incubate the slides consecutively with HRP-labeled antibody (1:500, 1 h, at room temperature) and streptavidin-biotinylated peroxidase complex (15 min, at room temperature).
7. Use as chromogen diaminobenzidine (5 min, at room temperature).
8. Counterstain and dehydrate the preparations with Mayer's hemalum.
9. Mount the slides with glycergel mounting medium.
10. Evaluate the slides in a microscope.

3.6. Determination of shRNA-Mediated Chemosensitization of Tumor Xenografts

1. In therapeutic experiments, monitor tumor growth during the entire experiment with the measurements of two perpendicular tumor diameters using the spheroid equation (*see* **Section 3.2**). The schedule for tumor volume determination may be adjusted depending on the in vivo tumor growth rates (*see* **Note 14**).
2. Evaluate the statistical significance of treatment influence on tumor growth in mice with the non-parametric Wilcoxon signed rank test.

4. Notes

1. The jet-injector prototype is a pneumatic needle-free injector using pressurized air or nitrogen for the injection of fluids. The headpiece (**Fig. 9.1**) can be filled with approximately 200 μL DNA-containing solutions and provides the possibility of multiple jet-injections of small volumes (5–10 μL per one jet-injection) with one filling. The ejected volumes are very constant with only small variations of less than 10%. The pressure of the jet-injector is adjustable from 1.5 to 3.0 bar that determines the force of the jet, the ejected volume, and penetration depths. In numerous previous studies, the best results were obtained for gene transfer into tumor tissues at pressures between 2.8 and 3.0 bar. Therefore, each parameter of jet-injection can be tightly controlled and adapted for the specific tissue type. This is important, since the pressure determines the energy for tissue penetration and the volume jet-injected into the tissue, which can also have impact on gene transfer efficiency.

   Jet-injection with plasmid DNA at concentrations of 1 μg/μL has shown reproducible results in in vivo gene transfer of different tumors. The solvent PBS had reproducibly shown stabilization of the plasmid DNA. Purification of plasmid DNA by affinity columns is sufficient for efficient gene transfer in pre-clinical studies. However, the DNA should be of endotoxin-free quality. We tested DNA solution in sterile water and in phosphate buffer with comparable results in gene transfer efficiency after jet-injection. However, phosphate buffer might contribute to better stabilization of the closed circular plasmid DNA (34).

Before filling of the jet-injector, the injector headpiece should be disinfected by washing in 70% ethanol or by autoclaving (all stainless steel and teflon components of the headpiece are autoclavable). Using these procedures, we never observed infections of the animals at the sites of jet-injection. During the filling of the chamber it is important to prevent air bubbles in the chamber, since this significantly decreases the jet-injection force and will negatively influence the transfer efficiency.

2. γ-[^{32}P]-ATP, 10 µCi/µL.
3. Oligodeoxynucleotide forward primers are antisense-MDR1 siRNA (5′-AGT GCT TGT CCA GAC AAC A-3′), anti-miR-16 (5′-TAG CAG CAC GTA AAT ATT GGC G-3′), and Universal reverse primer (SBI, Mountain View, CA) directed against the tagged miRNA. PCR program for detection of siRNAs and miRNAs is 10 min at 95°C, followed by 40 cycles of 15 s at 95°C, 10 s at 52°C, and 30 s at 72°C. RNA expression level can be calculated by threshold crossing point (CT) calculation using RelQuant software (Roche). CT values should be generated from three replicates for the target siRNA and the endogenous control (miR-16) amplification curves for each sample using the second derivate maximum mode of analysis.
4. α-[^{32}P]-dCTP, 10 µCi/µL.
5. Oligodeoxynucleotide primers used for amplification of the MDR1-specific template were MDR-fw 5′-TTG TTG AAA TGA AAA TGT TGT CTG G-3′ and MDR-rev 5′-GTC AAA GAA ACA ACG GTT CGG-3′. For gene-specific detection, the following labeled probes can be used: 5′-CAC TGA AAG ATA AGA AAG AAC TAG AAG GTG CT-3′-FITC and LCRed640-5′-GGA AGA TCG CTA CTG AAG CAA TAG AAA ACT-3′. As control, oligodeoxynucleotide primers and probes specific for glucose-6-phosphate-dehydrogenase were used (LightCycler h-G6PDH Housekeeping Gene Set, Roche Diagnostics). Reverse transcription can be performed with Murine Leukemia Virus Reverse Transcriptase (Applied

Biosystems) using arbitrary hexamers. Cycling conditions for MDR1 and (h-G6PDH) are as follows: initial enzyme activation at 95°C for 10 min, followed by 45 cycles at 95°C for 10 s, 60°C (61°C) for 30 s, and 72°C for 4 s. All cycling reactions should be performed in the presence of 4 μM $MgCl_2$.

6. In the past, many MDR1/P-gp-positive cell models were established from drug-sensitive cancer cells by in vitro exposure to drugs that are substrates of this pump protein, e.g., anthracyclines. The advantage of these models is that the degree of drug resistance and its modulation by RNAi can be directly compared to the corresponding parental cells as a control. Examples for well-known MDR1/P-gp-positive cell lines established by this procedure are the cell lines K562/ADR, EPG85-257RDB, EPP85-181RDB, MaTu/ADR, or MCF-7/ADR cells originally described to be derived from the breast cancer cell line MCF-7, but meanwhile identified as derived from OVCAR-8 ovarian adenocarcinoma cells. Thus, these cells were later re-designated as NCI/ADR-RES (35). To maintain the drug-resistant phenotype, cell culture medium for the MDR1/P-gp-expressing cell lines should be supplemented with appropriate concentration of the selection agent.
7. A nice overview of history, mechanism, and current recommendations for the selection of target sites for RNAi effectors are available at the homepage of Thomas Tuschl, a pioneer in this field (http://www.rockefeller.edu/labheads/tuschl/). Further information and commercial providers of siRNAs are listed below:
   General Information:
   - http://www.rnaiweb.com/RNAi/siRNA_Design/

   Design Tools:
   - http://www1.qiagen.com/Products/GeneSilencing/CustomSiRna/SiRnaDesigner.aspx
   - http://www.invitrogen.com/site/us/en/home/Products-and-Services/Applications/RNAi-Epigenetics-and-Gene-Regulation/RNAi.html?cid=invggl123000000007882s
   - http://www.dharmacon.com/DesignCenter/DesignCenterPage.aspx
8. Own experience and experience by others working in the field show that high effective siRNA target sites are in 50–75% also highly effective for shRNAs (7, 8).
9. In the experiments described here, the plasmid psiRNA-hH1zeo (InvivoGen) **(Fig. 9.2)** was used for the construction of the anti-MDR1 shRNA-encoding expression

vector. For amplification of psiRNA-hH1zeo-derived shRNA expression vectors, the *E. coli* strain GT116 is well suited. There are also well-established shRNA expression systems available from alternative suppliers like Ambion, Clontech, or Invitrogen.

10. NMRI:nu/nu mice are transplanted with 1×10^7 MaTu or MaTu/ADR cells/animal of (6 animals/group; Epo GmbH, Berlin, Germany).
11. Oligodeoxynucleotides for radioactive hybridization are anti-MDR1 (5′-AGT GCT TGT CCA GAC AAC A-3′; $T_M = 42°C$) (31) and anti-U6 snRNA (5′-TAT GGA ACG CTT CAC GAA TTT GC-3′; $T_M = 30°C$) (36).
12. The dilution series is performed using cDNA prepared from a MDR1/P-gp overexpressing cell line, for example, MaTu/ADR. Oligodeoxynucleotide primer sequences are mentioned in **Note 5**.
13. Slides should be suited for immunohistochemistry, e.g., SuperFrost Plus slides (Langenbrinck Labor und Medizintechnik, Emmendingen, Germany). As an alternative, slides may be coated by 15 s incubation in 2% 3-aminopropyltrimethoxysilane solution in acetone. Following washing in acetone (15 s) and water (2 × 15 s) the coated slides should be air-dried.
14. In a previous experiment, tumor volumes were determined at days 19, 22, 27, 30, 33, 36, and 40.

Acknowledgments

Own experiments for overcoming cancer MDR by RNAi were supported by grants LA 1039/2-1, LA 1039/2-3, and LA 1039/5-1 of the "Deutsche Forschungsgemeinschaft" (DFG), by the "RNA network" funded by the "Bundesministerium für Bildung und Forschung" (BMBF) and Berlin, by grant no. 01GU0615 of the BMBF, as well as by EMS Medical Systems, Nyon, Switzerland. Furthermore, we thank I. Fichtner for the support in the animal experiments. The excellent technical assistance of J. Aumann, P. Herrmann, M. Lemm, and B. Schaefer is gratefully acknowledged.

References

1. Fire, A., Xu, S., Montgomery, M.K., Kostas, S.A., Driver, S.E., and Mello, C.C. (1998) Potent and specific genetic interference by double-stranded RNA in Caenorhabditis elegans. *Nature*, **391**, 806–811.
2. Elbashir, S.M., Harborth, J., Lendeckel, W., Yalcin, A., Weber, K., and Tuschl, T. (2001) Duplexes of 21-nucleotide RNAs mediate RNA interference in cultured mammalian cells. *Nature*, **411**, 494–498.

3. Lage, H. (2005) Potential applications of RNA interference technology in the treatment of cancer. *Fut Oncol*, **1**, 103–113.
4. Lage, H. (2008) An overview of cancer multidrug resistance: a still unsolved problem. *Cell Mol Life Sci*, **65**, 3145–3167.
5. Stein, U. and Walther, W. (2006) Current Opinion article: Reversal of ABC transporter-mediated multidrug resistance in cancer: a realistic option?. *Am J Cancer*, **5**, 285–297.
6. Szakacs, G., Paterson, J.K., Ludwig, J.A., Booth-Genthe, C., and Gottesman, M.M. (2006) Targeting multidrug resistance in cancer. *Nat Rev Drug Discov*, **5**, 219–234.
7. Lage, H. (2006) MDR1/P-glycoprotein (ABCB1) as target for RNA interference-mediated reversal of multidrug resistance. *Curr Drug Targets*, 7, 813–821.
8. Lage, H. (2009) Therapeutric potential of RNA interference (RNAi) in drug-resistant cancers. *Fut Oncol*, **5**, 169–185.
9. Yamamoto, M. and Curiel, D.T. (2005) Cancer gene therapy. *Technol Cancer Res Treat*, **4**, 315–330.
10. Ledley, F.D. (1995) Nonviral gene therapy: the promise of genes as pharmaceutical products. *Hum Gene Ther*, **6**, 1129–1144.
11. Yang, N.-S., Burkholder, J., Roberts, B., Martinell, B., and McCabe, D. (1990) In vivo and in vitro gene transfer to mammalian cells by particle bombardment. *Proc Natl Acad Sci USA*, **87**, 9568–9572.
12. Furth, P.A., Shamay, A., Wall, R.J., and Hennighausen, L. (1992) Gene transfer into somatic tissue by jet injection. *Anal Biochem*, **205**, 365–368.
13. Sikes, M.L., O'Malley, B.W., Finegold, M.J., and Ledley, F.D. (1994) In vivo gene transfer into rabbit thyroid follicular cells by direct DNA injection. *Hum Gene Ther*, **6**, 837–844.
14. Aihara, H. and Miyazaki, J.-I. (1998) Gene transfer into muscle by electroporation in vivo. *Nat Biotechnol*, **16**, 867–870.
15. Vahlsing, H.L., Yankauckas, M., Sawdey, S.H., Gromkowski, M., and Manthorpe, M. (1994) Immunization with plasmid DNA using a pneumatic gun. *J Immunol Methods*, **175**, 11–22.
16. Rakhmilevich, A.L., Turner, J., Ford, M.J., McCabe, D., Sun, W.H., Sondel, P.H., Grota, K., and Yang, N.-S. (1996) Gene gun-mediated skin transfection with interleukin 12 gene results in regression of established primary and metastatic murine tumors. *Proc Natl Acad Sci USA*, **93**, 6291–6296.
17. Macklin, M.D., McCabe, D., McGregor, M.W., Neumann, V., Meyer, T., Callan, R., Hinshaw, V.S., and Swain, W.F. (1998) Immunization of pigs with a particle-mediated DNA vaccine to influenza A virus protects against challenge with homologous virus. *J Virol*, **72**, 1491–1496.
18. Turner, J.G., Tan, J., Crucian, B.E., Sullivan, D.M., Ballester, O.F., Dalton, W.S., Yang, N.-S., Burkholder, J.K., and Yu, H. (1998) Broadened clinical utility of gene gun-mediated granulocyte-macrophage colony-stimulating factor cDNA-based tumor cell vaccines as demonstrated with a mouse myeloma model. *Hum Gene Ther*, **9**, 1121–1130.
19. Furth, P.A., Kerr, D., and Wall, R. (1995) Gene transfer by jet injection into differentiated tissues of living animals and in organ culture. *Mol Biotechnol*, **4**, 121–127.
20. Cartier, R., Ren, S.V., Walther, W., Stein, U., Lewis, A., Schlag, P.M., Li, M., and Furth, P.A. (2000) In vivo gene transfer by low volume jet injection. *Anal Biochem*, **282**, 262–265.
21. Mitragotri, S. (2006) Current status and future prospects of needle-free liquid jet injectors. *Nat Rev Drug Discov*, 7, 543–548.
22. Walther, W., Stein, U., Fichtner, I., Malcherek, L., Lemm, M., and Schlag, P.M. (2001) Non-viral in vivo gene delivery into tumors using a novel low volume jet-injection technology. *Gene Ther*, **8**, 173–180.
23. Walther, W., Stein, U., Fichtner, I., and Schlag, P.M. (2004) Low-volume jet injection for efficient nonviral in vivo gene transfer. *Mol Biotechnol*, **28**, 121–128.
24. Walther, W., Stein, U., Fichtner, I., Kobelt, D., Aumann, J., Arlt, F., and Schlag, P.M. (2005) Nonviral jet-injection gene transfer for efficient in vivo cytosine deaminase suicide gene therapy of colon carcinoma. *Mol Ther*, **12**, 1176–1184.
25. Walther, W., Minow, T., Martin, R., Fichtner, I., Schlag, P.M., and Stein, U. (2006) Uptake, biodistribution, and time course of naked plasmid DNA trafficking after intratumoral in vivo jet injection. *Hum Gene Ther*, **17**, 611–624.
26. Walther, W., Siegel, R., Kobelt, D., Knösel, T., Dietel, M., Bembenek, A., Aumann, J., Schleef, M., Baier, R., Stein, U., and Schlag, P.M. (2008) Novel jet-injection technology for nonviral intratumoral gene transfer in patients with melanoma and breast cancer. *Clin Cancer Res*, **14**, 7545–7553.
27. Widmaier, R., Wildner, G.P., Papsdorf, G., and Graffi, I. (1974) Über eine neue, in vitro unbegrenzt wachsende Zellinie MaTu von Mamma-Tumorzellen des Menschen. *Arch Geschwulstforsch*, **44**, 1–10.

28. Stein, U., Walther, W., Lemm, M., Naundorf, H., and Fichtner, I. (1997) Development and characterisation of novel human multidrug resistant mammary carcinoma lines in vitro and in vivo. *Int J Cancer*, **72**, 885–891.
29. Stein, U., Walther, W., Stege, A., Kaszubiak, A., Fichtner, I , and Lage, H. (2008) Complete in vivo reversal of the multidrug resistance (MDR) phenotype by jet-injection of anti-MDR1 short hairpin RNA-encoding plasmid DNA. *Mol Ther*, **16**, 178–186.
30. Nieth, C., Priebsch, A., Stege, A., and Lage, H. (2003) Modulation of the classical multidrug resistance (MDR) phenotype by RNA interference (RNAi). *FEBS Lett*, **545**, 144–150.
31. Stege, A., Priebsch, A., Nieth, C., and Lage, H. (2004) Stable and complete overcoming of MDR1/P-glycoprotein-mediated multidrug resistance in human gastric carcinoma cells by RNA interference. *Cancer Gene Ther*, **11**, 699–706.
32. Kaszubiak, A., Holm, P.S., and Lage, H. (2007) Overcoming the classical multidrug resistance phenotype by adenoviral delivery of anti-MDR1 short hairpin RNAs and ribozymes. *Int J Oncol*, **31**, 419–430.
33. Zhou, D.C., Marie, J.P., Suberville, A.M., and Zittoun, R. (1992) Relevance of mdr1 gene expression in acute myeloid leukemia and comparison of different diagnostic methods. *Leukemia*, **6**, 879–885.
34. Hartikka, J., Bozoukova, V., Jones, D., Mahajan, R., Wloch, M.K., Sawdey, M., Buchner, C., Sukhu, L., Barnhart, K.M., Abai, A.M., Meek, J., Shen, M., and Manthorpe, M. (2000) Sodium phosphate enhances plasmid DNA expression in vivo. *Gene Ther*, 7, 1171–1182.
35. Liscovitch, M. and Ravid, D. (2007) A case study in misidentification of cancer cell lines: MCF-7/AdrR cells (re-designated NCI/ADR-RES) are derived from OVCAR-8 human ovarian carcinoma cells. *Cancer Lett*, **245**, 350–352.
36. Scherr, M., Battmer, K., Schultheis, B., Ganser, A., and Eder, M. (2005) Stable RNA interference (RNAi) as an option for anti-bcr-abl therapy. *Gene Ther*, **12**, 12–21.

Chapter 10

Short Hairpin RNA (shRNA): Design, Delivery, and Assessment of Gene Knockdown

Chris B. Moore, Elizabeth H. Guthrie, Max Tze-Han Huang, and Debra J. Taxman

Abstract

Shortly after the cellular mechanism of RNA interference (RNAi) was first described, scientists began using this powerful technique to study gene function. This included designing better methods for the successful delivery of small interfering RNAs (siRNAs) and short hairpin RNAs (shRNAs) into mammalian cells. While the simplest method for RNAi is the cytosolic delivery of siRNA oligonucleotides, this technique is limited to cells capable of transfection and is primarily utilized during transient in vitro studies. The introduction of shRNA into mammalian cells through infection with viral vectors allows for stable integration of shRNA and long-term knockdown of the targeted gene; however, several challenges exist with the implementation of this technology. Here we describe some well-tested protocols which should increase the chances of successful design, delivery, and assessment of gene knockdown by shRNA. We provide suggestions for designing shRNA targets and controls, a protocol for sequencing through the secondary structure of the shRNA hairpin structure, and protocols for packaging and delivery of shRNA lentiviral particles. Using real-time PCR and functional assays we demonstrate the successful knockdown of ASC, an inflammatory adaptor molecule. These studies demonstrate the practicality of including two shRNAs with different efficacies of knockdown to provide an additional level of control and to verify dose dependency of functional effects. Along with the methods described here, as new techniques and algorithms are designed in the future, shRNA is likely to include further promising application and continue to be a critical component of gene discovery.

Key words: RNA interference (RNAi), small interfering RNA (siRNA), short hairpin RNA (shRNA), lentivirus, design, delivery, ASC, *Porphyromonas gingivalis*, IL-1β, ELISA, THP1.

1. Introduction

In recent years, the use of RNA interference (RNAi) has emerged as a powerful tool for the study of gene function in mammalian cells. The mechanism of RNAi is based on the sequence-specific

M. Sioud (ed.), *RNA Therapeutics*, Methods in Molecular Biology 629,
DOI 10.1007/978-1-60761-657-3_10, © Springer Science+Business Media, LLC 2010

degradation of host mRNA through the cytoplasmic delivery of double-stranded RNA (dsRNA) identical to the target sequence (1). Degradation of target gene expression is achieved through an enzymatic pathway involving the endogenous RNA-induced silencing complex (RISC). One strand of the siRNA duplex (the guide strand) is loaded into the RISC with the assistance of Argonaute (Ago) proteins and double-stranded RNA-binding proteins. The RISC then localizes the guide strand to the complementary mRNA molecule, which is subsequently cleaved by Ago near the middle of the hybrid (2). The cleaved mRNA is further degraded by other endogenous nucleases. Likewise, the RISC also plays an important cellular role in inhibiting endogenously derived mRNA through a related micro-RNA (miRNA) mechanism (3).

Several methods of RNAi have evolved over time, with the simplest approach involving the transfection of chemically synthesized short interfering RNA oligonucleotides (siRNAs) directly into the cytosol (*see* **Chapters 4**, **5**, and **9**). While the delivery of siRNAs can be achieved in many cell types, variable transfection efficiencies have limited siRNA-mediated RNAi to only those cells capable of transfection. Another form of RNAi involves the use of short hairpin RNAs (shRNAs) synthesized within the cell by DNA vector-mediated production. Like siRNAs, shRNAs may be transfected as plasmid vectors encoding shRNAs transcribed by RNA pol III or modified pol II promoters, but can also be delivered into mammalian cells through infection of the cell with virally produced vectors. While siRNA delivers the siRNA duplex directly to the cytosol, shRNAs are capable of DNA integration and consist of two complementary 19–22 bp RNA sequences linked by a short loop of 4–11 nt similar to the hairpin found in naturally occurring miRNA. Following transcription, the shRNA sequence is exported to the cytosol where it is recognized by an endogenous enzyme, Dicer, which processes the shRNA into the siRNA duplexes (*see* **Chapter** 7). Like the exogenously delivered synthetic siRNA oligonucleotides, this endogenously derived siRNA binds to the target mRNA and is incorporated into the RISC complex for target-specific mRNA degradation (4).

Although siRNA and shRNA ultimately utilize a similar cellular mechanism (RISC), the choice of which method to use depends on several factors such as cell type, time demands, and the need for transient versus stable integration. There are a variety of reagents available for siRNA design and synthesis. Therefore, the efficiency of knockdown for each siRNA sequence can be rapidly determined and, in fact, there are several commercial sources for siRNA which have been functionally validated. In addition, siRNA delivery has benefited from the plethora of transfection reagents already in existence, yielding a potentially high level of gene silencing with minimal cellular toxicity. An increasing concern with siRNA, however, is the apparent increased probability of incurring off-target effects due to the high concentration of cytoplasmic siRNA. Another significant disadvantage to siRNA oligonucleotide delivery is that as the cells divide, the siRNA concentration becomes diluted, thereby rendering the generation of a long-term cell line with the desired target gene knockdown unfeasible. shRNA, on the other hand, may be used to generate stable knockdown cell lines, thereby eliminating the need for multiple rounds of transfection and greatly increasing reproducibility of results. However, the creation of a stable shRNA cell line is a time-consuming task as the construct preparation and the selection of shRNA-positive cells by drug resistance or fluorescent markers may take months. With this said, many cells cannot be transfected with siRNA at high levels, especially primary and non-adherent cells, such as immune cells and non-dividing cells. Transfection efficiency is a major issue for siRNA since incomplete transfection produces incomplete knockdown which may fail to ablate the function of the protein. For most untransfectable cells, adenoviral, retroviral, or lentiviral-based shRNA technology remains the only viable technology for the successful delivery of RNAi. For these reasons, this article focuses primarily on methodologies applicable to shRNA, though many of the suggestions may also be useful for siRNA.

The proper selection of a target sequence for a given gene of interest remains one of the most critical components of successful gene knockdown regardless of the RNAi methodology. Although target RNAi sequences have been constructed from 19 to 27 bp, most data on effective sequence selection involve the design of 19 bp targets. While there is no guarantee of effective gene silencing for a given siRNA until experimentally proven, numerous algorithms have been designed to predict these 19 bp targets with a nucleotide composition thought to confer the highest efficacy (5–8). In addition, as new algorithms are designed frequently, it is imperative that one should use the most modern design method available for selecting a target site. Protocol 3.1 provides general guidelines for how to use the available algorithms to design shRNA knockdowns and appropriate controls. A minimum of two target sequences should be designed for each gene, in order to increase the likelihood that at least one sequence results in significant gene knockdown. Additionally, two successful knockdowns can also provide a useful control for off-target knockdowns since it is statistically unlikely that different sequences will produce the same off-target knockdowns. **Figures 10.1** and **10.2** give an example of how two shRNAs with differing efficacies can also be extremely useful in showing dose dependency for functional studies.

Once the target sites are selected, shRNA vectors must be constructed. Two basic methods for constructing shRNA vectors,

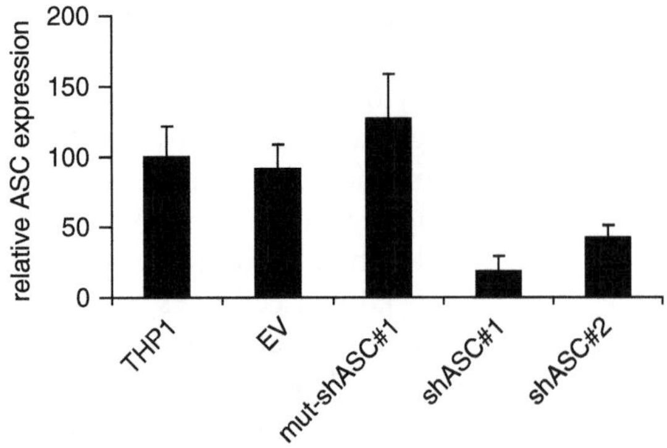

Fig. 10.1. Knockdown of ASC in THP1 cells transduced using lentiviral shRNA vector, FG12 (14). THP1 cells were transduced with lentivirus expressing shRNA against ASC. Our previous studies have shown efficacy for these same shRNAs in reducing ASC expression when expressed using a stem cell virus-based retroviral vector pHSPG (6, 13). Similar to our previous results, the shASCs transduced using FG12 reduce endogenous ASC levels in THP1 monocytic cells by approximately 80% (shASC#1) and 60% (shASC#2). Three control cell lines were also tested for comparison, untransfected THP1 cells (THP1), cells transduced with an empty lentiviral vector (EV), and cells transduced with a lentivirus expressing a scrambled target for shASC#1 (mut-shASC#1). Results represent the averages plus standard deviations of triplicates, are standardized to 18s rRNA expression, and are normalized to an average of 100 in THP1 cells.

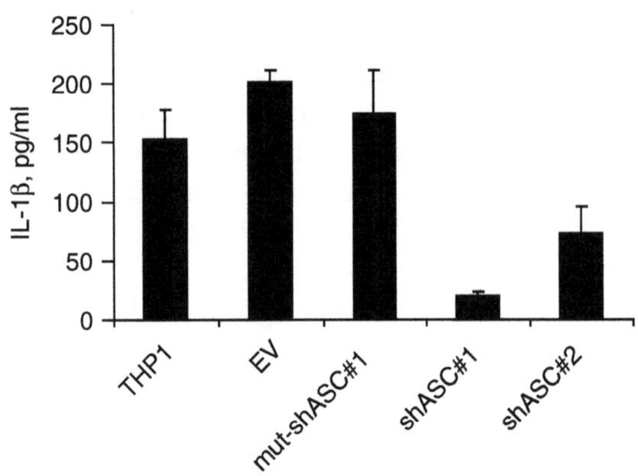

Fig. 10.2. ELISA of IL-1β in control and shASC knockdown cell lines following infection with 10 MOI *Porphyromonas gingivalis*. This figure demonstrates how a functional assay can be used to verify knockdowns. In this case of our protein of interest, ASC, has a well-established role in processing IL-1β following infection with bacteria (10, 11, 13). The reduced IL-1β that is observed for the shRNA cell lines following infection with *P. gingivalis* verifies the knockdowns. Additionally, the experiment shows dose dependency since the shASC#2 is less effective than shASC#1 in knocking down ASC (**Fig. 10.1**) and also has proportionally less efficacy in reducing IL-1β secretion levels. This figure, therefore, illustrates both the general utility of a functional assay and the advantage of having two different knockdowns of different efficacy to verify dose-dependent functional effects. The use of two shRNAs also provides an additional level of control for studies of ASC function since two shRNAs are statistically unlikely to promote the same off-target knockdowns.

oligonucleotide-based cloning and PCR-based cloning, have been provided elsewhere (9). Specific cloning vectors and protocols for constructing shRNA vectors for appropriate applications are also available commercially from multiple biotechnology companies. For any newly constructed shRNA vector, it is essential to confirm the sequence of the hairpin since single-base mismatches within the target can alter specificity. Though many shRNA plasmids will sequence sufficiently under standard sequencing conditions, a number of shRNAs will be problematic due to the intrinsic secondary structure of the hairpin. In **Section 3.2**, some basic steps are provided for sequencing even the most problematic shRNA hairpins. These recommendations are based on a detailed analysis of the effects of sequencing additives alone or in combination (6).

There are multiple methods of introducing siRNA and shRNA into cells. The method of choice depends on whether transient or stable expression is desired and the model system. Lentiviral-mediated transduction provides a convenient method of introducing shRNA into dividing or non-dividing cells and, in general, is less toxic to the cells than adenoviral-mediated transduction. **Sections 3.3** and **3.4** describe effective methods for preparation of lentiviral particles and transduction into adherent or non-adherent cells.

Once the shRNA plasmid is prepared and introduced into the cells, it is necessary to confirm effective knockdown. As an example of knockdown determination, Protocol 3.5 and **Fig. 10.1** describe the confirmation of knockdown of the innate immunity adaptor molecule, ASC (10, 11), by quantitative PCR analysis. These knockdowns are based on two shRNAs against ASC that have been previously produced by our laboratory using a murine stem cell-based retroviral vector, pHSPG (6, 12, 13). For the current study, the ASC shRNAs were re-cloned into a lentiviral vector, FG12 (14), and knockdown in this system was tested. Details of the PCR assay used to verify ASC knockdown by the FG12-shRNA vector are provided (**Section 3.5**).

If the gene targeted for knockdown has a known biological or physiological function, a functional assay can also be extremely useful in testing the efficacy of an shRNA. Protocol 3.6 illustrates how a functional assay can verify knockdown. In this example, ASC is known to play an integral role in the function of the inflammasome, a multiprotein cytosolic complex required for the cleavage and activation of IL-1β (10, 11). To test the function of the ASC shRNAs, we have performed ELISAs for IL-1β following infection of THP1 human monocytic cells with the periodontitis-associated pathogen, *Porphyromonas gingivalis* (**Fig. 10.2**). Similar to previous results using the pHSPG vector (6, 12, 13), the shRNAs produced in FG12 are functional in blocking IL-1β expression. The effects of the knockdowns are dose-dependent, providing further verification of the findings and a convenient additional control for the experiment (**Fig. 10.2**).

2. Materials

2.1. Selecting RNA Target Sites and Sequencing Through shRNA Hairpins

1. To select shRNA target sites and corresponding controls, internet access is required.
2. Dimethyl sulfoxide.
3. BigDye Terminator v1.1 cycle sequencing ready reaction mix (Applied Biosystems).
4. ABI Prism dGTP BigDye terminator ready reaction mix. (Applied Biosystems)
5. Betaine.
6. 1X PCRx Enhancer (Invitrogen; included with Pfx DNA Polymerase).
7. Centri-Sep 96-well spin plates (Princeton Separations).
8. 3730 series DNA Analyzer (Applied Biosystems).

2.2. Packaging shRNA-Encoding Lentivirus

1. Lentiviral vector, envelop vector (e.g., pMDL or other Gag-Pol vector), and packaging vector(s) (encoding VSV-G and Rev genes).
2. 2X BES-buffered saline (BBS)—0.5 M BES, 150 mM Na_2HPO_4, pH 6.95.
3. 1 M $CaCl_2$ dissolve in sterile water and stored at –20°C.
4. 293T cells.
5. Dulbecco's modified Eagle's medium.
6. Fetal calf serum (FCS).

2.3. Stable Transduction of Adherent or Non-adherent Cells with shRNA-Encoding Lentivirus

1. Viral supernatant from **Section 3.3**.
2. Cells to be transduced.
3. Growth media for the cells. Usually RPMI, 10% FCS or DMEM, 10% FCS, depending on the cell type.
4. 2.0 mL round-bottom microcentrifuge tubes.
5. 8.0 mg/mL polybrene stock solution dissolved in sterile water and stored at –20°C.
6. Table top centrifuge (for adherent cells) or swing-bucket microcentrifuge (for non-adherent cells).
7. Appropriate agent for drug selection for lentiviral vectors that contain a drug-resistance gene.

2.4. Confirmation of Knockdown of ASC by Real-Time PCR Analysis

1. RNeasy purification kit (Qiagen).
2. 100 μM oligo$(dT)_{15}$.
3. 10 mM dNTP mix: 10 mM each dATP, dCTP, dGTP, and dTTP.

4. MMLV reverse transcriptase.
5. 5X FS Buffer (Invitrogen; included with MMLV).
6. 0.1 M DTT (included with MMLV).
7. RNaseOUT (Invitrogen) or RNaseIN (Promega).
8. PCR grade water.
9. PCR pipette tips.
10. PCR primers (*see* **Section 3.5**).
11. 384-Well PCR plates.
12. Optical adhesive film.
13. 2X SYBR Green PCR Master Mix.
14. AB Prism 7700 thermocycler.

2.5. ELISA to Confirm Functional Knockdown of ASC

1. Stably transduced control and knockdown cells.
2. *P. gingivalis*, or other immunostimulatory agent.
3. RPMI 1640 supplemented with 10% FCS.
4. Human IL-1β ELISA set.

3. Methods

3.1. Selecting shRNA Target Sites and Corresponding Controls

1. Determine whether the gene of interest has one or multiple splice variants. Decide whether you want to target all potential forms of a gene or specific splice variants. Select exons for targeting accordingly (*see* **Note 1**).
2. Select several potential target sites within the exon or exons of interest within your gene or within the 5′ or 3′ UTR depending on which splice forms of the gene are to be targeted (*see* **Note 2**). If the function of an siRNA or shRNA against your gene of interest has been validated commercially or in a publication it may be useful to test the same target site in your system. An siRNA or shRNA that shows efficacy in one cell system in knocking down expression of its target is usually effective in other cell systems.
3. If a validated siRNA target is not available for your exon(s) of choice, you will need to design the shRNA target anew. There are several commercial and non-commercial web sites available for siRNA design (*see* **Note 3**). Use the most current design algorithm and web resource available. The following steps may or may not be included as an option depending on the algorithm and web resource, and their inclusion as design criteria depends upon your specific experimental model:

a. For shRNAs that are to be expressed from a plasmid or viral vector under control of a pol III promoter, avoid target sequences with runs of four or more A's or T's. These sequences may create potential problems with premature termination during transcription. Some of the common hairpin loop sequences begin with two T nucleotides, and if this is the case, be careful that the target sequence does not end with more than one T (*see* **Note 4**).

b. For the design of shRNA that will be expressed from a U6 or 7S K promoter, select target sequences beginning with a G.

4. Perform a BLAST search to eliminate potential target sequences that have a perfect match of 16 nt or more to an off-target gene of the same species (http://www.ncbi.nlm.nih.gov/BLAST/).
5. Eliminate potential target sites that overlap regions of single-nucleotide polymorphisms (SNPs) (http://www.ncbi.nlm.nih.gov/projects/SNP/).
6. Select two or more shRNAs targeting different regions within the same gene that have the fewest amounts of BLAST matches, that do not overlap a region of SNP, and that target different regions within the gene (*see* **Note 5**).
7. Also design control shRNAs, including a non-targeting shRNA with a fully or partially scrambled targeting sequence (*see* **Note 6**).

3.2. Sequencing Through shRNA Hairpins

1. Design two sequencing primers, one for either strand of the vector into which the shRNA is cloned. Primers should lie approximately 50–100 bp upstream (for the sense strand) or downstream (for the antisense strand) from the hairpin and be approximately 20 nt in length (*see* **Note 7**).
2. Set up a standard sequencing reaction (12.5 μL total volume) containing 1X BigDye Terminator v1.1 cycle sequencing ready reaction mix (Applied Biosystems), 0.26 μg of DNA, and 3.75 pmol of sequencing primer. Also add 5% DMSO. This will be sufficient for sequencing through the majority of shRNA hairpins.
3. For more problematic hairpins that do not sequence sufficiently with the conditions in step 2, sequencing reactions may be modified by substituting the standard BigDye Terminator v1.1 chemistry with a mixture of 10 parts BigDye v1.1 chemistry to 1 part ABI Prism dGTP BigDye Terminator ready reaction mix. Also add 0.83 M Betaine and 1X PCRx Enhancer (Invitrogen) to the sequencing reaction for read through of virtually any shRNA hairpin.

4. Perform PCR using the following standard thermal cycler program:
   95°C × 3 min

   98°C × 40s
   50°C × 5s
   60°C × 4 min

   98°C × 10s
   50°C × 5s for 24 cycles
   60°C × 4 min

   ---

   cool to 4°C

   ---

5. Purify the sequencing reactions using Centri-Sep columns or 96-well plates. Run the purified products on a 3730 series DNA Analyzer.

3.3. Packaging shRNA-Encoding Lentivirus

1. Day 0: Plate 293T cells in 10 cm plates in DMEM and 10% FCS and grow overnight. A 175 cm^2 flask at 50% confluency is sufficient for approximately 4–5 plates; however, the plating density will vary depending on the starting confluency and the growth of the cell culture (*see* **Note 8**).
2. Day 1: Cells should be about 70% confluent and evenly distributed on the plate. Aspirate off the medium and add 5 mL of DMEM without serum.
3. Aliquot the lentiviral vector and packaging vectors into a 5 mL snap cap tube as follows:
   a. 15 μg lentiviral vector.
   b. 5 μg envelop vector (e.g., pMDL or other Gag-Pol vector).
   c. 10 μg packaging vector(s) (*see* **Note 9**).
   d. Add water to yield 375 μL total volume.
4. Add 125 μL of 1 M $CaCl_2$ to the DNA mixture and vortex.
5. Add 500 μL of 2X BBS to the mixture dropwise, while vortexing, and then incubate the DNA/$CaCl_2$/BBS mixture for 30 min at room temperature.
6. Add the transfection mixture to the plate dropwise. Incubate in a CO_2 incubator (either 5% CO_2 or 3% CO_2) for 2–3 h and then add 0.5 mL serum to the culture. Incubate the cells overnight (*see* **Note 10** for steps 6–10 of this protocol and Protocol 3.4).
7. Day 2: Remove the media and replace with 7 mL DMEM and 10% FCS. Return cells to a 5% CO_2 incubator for 40–48 h. Virus production peaks at about 40 h (*see* **Note 11**).

8. Day 4: Collect viral supernatant and filter through a 0.45 μm syringe filter to remove any cells or cell debris. The 293T plates can now be bleached and discarded according to approved biosafety procedures.
9. Use viral supernatant fresh or aliquot and freeze at –80°C for at least 6 months to 1 year.

3.4. Stable Transduction of Adherent or Non-adherent Cells with shRNA-Encoding Lentivirus

1. Plate the cells to be transduced with the shRNA-encoding lentivirus as follows:
   a. For adherent cells, plate cells at approximately 60–70% confluency in a 24-well plate, one well per sample, and allow cells to attach overnight. Remove media and replace with 500 μL fresh growth media per well.
   b. For non-adherent cells, spin down 1×10^6 cells per sample and resuspend in 500 μL growth media in a 2 mL round-bottom microfuge tube.
2. Add 0.5 μL of polybrene to each sample to yield a final concentration of 8 μg/mL.
3. Add 1 mL of viral supernatant (*recall* **Note 10**).
4. Spin tubes or plates at 2,000 rpm for 1–3 h using a swing-bucket rotor if possible (*see* **Note 12**).
5. Remove the supernatant and replenish the cells with 1 mL of growth media for adherent cells, or for non-adherent cells replenish with 3 mL of growth media and plate in a 6-well plate in a CO_2 incubator overnight.
6. Repeat steps 1–4 for increased efficiency of infection.
7. Plate cells and culture as needed.
8. Depending on the marker gene in the lentiviral vector, you may be able to use the marker to assess percent transduction. If the virus has the green fluorescent protein (GFP) gene, for example, FACS analysis can be done to correlate GFP positivity and transduction levels. If cells are not 95–100% GFP positive, cell sorting could be done to increase levels of stable shRNA-expressing cells. If a drug resistance marker is contained within the viral vector, drug selection should be set up to eliminate any cells that did not receive the shRNA. For any of these assays it is necessary to wait at least 48–72 h to give the cells a chance to express the stably encoded marker gene.

3.5. Confirmation of Knockdown of ASC by Real-Time PCR (see Notes 13 and 14)

1. Design primers for analysis of mRNA levels by quantitative PCR. Primers should target the same splice forms that the shRNAs target, should span an intron/exon junction if possible (*see* **Note 15**), should lie approximately 100–150 nt apart, and should have a T_m of approximately 57°C. A

primer design program such as Primer Express or Primer Designer 4 can be used to assist in primer design (*see* **Note 16**).

2. Isolate RNA from knockdown cells on three different days (*see* **Note 17**). Also isolate RNA from several control cells, for example, untransfected cells, cells stably transfected with an empty vector, cells stably expressing a scrambled target sequence, and shRNA targeting an irrelevant gene. Use an RNeasy purification kit to purify total RNA from approximately 2×10^6 cells.
3. Prepare cDNA as follows:
   a. Combine 1 μL 100 μM oligo(dT), 1 μL 10 mM dNTP mix, and 1 μg of RNA. Add water to yield a final volume of 12 μL.
   b. Heat the mixture for 5 min at 65°C and then incubate on ice for at least 1 min.
   c. Prepare a master mix including the following components per sample (*see* **Note 18**):
      4 μL FS buffer,
      2 μL 0.1 M DTT,
      1 μL RNaseOUT or RNaseIN,
      1 μL MMLV reverse transcriptase.
   d. Add 8 μL of the master mix to each sample. Mix by pipetting up and down and then incubate at 42°C for 90 min.
4. Prepare a 1:10 dilution of the cDNA by combining 3 μL cDNA with 27 μL of PCR grade water. Also prepare a 1:5,000 dilution by combining 2 μL of the 1:10 diluted cDNA with 1 mL of PCR grade water.
5. Pipette 4.2 μL of either the 1:10 dilution of the cDNA for the gene of interest or the 1:5,000 dilution of the cDNA for the 18s rRNA (or other housekeeping gene; *see* **Note 19**) into duplicate wells of a 384-well plate.
6. Prepare a mastermix of SYBR green mix and primers (*see* **Note 18**). This should include 5 μL per sample of SYBR mix and 0.8 μL/sample of primer mix (5 μM each of forward and reverse primers from Step 1). Add 5.8 μL of mastermix per well.
7. Run on an AB Prism 7700 instrument (Applied Biosystems) or a similar thermocycler with the following program:
   48°C × 5 min

   95°C × 10 min

   95°C × 30s

58°C × 30s for 40 cycles

95°C × 15s
58°C × 15s dissociation stage
95°C × 15s

---

8. Determine the relative amounts of mRNA for your gene of interest using the comparative CT method (Applied Biosystems). Standardize values to the expression of the endogenous 18s rRNA or another endogenous housekeeping gene (*see* **Note 19**).

3.6. ELISA Analysis to Confirm Functional Knockdown of ASC (see Note 20)

1. Plate THP1 control and ASC knockdown cells lines at 10^6 cells/mL in a 24-well plate (1 mL/well) in a 37°C CO_2 incubator.
2. Add bacterial or other immunostimulatory agent that is known to activate IL-1β through the inflammasome complex (*see* **Note 21**).
3. Incubate cells for 2 h (for stronger inducers) to overnight (for weaker inducers) (*see* **Note 22**). For cells induced by infection with a bacterial pathogen, add antibiotics 1 h following infection to prevent subsequent bacterial growth in the culture.
4. Transfer supernatant to a microcentrifuge tube and spin for 5 min at high speed to remove any cells or cellular debris.
5. Recover 900 μL of supernatant to a new microcentrifuge tube and use immediately in an ELISA experiment or store at –20°C for up to several months.
6. Run the ELISA experiment according to manufacturer's recommendations and calculate IL-1β levels in control and knockdown cells by comparison to standard curve. Use log–log regression analysis as recommended by the manufacturer.

3.7. Future Challenges, Promise, and Scientific Developments

Since its inception, the use of RNAi technology has revolutionized how we perform research on gene function. However, the use of this technology is likely to include further challenges in addition to some exciting new applications. One major challenge is in the design of the RNAi. Although significant progress has been made over the past several years in predicting which siRNA target sequences are most effective in reducing gene expression (5–8, 16), currently the only way to ensure the efficacy of an siRNA is by direct experimentation. As our understanding of RNAi mechanisms improves, it is likely that we will be able to better predict the functionality of each siRNA through the development of more accurate algorithms. These new algorithms should include specific sequence requirements for each target and the ability to predict the putative effects of secondary mRNA structure. Also, the role of off-target knockdowns, including the degradation of non-identical mRNAs through the RISC pathway, as well as miRNA-type inhibition of translational elongation, is continually being understood in more detail. The more complete understanding of off-target knockdowns will further the effective design and implementation of RNAi. Other advances in the design of shRNA will likely include the identification of more effective RNA hairpin structures, including modified loop sequences; structural or chemical RNA modifications that can alter the mechanisms of action of the siRNA in a favorable and predictable manner; and the identification of additional proteins that can mimic or modulate the function of the Dicer and RISC complexes. Traditionally, shRNAs have been expressed using pol III promoters since they produce a shorter more predictable transcript; however, recent studies have identified modified pol II transcripts that can increase shRNA expression levels (17, 18). The ability to fine-tune expression levels of shRNAs will be important to the efficient use of shRNA since levels that are too low may not be effective, while levels too high can cause toxicity. The ability to control siRNA expression levels may be especially important for genes involved in cell survival, in which case the identification inducible shRNA promoters should be useful (19, 20).

An additional obvious challenge to RNAi is the effective delivery of siRNA or shRNA. The availability of improved commercial transfection reagents has improved siRNA, but still some cells remain difficult to transfect. Lentivirus and adenovirus have made it possible for cells that are refractory to transfection such as primary cells to become permissive to shRNA. However, as these methods are based on a viral backbone, each method harbors inherent dangers which would limit their use to in vitro studies.

In vivo delivery of RNAi also offers great promise for the future. Since current in vivo gene function studies involve the time-consuming development of transgenic mouse gene knockouts and double knockouts, a successful in vivo RNAi protocol would represent a tremendous step forward in terms of time allocation and likely lead to an explosion of knowledge obtained from such studies. Most current approaches to in vivo RNAi involve the systemic delivery of "naked" siRNAs. These so-called naked siRNAs are only moderately effective in the in vivo knockdown of a gene of interest and mostly are limited to genes expressed within the liver and kidney (21). In addition, since naked siRNAs exhibit poor pharmacokinetics, they are delivered at high concentrations, adding to their expense and putative off-target effects. However, there exists at least one in vivo siRNA delivery transfection reagent (Invivofectamine, Invitrogen, Inc.), which has shown promise in the delivery of much lower concentrations of siRNA to a mouse; however, this reagent remains cost prohibitive to most laboratories and these data remain to be reproduced readily outside of its commercial source (http://invitrogen.cnpg.com/Video/flatFiles/761/index.aspx). In addition, viral vectors encoding shRNAs have shown promise for in vivo delivery; however, most of these studies have utilized adenoviral delivery of shRNA, which has well-known toxic effects in the animal (22). Recently, adeno-associated viral vectors (AAV) have been designed with less toxicity and adequate shRNA delivery (23). Finally, there are numerous ongoing studies focused on virally mediated delivery of shRNA to hematopoietic stem cells (HSC) isolated from a mouse and re-implanted into an irradiated recipient mouse (24). These HSCs have been shown to give rise to cells with stable shRNA; however, the recipient mouse still retains gene expression within stromal cells. Nonetheless, this may prove an effective strategy for in vivo studies of gene expression in cells of an immune origin. Notwithstanding these advances in in vivo RNAi, there are still numerous challenges to methodology and application; however, with every new publication comes the exciting possibility of another breakthrough in RNAi technology which will likely advance this field far beyond what is conceivable today.

4. Notes

1. One of the most critical considerations in selecting a target site is the consideration of all splice forms of the targeted mRNA. For general knockdown of a gene, the site selected must target every splice form in order to yield interpretable results.
2. If the appropriate spice variants of the gene are targeted, the position of the target site (UTR, 5′, middle, or 3′) does not appear to have a general effect on the efficacy of a given siRNA or shRNA. However, local mRNA secondary structure has been postulated to play a role for certain genes. Since it is difficult to predict the local secondary structure effects at this time, it is best to design different shRNAs that target different regions of the gene if possible.
3. Web sites that offer algorithms for siRNA and shRNA target site selection include http://shRNAdesigner.med.unc.edu, http://www.dharmacom.com, and http://www.ambion.com. In most cases, a target site that is effective for siRNA will also be effective for shRNA. However, there are additional steps involved in the processing of shRNA by Dicer, and the design criteria for

shRNA may not be 100% identical to those for siRNA in all cases (6). For this reason we recommend our web algorithm which is specifically based on shRNA design (http://shRNAdesigner.med.unc.edu). Additionally, we have found that effective shRNA target sites may have overlap in which algorithms predict them, and looking for target sites predicted to be efficacious by two or more different algorithms could potentially provide an additional useful strategy.

4. For one of the most commonly used hairpin loops, based on a naturally occurring miRNA sequence (15), the complete hairpin sequence would read N19-TTCAAGAGA-rN19, where N19 is the target sequence and rN19 is the reverse complement of the target. If the N19 target were to end in two T's, the combination of the two T's from the target sequence together with the first two T's of the loop would constitute four T's in a row, which can comprise a termination signal for RNA pol III.
5. The use of two different siRNAs of shRNAs provides an extremely useful control against the possible effects of off-target knockdowns since it is statistically unlikely that two siRNAs will have the same off-target knockdowns. Also, since no algorithm can guarantee effective silencing, preparing two or more siRNAs or shRNAs increases the probability of obtaining knockdown. shRNA or siRNA with different extent of silencing also can be useful in verifying dose-dependent functional effects as shown in **Figs. 10.1** and **10.2**.
6. Experimental controls can include a mock-infected or mock-transfected sample, an empty vector, an shRNA encoding a scrambled target, or an shRNA targeting another gene entirely. If possible, use at least 2–3 controls. Many researchers prefer a control that is known to encode a functional shRNA against another gene. Genes within entirely different pathways or even of different species may be used. Some examples of common control genes for siRNA or shRNA are firefly luciferase and green fluorescence protein.
7. It is helpful to design a sequencing primer for each strand since the secondary structure of the hairpin often makes it selectively more difficult to read through one strand.
8. 293T cells should not be used after about passage 15 or if growth slows significantly.
9. In some packaging systems the VSV-G and Rev genes are combined into one plasmid and in other cases they are separated into two vectors. If the genes are provided in two separate vectors, use 5 μg of each vector.

10. Viral supernatant is infectious and should be treated with appropriate precautions.

11. For viruses encoding a fluorescent marker gene, such as GFP, 293T cells can be checked microscopically at this point to ensure transfection efficiency. The cells should be close to 100% GFP positive.

12. Spinoculation may not be essential for all non-adherent cells, but can greatly increase transduction efficiency depending on the cell type. A swing-bucket rotor is more effective than a fixed rotor at concentrating the virus onto the cells.

13. *See* **Fig. 10.1** for an example of a real-time PCR experiment used to verify knockdown of ASC in THP1 cells.

14. As a confirmation of real-time PCR results, immunoblotting may be used to assess knockdown at the level of the protein. Immunoblots should use either polyclonal antibodies targeted against the entire protein or polyclonal/monoclonal antibodies that target an epitope within the same splice forms targeted by the siRNA or shRNA.

15. Designing PCR primers to span an exon/intron junction reduces the possible background SYBR signal from contaminating genomic DNA within the sample. As an alternative, the optional on-the-column DNAse step can be performed during the RNeasy purification procedure.

16. Before using a new set of primers in a quantitative experiment, test the primers as follows: (1) add a dissociation step to the PCR profile and look at the dissociation curve after real-time PCR is performed to be sure that a distinct peak of SYBR activity is apparent; (2) run a titration of four 10-fold dilutions of a positive control cDNA sample and ensure that the SYBR activity accurately reflects the dilutions; and (3) recover the 384-well plate following real-time PCR and run the product of the PCR on an agarose gel. Make sure that the product appears as a discrete band that is approximately 100–150 bp in length and that the relative intensities of the bands in different lanes reflect the relative amounts of cDNA added to the PCR reaction.

17. We have found that knockdown levels for mRNA can fluctuate from day to day as assayed by real-time PCR. For that reason we assess knockdown on 3 days and calculate an average value.

18. Prepare enough master mix for all samples plus approximately 10% so that there will be enough to account for any pipetting errors.

19. We routinely use 18s rRNA as a standard for normalization of cDNA levels. We use the following forward and reverse primers for 18s rRNA (6, 13): FW-CGGCTACCACATCCAAGG; RV-GCTGCTGGCACCAGACTT. Other housekeeping genes, such as GAPDH or cyclophilin-b, can also be used in place of the 18s gene.
20. *See* **Fig. 10.2** for an example of the application of ELISA to verify ASC knockdown.
21. 1.0 μg/mL lipopolysaccharide is a common immunostimulant that can be used for inducing inflammasome activation (10, 11); however, there are many inflammasome inducers. In **Fig. 10.2** we have chosen to use 10 MOI *P. gingivalis*, an oral pathogen that has previously been shown to induce IL-1β in an ASC-dependent manner (12, 13).
22. Strong immunostimulants can also promote toxicity, requiring a shortened length of induction.

Acknowledgments

We thank Dr. David Baltimore for supplying us with FG12. We thank Dr. Jenny Ting for her support and guidance. These studies were supported by grants 1-RO1-DE016326 and 1-U54-AI057157.

References

1. Fire, A., Xu, S., Montgomery, M.K., Kostas, S.A., Driver, S.E., and Mello, C.C. (1998) Potent and specific genetic interference by double-stranded RNA in Caenorhabditis elegans. *Nature*, **391**, 806–811.
2. Elbashir, S.M., Lendeckel, W., and Tuschl, T. (2001) RNA interference is mediated by 21- and 22-nucleotide RNAs. *Genes Dev*, **15**, 188–200.
3. Maniataki, E. and Mourelatos, Z. (2005) A human, ATP-independent, RISC assembly machine fueled by pre-miRNA. *Genes Dev*, **19**, 2979–2990.
4. Kutter, C. and Svoboda, P. (2008) miRNA, siRNA, piRNA: knowns of the unknown. *RNA Biol*, **5**, 181–188.
5. Ui-Tei, K., Naito, Y., Takahashi, F., Haraguchi, T., Ohki-Hamazaki, H., Juni, A., Ueda, R., and Saigo, K. (2004) Guidelines for the selection of highly effective siRNA sequences for mammalian and chick RNA interference. *Nucl Acids Res*, **32**, 936–948.
6. Taxman, D.J., Livingstone, L.R., Zhang, J., Conti, B.J., Iocca, H.A., Williams, K.L., Lich, J.D., Ting, J.P., and Reed, W. (2006) Criteria for effective design, construction, and gene knockdown by shRNA vectors. *BMC Biotechnol*, **6**, 7.
7. Reynolds, A., Leake, D., Boese, Q., Scaringe, S., Marshall, W.S., and Khvorova, A. (2004) Rational siRNA design for RNA interference. *Nat Biotechnol*, **22**, 326–330.
8. Amarzguioui, M. and Prydz, H. (2004) An algorithm for selection of functional siRNA sequences. *Biochem Biophys Res Commun*, **316**, 1050–1058.
9. Taxman, D.J. (2009) siRNA and shRNA design. In: Helliwell, T.D.C. (Ed.) *RNA Interference Methods for Plants and Animals*. CABI, Oxfordshire, UK, Chapter 10, pp. 228–253.

10. Mariathasan, S., Newton, K., Monack, D.M., Vucic, D., French, D.M., Lee, W.P., Roose-Girma, M., Erickson, S., and Dixit, V.M. (2004) Differential activation of the inflammasome by caspase-1 adaptors ASC and Ipaf. *Nature*, **430**, 213–218.
11. Martinon, F., Burns, K., and Tschopp, J. (2002) The inflammasome: a molecular platform triggering activation of inflammatory caspases and processing of proIL-beta. *Mol Cell*, **10**, 417–426.
12. Huang, M.T., Taxman, D.J., Holley-Guthrie, E.A., Moore, C.B., Willingham, S.B., Madden, V., Parsons, R.K., Featherstone, G.L., Arnold, R.R., O'Connor, B.P. et al. (2009) Critical role of apoptotic speck protein containing a caspase recruitment domain (ASC) and NLRP3 in causing necrosis and ASC speck formation induced by Porphyromonas gingivalis in human cells. *J Immunol*, **182**, 2395–2404.
13. Taxman, D.J., Zhang, J., Champagne, C., Bergstralh, D.T., Iocca, H.A., Lich, J.D., and Ting, J.P. (2006) Cutting edge: ASC mediates the induction of multiple cytokines by Porphyromonas gingivalis via caspase-1-dependent and -independent pathways. *J Immunol*, **177**, 4252–4256.
14. An, D.S., Donahue, R.E., Kamata, M., Poon, B., Metzger, M., Mao, S.H., Bonifacino, A., Krouse, A.E., Darlix, J.L., Baltimore, D. et al. (2007) Stable reduction of CCR5 by RNAi through hematopoietic stem cell transplant in non-human primates. *Proc Natl Acad Sci USA*, **104**, 13110–13115.
15. Brummelkamp, T.R., Bernards, R., and Agami, R. (2002) A system for stable expression of short interfering RNAs in mammalian cells. *Science*, **296**, 550–553.
16. Schwarz, D.S., Hutvagner, G., Du, T., Xu, Z., Aronin, N., and Zamore, P.D. (2003) Asymmetry in the assembly of the RNAi enzyme complex. *Cell*, **115**, 199–208.
17. Xia, H., Mao, Q., Paulson, H.L., and Davidson, B.L. (2002) siRNA-mediated gene silencing in vitro and in vivo. *Nat Biotechnol*, **20**, 1006–1010.
18. Denti, M.A., Rosa, A., Sthandier, O., De Angelis, F.G., and Bozzoni, I. (2004) A new vector, based on the PolII promoter of the U1 snRNA gene, for the expression of siRNAs in mammalian cells. *Mol Ther*, **10**, 191–199.
19. Gupta, S., Schoer, R.A., Egan, J.E., Hannon, G.J., and Mittal, V. (2004) Inducible, reversible, and stable RNA interference in mammalian cells. *Proc Natl Acad Sci USA*, **101**, 1927–1932.
20. Unwalla, H.J., Li, M.J., Kim, J.D., Li, H.T., Ehsani, A., Alluin, J., and Rossi, J.J. (2004) Negative feedback inhibition of HIV-1 by TAT-inducible expression of siRNA. *Nat Biotechnol*, **22**, 1573–1578.
21. Lewis, J., Melrose, H., Bumcrot, D., Hope, A., Zehr, C., Lincoln, S., Braithwaite, A., He, Z., Ogholikhan, S., Hinkle, K. et al. (2008) In vivo silencing of alpha-synuclein using naked siRNA. *Mol Neurodegener*, **3**, 19.
22. Vlachaki, M.T., Hernandez-Garcia, A., Ittmann, M., Chhikara, M., Aguilar, L.K., Zhu, X., Teh, B.S., Butler, E.B., Woo, S., Thompson, T.C. et al. (2002) Impact of preimmunization on adenoviral vector expression and toxicity in a subcutaneous mouse cancer model. *Mol Ther*, **6**, 342–348.
23. Monahan, P.E., Jooss, K., and Sands, M.S. (2002) Safety of adeno-associated virus gene therapy vectors: a current evaluation. *Expert Opin Drug Saf*, **1**, 79–91.
24. Bot, I., Guo, J., Van Eck, M., Van Santbrink, P.J., Groot, P.H., Hildebrand, R.B., Seppen, J., Van Berkel, T.J., and Biessen, E.A. (2005) Lentiviral shRNA silencing of murine bone marrow cell CCR2 leads to persistent knockdown of CCR2 function in vivo. *Blood*, **106**, 1147–1153.

Chapter 11

Effective Pol III-Expressed Long Hairpin RNAs Targeted to Multiple Unique Sites of HIV-1

Sheena M. Saayman, Patrick Arbuthnot, and Marc S. Weinberg

Abstract

The RNA interference (RNAi) pathway has in recent years been exploited for the development of novel antiviral therapies. The emergence of viral escape mutants, however, is a major impediment to the use of RNAi effectors to treat highly mutable viruses such as HIV-1. A combinatorial approach is therefore required for long-term inhibition of gene expression. RNA Pol III-driven long hairpin RNA (lhRNA) duplexes can be cleaved several times by Dicer, yielding multiple functional siRNAs from a single construct. Here we describe a method for the generation of ectopically expressed U6-lhRNAs encoding three separate siRNA sequences targeting unique sites in HIV-1. This methodological overview explains some crucial aspects of lhRNA design and cloning as well as facile experiments to determine their efficacy in cell culture.

Key words: Long hairpin RNA, lhRNA, RNA interference, Dicer, HIV-1, therapeutics.

1. Introduction

RNA interference (RNAi) is an evolutionary conserved pathway in eukaryotes whereby double-stranded RNA acts as an intracellular trigger to regulate gene expression (1). The RNAi pathway has been popularly used as a tool to silence genes and holds much promise as novel therapeutic approach aimed at suppressing specific cellular and viral genes at the post-transcriptional level. The therapeutic development of RNAi has been made possible by usurping elements of the endogenous mammalian microRNA (miRNA) biogenesis pathway (*see* **Chapter 14**) through exogenously introduced synthetic short interfering RNAs (siRNAs) (2)

M. Sioud (ed.), *RNA Therapeutics*, Methods in Molecular Biology 629,
DOI 10.1007/978-1-60761-657-3_11, © Springer Science+Business Media, LLC 2010

or through gene expression constructs which produce 21–29 bp short hairpin RNAs (shRNAs) (3, 4). These shRNA mimics of precursor miRNA are cleaved by the RNase III endonuclease Dicer to form effective siRNAs which unwind to load single-stranded RNA guide strands into an Argonaute-containing RNA-induced silencing complex (RISC). Loaded RISC complexes target cognate complementary mRNA sequences to mediate post-transcriptional gene silencing by Argonaute-2-mediated cleavage or translational suppression of targeted mRNAs (5).

A significant hurdle for the use of RNAi-based therapeutics has been the targeting of highly mutable sequences, such as genomic and sub-genomic RNAs from the human immunodeficiency virus type 1 (HIV-1). HIV replicates using an error-prone reverse transcriptase and has been shown to escape the silencing effects of expressed shRNAs by rapidly developing resistant viral variants (6–8). The emergence of such viral escape mutants is not unexpected since single-base mismatches between guide strand and target is enough to prevent silencing (6, 9). To circumvent this problem, efforts are underway to generate expressed RNAi effectors, which simultaneously target multiple sites within the viral genome (10–13). While a number of approaches are available to induce combinatorial silencing of viral sequences [reviewed in (14–16)], we have made use of RNA Pol III-expressed long hairpin RNAs (lhRNAs) with duplex stems encoding more than one putative siRNA sequence (17). In mammalian cells, introduction of synthetic dsRNA of greater than 30 bp leads to a strong innate immunostimulatory response (18). Although certain synthetic siRNA sequences were found to activate innate immunity (*see* **Chapter 3**), studies using expressed sequences have shown that safe and effective gene-specific silencing can be achieved and that these duplexes evade cytoplasmic activators of the type 1 interferon response (19, 20).

Long hairpin RNAs generated from Pol III promoters include 2–3 nt 3′-terminal uridine overhangs, which are produced by transcriptional termination. These 3′-overhangs facilitate export to the cytoplasm and allow binding of the Paz domain of human Dicer (hDicer) (21, 22). Processive cleavage of the lhRNA by hDicer then occurs from the open-ended stem to the loop of the hairpin duplex (23). Multiple successive siRNAs are produced by the intracellular processivity of hDicer in decreasing order of efficiency along the lhRNA duplex, resulting in the production of at least three non-overlapping functional siRNAs per lhRNA (17, 23–26). Long hairpin RNAs can be designed to target one contiguous sequence within the genome or, alternatively, can be made to incorporate multiple independent target sequences. The advantage of the latter is that separate mRNAs can be silenced simultaneously and previously characterized effective shRNA/siRNA sequences can be incorporated into a single lhRNA. The optimal positioning of different 19 bp siRNA within an lhRNA duplex for efficient processing of successive siRNAs often requires empirically testing different sequence spacing arrangements at the junction of each siRNA. However, while no clear rules exist, we provide some design guidelines for optimally spacing effective 19 bp siRNA sequences along the length of an lhRNA duplex.

Here we present a method for cloning and generating 69 bp long hairpin RNAs expressed from the U6 snRNA promoter which encodes three separate anti-HIV siRNA duplexes (**Fig. 11.1a**). This method can be generally adapted for applications where multiple siRNAs are required for the simultaneous targeting of up to three separate and unique RNA sites. A number of different methodologies are available for producing shorter Pol III-expressed hairpin duplexes such as shRNAs. However, many of these are not easily adapted for generating lhRNAs. Owing to the increased length of the dsRNA duplex for lhRNAs and associated problems with subsequent PCR, cloning and sequencing of long inverted repeat sequences, we have relied on a modification of the two-round PCR protocol originally used by Castanotto et al. (27) to generate Pol III-expressed shRNAs. In addition

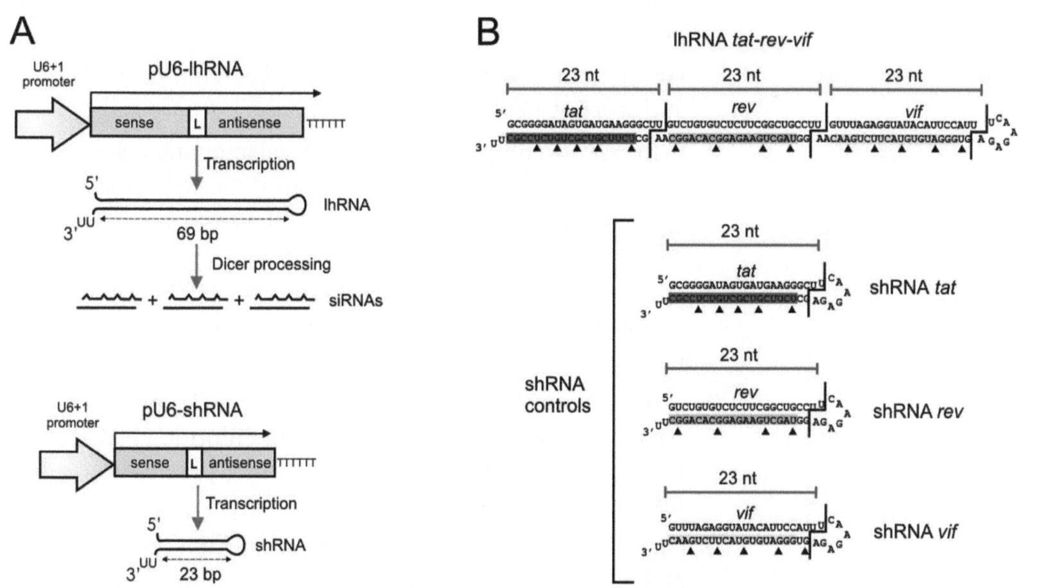

Fig. 11.1. Design of U6-expressed lhRNAs and shRNA controls targeting three sites of HIV-1. (**a**) Schematic illustration of lhRNAs and shRNAs comprising 69 and 23 bp duplexes, respectively. G:U pairings are indicated as a *corrugated* sense strand. A sequence of two terminal 3′-U residues is derived from the Pol III (U6) transcriptional termination signal. The intended mechanism of transcription and processing by Dicer of the lhRNAs to form three anti-HIV siRNAs is illustrated. (**b**) The sequence and predictive structure of the lhRNA, lhRNA *tat-rev-vif*, and each shRNA are shown. The order and special arrangement of the siRNA-encoded sequences within the lhRNA is indicated along the extent of the duplex. G:U and U:G pairings are indicated with an *arrowhead.* The putative guide strands are *shaded.*

to describing the design, PCR and cloning of U6-lhRNA cassettes, we describe a simple luciferase-based reporter gene assay for detecting the efficacy of each siRNA generated from the parental lhRNA duplex. Moreover, we provide a standard polyacrylamide gel electrophoresis (PAGE) northern blot protocol, which allows convenient quantitative detection of each siRNA and associated precursors generated from an lhRNA expression cassette.

2. Materials

2.1. PCR of lhRNA-Encoding DNA Templates

1. Expand High Fidelity^{PLUS} PCR kit (Roche) (*see* **Note 1**).
2. pTZU6+1 template plasmid (28) containing the human U6 snRNA RNA Pol III promoter.
3. PAGE-purified, synthesized oligodeoxynucleotide primers for PCR.
4. 1X Tris-acetate-EDTA (TAE) electrophoresis running buffer: 40 mM Tris, 1 mM ethylenediaminetetraacetic acid (EDTA), 20 mM glacial acetic acid. Store at room temperature.
5. InsTAclone^{TM} PCR Cloning Kit (Fermentas, WI, USA) which includes the plasmid pTZ57R/T. Store all kit components at –20°C.
6. Competent *Escherichia coli* DH5α bacterial cells. Aliquot and store at –70°C.
7. 5-Bromo-4-chloro-3-indolyl-β-D-galactopyranoside (X-Gal). X-Gal should be dissolved in dimethylformamide at a concentration of 20 mg/mL and stored in the dark at –20°C.
8. Isopropyl-β-D-1-thiogalactopyranoside (IPTG). IPTG is dissolved in ddH_2O at a concentration of 100 mg/mL, filter sterilized and stored at –20°C.
9. Luria Bertani broth (LB): 10 g/L Bacto-Tryptone, 5 g/L bacto yeast extract, 10 g/L sodium chloride (NaCl) containing 1 μg/mL ampicillin. Autoclave prior to the addition of antibiotic and store at 4°C.
10. Agar plates containing 1 μg/mL ampicillin. Store inverted at 4°C.
11. High Pure Plasmid Isolation Kit for small-scale (mini) preparations of purified plasmid DNA.
12. *Spe*I, *Not*I, *Eco*RI and *Hind*III restriction enzymes (10 U/μL), supplied with recommended 10X buffers. Store at –20°C.
13. QIAGEN Plasmid Maxi Kit (for transfections).

14. Common restriction enzymes, each 5–10 U/μL, supplied with recommended 10X buffers. Store at –20°C.

2.2. Generation of Target Reporter Plasmids

1. psiCHECK™-2 vector.
2. Complementary forward and reverse oligodeoxynucleotides for annealing and directional cloning.
3. *Xho*I and *Not*I restriction enzymes (10 U/μL) supplied with recommended 10X buffers. Store at –20°C.
4. MinElute™ Gel Extraction Kit.
5. Antarctic phosphatase.

2.3. Cell Culture and Determination of Knockdown Efficiency by Dual Luciferase Reporter Assays

1. Human embryonic kidney 293 (HEK293) cell line.
2. Dulbecco's modified Eagle's medium (DMEM) supplemented with 10% heat-inactivated fetal calf serum (FCS). Store at 4°C.
3. Phosphate-buffered saline containing 0.01% EDTA.
4. Lipofectamine™ 2000 transfection reagent.
5. OptiMEM.
6. Plasmid pCI-eGFP (29), a GFP-expression plasmid under control of the CMV promoter.
7. Dual-Luciferase® Reporter Assay System (Promega, WI, USA). Store the kit at –20°C, and once the luciferase assay substrate has been reconstituted, aliquot and store in the dark at –70°C.
8. Costar® 96 well flat bottom assay plates.
9. Veritas™ Microplate Luminometer.

2.4. Detection of lhRNA Processing by PAGE Northern Blots

1. TriReagent™ (Sigma, MO, USA). Store at 4°C. TriReagent™ should only be used in a fume hood.
2. A 15% polyacrylamide gel with 8 M urea (20 × 20 × 0.8 cm): 40% acrylamide solution (19:1 acrylamide:bis-acrylamide) 18.5 mL; 10X TBE 5 mL; urea 24.03 g; ammonium persulphate 20 mg; *N,N,N,N'*-tetramethylethylenediamine (TEMED) 15 μL (*see* **Note 2**).
3. Loading buffer: 95% formamide, 0.025% xylene cyanol, 0.025% bromophenol blue, 0.025% sodium dodecyl sulphate (SDS), 18 mM EDTA.
4. 10X Tris-borate-EDTA buffer (TBE): 890 mM Tris–HCl, 890 mM boric acid and 20 mM EDTA, pH 8.0. Store at room temperature and discard if a precipitate begins to form.
5. Decade™ Marker kit (Ambion, TX, USA). Components include Decade Marker RNA (100 ng/μL in 10 mM

Tris–HCl, pH 7.0), T4 polynucleotide kinase (10 U/μL), 10X kinase reaction buffer, 10X cleavage reagent, gel loading buffer and nuclease-free water. Store the Decade Marker RNA at –70°C, the cleavage reagent at room temperature and all the other reagents at –20°C.

6. Semi-dry blotter.
7. Hybond-N$^+$ nylon membrane.
8. Ultra-Violet Products (UVP) UV cross-linker.
9. Rapid-hybridization buffer.
10. γ^{32}P-ATP (6,000 Ci/mmol).
11. T4 polynucleotide kinase (PNK), supplied with 10X PNK buffer.
12. DNA oligonucleotides (approximately 18–21 nt) complementary to the antisense strand of each siRNA sequence (*see* **Note 3**).
13. Sephadex columns. These can be prepared using 5 g Sephadex® G-25 in 50 mL TE buffer. Insert 0.5 cm nylon fibre into a 1 mL syringe and then add 1 mL of sephadex/TE solution. Spin at 2,000*g* for 2 min (*see* **Note 4**).
14. Hybridization incubator with a rotisserie.
15. Sodium dodecyl sulphate, 1% solution in water. Store at room temperature. However if a precipitate begins to form, the solution may be heated to re-dissolve precipitate.
16. 20X Sodium chloride sodium citrate (SSC) buffer: 3 M NaCl, 0.3 M sodium Citrate dehydrate, pH 7.0.
17. Medical x-ray film.
18. Phosphorimaging plates.
19. Phosphorscanner.

3. Methods

3.1. Generating U6-Driven lhRNA Vectors Targeting Three Unique Sequences Within HIV-1

3.1.1. The Design of U6-Driven lhRNA Cassettes

Here we provide some design guidelines for optimally spacing three effective anti-HIV 19 bp siRNA sequences along the length of a U6-driven lhRNA duplex. We also describe important design features which need to be considered for constructing and cloning a plasmid containing an lhRNA expression cassette. The lhRNA comprises a stem of 69 bp and a loop of seven nucleotides. The lhRNA is designed to be transcribed from a U6 RNA Pol III promoter such that three 21–23 bp siRNAs can potentially be generated by hDicer cleavage (**Fig. 11.1a, b**). We chose to incorporate the sequences of three unique anti-HIV-1 shRNAs targeting Tat/Rev (referred to as *tat*) (30), Rev/Env (*rev*) (30) and Vif

(*vif*) open reading frames (31) (**Fig. 11.1b**). Some of the design criteria are described below:

1. A single 69 bp lhRNA duplex encodes three separate 19 bp siRNAs (for *tat, rev* and *vif*, respectively). However, since intracellular Dicer cleavage occurs approximately every 22–23 nucleotides along the dsRNA duplex, for optimal Dicer cleavage, each 19 bp siRNAs with two 3′-terminal nucleotides is spaced every 23 nt from the base of the stem (*see* **Fig. 11.1b**) (*see* **Note 5**).
2. The guide strands for each siRNA are placed sequentially in the 3′-arm of the lhRNA duplex. To avoid confusion, the sense strand always represents the 5′-arm such that transcription occurs in the following direction: sense–loop–antisense.
3. Wobble base mismatches (G:U or U:G bp) are introduced into the sense strand of the lhRNA duplex (C is replaced by a T; A is replaced by a G in the DNA sequence) at regular intervals (every 4–8 bp) (**Fig. 11.1b**). These mismatches greatly facilitate PCR of the lhRNA expression cassette, cloning of inverted repeat sequences in *E. coli* and later sequencing of clones (*see* **Note 6**).
4. Individual short hairpin RNAs (shRNAs) with corresponding G:U mismatches serve as positive controls for each siRNA generated from the lhRNA duplex (**Fig. 11.1b**).
5. We chose a random 7 nt loop sequence: 5′-UCAAGAG-3′. A longer loop provides a unique anchor for two rounds of PCR (**Fig. 11.1b** and **Section 3.1.2**) (see **Note 7**).

3.1.2. Generation of Expressed lhRNAs by a Two-Step PCR

1. U6-expressing lhRNAs are constructed using a two-step PCR which was adapted from Castanotto et al. (27) (**Fig. 11.2**). In the first round of PCR, 10 pg of pTZU6+1 is used as a template.
2. The universal U6 forward primer, 5′-CTA ACT AGT GGC GCG CCA AGG TCG GGC AGG AAG AGG G-3′, is complementary to the 5′-end of the U6 promoter and is used for both rounds of PCR. It includes *Spe*I and *Not*I sites to facilitate later screening of correctly inserted clones (*see* **Section 3.2**).
3. The reverse primer for the first round (R1) of PCR for the lhRNA (*tat-rev-vif*) is complementary to 18–21 nt of the 3′-end of the U6 promoter and contains a linker encoding the *tat-rev-vif* siRNA sequences: lhRNA *tat-rev-vif* (R1) 5′-CTT GAA ATG GAA TGT ATA CCT CTA AAC AAG GCA GCC GAA GAG ACA CAG ACA AGC CCT TCA TCA CTA TCC CCG CGG TGT TTC GTC CTT TCC ACA A-3′.

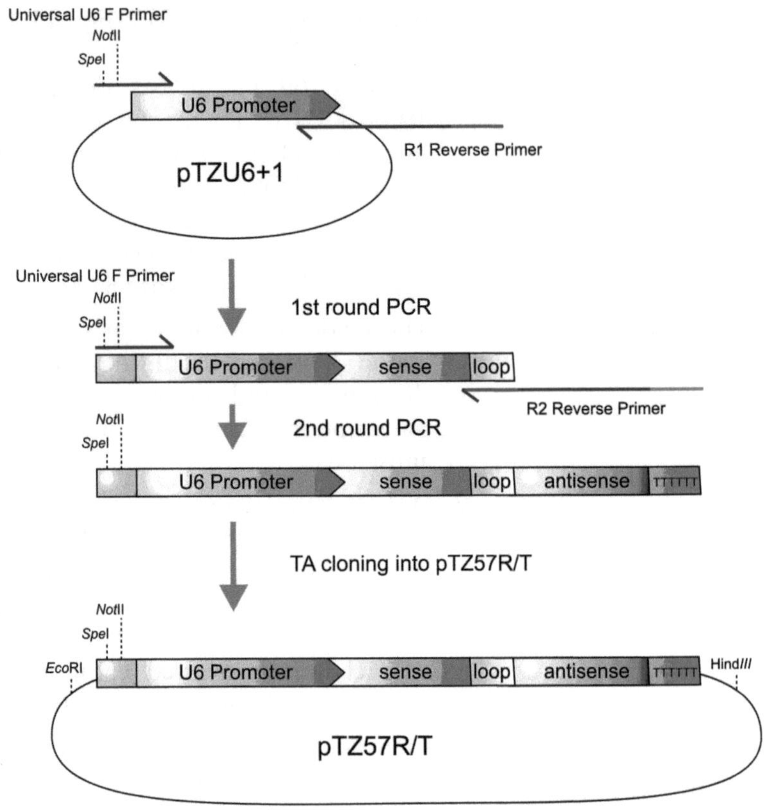

Fig. 11.2. A two-step PCR strategy for producing a U6-driven lhRNA cassette for cloning in the TA vector pTZ57R/T.

4. We use standard thermocycling conditions: 95°C for 5 min, followed by 30 cycles of denaturing at 95°C for 30 s, annealing at 55°C for 30 s, extension at 72°C for 30 s. With the Expand High Fidelity Taq polymerase kit, 10 pmol of each primer is used in a 50 μL reaction. The first round of PCR produces the complementary sequence of the lhRNA sense strand as well as of the loop sequence.
5. For the second round of PCR, a 1:500 dilution of the round one PCR product (approximately 10 pg) is used as the template. The U6 forward primer is used again and the reverse primer sequence (R2) is designed to have 18 nt of overlapping sequence with the loop region of the round one reverse primer. The R2 primer encodes the complementary sequence of the antisense strand of the lhRNA as well as six thymidine (uridine) residues for Pol III-transcriptional termination (**Fig. 11.2**). The primer for round two is lhRNA *tat-rev-vif* (R2) 5′-AAA AAA GCG GAG ACA GCG ACG AAG AGC TTG CCT GTG CCT CTT CAG CTA CCT TGT TCA GAA GTA CAC ATC

CCA CTC TCT TGA AAT GGA ATG TAT A-3′. This overlap between each pair of reverse primers enables the extension of the PCR product to generate a U6-expressed lhRNA cassette with a transcription termination signal.

6. The conditions for the second round PCR are the same as the first, except for the additional 10 min extension step at completion of thermal cycling. This facilitates cloning into the TA vector, pTZ57R/T.
7. Load 5 μL of each PCR product on a 1.5% agarose gel in 1X TAE buffer, resolve electrophoretically and visualize on a UV transilluminator.

3.1.3. Plasmid Propagation of lhRNA Expression Cassettes

1. If a single band is visualized representing the expected size of the round two PCR product on the agarose gel (*see* **Section 3.1.2**), it can be ligated directly into the TA cloning vector pTZ57R/T (*see* **Note 8**). The ligation reaction is set up by adding together: 4 μL round two PCR product, 1 μL pTZ57R/T, 4 μL 5X ligation buffer and 1 μL (5 U) T4 DNA ligase. The volume is made up to 20 μL with water. The reaction can be left at room temperature overnight.
2. The ligation reaction (7 μL) is used to transform 100 μL competent DH5α cells before plating on ampicillin-containing agar plates.
3. Agar plates contain 40 μL of X-Gal stock and 8 μL of IPTG stock for blue-white screening. Plates containing transformed bacteria are left at 37°C overnight.
4. Pick white colonies and grow in 3 mL of ampicillin-containing LB at 37°C overnight for mini-preparation of plasmid DNA using the Roche high-pure miniprep kit.
5. Screen plasmids for presence of inserts and orientation by digesting plasmid DNA with *Eco*RI and *Spe*I, and *Hin*dIII and *Spe*I, respectively (**Fig. 11.2**). Run digested plasmids on a 3% agarose gel (*see* **Note 9**).
6. Sequence positive clones with M13 forward and reverse primers. Since errors are commonly incorporated in the synthesis of long oligonucleotides, a number of clones may need to be sequenced.

3.2. Generation of Target Reporter Plasmids

1. A standard PCR reaction is used to amplify approximately 200 bp of the gene containing the desired target sequences for *tat*, *rev* and *vif* using 10 pmol each of gene-specific primers and 10 pg of HIV-1 subtype B plasmid, pNL4-3, as the template. The forward primer contains a 5′-*Xho*I restriction site linker, while the reverse primer contains a 5′-*Not*I restriction site linker.

2. PCR products for *tat*, *rev* and *vif* are individually ligated into pTZ57R/T as described above (**Section 3.2**). Plasmid pTZ57R/T, containing the target sequence, is screened by digestion with *Xho*I and *Not*I and the insert is then excised from an agarose gel and purified using the MinElute Gel Extraction kit according to the manufacturer's instructions. Fragments are eluted from the column in 20 μL of TE buffer.
3. To insert target sequences in the 3′-UTR of the *Renilla* luciferase open reading frame, the psiCHECK plasmid backbone is prepared by digestion with 15 units (1.5 μL) each of *Xho*I and *Not*I for 1.5 h at 37°C in a 50 μL volume. One microlitre (5 U) of Antarctic phosphatase (AP) is added to the digestion reaction together with 10X AP buffer and water to 60 μL total volume. The reaction is incubated at 37°C for a further 10 min. The phosphatase reaction is heat-inactivated for 15 min at 65°C followed by resolving the linear psiCHECK vector backbone DNA on a 0.8% agarose gel. The double-digested vector backbone band is excised and purified using the MinElute Gel Extraction kit. Fragments are eluted from the column in 20 μL of TE buffer.
4. To ligate vector backbone to digested PCR fragments, 60 fmol (approximately 50 ng) of purified psiCHECK backbone fragment is ligated with 180 fmol of each *tat*, *rev* and *vif* fragment in a 20 μL reaction volume containing 1 μL (10 U) of T4 DNA ligase and 1X ligase buffer.
5. The ligase reaction is used to transform bacteria as described previously: 10 μL of ligation reaction is used to transform 100 μL of chemically competent *E. coli*; the reaction is plated onto ampicillin-containing agar plates and incubated overnight at 37°C.
6. Colonies are screened by digestion with *Xho*I and *Not*I and clones with the correct sized insert are subsequently sequenced using a forward primer specific to the *Renilla* luciferase ORF: 5′-GAG GAC GCT CCA GAT GAA ATG-3′.

3.3. Cell Culture and Knockdown Assays

1. The HEK293 cell line is maintained with DMEM supplemented with 10% heat-inactivated FCS at 37°C and 5% CO_2. Cells are passaged using PBS containing EDTA.
2. Seeding of 120,000 cells per well in 24 well plates is carried out 24 h prior to transfection. lhRNA-encoding plasmid (750 ng) together with target reporter plasmid (150 ng) are co-transfected in a 5:1 ratio (*see* **Note 10**). A 100 ng of plasmid pCI-eGFP is also co-transfected to control for

transfection efficiency. Each transfection should be performed in triplicate. Transfections are carried out using 1 μL lipofectamine 2000 per 1 μg DNA and 0.2X optiMEM.

3. Twenty-four hours post-transfection media should be replaced and the dual luciferase reporter assay is carried out a further 24 h later according to the manufacturer's instructions. All *Renilla* luciferase values are normalized against background firefly luciferase values. The average expression ratio for a control plasmid containing the U6 promoter with no hairpins is set to 100%, and relative expression levels of other samples are calculated accordingly (**Fig. 11.3a**).

3.4. PAGE Northern Blot Hybridization

1. Cells are cultured and transfected as previously described. For northern blot analysis cells are seeded to approximately 70% confluence in 10 cm culture plates and transfected 24 h later with 18 μg hairpin construct. Forty-eight hours post-transfection, total RNA is extracted using TriReagent according to the manufacturer's instructions. Standard RNA-handling procedures to avoid RNase contamination should be followed.
2. Prepare a 15% polyacrylamide gel with a 1:19 ratio of bis:acrylamide with 8 M urea and 1X TBE.
3. Ambion Decade Marker should be prepared as per the kit's instructions.
4. Add an equal volume of loading dye to 30 μg of each RNA sample, heat at 80°C for 5 min and then return sample to ice before loading.
5. To warm the apparatus prior to the loading of RNA samples, the gel should be pre-electrophoresed in 0.5X TBE buffer at 200 V for 30 min.
6. Resolve RNA samples and labelled markers at constant voltage (200–300 V) and run the gel until the bromophenol blue band migrates to within 1 cm from the bottom of the gel.
7. Transfer to a positively charged membrane (Hybond-N$^+$) using a semi-dry blotter. For the transfer, the gel and membrane are tightly placed between six layers of 0.5X TBE-soaked chromatography paper and the current applied for 1 h.
8. UV cross-link the RNA to the membrane at 2,000 × 100 μJ/cm^2.
9. Bake the membrane at 80°C for 1 h.
10. Prehybridize the membrane in 10 mL of pre-warmed rapid-hyb buffer at 42°C for 20 min.

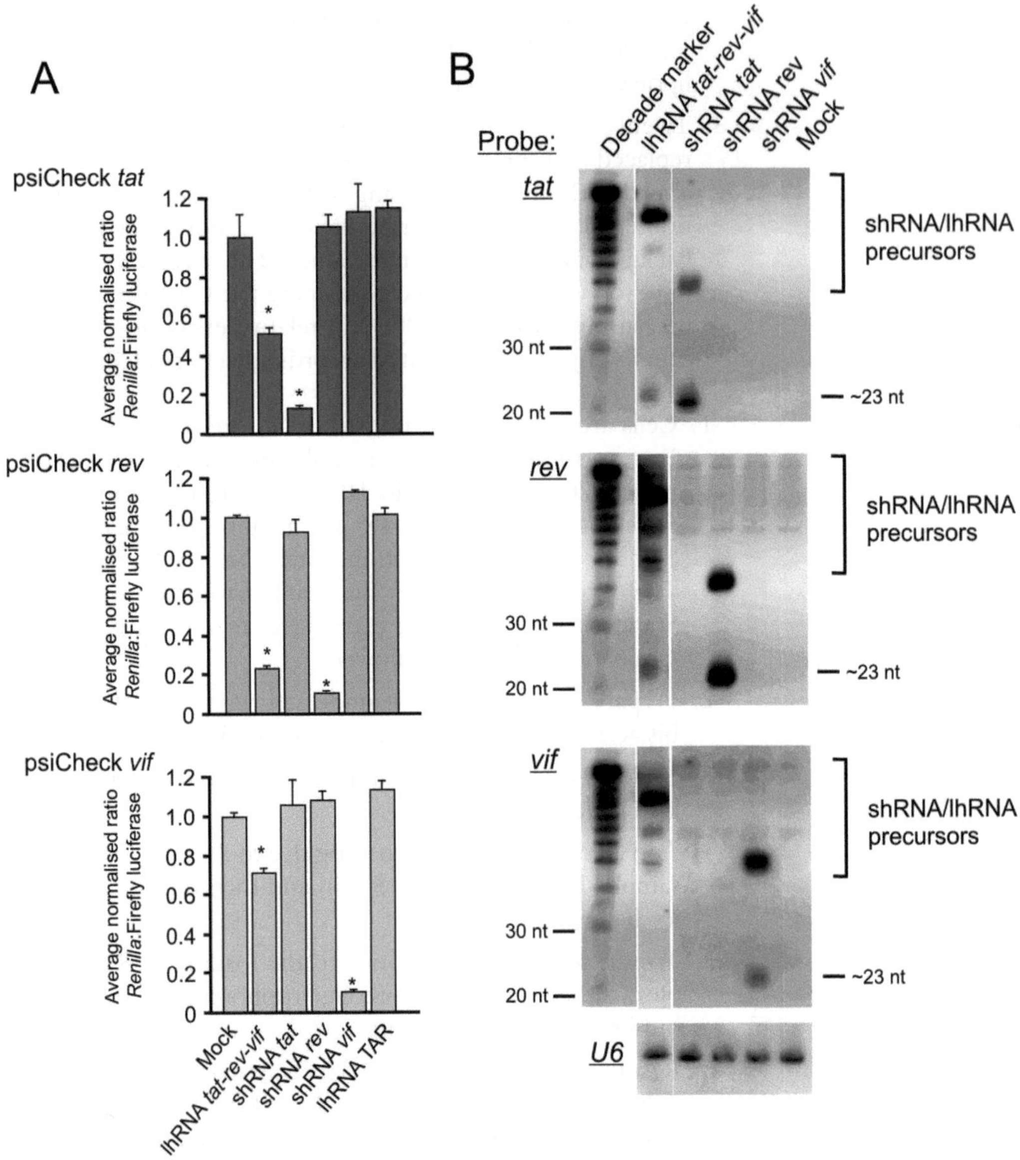

Fig. 11.3. Knockdown of individual reporter gene targets and intracellular processing of lhRNA and shRNA-expressing plasmids in transfected HEK293 cells. (**a**) Average normalized *Renilla*:firefly luciferase activity determined 48 h after transfecting HEK293 cells with the psiCHECK *tat-rev-vif* target together with each individual shRNA control and an irrelevant lhRNA control (lhRNA TAR) (*$p < 0.5$, for experiments conducted in triplicate). (**b**) Northern blot analysis of RNA extracted from HEK293 cells that had been transfected with lhRNA *tat-rev-vif* and shRNA-expressing plasmids. The blot was probed with an oligonucleotide that was complementary to putative *tat*, *rev* and *vif* guide sequence. The blot was stripped and re-probed with an oligonucleotide complementary to U6 snRNA to control for equal RNA loading.

11. Label 2 μL oligo probe (20 μM) with fresh γ-^{32}P-ATP in a 20 μL total volume using 1 μL PNK (5 U) and 2 μL 10X PNK buffer. Dilute to 50 μL and spin through a freshly prepared sephadex G-25 column. Add the labelled

probe to the hybridization buffer and leave rotating at 42°C overnight.

12. Wash membrane once with 0.1% SDS and 5X SSC in a volume of 50 mL at room temperature for 20 min. Wash membrane twice with 0.1% SDS and 1X SSC in a volume of 50 mL at 42°C for 15 min each (*see* **Note 11**).
13. Place membrane in cling wrap and expose membrane to x-ray film for 24–72 h before developing film. The membrane can also be exposed to a phosphor plate and scanned using a phosphor-imager (e.g. Fuji FLA-7000).
14. Strip membrane in 50 mL 1% SDS at 80°C for 30 min. The membrane can then be re-probed as described above.
15. Once stripped, the membrane is re-probed using a 20-mer oligonucleotide antisense to the U6 snRNA. This serves as an ideal loading control.

4. Notes

1. We also find standard Taq polymerase easily amplifies long hairpins but does introduce mismatches which need to be screened by sequencing. The use of high-fidelity thermostable polymerases may improve sequence fidelity but often hampers the generation PCR products useful for subsequent cloning. Nevertheless, we have used the Expand High Fidelity^{PLUS} PCR System (Roche) to produce lhRNA-encoded PCR fragments.
2. Ammonium persulphate should be made up fresh each time. Therefore make up small quantities at a time. Alternatively, small 500 μL aliquots can be frozen for up to 3 months.
3. Locked nucleic acid (LNA) oligonucleotide probes may be used for increased specificity.
4. The sephadex G-25 should be hydrated by overnight agitation in 50 mL TE buffer. Briefly centrifuge the solution at 4,000 rpm for 2 min, discard the supernatant and add another 50 mL TE buffer. Repeat this two to three times before adding a final volume of TE buffer.
5. If the lhRNA is designed to encode more than two siRNAs, the third siRNA may not be processed efficiently or may be present at too low a concentration to be an effective inhibitor. Although efficient production of all siRNAs is not always guaranteed, there are modifications that may be employed to improve the yield of the third siRNA. The

processing of multiple siRNAs in a single duplex may be augmented by empirically testing different spatial arrangement of siRNAs along the hairpin stem. Although gross generalizations regarding the most favourable siRNA spacing cannot yet be made, improvements can be achieved by inserting or deleting random base pairs at each siRNA junction as well as before the loop sequence.

6. Wobble mismatches can be strategically used to improve the thermodynamic asymmetry of each siRNA, thereby facilitating correct guide strand incorporation into RISC. Moreover, there is some anecdotal evidence to suggest that wobble base pairs help prevention of the induction of the IFN response by masking protein kinase R (PKR) recognition (32).
7. There is some evidence to suggest that increasing the duplex length at the loop side of the duplex (by 5 bp or more) may improve cleavage of the third siRNA (26). However, this has not been tested by us.
8. If non-specific bands are visible on the agarose gel, purify the desired PCR fragment from the gel and then ligate 10 μL gel-purified DNA into the TA cloning vector.
9. Resolve bands properly on a high percentage agarose gel (e.g. 3% gel). Clones with minor deletions and insertions can be detected on the gel and screened out.
10. Lower ratios of hairpin to target can be used without significantly affecting efficient target knockdown.
11. If high levels of background are present on the film, wash steps can be repeated.

References

1. Fire, A., Xu, S., Montgomery, M.K., Kostas, S.A., Driver, S.E., and Mello, C.C. (1998) Potent and specific genetic interference by double-stranded RNA in Caenorhabditis elegans. *Nature*, **391**, 806–811.
2. Elbashir, S.M., Lendeckel, W., and Tuschl., T. (2001) RNA interference is mediated by 21- and 22-nucleotide RNAs. *Genes Develop*, **15**, 188–200.
3. Paddison, P.J., Caudy, A.A., Bernstein, E., Hannon, G.J., and Conklin, D.S. (2002) Short hairpin RNAs (shRNAs) induce sequence-specific silencing in mammalian cells. *Genes Dev*, **16**, 948–958.
4. Brummelkamp, T.R., Bernards, R., and Agami, R. (2002) A system for stable expression of short interfering RNAs in mammalian cells. *Science*, **296**, 550–553.
5. Filipowicz, W., Bhattacharyya, S.N., and Sonenberg, N. (2008) Mechanisms of post-transcriptional regulation by microRNAs: are the answers in sight? *Nat Rev Genet*, **9**, 102–114.
6. von Eije, K.J., ter Brake, O., and Berkhout, B. (2008) Human immunodeficiency virus type 1 escape is restricted when conserved genome sequences are targeted by RNA interference. *J Virol*, **82**, 2895–2903.
7. Boden, D., Pusch, O., Lee, F., Tucker, L., Shank, P.R., and Ramratnam, B. (2003) Promoter choice affects the potency of HIV-1 specific RNA interference. *Nucl Acids Res*, **31**, 5033–5038.
8. Das, A.T., Brummelkamp, T.R., Westerhout, E.M., Vink, M., Madiredjo, M., Bernards,

R., and Berkhout, B. (2004) Human immunodeficiency virus type 1 escapes from RNA interference-mediated inhibition. *J Virol*, **78**, 2601–2605.
9. Schwarz, D.S., Ding, H., Kennington, L., Moore, J.T., Schelter, J., Burchard, J., Linsley, P.S., Aronin, N., Xu, Z., and Zamore, P.D. (2006) Designing siRNA that distinguish between genes that differ by a single nucleotide. *PLoS Genet*, **2**, e140.
10. ter Brake, O., 't hooft, K., Liu, Y.P., Centlivre, M., von Eije, K.J., and Berkhout, B. (2008) Lentiviral vector design for multiple shRNA expression and durable HIV-1 inhibition. *Mol Ther*, **16**, 557–564.
11. Liu, Y.P., Haasnoot, J., Ter Brake, O., Berkhout, B., and Konstantinova, P. (2008) Inhibition of HIV-1 by multiple siRNAs expressed from a single microRNA polycistron. *Nucl Acids Res*, **36**, 2811–2824.
12. Aagaard, L.A., Zhang, J., von Eije, K.J., Li, H., Saetrom, P., Amarzguioui, M., and Rossi, J.J. (2008) Engineering and optimization of the miR-106b cluster for ectopic expression of multiplexed anti-HIV RNAs. *Gene Ther*, **53**, 1536–1549.
13. Li, M., Li, H., and Rossi, J.J. (2006) RNAi in combination with a ribozyme and TAR decoy for treatment of HIV infection in hematopoietic cell gene therapy. *Ann NY Acad Sci*, **1082**, 172–179.
14. Grimm, D. and Kay, M.A. (2007) Combinatorial RNAi: a winning strategy for the race against evolving targets? *Mol Ther*, **15**, 878–888.
15. Haasnoot, J. and Berkhout, B. (2009) Nucleic acids-based therapeutics in the battle against pathogenic viruses. *Handb Exp Pharmacol*, **189**, 243–263.
16. Scherer, L., Rossi, J.J., and Weinberg, M.S. (2007) Progress and prospects: RNA-based therapies for treatment of HIV infection. *Gene Ther*, **14**, 1057–1064.
17. Saayman, S., Barichievy, S., Capovilla, A., Morris, K.V., Arbuthnot, P., and Weinberg, M.S. (2008) The efficacy of generating three independent anti-HIV-1 siRNAs from a single U6 RNA Pol III-expressed long hairpin RNA. *PLoS ONE*, **3**, e2602.
18. Manche, L., Green, S.R., Schmedt, C., and Mathews, M.B. (1992) Interactions between double-stranded RNA regulators and the protein kinase DAI. *Mol Cell Biol*, **12**, 5238–5248.
19. Marques, J.T., Devosse, T., Wang, D., Zamanian-Daryoush, M., Serbinowski, P., Hartmann, R., Fujita, T., Behlke, M.A., and Williams, B.R. (2006) A structural basis for discriminating between self and nonself double-stranded RNAs in mammalian cells. *Nat Biotechnol*, **24**, 559–565.
20. Robbins, M.A., Li, M., Leung, I., Li, H., Boyer, D.V., Song, Y., Behlke, M.A., and Rossi, J.J. (2006) Stable expression of shRNAs in human CD34$^+$ progenitor cells can avoid induction of interferon responses to siRNAs in vitro. *Nat Biotechnol*, **24**, 566–571.
21. Lund, E. and Dahlberg, J.E. (2006) Substrate selectivity of exportin 5 and Dicer in the biogenesis of microRNAs. *Cold Spring Harb Symp Quant Biol*, **71**, 59–66.
22. Macrae, I.J., Zhou, K., Li, F., Repic, A., Brooks, A.N., Cande, W.Z., Adams, P.D., and Doudna, J.A. (2006) Structural basis for double-stranded RNA processing by Dicer. *Science*, **311**, 195–198.
23. Weinberg, M.S., Ely, A., Barichievy, S., Crowther, C., Mufamadi, S., Carmona, S., and Arbuthnot, P. (2007) Specific inhibition of HBV replication in vitro and in vivo with expressed long hairpin RNA. *Mol Ther*, **15**, 534–541.
24. Barichievy, S., Saayman, S., von Eije, K.J., Morris, K.V., Arbuthnot, P., and Weinberg, M.S. (2007) The inhibitory efficacy of RNA POL III-expressed long hairpin RNAs targeted to untranslated regions of the HIV-1 5′ long terminal repeat. *Oligonucleotides*, **17**, 419–431.
25. Soifer, H.S., Sano, M., Sakurai, K., Chomchan, P., Saetrom, P., Sherman, M.A., Collingwood, M.A., Behlke, M.A., and Rossi, J.J. (2008) Role for the Dicer helicase domain in the processing of thermodynamically unstable hairpin RNAs.. *Nucl Acids Res*, **36**, 6511–6522.
26. Sano, M., Li, H., Nakanishi, M., and Rossi, J.J. (2008) Expression of long anti-HIV1 hairpin RNAs for the generation of multiple siRNAs: advantages and limitations. *Mol Ther*, **16**, 170–177.
27. Castanotto, D., Li, H., and Rossi, J.J. (2002) Functional siRNA expression from transfected PCR products. *RNA*, **8**, 1454–1460.
28. Bertrand, E., Castanotto, D., Zhou, C., Carbonnelle, C., Lee, N.S., Good, P., Chatterjee, S., Grange, T., Pictet, R., Kohn, D., Engelke, D., and Rossi, J.J. (1997) The expression cassette determines the functional activity of ribozymes in mammalian cells by controlling their intracellular localization. *RNA*, **3**, 75–88.
29. Passman, M., Weinberg, M., Kew, M., and Arbuthnot, P. (2000) In situ demonstration of inhibitory effects of hammerhead ribozymes that are targeted to the hepatitis

Bx sequence in cultured cells. *Biochem Biophys Res Commun*, **268**, 728–733.
30. Lee, N.S., Dohjima, T., Bauer, G., Li, H., Li, M.J., Ehsani, A., Salvaterra, P., and Rossi, J. (2002) Expression of small interfering RNAs targeted against HIV-1 rev transcripts in human cells. *Nat Biotechnol*, **20**, 500–505.
31. Lee, S.K., Dykxhoorn, D.M., Kumar, P., Ranjbar, S., Song, E., Maliszewski, L.E., Francois-Bongarcon, V., Goldfeld, A., Swamy, N.M., Lieberman, J., and Shankar, P. (2005) Lentiviral delivery of short hairpin RNAs protects CD4 T cells from multiple clades and primary isolates of HIV. *Blood*, **106**, 818–826.
32. Akashi, H., Miyagishi, M., Yokota, T., Watanabe, T., Hino, T., Nishina, K., Kohara, M., and Taira, K. (2005) Escape from the interferon response associated with RNA interference using vectors that encode long modified hairpin-RNA. *Mol Biosyst*, **1**, 382–390.

Chapter 12

Functional Studies on RNA-Transfected Cell Microarrays

Christina Sæten Fjeldbo, Kristine Misund, Clara-Cecilie Günther, Mette Langaas, Tonje Strømmen Steigedal, Liv Thommesen, Astrid Lægreid, and Torunn Bruland

Abstract

RNA-transfected cell microarray shows great promise in functional genomics. By printing siRNA complexed with transfection reagent on glass slides, arrays of transfected cells are formed in which the phenotypic consequences of gene suppression can be investigated. Using reporter plasmids with fluorescence intensity as output data, we have developed a strategy for statistical analysis of the intensity data from medium-scale functional studies using data from several experimental replicates.

Key words: Reverse transfection, RNA interference, small interfering RNA (siRNA), linear regression (ANOVA) modeling.

1. Introduction

Transfected cell microarray is a versatile, efficient, and cost-reducing technology that can contribute to further understanding of gene functions and regulatory mechanisms governing gene expression by enabling measurements of biological responses in cells with overexpression or downregulation of specific gene products combined with the possibility of assaying effects of external stimuli (1, 2). This technology allows spatially restricted transfection without the use of wells by immobilizing nucleic acids formulated with a transfection reagent in a gel, from which it is only accessible to nearby cells. Adherent cells growing on top of such printed spots will take up the nucleic acids deposited in the spot, while cells growing between the spots will not be transfected (*see* **Fig. 12.1e, f**). By printing expression plasmids, specific overexpression can be achieved. Recently RNA interference has become

M. Sioud (ed.), *RNA Therapeutics*, Methods in Molecular Biology 629,
DOI 10.1007/978-1-60761-657-3_12, © Springer Science+Business Media, LLC 2010

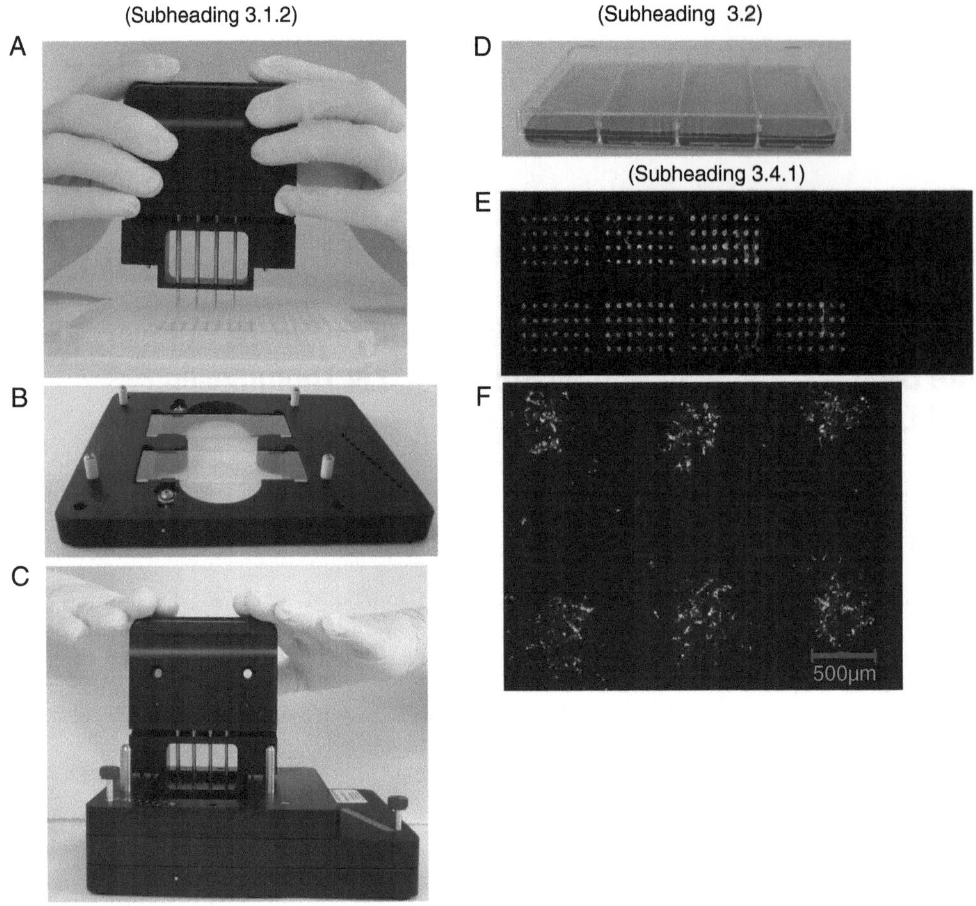

Fig. 12.1. Workflow for transfected cell microarrays. (**A**) The printing solutions are mixed and transferred to a 384-well plate. The eight pins in the arrayer tool of the hand-held microarrayer MicroCaster^{TM} are carefully lowered into the wells. (**B**) UltraGAPS^{TM} slides are placed in the base unit of the slide holder. (**C**) The indexing deck is placed on top of the base unit and the printing solutions are spotted onto the slides using the arrayer tool. (**D**) Dried arrays are placed into QuadriPERM chambers and cells are seeded out onto each array for transfection. (**E**) Image of the fluorescence intensities from spots printed with pEGFP acquired using a laser scanner (6 μm resolution). (**F**) Fluorescent microscope image of six spots showing the spot diameter to be approximately 500 μm (image taken at 40× magnification). (Modified from (7) with permission from *Nucleic Acids Research*).

a key method for inhibiting gene expression in vitro and in vivo (*see* **Chapters 9–11**). By printing siRNAs or shRNAs, gene silencing is achieved on transfected cell microarrays (3–6). Applying the appropriate assays, the phenotypic consequences of gene inactivation can be monitored (*see* **Chapter 13**).

Although the technology seems straightforward, one of the main challenges with cell microarray is the analysis of output data. There is a huge variety in the assay outputs, with the possibility of measuring many parameters per spot or per cell. Transfected cell microarrays can be applied for high-throughput screening. However, it is an attractive method also for more medium-scale studies addressing defined biological questions with fewer genes due to the possibility of printing many replicate spots on one array, as well as the possibility of multiplexing, e.g., using different biological assays on identical arrays. Moreover, assessment of the role of particular gene products in gene regulation can be performed by overexpressing (with expression plasmids) or inhibiting (with siRNA) the gene product of interests in the presence of fluorescent protein reporter plasmids, containing specific promoter elements. We developed a strategy for analysis of fluorescence intensity data from each spot in such medium-scale studies including both technical and experimental replicates in the statistical analysis (7). Screening data containing few, if any replicates will need other statistical approaches (8). In this chapter, we describe the protocol for making the printing solution, printing the arrays, reverse transfection, laser scanning and image analysis, and statistical analysis using linear regression (ANOVA) modeling for estimating biological effects and calculating *p*-values for comparisons of interests.

2. Materials

2.1. Array Printing

1. X-tremeGENE siRNA transfection reagent (Roche) (*see* **Note 1**).
2. Sucrose solution: 1.5 M in sterile water.
3. Gelatine solution, 0.4% (w/v): Dissolve gelatine powder, Type B: 225 Bloom (Sigma), in sterile water by swirling the solution gently for 15 min in a 60°C water bath. Cool down to approximately 37–40°C and sterile filtrate using a 0.45 μm filter. Store at 4°C (stable at least for 1 month) (1).
4. Expression plasmids. We have used the following:
   a. pEGFP-N1 and pDsRed-express-N1 (BD Bioscience Clontech).
   b. pCRE-EGFP and pNFκB-EGFP (7).
   c. Negative control plasmid (i.e., a plasmid without a gene encoding a fluorescent protein): pCRE-Luc (Stratagene).
5. siRNAs. We have used the following sequences:
   a. siEGFP: sense, 5′-GCAAGCUGACCCUGAAGUUCAU-3′; antisense, 5′-GAACUUCAGGGUCAGCUUGCCG-3′ (4).
   b. siICER: sense, 5′-CAUUAUGGCUGUAACUGGATT-3′; antisense, 5′-UCCAGUUACAGCCAUAAUGGG-3′ (9).

c. Control siRNA: siCAT (CAT, chloramphenicol acetyl transferase): sense, 5′-GAGUGAAUACCACGACGAUUUC-3′; antisense, 5′-AAUCGUCGUGGUAUUCACUCCA-3′ (4).

6. Notably, the optimal plasmid and siRNA concentrations will vary with the experimental setting. For plasmids with strong promoters use 25–75 ng/μL. However, for siRNA several concentrations should be tested. A good starting point would be 15–30 ng/μL.
7. UltraGAPS™ coated slides (Corning).
8. Hand-held arrayer: MicroCaster™ from Schleicher & Schuell (S&S), consisting of a slide holder and an arrayer with eight printing pins, giving up to 768 spots (about 500 μm in diameter) per slide.
9. S&S MicroCaster™ Pin Conditioner. Working solution: 1:5 dilution in distilled water. This solution can be used several times. Store at room temperature.
10. 384-well plate.

2.2. Reverse Transfection

1. HEK 293 cells. We have used HEK 293 cells stably transfected with a gene encoding the ICER IIγ splice variant driven by a tetracycline-inducible promoter: HEK 293_{ind}-ICER IIγ cells (7).
2. Dulbecco's modified Eagle's medium (DMEM) containing 4.5 g/L glucose and supplemented with 1% (v/v) penicillin–streptomycin, 0.1 mg/mL L-glutamine, 150 μg/mL hygromycin B, 15 μg/mL blasticidin, and 10% FCS.
3. QuadriPERM plates (Vivascience) (*See* **Note 2**).
4. Trypsin–EDTA solution: 0.25% Trypsin and 1 mM EDTA.
5. External treatment. We have used the following:
   a. Tetracycline: Dissolve in sterile water at 1 mg/mL and add to culture medium at a final concentration of 1 μg/mL. The working solution should be stored at 4°C.
   b. Forskolin: Dissolve in DMSO at 50 mM and store in aliquots at –20°C until use. In the present experiments we added forskolin to the cell culture at a final concentration of 10 μM.
   c. Human recombinant TNFα: Dissolve in PBS with 0.2% BSA at 50 ng/μL and store in aliquots at –20°C until use. In the present experiments we added TNFα to the cell culture at a final concentration of 20 ng/mL.

2.3. Fixation

1. PBS (phosphate-buffered saline): Phosphate Buffered Saline Tablets Dulbecco A' (Oxoid). To obtain 2X solution, dissolve two tablets in 100 mL sterile water.

2. Fixation solution: 3.7% (w/v) paraformaldehyde and 4% (w/v) sucrose in PBS. Prepare the solution fresh for each experiment (in a fume hood). Heat water to 59°C (not above 60°C), add paraformaldehyde and 10–20 droplets of 1 M NaOH until the paraformaldehyde is dissolved (use a stirring hot-plate). Subsequently, adjust the solution to 1X PBS by adding 0.5 of the total volume 2X PBS and add sucrose. Adjust the pH to 7.2–7.4 with HCl. Sterile filtrate the solution and cool to room temperature before use.
3. Nuclear stain: 1 μL/mL DAPI (4′,6′-diamidino-2-phenylindole) (Invitrogen) in 1X PBS. Store at 4°C.
4. Mowiol 4-88 mounting medium pH 8.5 (*see* **Note 3**): 6 g glycerol, 2.4 g Mowiol 4-88 (Hoechst), 6 mL dH_2O, and 12 mL Tris–HCl, pH 8.5. This solution should be prepared as follows:
   a. Add glycerol to a 50 mL centrifuge tube.
   b. Add Mowiol 4-88 and mix well.
   c. Add dH_2O and incubate for 2 h at room temperature.
   d. Add Tris-HCl buffer pH 8.5 and incubate in a 50°C water bath for 10 min. Shake the solution regularly to dissolve Mowiol 4-88.
   e. Centrifuge at 5,000×*g* for 15 min.
   f. Make 1,000 μL aliquots and store at –20°C. (stable at least 1 month at room temperature).
5. Cover glass (Corning): 24 × 60 mm, No. 1 $^1/_2$.

2.4. Scanning and Analysis

1. Microarray laser scanner: Tecan's LS Reloaded™ scanner using 6 μm resolution (*see* **Note 4**). The EGFP and DsRed emissions are visualized using lasers with wavelength 488 and 532 nm, and the filters 535/25 and 575/50 nm, respectively.
2. Image analysis software such as the GenePix Pro software (Axon Instruments).
3. The free software environment for statistical computing and graphics R. The R code used in the statistical analysis is available upon request: mettela@math.ntnu.no

3. Methods

The following subsections describe the process of printing the arrays with the desired nucleic acids, reverse transfection, fixation, and analysis of the spot intensities to detect biological effects.

3.1. Array Printing

3.1.1. Making the Printing Solution

1. In a 1.5 mL microcentrifuge tube, mix the reporter plasmid, and if possible, an internal control plasmid (*see* **Note 5**) and the siRNA with serum-free medium. Add 0.5 μL 1.5 M sucrose to yield 25 mM sucrose in the final printing solution (*see* **Note 6**). Subsequently, add 3 μL X-tremeGENE per μg nucleic acid to a final volume of 22.5 μL (*see* **Note 7**).
2. Incubate the solution for 15–20 min at room temperature.
3. Add 7.5 μL 0.4% gelatine to give 30 μL printing solution, yielding 0.2% gelatine in the final printing solution (*see* **Note 6**).

The printing solutions can be stored at 4°C for at least 2 months for re-arraying purposes.

3.1.2. Printing the Arrays

The requirement of an expensive robotic arrayer for printing the microarrays can be an obstacle for many research groups to adopt this method in their lab. Alternatively, a hand-held microarrayer like MicroCaster^{TM} can be used. *See* **Fig. 12.1** for the illustration of the printing protocol and **Note 8** for some considerations about the experimental design.

1. Prepare the printing pins by coating with Pin Conditioner according to the manufacturer's description.
2. Add the printing solutions to wells in a 384-well plate; 15 μL in each well.
3. Place UltraGAPS^{TM} slides in the slide holder of MicroCaster^{TM}.
4. Dip the printing pins into the wells of the 384-well plate (*see* **Note 9**) and print the arrays according to the manufacturer's description.

Array printing can also be performed using other equipment (*see* **Note 10**). After printing, the slides are dried for at least 1 h in room temperature before placed at 4°C together with a desiccant until use. The printed slides may be stored for several weeks before use.

3.2. Reverse Transfection

1. About 24 h before reverse transfection, split HEK 293 cell culture in order to obtain exponentially growing cells on the printed slides (*See* **Note 11**).
2. At the day of reverse transfection, place the printed slide(s) in QuadriPERM plates so that they reach room temperature before cells are added.
3. Immediately before transfection, trypsinize the actively growing cells and resuspend in growth medium to a density of 37.5×10^4 cells/mL.

4. Gently pipette 8 mL cell suspension to each slide (i.e., 3 × 10^6 HEK 293 cells per slide), avoiding direct pouring on the printed areas.
5. Incubate the cells in a 37°C, 5% CO_2 humidified incubator.
6. External treatment can now be added to the cells according to the protocol of interest. To compare treated and untreated cells, use two identical printed slides. Alternatively the slides can be divided into wells (*see* **Note 12**).
7. Incubate the cells on the slides for 48 h before fixation to achieve near confluent layer. However, the optimal incubation time is dependent on the used cell line, plasmid, and siRNA (e.g., 24–72 h).

3.3. Fixation

1. Remove the medium carefully.
2. Wash the slides gently by dipping them carefully into a well in a QuadriPERM plate containing cold PBS.
3. Transfer the slides to another QuadriPERM plate well containing fixation solution, and incubate the slides in fixation solution for 20 min at room temperature.
4. Wash the slides gently in cold PBS.
5. Stain the nuclei by incubating the slides for 5 min with 500 μL of 1 μL/mL DAPI in PBS per slide.
6. Wash the slides gently three times in cold PBS.
7. Mount coverslip using 50 μL Mowiol 4-88 mounting medium pH 8.5.
8. Place the slides at 4°C overnight for hardening before image acquisition (*see* **Note 13**). Keep the slides in the dark to minimize photobleaching.

3.4. Scanning and Data Analysis

The analysis pipeline is shown in **Fig. 12.2**. This protocol describes the analysis of spot intensities. Transfected cell microarrays can also be used to detect and measure phenotypes visible on single-cell level (4, 10–14).

3.4.1. Scanning and Quantification of Spot Intensities

1. Obtain an image of the fluorescence signals in the spots using a microarray laser scanner. *See* **Fig. 12.1e** for an example of such an image. The scanner settings have to be adjusted according to the type of fluorochrome used and the degree of fluorescence intensity.
2. Quantify the level of fluorescence protein expression in each spot using, e.g., the GenePix Pro software. Measure the mean grayscale values in each spot using a grid with

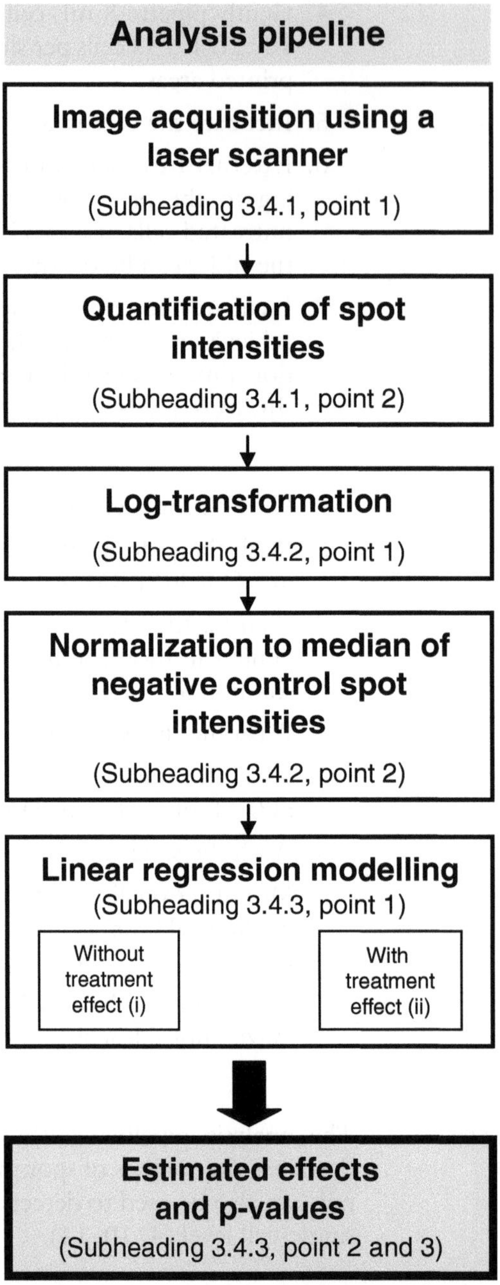

Fig. 12.2. Analysis pipeline. An image of the fluorescence intensities on the slides is obtained using a laser scanner. Image analysis software like the GenePix Pro software is used to quantify the mean intensity in each spot. The data is log transformed (base 2), and the log-transformed data is normalized using the median of negative control spots printed on the same array. The log-transformed and normalized data is used to fit a linear model to obtain estimated effects of each condition, and *p*-values for comparisons of interests are calculated.

circles of diameter corresponding to 600 μm in the scanning image. Eliminate spots with visual defects from the analysis. Observe the array using a fluorescence microscope to ensure reasonable results obtained in the scanning image. The rest of the analysis described below is performed using an R code written for this purpose (7), with text files containing the quantified intensity data as input files.

3.4.2. Log-Transformation and Normalization

1. Log-transform (base 2) the intensity data.
2. Normalize the log-transformed data to negative control spots (spots without fluorescent protein expression, i.e., spots printed with a non-fluorescent reporter plasmid) by subtracting the median of the log-transformed intensities in the negative control spots from all the spots printed on the same array, i.e., as a means of background correction on the log scale. This correction makes it possible to compare spot intensities on different arrays.

3.4.3. Linear Regression Method

Use linear regression to model the observed log-transformed and normalized spot intensities on the basis of different conditions printed on the slide (different printing solutions with plasmids and siRNAs), the different treatments (external stimulus added to the cells), and the different independent experiments (experimental replicates). Each condition is represented by several replicate spots per treatment (technical replicates) in each experimental replicate.

1. Fit the model to the log-transformed and normalized data:
   a. If no external treatment is added to the cells:

$$Y_{ikl} = \mu + e_i + c_k + (ec)_{ik} + \varepsilon_{ikl} \qquad [1]$$

Y_{ikl}: the normalized log-transformed intensity signal in the lth spot in experimental replicate i for condition k.

μ: the overall level of all spots in all experimental replicates and for all conditions.

$e_i,\ i = 1,\ldots,n_e$: the effect of the ith experimental replicate.

n_e: the number of experimental replicates.

$c_k,\ k = 1,\ldots,n_c$: the effect of the kth condition.

n_c: the number of conditions.

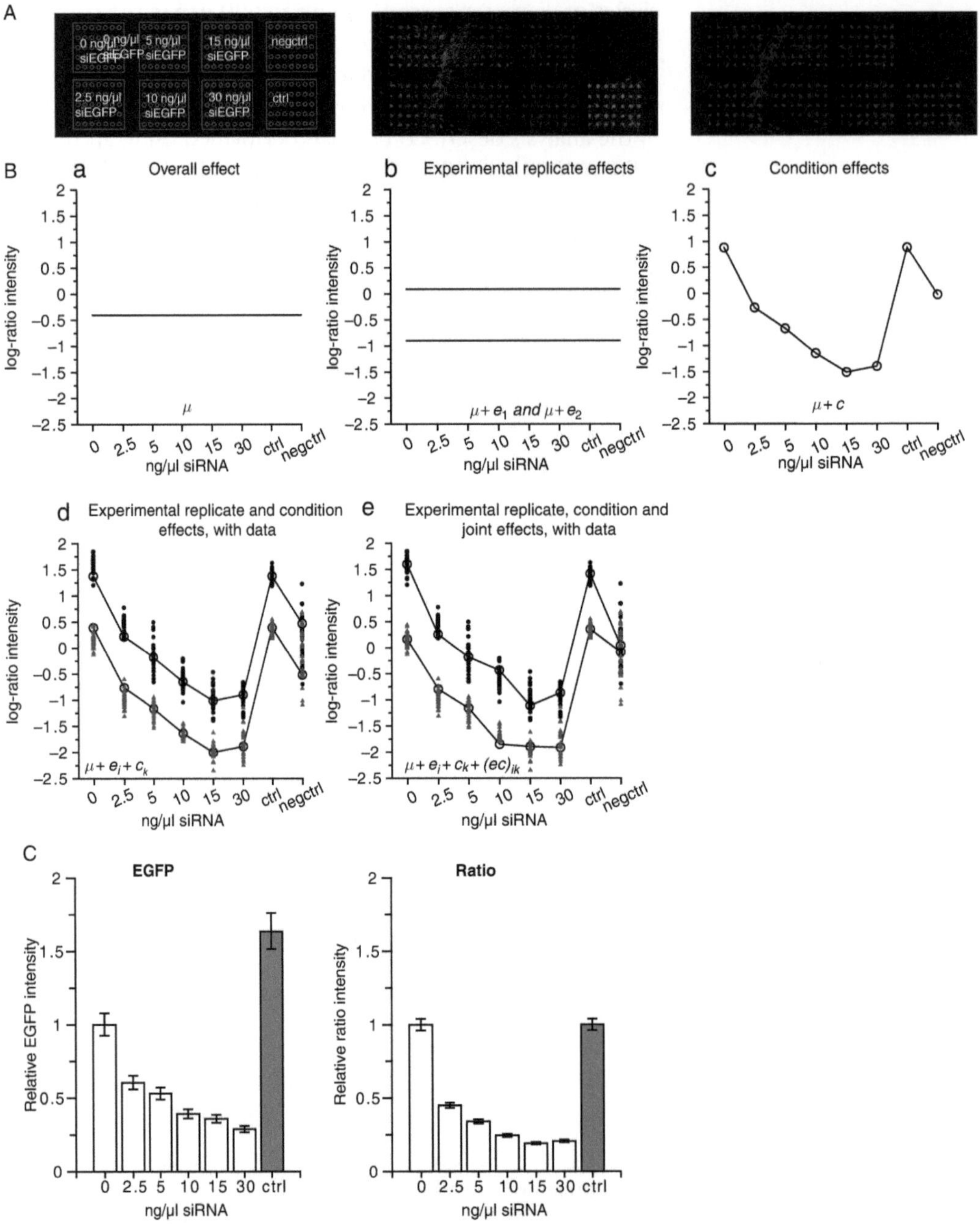

Fig. 12.3. Linear regression modeling without a treatment effect. Cells were cotransfected with pEGFP (30 ng/µl), pDsRed (30 ng/µl), and different concentrations of siEGFP (0–30 ng/µl). (**A**) *Left diagram* indicates the placement of the cell clusters for eight conditions. pEGFP, pDsRed, and 0, 2.5, 5, 10, 15, or 30 ng/µl siEGFP. The non-fluorescent reporter pCRE-Luc (60 ng/µl) as a control of the background fluorescence in the spots (negctrl). pEGFP, pDsRed, and 30 ng/µl siCAT as a siRNA control (ctrl). *Middle diagram:* Scanning image for EGFP-detection. *Right diagram:* Scanning image for DsRed-detection. (**B**) Illustration of the different terms in the linear regression model in Eq1. The ratio fluorescence signal intensities (EGFP intensities normalized to DsRed intensities) from each spot (log transformed and normalized to negative control spot intensities) in two experimental replicates (28–35 technical replicates after removing flagged spots) were used to fit the model. (**a**) The overall effect μ. (**b**) The overall effect μ and the experimental replicate effects e_1 and e_2.

$(ec)_{ik}$: models the interaction between experimental replicate i and condition k.

ε_{ikl}: the remaining unexplained variation in the data. We assume that ε_{ikl} is normally distributed with mean 0 and variance σ^2. We assume that this unexplained variation is independent between the spots.

b. If an effect of one external stimulus is to be studied, a treatment effect is added to the previous model, t_j:

$$Y_{ijkl} = \mu + e_i + t_j + c_k + (et)_{ij} + (ec)_{ik} + (tc)_{jk} + (etc)_{ijk} + \varepsilon_{ijkl} \quad [2]$$

Y_{ijkl}: the log-transformed and normalized signal in the lth spot in experimental replicate i for treatment j and condition k.

We have assumed only two treatment levels, thus $j = 1,2$.

In the estimation of the parameters we use the following sum-to-zero constraints:

$$\sum_{i=1}^{n_e} e_i = 0, \quad \sum_{j=1}^{n_t} t_j = 0, \quad \sum_{k=1}^{n_c} c_k = 0,$$
$$\sum_{i=1}^{n_e} (et)_{ij} = \sum_{j=1}^{n_t} (et)_{ij} = 0, \quad \sum_{i=1}^{n_e} (ec)_{ik} = \sum_{k=1}^{n_c} (ec)_{ik} = 0,$$
$$\sum_{j=1}^{n_t} (tc)_{jk} = \sum_{k=1}^{n_c} (tc)_{jk} = 0, \text{ and}$$
$$\sum_{i=1}^{n_e} (etc)_{ijk} = \sum_{j=1}^{n_t} (etc)_{ijk} = \sum_{k=1}^{n_c} (etc)_{ijk} = 0$$

2. The fitted model gives an estimate of the effects:
For Eq 1: $\mu + c_k$.

For Eq 2: $\mu + c_k + t_j + (tc)_{jk}$.

3. Compare conditions within or between treatments by testing the null hypothesis that their difference on log scale is zero. The test statistic is the estimated difference between

Fig. 12.3. (continued) (**c**) The overall effect μ and the condition effects c_1 through c_8. (**d**) Combining μ, e_i, and c_k. (**e**) Addition of the interaction effect term $(ec)_{ik}$ to the terms in (**d**). The observed data from each spot are shown together with the fitted model in (**d**) and (**e**) for experimental replicate 1 (*circles*) and experimental replicate 2 (*triangles*). (**C**) Estimated effects of the different conditions are shown based on the EGFP fluorescence intensity data (*left*) or the EGFP fluorescence intensity data normalized to DsRed control plasmid intensities (*right*). The intensities are shown relative to 0 ng/μL siRNA, error bars are 95% confidence interval. The arrays were printed using MicroCaster^{TM}. (Reproduced from (7) with permission from *Nucleic Acids Research*).

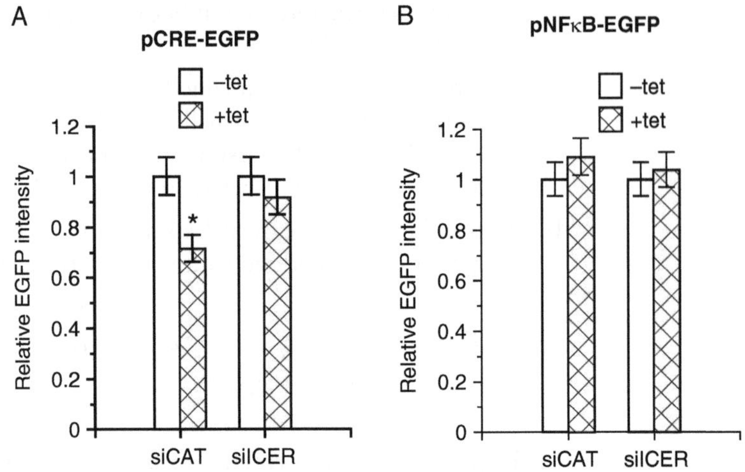

Fig. 12.4. Linear regression modeling with a treatment effect for analysis of transcriptional repression. (**A**) Effect of ICER IIγ overexpression on forskolin stimulated CRE-driven transcription. Duplicate arrays with spots printed with pCRE-EGFP (50 ng/μL) together with siCAT or siICER (25 ng/μL), and negative control spots (pCRE-Luc, 50 ng/μL) were overlaid with $HEK293_{ind}$-ICER IIγ cells. The cells on one of the two arrays were treated with 1 μg/mL tetracycline (tet) for 20 h before fixation to induce overexpression of ICER IIγ, and both arrays were stimulated with 10 μM forskolin 6 h before fixation to induce CRE-driven expression. The resulting estimated effect of each condition and treatment based on the EGFP fluorescence intensities from four experimental replicates (33–36 technical replicates after removing flagged spots) using the linear regression model in Eq. 2. The EGFP fluorescence intensities are expressed relative to tetracycline untreated state. Error bars are 95% confidence intervals. $*p < 0.01$; significant difference from tetracycline untreated cells. (**B**) Effect of ICER IIγ overexpression on TNFα–stimulated NFκB-driven transcription. The experiment was performed as in (**A**), with pCRE-EGFP substituted by pNFκB-EGFP, and stimulation with TNFα (20 ng/mL) instead of forskolin. EGFP fluorescence intensity relative to the tetracycline untreated state estimated from three experimental replicates using the linear regression model in Eq. 2. Error bars are 95% confidence intervals (24–36 technical replicates). The arrays were printed using a 10 μL pipette tip. (Reproduced from (7) with permission from *Nucleic Acids Research*).

the conditions or treatments of interest divided by the estimated standard error of this difference. The standard error is found from the covariance matrix of the estimated effects, and the test statistic is t-distributed. The p-value is then calculated from the test statistic. **Figures 12.3** and **12.4** show an example of the estimated effects using the model without and with a treatment effect, respectively.

4. Notes

1. Other transfection reagents can be used. It is important to find a reagent that give high transfection efficiency for the type of nucleic acid and cell line used. In our hand the X-tremeGENE siRNA transfection reagent (Roche) was found to be suitable. Other transfection reagents that have been used on transfected cell microarrays include Effectene (1) and Lipofectamine 2000 (10) (*see* **Chapter 10**).

2. The QuadriPERM plate contains four chambers, each with room for one slide. Other less expensive plates like Petri dishes may be used.
3. Other mounting medium may be used. In our experience, the used mounting medium conserves the fluorescence well [reduces light-induced fading (photobleaching) of the fluorophore]. We have observed that the fluorescence is still detectable 1 year after mounting the slide.
4. An advantage with this scanner is that the slide can be mounted before scanning. For other scanners, like ScanArray Express HT (PerkinElmer), the slides have to be scanned without mounting. The advantage of mounting the slides before scanning is a better preservation of the fluorescence signals, and the results can be observed in a microscope before scanning.
5. An internal control plasmid that gives constitutively active expression of a fluorescent protein is beneficial for normalizing the fluorescence intensity in each spot. It minimizes the effect of differences in transfection efficiency between spots. *See* **Fig. 12.3** where pDsRed is used as an internal control plasmid.
6. The sucrose and gelatine concentrations in the printing solution influence both transfection efficiency and spot integrity (7, 14, 15). *See* **Fig. 12.5** for an example. If cross-contamination between spots is a problem, try to reduce the concentrations of sucrose and/or gelatine. Sucrose is added to the printing solution to stabilize the complex between the lipid (transfection reagent) and nucleic acids and is especially beneficial when storing the printed slides several weeks before use.
7. The protocol is based on the so-called lipid–DNA method. Arrays can also be printed without transfection reagent (gelatine–DNA method), by including an incubation step of the slide with transfection reagent immediately before transfection (1). If low transfection efficiency is a problem, the amount of transfection reagent can be increased, in addition to adjusting the sucrose and gelatine concentrations. However, we have observed that too much X-tremeGENE may result in less localized spots.
8. When designing how to print the arrays, consider the following points:
   - Spots printed with a non-fluorescent reporter plasmid (e.g., pCRE-Luc) have higher fluorescence intensity than the cell layer between the spots. Such a negative (or background) control is valuable for the assessment of exogenous expression of fluorescence protein as spots

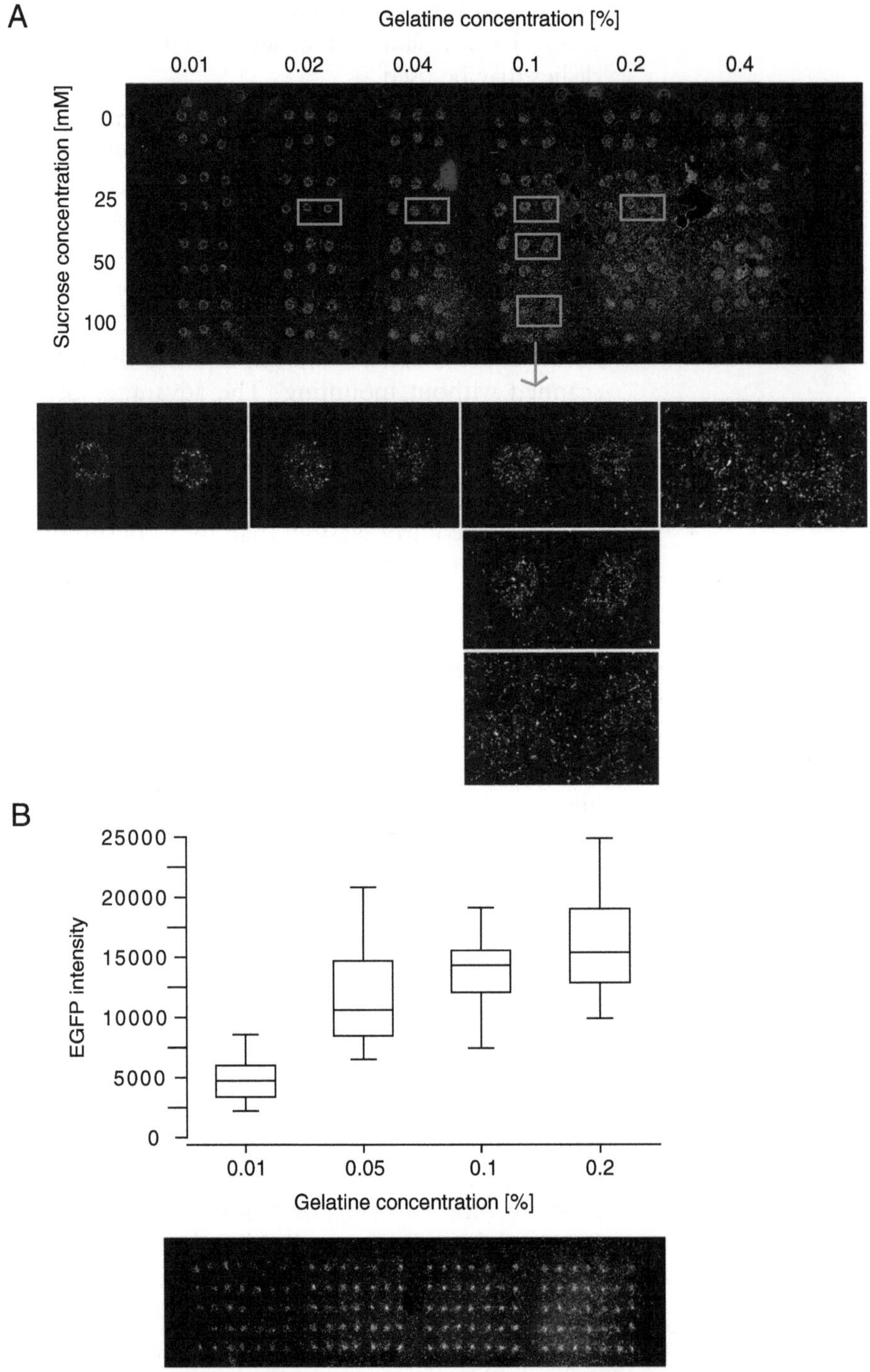

Fig. 12.5. Effects of sucrose and gelatine concentrations on spot integrity and transfection efficiency. (**A**) Array printed with pDsRed (50 ng/μL) in a printing solution with different gelatine and sucrose concentrations. Scanning image of the whole array and magnifications of specific spots. (**B**) Array printed with pEGFP (50 ng/μL) in a printing solution with 25 mM sucrose and four different concentrations of gelatine. *Top:* Box plot of the fluorescence intensities in each spot ($n = 32–34$). *Bottom*: Scanning image showing squares of seven times five spots for the four gelatine concentrations. From left to right: 0.01%, 0.05%, 0.1%, and 0.2% gelatine. The DNA–lipid–gelatine–sucrose solutions were printed manually with a 10 μL pipette tip (**A**) or by MicroCaster^{TM} manual arrayer system (**B**). (Reproduced from (7) with permission from *Nucleic Acids Research*).

expressing such proteins should yield higher intensity signals. As this background signal varies between arrays, these spots are valuable for normalization of the data for differences in the global background signal on the array to allow comparisons of spot intensities between arrays (*see* **Section 3.4.2**).

- It is important to include both positive and negative controls for the assay being performed in order to be sure that the assay is able to detect the effects under study and to be able to set the *p*-value cutoff for significance properly.
- The inclusion of both technical (replicate spots on an array) and experimental replicates in transfected cell microarray experiments is important in order to achieve results with a high degree of confidence. The number of replicates that is sufficient will vary with the biological problem to be addressed (7). Positive and negative controls can be used to indicate how many replicates that has to be included in order to detect the biological effect.

9. Small spirit levels may be useful to guide the dipping of the eight printing pins into the wells of a 384-well plate to ensure equal exposure to the printing solution.
10. Array printing may also be performed using a microarray printing robot (1). For small or medium-scale studies a pipette tip may be used (7, 16).

    Printing with a small pipette tip: 10 μL pipette tip (Biosphere Filter Tips, type Gilson/Biohit, Sarstedt) giving spots of about 800 μm in diameter:

    1. Make a template of the printing pattern and place it inside a plastic pocket.
    2. Attach the slide to the plastic pocket over the template using tape.
    3. Make small marks with a black pen for orientation on the slide.
    4. Pipette 2 μL printing solution in the pipette tip, remove the tip from the pipette (ensure that the printing solution is in the tip of the pipette without air bubbles) and use it to print the array by gently touching the slide surface. Printing is performed at room temperature.
11. Other cell lines can be used, but the protocol presented here is optimized for the HEK 293 cell line.
12. To be able to observe a treatment effect on the different conditions in the spots, it is necessary to use either two identical printed slides or divide the slide into separate wells. A cell culture accessory made of silicone (e.g.,

FlexiPERM from Vivascience) can be used to divide the slides into separate wells with several spots in each well.

13. The cover slip may be sealed to the slide with, e.g., nail polish for observation of the result the same day.

Acknowledgments

We thank Hallgeir Bergum and Atle van Beelen Granlund at the National Technology Microarray Platform (Norwegian Microarray Consortium) at NTNU, which is founded by the Functional Genomics Programme (FUGE) of the Norwegian Research Counsil for valuable input and help in the work of establishing transfected cell microarrays in our lab. Funding was provided by Liaison Committee between the Central Norwegian Region Health Authority (RHA) and NTNU.

References

1. Ziauddin, J. and Sabatini, D.M. (2001) Microarrays of cells expressing defined cDNAs. *Nature*, **411**, 107–110.
2. Starkuviene, V., Pepperkok, R., and Erfle, H. (2007) Transfected cell microarrays: an efficient tool for high-throughput functional analysis. *Expert Rev Proteomics*, **4**, 479–489.
3. Kumar, R., Conklin, D.S., and Mittal, V. (2003) High-throughput selection of effective RNAi probes for gene silencing. *Genome Res*, **13**, 2333–2340.
4. Mousses, S., Caplen, N.J., Cornelison, R., Weaver, D., Basik, M., Hautaniemi, S., Elkahloun, A.G., Lotufo, R.A., Choudary, A., Dougherty, E.R. et al. (2003) RNAi microarray analysis in cultured mammalian cells. *Genome Res*, **13**, 2341–2347.
5. Silva, J.M., Mizuno, H., Brady, A., Lucito, R., and Hannon, G.J. (2004) RNA interference microarrays: high-throughput loss-of-function genetics in mammalian cells. *Proc Natl Acad Sci U S A*, **101**, 6548–6552.
6. Wheeler, D.B., Carpenter, A.E., and Sabatini, D.M. (2005) Cell microarrays and RNA interference chip away at gene function. *Nat Genet*, **37 Suppl**, S25–S30.
7. Fjeldbo, C.S., Misund, K., Gunther, C.C., Langaas, M., Steigedal, T.S., Thommesen, L., Laegreid, A., and Bruland, T. (2008) Functional studies on transfected cell microarray analysed by linear regression modelling. *Nucleic Acids Res*, **36**, e97.
8. Boutros, M., Bras, L.P., and Huber, W. (2006) Analysis of cell-based RNAi screens. *Genome Biol*, 7, R66.
9. Misund, K., Steigedal, T.S., Laegreid, A., and Thommesen, L. (2007) Inducible cAMP early repressor splice variants ICER I and IIgamma both repress transcription of c-fos and chromogranin A. *J Cell Biochem*, **101**, 1532–1544.
10. Neumann, B., Held, M., Liebel, U., Erfle, H., Rogers, P., Pepperkok, R., and Ellenberg, J. (2006) High-throughput RNAi screening by time-lapse imaging of live human cells. *Nat Methods*, **3**, 385–390.
11. Conrad, C., Erfle, H., Warnat, P., Daigle, N., Lorch, T., Ellenberg, J., Pepperkok, R., and Eils, R. (2004) Automatic identification of subcellular phenotypes on human cell arrays. *Genome Res*, **14**, 1130–1136.
12. Carpenter, A.E., Jones, T.R., Lamprecht, M.R., Clarke, C., Kang, I.H., Friman, O., Guertin, D.A., Chang, J.H., Lindquist, R.A., Moffat, J. et al. (2006) CellProfiler: image analysis software for identifying and quantifying cell phenotypes. *Genome Biol*, 7, R100.
13. Pepperkok, R. and Ellenberg, J. (2006) High-throughput fluorescence microscopy for systems biology. *Nat Rev Mol Cell Biol*, 7, 690–696.
14. Baghdoyan, S., Roupioz, Y., Pitaval, A., Castel, D., Khomyakova, E., Papine, A., Soussaline, F., and Gidrol, X. (2004) Quantitative analysis of highly parallel transfection in cell microarrays. *Nucleic Acids Res*, **32**, e77.
15. Mannherz, O., Mertens, D., Hahn, M., and Lichter, P. (2006) Functional screening for proapoptotic genes by reverse transfection

cell array technology. *Genomics*, **87**, 665–672.

16. Hu, Y.H., Warnatz, H.J., Vanhecke, D., Wagner, F., Fiebitz, A., Thamm, S., Kahlem, P., Lehrach, H., Yaspo, M.L., and Janitz, M. (2006) Cell array-based intracellular localization screening reveals novel functional features of human chromosome 21 proteins. *BMC Genomics*, 7, 155.

Chapter 13

Transfection Microarrays for High-Throughput Phenotypic Screening of Genes Involved in Cell Migration

Reiko Onuki-Nagasaki, Akira Nagasaki, Kazumi Hakamada, Taro Q.P. Uyeda, Satoshi Fujita, Masato Miyake, and Jun Miyake

Abstract

Cell migration is important in several biological phenomena, such as cancer metastasis. Therefore, the identification of genes involved in cell migration might facilitate the discovery of antimetastatic drugs. However, screening of genes by the current methods can be complicated by factors related to cell stimulation, for example, abolition of contact inhibition and the release inflammatory cytokines from wounded cells during examinations of wound healing in vitro. To overcome these problems and identify genes involved in cell migration, in this chapter we describe the use of transfection microarrays for high-throughput phenotypic screening.

Key words: Cell migration, transfection microarray, cell chip, siRNA library, time-lapse imaging, high-throughput analysis.

1. Introduction

Cell migration is essential for many physiological and pathological processes, such as tissue remodeling, wound healing, inflammation, and tissue invasion by tumor cells (1). To date, most analyses of chemotaxis and invasion involve the so-called Boyden chamber (2) or the healing of "wounds" in culture dishes (3). Screening for cell-migration genes in a Boyden chamber generally involves the use of chemoattractants, and their downstream effects that might interfere with the identification of candidate genes. Moreover, the release of inflammatory cytokines from wounded cells can obviously influence rates of cell migration. As a result, it is difficult to postulate roles for candidate genes since they might or might not be involved directly in cell migration. Classical

M. Sioud (ed.), *RNA Therapeutics*, Methods in Molecular Biology 629,
DOI 10.1007/978-1-60761-657-3_13, © Springer Science+Business Media, LLC 2010

methods are useful for the analysis of individual target genes but not for high-throughput screening, because of these serious problems described above.

Microarrays of cells for transfection represent effective tools for high-throughput phenotypic screening (4). We have improved the efficiency of transfection in such microarrays by solid-phase reverse transfection (5). In this respect, we found that inclusion of fibronectin and optimization of the ratio of transfection reagents can enhance efficiency of transfection by facilitating the contact between cells and siRNA attached to a solid surface. Moreover, there is no cross-transfection between the respective spotted areas even in the absence of physical barriers between spots of siRNA because the attached siRNA to the solid surface is localized on an area of the surface that has been treated with an agent that ensures the slow release of the siRNA without diffusion into the medium. Therefore, it is easy to make a dense microarray by miniaturizing and concentrating each spot of siRNA to yield what we call a transfection microarray (TMA) (4, 5)

To use TMAs for the identification of genes involved in cell migration, we have also refined the system by selecting appropriate cell lines because the growth rate and the speed of migration are very important parameters for the development of a reliable screening method using TMA technology. Cell lines associated with the unidirectional and rapid migration are the best suitable for rapid screening on TMAs. Therefore, we have evaluated the migration characteristics of five common cell lines and selected NBT-L2b cells for our study. To quantify migration distances rapidly, we also developed a novel and simple method of the detection of siRNA-transfected cells and for determining the starting and end points of the migration of individual cells. This method is based on labeling the printed area to which siRNA was attached with rhodamine and using cells that express green fluorescent protein (GFP).

2. Materials

2.1. Cell Culture

1. NBT-L2b cells (RIKEN Cell Bank, Tsukuba, Ibaraki, Japan).
2. Minimum essential medium supplemented with 10% fetal bovine, nonessential amino acids, and antibiotics (streptomycin and kanamycin 100 μg/mL). Store at 4°C.
3. 0.05% (w/v) trypsin solution in water containing 0.53 mM ethylenediamine tetraacetic acid (EDTA). Store at 4°C.
4. Dulbecco's phosphate-buffered saline without calcium and magnesium (D-PBS-). Store at 4°C.

2.2. Glass Slides Coated with Type I Collagen and Rhodamine-Labeled Fibronectin

1. 0.01% (w/v) type I collagen solution. Store at 4°C.
2. VWR Plane Slides (thickness, 1 mm; size, 3×1 in.)
3. 1,1,1,3,3,3-Hexamethyldisilazane.
4. Absolute ethanol.
5. Fibronectin.
6. EZ-Label™ Rhodamine Protein-Labeling Kit (Pierce Biotechnology).

2.3. Preparation of a Cell Chip

1. Dulbecco's modified Eagle's Medium (DMEM) without serum and phenol red. Stored in 1-mL aliquots at 4°C.
2. Endotoxin-tested water.
3. Solution of siRNA (20 μM). In the case of freeze-dried samples, dissolve siRNA at 20 μM in endotoxin-free water and store in single-use aliquots (10 μL) at –80°C. As controls, we used AllStar RNAi Controls (Qiagen). The following specific siRNAs were obtained from Qiagen: *actr3*-specific siRNAs, *actr2*-specific siRNAs, *wasf2*-specific siRNAs, and the ***paxilin*** (***pxn***)-specific siRNAs. When siRNAs from another supplier are used, it is necessary to identify suitable conditions for their use because some components, such as salts, differ among freeze-dried siRNAs from different suppliers.
4. Lipofectamine™ 2000 transfection reagent.
5. Solution of rhodamine-labeled fibronectin (2 mg/mL). Store at 4°C.
6. A solution of 0.1% (w/v) gelatin dissolved in endotoxin-free water and autoclaved. Store at 4°C.
7. Type I collagen-coated glass slide.
8. An ink-jet microarray printer.
9. The pEGFP-N1 expression vector (1 μg/μL).

2.4. Motility Assay

1. DMEM/F12 Ham's medium without phenol red, supplemented with 10% fetal bovine serum. Store at 4°C.
2. Printed glass slides.
3. Programmable Cellular Image Tracer (Olympus, Tokyo, Japan).

3. Methods

3.1. Methodological Consideration

We have developed a TMA-based cell-motility chip that allows the identification of genes that are involved in cell migration (6). In order to evaluate cell migration on such a chip, we used

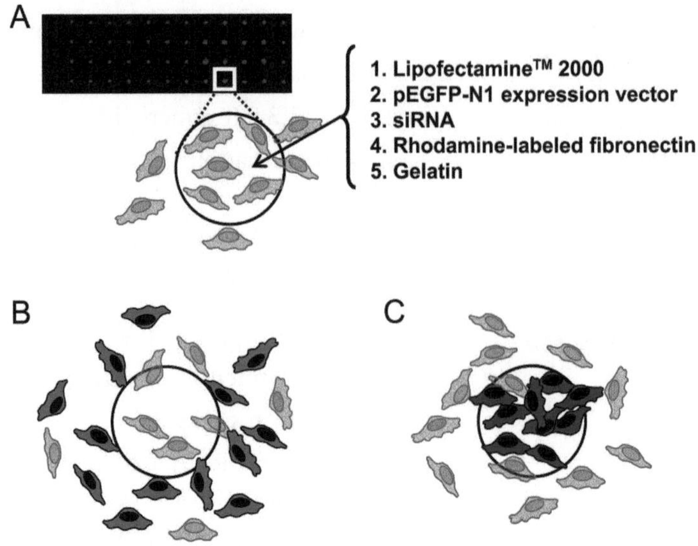

Fig. 13.1. Schematic illustration of the cell-motility chip. (**a**) The transfection mixture, consisting of Lipofectamine^{TM} 2000, the pEGFP-N1 expression vector, siRNA, rhodamine-labeled fibronectin and gelatin, was printed as a 4×12 grid on the surface of a type I collagen-coated glass slide by a high-precision ink-jet microarrayer. *Red signals* due to rhodamine-labeled fibronectin revealed the spots of the transfection mixture on the slide. For analysis, NBT-L2b cells (orange cells; 2×10^5 cells/array) were seeded on the printed chip, which was incubated at 37°C in an atmosphere of 5% CO_2 in air for 24–48 h. Fluorescence due to rhodamine (*white circle*) and GFP (*dark gray*) was monitored with an improved fluorescence microscope for TMA assays. (**b**) GFP-expressing cells transfected with siRNA that had no significant effect on cell migration (*dark gray cells*) migrated freely around each printed spot. (**c**) When the siRNA suppressed the expression of migration-related gene, the migration of GFP-expressing cells was inhibited and the distribution of these cells was limited to the printed spot. (Reproduced from 6 with permission from the Royal Society of Chemistry.)

rhodamine-labeled fibronectin to identify the individual spots on the glass slide that corresponded to the transfection complex, and we used the GFP as a marker of transfectants (**Fig. 13.1a**). When a transfected siRNA had no negative affect on cell migration, we anticipated that those GFP-positive cells that harbored siRNA would migrate normally and fluoresce due to GFP and would be scattered over a wide area (**Fig. 13.1b**). By contrast, when an siRNA suppresses the expression of a migration-related gene, we anticipated that the migration of cells that express GFP would be inhibited and the distribution of such cells would be limited to the rhodamine-labeled printed region (**Fig. 13.1c**). For easy visual discrimination of significant results in our assay, we chose NBT-L2b cells, which migrate rapidly and unidirectionally (**Fig. 13.2**). By contrast, HeLa cells were immobile.

To evaluate the effects of siRNAs on the cell chip, we prepared two siRNAs for each targeted gene, namely, the *pxn* (for *paxillin*), *actr2* (for *arp2*, a homolog of actin-related

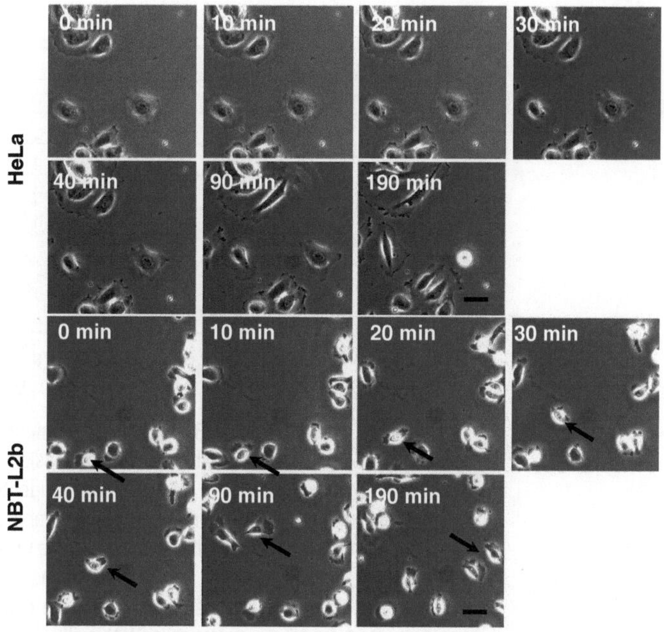

Fig. 13.2. Migration of HeLa cells and NBT-L2b cells on type I collagen-coated glass dishes. Cells were incubated for 190 min on type I collagen-coated dishes and phase-contrast images were recorded at 10-min intervals. *Arrows* indicate migrating cells; HeLa cells remained stationary. Scale bar, 100 μm.

protein 2), *actr3* (for *arp3*, a homolog of actin-related protein 3), and *wasf2* (for member 2 of the WAS protein family) genes. All these genes are known to be involved in cell migration (7, 8). We analyzed captured phase-contrast and fluorescence time-lapse images by manually tracking migrating cells on a large number of time-lapse images (**Fig. 13.3a**). However, cell tracking with image analysis software, such as ImageJ (http://rsb.info.nih.gov/ij/), is very time-consuming. Therefore, we also used an analytic method, using snapshots instead of time-lapse images. To distinguish transfectants that harbored an siRNA targeted to a migration-related gene from those that harbored an ineffective siRNA, we posited that the distance migrated by each cell corresponded to the distance from the center of the printed area to the location of the GFP-positive cell, and we used a permutation test to compare distances (Data not shown and see ref (6) and **Note 1**).

3.2. Cell Culture

1. Culture the NBT-L2b cells in MEM supplemented with 10% fetal bovine serum and nonessential amino acids at 37°C in an atmosphere of 5% CO_2 in air. To avoid contamination, add antibiotics.
2. Usually, cells are subcultured when they approach approximately 80% confluence on 100-mm tissue-culture dishes

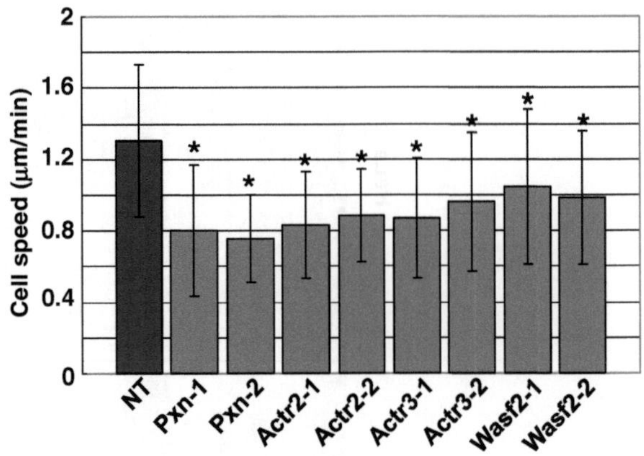

Fig. 13.3. Data analysis. siRNAs targeted to migration-related genes or nontarget siRNA (NT) were cotransfected with the GFP expression vector into NBT-L2b cells. First ($t = 0$) fluorescent images were recorded for the identification of transfectants and phase-contrast images were recorded at 5-min intervals for 3 h to follow the migration of cells. The speed of migration of individual cells was analyzed with ImageJ software with a manual tracking plug-in. The speeds of movement of nontarget siRNA-transfected cells and other siRNA-transfected cells were compared by the Mann–Whitney *U*-test ($^*p < 0.01$ for NT versus each siRNA), with $n = 25$–100 cells for each assay of gene knockdown. (Reproduced from 6 with permission from the Royal Society of Chemistry.)

(*see* **Note 2**). Prepare prewarmed 0.05% trypsin/0.50 mM EDTA solution, and MEM supplemented with 10% fetal bovine serum and nonessential amino acids at 37°C.

3. After aspiration of the culture medium, gently wash the cell with 8 mL D-PBS (−), add 2 mL of prewarmed 0.05% trypsin/0.53 mM EDTA solution and then incubate the dish at 37°C for 10 min (*see* **Note 3**).
4. Add 8 mL of MEM to neutralize the effects of trypsin, transfer the cell suspension to a 15-mL tube and then centrifuge at 800–1, 200×*g* for 3 min.
5. Remove the supernatant, resuspend the cell in 8 mL of medium, and then place the cells in a 100-mm tissue-culture dish.

3.3. Glass Slides Coated with Type I Collagen

1. Soak glass slides in HMDS at room temperature for 3 h (*see* **Note 4**).
2. Place the glass slides in a beaker of absolute ethanol and wash them for 10 min in an ultrasonic washing machine.
3. Discard the ethanol and wash the slides with distilled water for 10 min in the ultrasonic washing machine.
4. Dry the glass slides completely, one by one, at room temperature and sterilize slides by exposure to a UV germicidal lamp in a clean bench for 30 min.

5. Dilute the 0.01% (w/v) solution of type I collagen to 0.001% with PBS (pH 7.0).
6. Place the glass slides in the 0.001% solution of type I collagen and incubate them at 37°C for more than 3 h but less than 10 h.
7. Wash the incubated glass slides twice with an excess of distilled water, shake off the excess water, dry at room temperature, and store with a desiccant at room temperature.

3.4. Rhodamine-Labeled Fibronectin

1. Dissolve one vial of fibronectin (1 mg/vial) in 250 μL of BupH™ borate buffer from the EZ-Label™ Rhodamine Protein-Labeling Kit. Transfer the solution of fibronectin to a light-shielded reaction tube for labeling.
2. Reconstitute the "Rhodamine Labeling Reagent" by puncturing the foil and adding 100 μL of dimethylformamide (DMF). Pipette the mixture up and down until the dye is completely dissolved.
3. Transfer 25 μL of the solution of the dye to the light-shielded reaction tube.
4. Mix well and incubate at room temperature for 1 h.
5. Place aliquots of 220 μL of the mixture in a Slide-A-Lyzer® Mini Dialysis unit from the EZ-Label™ Rhodamine Protein-Labeling Kit. Cap the dialysis unit and place in the flotation device for dialysis.
6. Float the dialysis unit in a beaker with 1 L of PBS.
7. Place the beaker on a stirring plate and stir at low speed such that the flotation device is not submerged.
8. Dialyze the samples overnight at room temperature.
9. Remove each sample and bring the volume to 500 μL with PBS (final concentration, 2 mg/mL). Store samples at 4°C.

3.5. Preparation of a Cell Chip

1. Prepare DMEM without serum and phenol red, siRNA, Lipofectamine™ 2000 transfection reagent, rhodamine-labeled fibronectin (2 mg/mL), and a 0.1% (w/v) solution of gelatin. Subsequently, mix the following in an Eppendorf tube on ice:
   - 7.5 μL DMEM
   - 1 μL of pEGFP-N1 vector (1 μg/μL)
   - 7 μL of a 20 μM solution of siRNA
2. Spin the solutions down and mix by pipetting up and down 20 times (*see* **Note 5**).
3. Add 2 μL of Lipofectamine™ 2000 transfection reagent to the Eppendorf tube, spin down, and mix by pipetting up and down 20 times.

4. Incubate for 30 min at room temperature (15–25°C) to allow formation of the transfection complex.
5. Add 5 μL of the solution of rhodamine-labeled fibronectin and 2.5 μL of the 0.1% solution of gelatin to the sample, spin down, and then mix by pipetting up and down 20 times (*see* **Note 6**).
6. Spot the sample(s) on a collagen-coated glass slide using the ink-jet printer according to the supplier's instructions (*see* **Notes 7** and **8**). Spots of 15–25 nL, which are suitable for effective transfection, dry immediately in air as circles of 600–800 nm diameter (*see* **Notes 9** and **10**), and the amount of siRNA per spot is approximately 112 fmol.

3.6. Motility Assay

1. Collect NBT-L2b cells as described in **Section 3.1** and suspend in complete DMEM/F12 Ham's medium at 1×10^5 cells/mL. Place the printed glass slides in a multiwell dish (*see* **Note 11**).
2. Drop 2 mL of the cell suspension of cells slowly onto each printed glass slide, and then incubate the slides at room temperature for 1 h (*see* **Note 12**).
3. Slowly add 5 mL of prewarmed complete DMEM/F12 Ham's medium to each well in the dish.
4. Incubate the dish at 37°C for 24 h. Depending on the siRNA effects, incubation time can be extended by several hours.
5. Place the glass slides in the prewarmed culture chamber that fits the Programmable Cellular Image Tracer (Olympus) and add 10 mL of prewarmed fresh complete DMEM/F12 Ham's medium (*see* **Notes 13** and **14**).
6. Set the culture chamber in the image tracer. After acquisition of fluorescent images due to rhodamine, record both phase-contrast images and images due to the fluorescence of GFP in time-lapse mode at 5-min intervals for 3 h. After 21 h cell incubation in the Programmable Cellular Image Tracer, record final phase-contrast images and images due to the fluorescence of GFP and rhodamine.

3.7. Analysis of Data

1. The speed of migration of individual cells is analyzed with ImageJ software with a manual-tracking plug-in.
2. After incubation of cells for 24 or 48 h on collagen-coated slides with siRNAs, phase-contrast and fluorescence images are recorded as snapshots, as described in **Section 3.5**. The distances migrated by individual cells from the center of each printed spot to the sites of their final fluorescent signals are calculated with ImageJ software. A permutation test is applied to data sets for comparison of migrated distances (*see* **Note 1**)

4. Notes

1. Two independent siRNAs directed against the paxillin gene (Pxn-1 and Pxn-2 siRNAs) were cotransfected into NBT-L2b cells with the GFP expression vector. After incubation for 24 or 48 h, fluorescence due to GFP was monitored. We applied a permutation test to data sets as follows: nontarget siRNA, 24 h ($n = 57$) and nontarget siRNA, 48 h ($n = 61$); Pxn-1 siRNA, 24 h ($n = 99$) and Pxn-1 siRNA, 48 h ($n = 133$); and Pxn-2 siRNA, 24 h ($n = 91$) and Pxn-2 siRNA, 48 h ($n = 120$) (for details, *see* **Fig. 13.3**). For each analysis, a small number of results were selected at random from the large data set and 20 data sets were prepared. For example, to compare nontarget siRNA-transfected cells at 24 h with those at 48 h, we selected 57 samples at random from the data set for nontarget siRNA-transfected cells at 48 h. Then, using the data set mentioned above, we applied the permutation test to examine the statistical significance of differences.
2. The migratory ability of NBT-L2b cells is influenced by culture conditions. Over-growth and long-term culture result in dramatic decreases in the speed of cell migration. We recommend the preparation of numerous cell stocks prior to the start of motility assays and the use of a fresh stock for each assay.
3. The attachment of NBT-L2b cells to dishes is very strong when compared to other cell lines. Therefore, cover the cells well with the trypsin/EDTA solution and incubate at 37°C for at least 10 min.
4. HMDS is sensitive to humidity and should be stored with an aliquot of Molecular Sieves 4A (Wako) with 25 g of Molecular Sieves 4A per 100 mL of HMDS. HMDS should be handled in a fume hood.
5. Differences among cell lines and target genes can affect the efficiency of gene knockdown by siRNA. Therefore, appropriate siRNA concentrations should be determined for each experiment. The volume of siRNA can range from 1 to 10.5 μL. Add DMEM to bring the total volume up to 13.5 μL.
6. The purity and stability of fibronectin influence the efficiency of transfection. Moreover, some batches of fibronectin allow cross-transfections among different spots of siRNA. We recommend the addition of 0.1% gelatin to prevent such cross-transfection (4).

7. An ink-jet microarray printing device is preferable to a pin-type microarrayer because droplets made by a pin-type microarrayer dry as torus-shaped spots.
8. In the absence of a printing device, spots on glass slides can be made by hand. In this case, immerse the tip of a micropipette in the sample and take up the solution by capillary action. A droplet is spotted when the tip touches against the surface of the glass slide.
9. When printing many spots, please pay attention to possible evaporation of the sample.
10. Slides are stored in a vacuum-sealed plastic bag with silica gel at –20°C.
11. In our study with NBT-L2b cells, we seeded 2×10^5 cells (1×10^5 cells/mL in 2 mL of DMEM/F12 Ham's medium) on each slide. However, it is necessary to determine optimal conditions for each experiment.
12. Cell migration is inhibited in a confined environment and, thus, an appropriate seeding procedure is important for reproducible results. For uniform seeding of cells, the minimum volume of cell suspension should be used. Do not move the multiwell dish until cell adhesion is complete.
13. We use a Programmable Cellular Image Tracer, a prototype of CELAVIEW RS100 (Olympus) to capture images. The chamber is provided as ancillary equipment for the Programmable Cellular Image Tracer.
14. To avoid out-of-focus images, the culture chamber and medium should be warmed before they are placed in the Programmable Cellular Image Tracer.

Acknowledgment

This study was performed as part of "Analytical technology development of dynamic networks inside living cells/Technology development for the real time analysis of the network in the living cells," funded by the New Energy and Industrial Technology Development Organization (NEDO) of Japan.

References

1. Yamaguchi, H. and Condeelis, J. (2007) Regulation of the actin cytoskeleton in cancer cell migration and invasion. *Biochim Biophys Acta*, **1773**, 642–652.
2. Boyden, S. (1962) The chemotactic effect of mixtures of antibody and antigen on polymorphonuclear leucocytes. *J Exp Med*, **115**, 453–466.

3. Nobes, C.D. and Hall, A. (1999) Rho GTPases control polarity, protrusion, and adhesion during cell movement. *J Cell Biol*, **144**, 1235–1244.
4. Ziauddin, J. and Sabatini, D.M. (2001) Microarrays of cells expressing defined cDNAs. *Nature*, **411**, 107–110.
5. Yoshikawa, T., Uchimura, E., Kishi, M., Funeriu, D.P., Miyake, M., and Miyake, J. (2004) Transfection microarray of human mesenchymal stem cells and on-chip siRNA gene knockdown. *J Contol Release*, **96**, 227–232.
6. Onuki-Nagasaki, R., Nagasaki, A., Hakamada, K., Fujita, S., Miyake, M., and Miyake, J. (2008) On-chip screening method for cell migration genes based on a transfection microarray. *Lab Chip*, **8**, 1502–1506.
7. Takenawa, T. and Miki, H. (2001) WASP and WAVE family proteins: key molecules for rapid rearrangement of cortical actin filaments and cell movement. *J Cell Sci*, **114**, 1801–1809.
8. Schaller, M. (2001) Paxillin: a focal adhesion-associated adaptor protein. *Oncogene*, **20**, 6459–6472.

Chapter 14

Intron-Mediated RNA Interference, Intronic MicroRNAs, and Applications

Shao-Yao Ying, Chen Pu Chang, and Shi-Lung Lin

Abstract

Nearly 97% of the human genome is non-coding DNA. The intron occupies most of it around the gene-coding regions. Numerous intronic sequences have been recently found to encode microRNAs (miRNAs), responsible for RNA-mediated gene silencing through RNA interference (RNAi)-like pathways. miRNAs, small single-stranded regulatory RNAs capable of interfering with intracellular messenger RNAs (mRNAs) that contain either complete or partial complementarity, are useful for the design of new therapies against cancer polymorphism and viral mutation. This flexible characteristic differs from double-stranded siRNAs (small interfering RNAs) because more rigid complementarity is required for siRNA-induced RNAi gene silencing. miRNAs were firstly discovered in *Caenorhabditis elegans* as native RNA fragments that modulate a wide range of genetic regulatory pathways during embryonic development. Currently, varieties of miRNAs are widely reported in plants, animals, and even microorganisms. Intronic miRNA is a new class of miRNAs derived from the processing of gene introns. The intronic miRNAs differ from previously described intergenic miRNAs due to the requirement of type II RNA polymerases (Pol-II) and spliceosomal components for their biogenesis. Several kinds of intronic miRNAs have been identified in *C. elegans*, mouse, and human cells. However, neither function nor application has been reported. Here, we show that, for the first time, intron-derived miRNAs are able to induce RNA interference not only in human and mouse cell lines but also in zebrafish, chicken, and mouse, which demonstrates the evolutionary preservation of the intron-mediated gene silencing through miRNA functionality in cell and in vivo. Based on this novel mechanism, numerous biomedical applications have been developed, including cosmetic skin whitening, transgenic animal generation, anti-viral vaccination and therapy, and somatic cell reprogramming into induced pluripotent stem (iPS) cells. These findings suggest an important miRNA-mediated gene regulatory system, which fine-tunes a variety of cellular and developmental events through the mechanism of RNAi-like gene silencing.

Key words: RNA interference, intronic microRNA, microRNA, nonsense-mediated RNA decay, RNA-induced gene silencing complex, stem cells.

M. Sioud (ed.), *RNA Therapeutics*, Methods in Molecular Biology 629,
DOI 10.1007/978-1-60761-657-3_14, © Springer Science+Business Media, LLC 2010

1. Introduction

The first microRNA (miRNA) molecules, *lin-4* and *let-7*, were identified in 1993 (1). Subsequently, the biogenesis, functional significance, and target genes of many miRNAs were reported. Most of these early miRNAs were located in the non-coding regions between the genes and transcribed by un-identified promoters; these are intergenic miRNA. All miRNAs studied at this stage were recognized as the intergenic miRNA. Ambros et al. (1) discovered some tiny non-coding RNA derived from the intronic regions of gene transcripts. Meanwhile, Lin et al. (2) proved the biogenesis and gene silencing mechanism of these intron-derived miRNAs, which provides the first functional evidence for a new miRNA category, the intronic miRNA. As shown in **Table 14.1**, several intronic miRNA molecules have been identified in *Caenorhabditis elegans*, mouse, and human genomes (1–3). Some of their functions have been related to RNA interference (RNAi).

Table 14.1
Currently studied intronic microRNA (Id-miRNA)

miRNA	Species	Host gene (intron) [#]	Target gene(s)
miR-2a, -b2	Worm	Spi	
miR-7b	Mammal	Pituitary gland-specific factor 1A (2) [NM174947]	Paired mesoderm homeobox protein 2b; HLHm5
miR-10b	Mammal	Homeobox protein HOX-4 (4)	
miR-11	*Drosophila*	E2F	
miR-13b2	*Drosophila*	CG7033	
miR-15b, -16-2	Mammal	Chromosome-associated polypeptide C	
miR-17-92	Human	C13orf25	
miR-25, -93, -106b	Mammal	CDC47 homologue (13)	
miR-26a1, -26a2, -26b	Vertebrate	Nuclear LIM interactor-interacting factor 1, 2, 3	
miR-28	Human	LIM domain-containing preferred translocation partner in lipoma [NM005578]	
miR-30c1, -30e	Mammal	Nuclear transcription factor Y subunit γ (5)	Transcription factor HES-1; PAI-1 mRNA-binding protein

Table 14.1 (continued)

miRNA	Species	Host gene (intron) [#]	Target gene(s)
miR-33	Vertebrate	Sterol regulatory element binding protein-2 (15)	RNA-dependent helicase p68; NAG14 protein
miR-101b	Human	RNA 3′-terminal phospate cyclase-like protein (8)	
miR-103, -107	Human	Pantothenate kinase 1, 2, 3	
miR-105-1, -105-2, -224	Mammal	γ-aminobutyric-acid receptor α-3 subunit precursor, epsilon subunit precursor	
miR-126, -126*	Mammal	EGF-like, Notch4-like, NEU1 protein (6) [NM178444]	
miR-128b	Mammal	cAMP-regulated phosphoprotein 21 (11)	
miR-139	Mammal	cGMP-dependent 3′,5′-cyclic phosphodiesterase (2)	
miR-140	Human	NEDD4-like ubiquitin-protein ligase WWP2 (15)	
miR-148b	Mammal	Coatomer ζ-1 subunit	
miR-151	Mammal		
miR-152	Human	Coatomer ζ-2 subunit	N-myc proto-oncogene protein; noggin precursor
miR-153-1, -153-2	Human	Protein–tyrosine phosphatase N precursors	
miR-208	Mammal	Myosin heavy chain, cardiac muscle α isoform (28)	
miR-218-1, -218-2	Human	Slit homologue proteins [NM003062]	

Intron occupies the largest proportion of non-coding sequences in the protein-coding DNA of a genome. The transcription of the genomic protein-coding DNA generates precursor messenger RNA (pre-mRNA), which contains four major parts: the 5′-untranslational region (UTR), protein-coding exon, non-coding intron, and the 3′-UTR. Generally, both 5′- and 3′-UTR can be seen as a kind of intron extension. However, their processing during mRNA translation is different from the intron located between two protein-coding exons, called the in-frame intron. The in-frame intron can size up to several 10 kb nucleotides and was thought to be a huge genetic waste in gene transcripts. Recently, this misunderstanding was changed after the finding of the intronic miRNA. miRNA is usually about 18–27 oligonucleotide long, capable of either directly degrading its intracellular messenger RNA (mRNA) target or suppressing the protein translation of its targeted mRNA, depending on the complementary between the miRNA and its target. In this way, the intronic miRNA is similar to the previously delineated intergenic miRNAs, structurally and functionally. But slightly differs from intergenic miRNAs due to the requirement of Pol-II and RNA splicing components for biogenesis (2, 4, 5). Approximately, 10–30% of a spliced intron are exported into cytoplasm with a moderate half-life (6).

RNA interference (RNAi) is a post-transcriptional gene silencing mechanism in eukaryotes, which can be triggered by small RNA molecules such as miRNAs and siRNAs. These small RNA molecules usually function as a gene silencer that interferes with intracellular expression of genes either completely or partially complementary to the small RNAs. In principle, siRNAs are double-stranded RNAs capable of degrading target gene transcripts with almost perfect complementarity (7, 8). Unlike the stringent complementarity of siRNAs to their RNA targets, miRNAs are single stranded and able to pair with target RNAs that have partial complementarity to the miRNAs (9, 10). Numerous natural miRNAs are derived from hairpin-like RNA precursors in almost all eukaryotes, including yeast (*Schizosaccharomyces pombe*), plant (*Arabidopsis* spp.), nematode (*C. elegans*), fly (*Drosophila melanogaster*), mouse, and human; they function in defense against viral infections and regulation of certain gene expressions during development (11–21). In contrast, natural siRNAs were abundantly discovered in plants and low-level animals (worms and flies), but rarely in mammals (10). Given that miRNAs are abundant in eukaryotes, they have been used to design novel therapeutics against cancers and viral infections (22, 23). In fact, gene silencing mechanisms involving miRNAs have been proposed to be an intracellular defense system for eliminating undesired transgenes and foreign RNAs, such as viral infections and retrotransposon activities (22, 24).

2. Biogenesis and Definition

The definition of intronic miRNA is based on two factors; first, they must share the same promoter with their encoded genes, and second, they are spliced out of the transcript of their encoded genes and further processed into mature miRNAs (**Fig. 14.1**). Although some of the currently identified miRNAs are encoded in the genomic intron region of a gene, but in the opposite orientation to the gene transcript. These miRNAs are not intronic miRNAs because they neither share the same promoter with the

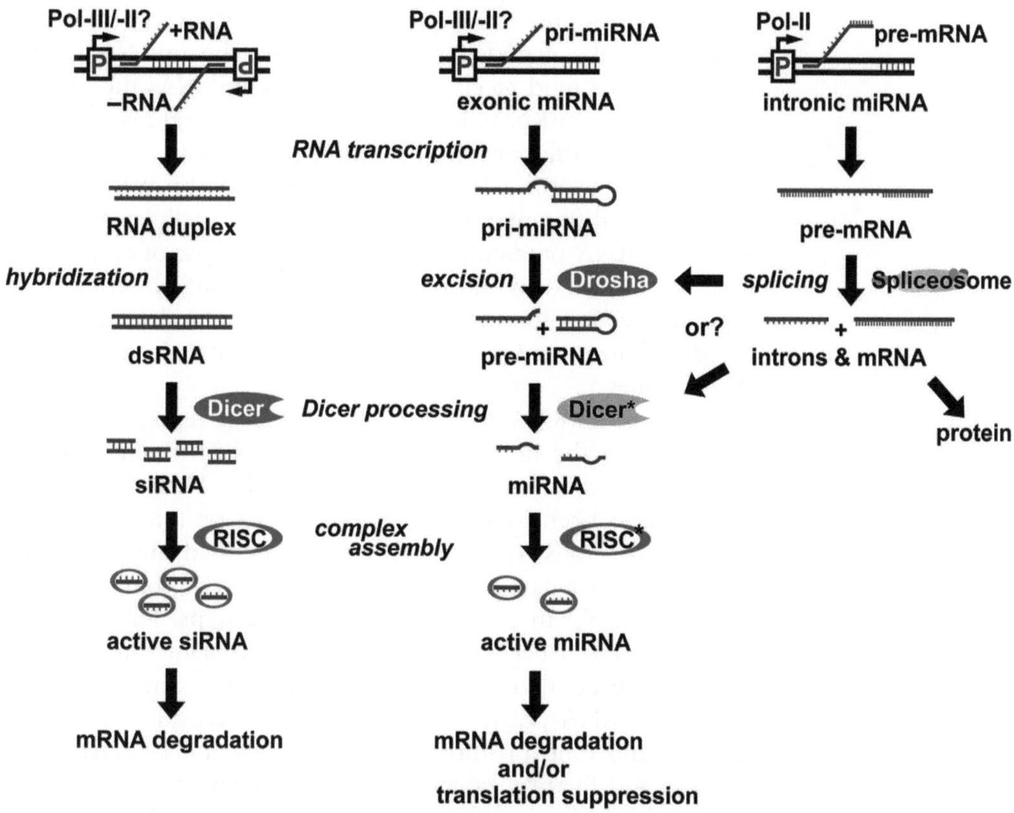

Fig. 14.1. Comparison of biogenesis and RNAi mechanisms among siRNA, intergenic (exonic) miRNA, and intronic miRNA. SiRNA is likely formed by two perfectly complementary RNAs transcribed from two different promoters (remains to be determined) and further processing into 19–22 bp duplexes by RNase III-familial endonuclease, Dicer. The biogenesis of intergenic miRNAs, e.g., *lin*-4 and *let*-7, involves a long transcript precursor (pri-miRNA), which is probably generated by Pol-II or Pol-III RNA promoters, while intronic miRNAs are transcribed by the Pol-II promoters of its encoded genes and co-expressed in the intron regions of the gene transcripts (pre-mRNA). After RNA splicing and further processing, the spliced intron may function as a pri-miRNA for intronic miRNA generation. In the nucleus, the pri-miRNA is excised by Drosha RNase to form a hairpin-like pre-miRNA template and then exported to cytoplasm for further processing by Dicer* to form mature miRNAs. The Dicers for siRNA and miRNA pathways are different. All three small regulatory RNAs are finally incorporated into a RNA-induced silencing complex (RISC), which contains either strand of siRNA or the single-strand of miRNA. The effect of miRNA is considered to be more specific and less adverse than that of siRNA because only one strand is involved. On the other hand, siRNAs primarily trigger mRNA degradation, whereas miRNAs can induce either mRNA degradation or suppression of protein synthesis. It depends on the sequence complementarity to the target gene transcripts.

gene nor need to be released from the gene transcript by RNA splicing. For the transcription of those miRNAs, their promoters are located in the antisense direction to the gene probably using the gene transcript as a potential target for the antisense miRNAs. A good example is *let-7c*, which was found to be an intergenic miRNA located in the antisense region of a gene intron. Current computer programs specialized for miRNA prediction cannot distinguish the intronic miRNAs from the intergenic miRNAs. Because intronic miRNAs are encoded in the gene transcript precursors (pre-mRNAs) and share the same promoter with the encoded gene transcripts, the miRNA prediction programs tend to classify the intronic miRNAs the same as the intergenic miRNAs located in the exonic regions of genes. Due to the difference in biogenic mechanisms between the intronic and exonic miRNAs, these two types of miRNAs may function as different gene regulators in the cellular function. Thus, a miRNA-prediction program, with database of non-coding sequences located in the protein-coding pre-mRNA regions, is urgently needed for thorough screening and understanding of the distribution and variety of hairpin-like intronic miRNAs in the genomes.

Intronic miRNA is a new class of small regulatory RNAs derived from the non-coding DNA regions of a gene, such as intron, 5′- and 3′-untranslated region (UTR). Many introns and UTR's have been identified to contain tri- or tetra-nucleotide repeat expansions, capable of being transcribed and processed into repeat-associated miRNAs (25, 26). The biogenesis of intronic miRNA in eukaryotes involves five steps. First, miRNA is generated as a part of a long primary precursor miRNA (pri-miRNA), located in the intron or UTR of a gene transcript, by type II RNA polymerases (Pol-II) (2). Second, after intron splicing, the long pri-miRNA is excised by spliceosomal components and further processed by certain non-*Drosha* RNaseIII endonucleases to form precursor miRNA (pre-miRNA) (2, 27, 28). Third, the pre-miRNA is exported out of the nucleus by Ran-GTP and a receptor of exportins (29). Fourth, once in the cytoplasm, *Dicer*-like nucleases cleave the pre-miRNA to form mature miRNA. Lastly, the mature miRNA is assembled into a ribonuclear particle (RNP) to form a RNA-induced silencing complex (RISC) or RNA-induced transcriptional silencing (RITS) complex for executing the RNA interference (RNAi)-like gene silencing mechanisms (27, 30). Although the in vitro model of siRNA-associated RISC assembly has been reported, the link between the final miRNA maturation and RISC assembly remains to be determined. Nevertheless, there were observations that the characteristics of Dicer and RISC in the biogenesis and degradation of siRNA and miRNA were different (31, 32). In zebrafish, we have recently observed that the stemloop structure of pre-miRNAs is involved in the strand selection for mature miRNA, during RISC assembly (27). These findings suggest that the duplex structure of siRNA may be not essential for the assembly of miRNA-associated RISC in vivo. The biogenesis of miRNA and siRNA seems to share certain similarities. However, the miRNA mechanisms previously proposed were based on the model of siRNA, and thus one must distinguish their individual properties and differences in these two types of RNAs to understand the evolutional and functional relationships of these gene silencing pathways.

In addition, the differences may provide insight into the prevalence of native siRNAs in invertebrates and, sporadically, in mammals.

3. Intronic MicroRNA and Diseases

The majority of human gene transcripts contain introns which changes frequently as observed in the clinical and circumstantial malfunction. Numerous introns encode miRNAs, which are involved in RNAi-related chromatin silencing mechanisms. Over 90 intronic miRNAs have been identified to use the bioinformatic approaches to date, but the functions of the vast majority of these RNA agents remain unclear (3). Based on the strictly expressive correlation between intronic miRNAs and their encoded genes, it is possible to speculate the levels of condition-specific, time-specific, and tissue-specific expression of miRNAs as determined by the intracellular modulation of the encoded gene. This interpretation accounts for more accurate genetic expression of various traits, and any dysregulation of this miRNA-encoded gene correlation will result in genetic diseases. For instance, many introns contain tri- or tetranucleotide repeat expansions capable of being transcribed and processed into repeat-associated miRNAs (25). They play an important role in the modulation of epigenetic alterations of genes, which are involved in several genetic diseases that have been identified to be caused by triplet repeat expansions in specific single genes, collectively termed triplet repeat expansion diseases (TREDs). TREDs include fragile X syndrome (FXS), Huntington's disease (HD), myotonic dystrophy (DM), and a number of spinocerebellar ataxias (SCAs).

One of the early intron-related TREDs was observed in monozygotic twins who frequently demonstrate slightly, but definitely, distinguishing disease susceptibility and physiological behavior. It was determined that a long CCTG expansion in the intron 1 of a zinc finger protein ZNF9 gene of the twin with higher susceptibility was correlated to type 2 myotonic dystrophy (DM2) (33). Since the repeat expansion motifs often obtain high affinity to certain RNA-binding proteins, the interfering role of intron-derived repeat expansion fragments in DM and HD has been recently found to trigger RNAi-like gene silencing (26). However, the mechanism involved is still unclear. Another more established example is fragile X syndrome (FXS), which represents about 30% of human inherited mental retardation, and affects approximately 1 in every 2,000 males and 1 in every 4,000 females. This disease is caused by a dynamic pre-mutation [expansion of microsatellite-like trinucleotide—(cytosine–guanine–guanine)—repeats or r(CGG)] at an inherited fragile site on the long arm of the X chromosome, where the *FMR1* gene is located (34–36). Because this pre-mutation is dynamic, it can change in length and severity from generation to generation, from person to person, and even within a given person. Patients with the FXS usually have an increased number of r(CGG) (>230) copies in the 5′-UTR of the *FMR1* gene, and this CpG-rich r(CGG) expansion region is often heavily methylated (37). Such r(CGG) expansion and methylation lead to physical, neurocognitive, and emotional characteristics linked to the inactivation of the *FMR1* gene and the deficiency of its protein product (30). Two theories have been proposed to explain this *FMR1* methylation mechanism in FXS. First, Handa et al. (25) suggested that non-coding RNA transcripts, transcribed from the *FMR1* 5′-UTR r(CGG) expansion, can be folded into RNA hairpins and further processed by *Dicer* RNases to form microRNA (miRNA)-like molecules directed against the *FMR1* gene expression. Second, Jin et al. (38) proposed that RNA-mediated gene methylation may occur in the CpG regions of the *FMR1* r(CGG) expansion, which are targeted by hairpin RNAs derived from the 3′-UTR of the *FMR1* expanded allele transcript. The *Dicer*-processed hairpin RNA may trigger the formation of RNA-induced initiator of transcriptional silencing (RITS) on homologous r(CGG) sequences and leads to hypermethylation of the chromosomal *FMR1* gene locus in 99% of FXS patients.

Using a fish-compatible retrovector, *pGABAR2-rT* containing the *fmr1* 5′-UTR r(CGG) expansion, Lin et al. (30) were able to deliver an intronic transgene, *SpRNAi-rGFP* in zebrafish. One of the most abundant ramRNA species is derived from the *fmr1* 5′-untranslational r(CGG) expansion region approximately 65-nucleotide upstream of the translational start codon (accession number NM152963). It is capable of causing the r(CGG)-methylation and inactivation of the *FMR1* gene. This ramRNA is specially characterized by its unique pre-miRNA structure, consisting of (a) multiple loops and matched CGG sites in the stem of a relatively long hairpin precursor, (b) a nuclear import signal (NIS) motif, and (c) multiple CCG-rich DNA-binding motifs. Recent FXS studies using this ramRNA-induced disease model have found that formation of synaptic connection was markedly reduced among the dendrites of the *FMR1*-deficient neurons, similar to the diseased hippocampal neurons in human FXS. *FMR1* deficiency often caused synaptic deformity in the neurons essential for cognition and memory activities, damaging the activity-dependent synaptic neuron plasticity. Furthermore, the metabotropic glutamate receptor (mGluR)-activated long-term depression (LTD) was augmented after *FMR1* inactivation, suggesting that exaggerated LTD may be responsible for aspects of abnormal neuronal responses in FXS, such as autism. These findings support a novel disease model in which mature ramR-

NAs, originating from the triplet repeat expansion of a gene, can reversely bind back to the corresponding triplet repeat regions of the gene. More triplet repeats in the gene generate more mature ramRNAs. DNA methylation takes place in the triplet repeat regions when ramRNAs bind to their targeted gene, consequently inactivate the expression of this gene. The animal model may provide a useful tool for studying TREDs involving epigenetic alterations.

4. Man-Made Intronic MicroRNA

To understand the disease caused by dysregulation of miRNAs, an artificial expression system is needed to recapitulate the function and mechanism of the miRNA in vitro and in vivo. The same strategy may be used to design and develop therapies for the disease. Previously, several vector-based RNAi expression systems have been developed, using type III RNA polymerase (Pol-III)-directed transcription activities, to generate more stable RNAi efficacy in vitro and in vivo (39–43). Although these approaches have succeeded in maintaining constant gene silencing efficacy in vivo, their delivery strategies failed to target a specific cell population due to the ubiquitous existence of Pol-III activity in all cell types. Moreover, the requirement of using Pol-III RNA promoters, e.g., U6 and H1, for small RNA expression causes another drawback. The read-through side effect of Pol-III occurs on a short transcription template. If without proper termination, large RNA products longer than desired (18–25 bp) can be synthesized and subsequently cause unexpected interferon cytotoxicity (44, 45). Such a problem can also result from the competitive conflict between the Pol-III promoter and another vector promoter (i.e., LTR and CMV promoters). Sledz et al. and we have found that high dosage of siRNAs (e.g., >250 nM in human T cells) was able to cause strong cytotoxicity similar to that of long double-stranded dsRNAs (46, 47). This toxicity is due to the double-stranded structure of siRNAs and dsRNAs, which activates the interferon-mediated non-specific RNA degradation and programmed cell death through the signaling of PKR and 2-5A systems. It is known that interferon-induced protein kinase PKR can trigger cell apoptosis, while activation of interferon-induced 2′,5′-oligoadenylate synthetase (2-5A) system leads to extensive cleavage of single-stranded RNAs (i.e., mRNAs) (48). High siRNA/shRNA concentrations generated by the Pol-III-directed RNAi systems can also over-saturate the cellular microRNA pathway and therefore caused global miRNA inhibition and cell death (49). In contrast, a Pol-II-directed intronic miRNA expression system does not show these problems due to their precise regulation under cellular RNA splicing and nonsense-mediated decay (NMD) mechanisms (50–54). These degrade excessive RNA accumulation to prevent potential cytotoxicity. For the therapeutical purpose in vivo, the Pol-II-directed intronic miRNA expression system is likely a better solution than the Pol-III-based siRNA/shRNA expression systems.

The intron-derived miRNA system is activated in a specific cell type under the control of a type II RNA polymerases (Pol-II)-directed transcriptional machinery. To overcome the Pol-III-mediated siRNA side effects, we have successfully developed a new Pol-II-based miRNA biogenesis strategy, employing intronic miRNA molecules (2) to knock down more than 85% of selected oncogene function or viral genome replication (22, 23, 47). Because of its flexibility in binding with partially complementary mRNA targets, miRNA can serve as an anti-cancer drug or vaccine to achieve a major breakthrough in the treatments of cancer polymorphisms and viral mutations. We are the first research group to discover the biogenesis of miRNA-like precursors from the 5′-proximal intron regions of gene transcripts (pre-mRNAs) produced by the mammalian Pol-II. The intronic miRNA is co-expressed with its encoding gene in specific cells, which activates the promoter and expresses the gene. It has been noted that a spliced intron was not completely digested into monoribonucleotides for transcriptional recycling since approximately 10–30% of the intron was found in the cytoplasm with a moderate half-life (6, 55). This type of miRNA generation has been found to rely on the coupled interaction of nascent Pol-II-mediated pre-mRNA transcription and intron excision, occurring within certain nuclear regions proximal to genomic perichromatin fibrils (22, 56–58). After Pol-II RNA processing and splicing excision, some of the intron-derived miRNA fragments can form mature miRNAs and effectively silence the target genes through the RNAi mechanism, while the exons of pre-mRNA ligate together to form a mature mRNA for protein synthesis(**Fig. 14.2a**). Because miRNAs are single-stranded molecules insensitive to PKR- and 2-5A-induced interferon systems, the utilization of this Pol-II-mediated miRNA generation can be safe in vitro and in vivo without the cytotoxic effects of dsRNAs and siRNAs. These findings signify new functions for mammalian introns in intracellular miRNA generation and gene silencing, and it can be used as a tool for analysis of gene functions and development of gene-specific therapeutics against cancers and viral infections.

Using artificial introns carrying hairpin-like miRNA precursors (pre-miRNA), we have successfully generated mature miRNA molecules with full capacity in triggering RNAi-like gene silencing in human prostate cancer LNCaP, human cervical cancer HeLa, rat neuronal stem HCN-A94-2 cells (2, 59, 60), zebrafish, chicken, and mouse in vivo (30, 61). As shown in **Fig. 14.2b**,

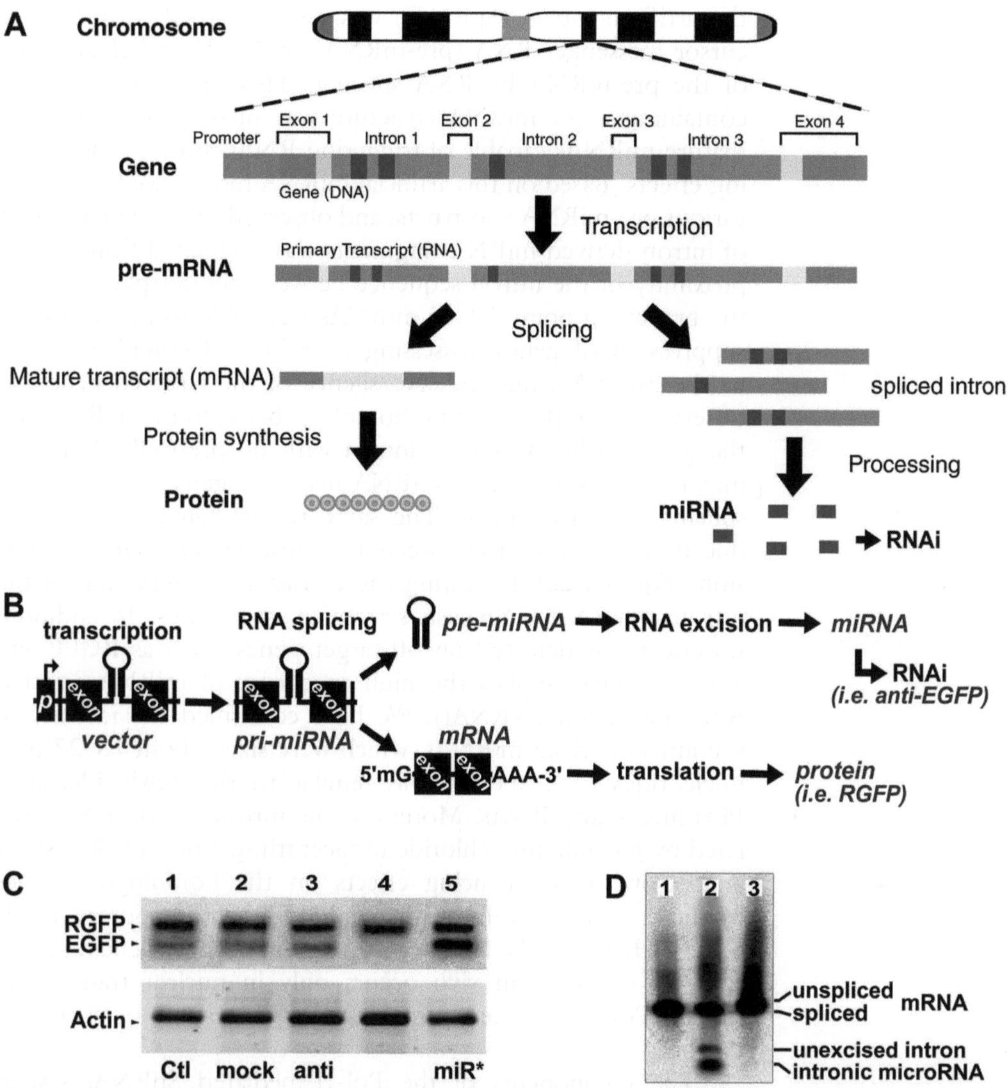

Fig. 14.2. Biogenesis and function of intronic miRNAs. (**a**) The native intronic miRNA is co-transcribed with a precursor messenger RNA (pre-mRNA) by Pol-II and cleaved out of the pre-mRNA by a RNA splicing machinery, spliceosome. The spliced intron with hairpin-like secondary structures is further processed into mature miRNAs capable of triggering RNAi effects, while the ligated exons become a mature messenger RNA (mRNA) for protein synthesis. (**b**) We designed an artificial intron containing pre-miRNA, namely SpRNAi, mimicking the biogenesis processes of the native intronic miRNAs. (**c**) When a designed miR-EGFP(280–302)-stemloop RNA construct was tested in the EGFP-expressing Tg(UAS:gfp) zebrafishes, we detected a strong RNAi effect only on the target EGFP (*lane* 4). No detectable gene silencing effect was observed in other lanes from *left* to *right*: 1, blank vector control (Ctl); 2, miRNA-stemloop targeting HIV-p24 (mock); 3, miRNA without stemloop (anti); and 5, stemloop-miRNA* complementary to the miR-EGFP(280–302) sequence (miR*). The off-target genes such as vector RGFP and fish actin were not affected, indicating the high target specificity of miRNA-mediated gene silencing. (**d**) Three different miR-EGFP(280–302) expression systems were tested for miRNA biogenesis from *left* to *right*: 1, vector expressing intron-free RGFP, no pre-miRNA insert; 2, vector expressing RGFP with an intronic 5′-miRNA-stemloop-miRNA*-3′ insert; and 3, vector similar to the 2 construct but with a defected 5′-splice site in the intron. In Northern bolt analysis probing the miR-EGFP(280–302) sequence, the mature miRNA was released only from the spliced intron resulted from the vector 2 construct in the cell cytoplasm.

the artificial intron (SpRNAi) was co-transcribed within a precursor messenger RNA (pre-mRNA) by Pol-II and cleaved out of the pre-mRNA by RNA splicing. Then, the spliced intron containing a pre-miRNA structure was further processed into mature miRNAs capable of triggering RNAi-related gene silencing effects. Based on this artificial miRNA model, we have tested various pre-miRNA constructs, and observed that the production of intron-derived miRNA fragments was originated from the 5′-proximity of the intron sequence between the 5′-splice site and the branching point. These miRNAs were able to trigger strong suppression of genes possessing over 70% of complementarity to the miRNA sequences. No silence of the targeted gene was observed when there was no homology between the miRNA and the gene, including empty intron with no pre-miRNA inserts, intron with an off-target miRNA insert (negative control), and splicing-defective intron. The same results can also be reproduced in the zebrafish directed against target EGFP expression (**Fig. 14.2c**), indicating the consistent preservation of the intronic miRNA biogenesis system in vertebrates. In addition, no effect was detected on off-target genes, such as RGFP and β-actin, which implies the high specificity of miRNA-directed RNA interference (RNAi). We have confirmed the identity of the intron-derived miRNAs, which were sized about 18–27 base nucleotides (nt), and roughly similar to the newly identified Piwi-interacting RNAs. Moreover, the intronic small RNAs isolated by guanidinium chloride ultracentrifugation can elicit short and strong gene silencing effects on the homologous genes in the transfected cells, which indicate their temporary RNAi effects. However, the long-term (>1 month) gene silencing effect that we observed in vivo occurs only in nuclear transfection of the Pol-II-mediated intronic miRNA system by retroviral vectors.

The components of the Pol-II-mediated SpRNAi system include several consensus nucleotide elements that consist the following: a 5′-splice site, a branch-point domain, a poly-pyrimidine tract, and a 3′-splice site (**Fig. 14.3**). Additionally, a pre-miRNA insert sequence is placed within the artificial intron between the 5′-splice site and the branch-point domain. This portion of the intron normally forms a lariat structure during RNA splicing and processing. We currently know that spliceosomal U2 and U6 snRNPs (helicases) may be involved in the unwinding and excision of the lariat RNA fragment into pre-miRNA; however, specific details remain to be elucidated. Moreover, the SpRNAi contains a translation stop codon domain (T codon) in its 3′-proximal region to facilitate the accuracy of RNA splicing, which if present in a cytoplasmic mRNA would signal the diversion of a splicing-defective pre-mRNA to the nonsense-mediated decay (NMD) pathway and cause the elimination of any unspliced pre-mRNA in the cell. For intracellular expression of the SpRNAi, we need to

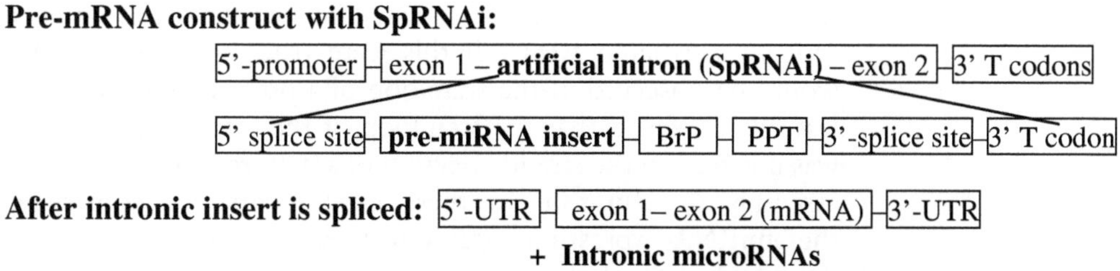

Fig. 14.3. Schematic construct of the artificial SpRNAi intron in a recombinant gene SpRNAi-RGFP for intracellular expression and processing. The components of the Pol-II-mediated SpRNAi system include several consensus nucleotide elements consisting of a 5′-splice site, a branch-point domain (BrP), a poly-pyrimidine tract (PPT), a 3′-splice site, and a pre-miRNA insert located between the 5′-splice site and the BrP domain. The expression of the recombinant gene is under the regulation of either a mammalian Pol-II RNA promoter or a compatible viral promoter for cell-type-specific effectiveness. Mature miRNAs are released from the intron by RNA splicing and further Dicer processing.

insert the SpRNAi construct into the *Dra*II cleavage site of a red fluorescent membrane protein (RGFP) gene from mutated chromoproteins of coral reef *Heteractis crispa*. The cleavage of RGFP at its 208th nucleotide site by restriction enzyme *Dra*II generates an AG–GN nucleotide break with three recessing nucleotides in each end, forming 5′- and 3′-splice sites, respectively, after the SpRNAi insertion. Because this intronic insertion disrupts the expression of functional RGFP, it is then possible to determine the occurrence of intron splicing and RGFP-mRNA maturation through the appearance of red fluorescent emission around the membrane surface of the transfected cells. The RGFP also provides multiple exonic splicing enhancers (ESEs) to increase RNA splicing efficiency.

5. Strand-Specific Gene Silencing in Zebrafish

The foregoing establishes the fact that intronic miRNAs can be used as an effectual strategy to silence specific target gene in vivo (27). We first tried to decide the structural design of pre-miRNA inserts for the best gene silencing effect and found out that a strong structural bias exists in the selection of a mature miRNA strand during assembly of the RNAi effector, RNA-induced gene silencing complex (RISC). RISC is a protein–RNA complex that directs either target gene transcript degradation or translational repression through the RNAi mechanism. Formation of siRNA duplexes has been reported to play a key role in assembly of the siRNA-associated RISC. The two strands of the siRNA duplex are functionally asymmetric, but assembly into the RISC complex is preferential for only one strand. Such preference is determined by the thermodynamic stability of each 5′-end base pairing in the strand. Based on this siRNA model, the formation of miRNA and its complementary miRNA (miRNA*) duplexes was thought to be an essential step for the assembly of miRNA-associated RISC. If this were true, no functional bias would be observed in the stemloop of a pre-miRNA. Nevertheless, we observed that the stemloop of the intronic pre-miRNA was involved in the strand selection of a mature miRNA for RISC assembly in zebrafish. In these experiments, we constructed the miRNA-expressing *SpRNAi-rGFP* vectors as previously described (2). Subsequently, two symmetric pre-miRNAs, miRNA-stemloop-miRNA* [**1**] and miRNA*-stemloop-miRNA [**2**], were synthesized and inserted into the vectors, respectively. Both pre-miRNAs contained the same double-stranded stem arm region, which was directed against the EGFP nts 280–302 sequence. Because the intronic insert region of the SpRNAi-RGFP recombined gene is flanked with a *PvuI* and an *MluI* restriction site at the 5′- and 3′-ends, respectively, the primary insert can be easily removed and replaced by various gene-specific inserts (e.g., anti-EGFP) possessing cohesive ends. By changing the pre-miRNA inserts directed against different gene transcripts, this intronic miRNA generation system provides a valuable tool for genetic and miRNA research in vivo.

To determine the structural preference of the designed pre-miRNAs, we have isolated the zebrafish RNAs by mirVana miRNA isolation columns (*see* **Chapter 10**). Then, we precipitated all potential miRNAs complementary to the target EGFP region by latex beads containing the target RNA sequence. One full-length miRNA identity, miR-EGFP(280–302), was verified to be active in the transfections of the 5′-miRNA-stemloop-miRNA*-3′ construct, as shown in **Fig. 14.4a** (gray-shading

Fig. 14.4. (continued) [**2**]- transfected zebrafishes, whereas the [**1**]-transfection produced another kind of miRNA, miR*-EGFP(301-281), which was partially complementary to the miR-EGFP(280/302). (**b**) The RNAi effect was only observed in the transfection of the [**2**] pre-miRNA, showing less EGFP expression in the transfectant [**2**] than [**1**], while the miRNA indicator RGFP was evenly present in all vector transfections. (**c**) Western blot analysis of the EGFP protein levels confirmed the specific silencing result of (**b**). No detectable gene silencing was observed in fishes without (Ctl) and with liposome only (Lipo) treatments. The transfection of either a U6-driven siRNA vector (siR) or an empty vector (Vctr) without the designed pre-miRNA insert resulted in no gene silencing significance. (**d**)–(**g**) Silencing of endogenous β-catenin and noggin genes in chicken embryos. (**d**) The pre-miRNA construct and fast *green dye* mixtures were injected into the ventral side of chicken embryos near the liver primordia below the heart. (**e**) Northern blot analysis of extracted RNAs from chicken embryonic livers with anti-β-catenin miRNA transfections (*lanes* 4–6) in comparison with wild types (*lanes* 1–3) showed a more than 98% silencing effect on β-catenin mRNA expression, while the housekeeping gene, GAPDH, was not affected. (**f**) Liver formation of the β-catenin knockouts was significantly hindered (*upper right* 2 panels). Microscopic examination revealed a loose structure of hepatocytes, indicating the loss of cell–cell adhesion due to breaks in adherins junctions formed between β-catenin and cell membrane E-cadherin in early liver development. In severely affected regions, feather growth in the skin close to the injection area was also inhibited (*lower right* 2 panels). Immunohistochemistry staining of β-catenin protein expression showed a significant decrease in the feather follicle sheaths. (**g**) The lower beak development was increased by the mandible injection of the anti-noggin pre-miRNA construct (*down panel*) in comparison to the wild type (*up panel*). *Right panels* showed bone (*alizarin red*) and cartilage staining to demonstrate the out growth of bone tissues in the lower beak of the noggin knockout. Northern blot analysis (*small windows*) confirmed a ~60% decrease of noggin mRNA expression in the lower beak area.

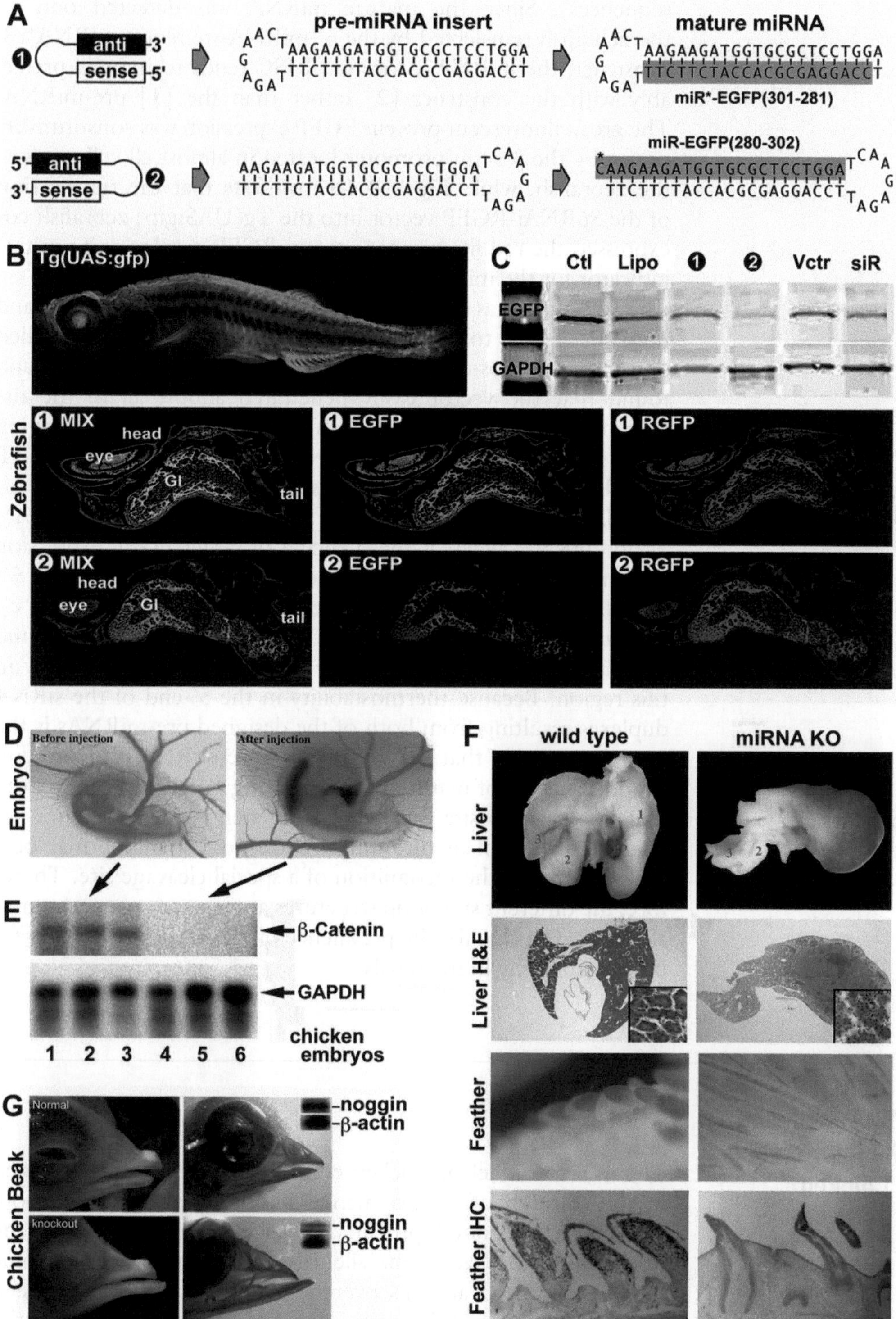

Fig. 14.4. Intronic miRNA-mediated gene silencing effects in vivo. (**a**)–(**c**) Different preferences of RISC assembly were observed by transfection of 5′-miRNA*-stemloop-miRNA-3′ [**1**] and 5′-miRNA-stemloop-miRNA*-3′ [**2**] pre-miRNA structures in zebrafish, respectively. (**a**) One mature miRNA, namely miR-EGFP(280/302), was detected in the

sequences). Since the mature miRNA was detected only in the zebrafish transfected by the 5′-miRNA-stemloop-miRNA*-3′ construct, the miRNA-associated RISC tends to interact preferably with the construct [**2**] rather than the [**1**] pre-miRNA. The green fluorescent protein EGFP expression was constitutively driven by the β-actin promoter located in almost all cell types of the zebrafish, while **Fig. 14.4b** manifests that the transfection of the SpRNAi-RGFP vector into the Tg(UAS:gfp) zebrafish co-expressed the red fluorescent protein RGFP, serving as a positive indicator for the miRNA generation in the transfected cells.

This approach has been successfully used in several mouse and human cell lines to show the RNAi effects (59, 60). We applied the liposome-capsulated vector (total 60 μg) to the fishes and found that the vector easily penetrated almost all of the tissues of the 2-week-old zebrafish larvae within 24 h, reaching full systemic delivery of the miRNA effect. The indicator RGFP was detected in both of the fishes transfected by either 5′-miRNA*-stemloop-miRNA-3′ or 5′-miRNA-stemloop-miRNA*-3′ pre-miRNA, whereas the silencing of target EGFP expression (green) was observed only in the fish transfected with the 5′-miRNA-stemloop-miRNA*-3′ pre-miRNA (**Fig. 14.4b and c**). The suppression level in the gastrointestinal (GI) tract was found to be less effective, probably due to the high RNase activity in this region. Because thermostability in the 5′-end of the siRNA duplexes resulting from both of the designed pre-miRNAs is the same, we suggest that the stemloop of pre-miRNA is involved in strand selection of mature miRNA during RISC assembly. Given that the cleavage site of Dicer in the stem arm determines the strand selection of mature miRNA (62), the stemloop may be a determinant for the recognition of a special cleavage site. Therefore, the different stemloop structures among various species may also provide a clue for the prevalence of native siRNAs in invertebrates and rarely in mammals.

6. Intronic piRNA-Mediated RNA Interference in Chicken

The in vivo model of chicken embryos has been widely utilized in the research of developmental biology, signal transduction, and flu vaccine development. We have successfully tested the feasibility of localized gene silencing in vivo using the intronic miRNA approach and discovered that the interaction between pre-mRNA and perichromatin DNA may be essential for the intronic miRNA biogenesis. For example, the β-catenin gene was selected because its products play a critical role in the biological development and ontogenesis (63). β-Catenin plays a role in the growth control of skin and liver tissues in chicken embryos. The loss-of-function of β-catenin is lethal in transgenic animals. Experimental results demonstrated that the miRNAs derived from a long RNA–DNA hybrid construct ($\geq$150 bp) were capable of inhibiting β-catenin gene expression in the liver and skin of developing chicken embryos (**Fig. 14.4d–g**). Again, the interaction between the intronic miRNA precursor and genomic DNA may account for the specific gene silencing (22, 23, 51). We have demonstrated that the [P^{32}]-labeled DNA component of a long RNA–DNA duplex construct in cell nuclear lysates was intact during the effective period of miRNA-induced RNA interference (RNAi) phenomena, while the labeled RNA part was replaced by cold homologues and excised into small 18–27-nt RNA fragments in a 3-day-incubation period (64). Since the intronic miRNA generation relies on a coupled interaction of nascent Pol-II-directed pre-mRNA transcription and the intron excision in proximity to the genomic perichromatin fibrils, the above observation indicates that the pre-mRNA–perichromatin DNA interaction may facilitate new intronic miRNA generation. Thus, Pol-II may actually function as an RNA-dependent RNA polymerase (RdRp) for producing more small miRNAs.

Recent studies have shown that the pre-mRNA–perichromatin DNA interaction results in the generation of Piwi-interacting RNAs (piRNA), which are similar to intronic miRNAs, but distinct from other small double-stranded siRNAs and shRNAs by their relatively larger size (approximately 26–31 nucleotides), single-strandedness, strand-specificity, and the clustered arrangement of their origins (58). The piRNA class of small RNAs is likely transcribed by an intracellular RNA polymerase, similar to RdRp, from the pre-mRNA–perichromatin DNA duplex region of a replicating cell genome during mitosis or meiosis. Mammalian type II RNA polymerases (Pol-II) have been observed to possess RdRp-like activities (4, 65, 66). Nuclear transfection of long DNA–RNA duplex templates has also been shown to trigger piRNA-like gene silencing effects against viral infection and retrotransposon activity (22). In *Drosophila* and zebrafish, Piwi proteins are recently found to directly implicate in piRNA biogenesis to maintain transposon silencing in the germline genome (24, 67). This function may be conserved in mice as loss of Miwi2, a mouse Piwi homologue, leads to germline stem cell and meiotic defects correlated with increased transposon activity (68). Because the RNAi effector of the piRNA-mediated gene silencing requires Piwi proteins rather than siRNA/shRNA-associated Dicer RNases, this suggests that the piRNA-mediated RNAi mechanism is slightly different from the siRNA/shRNA-mediated RNAi pathway.

In an effort to examine the pre-mRNA–perichromatin DNA interaction theory, we tested intracellular transfection of a long RNA–DNA hybrid construct containing a hairpin anti-β-catenin intronic pre-miRNA, which was directed against the central region of the β-catenin coding sequence (aa 306–644) with perfect complementarity. A perfectly complementary miRNA theoretically directs target mRNA degradation more efficient than translational repression. Using embryonic day-3 chicken embryos, a 25 nM dose of the pre-miRNA construct was injected into the ventral body cavity, which is close to where the liver primordia would form (**Fig. 14.4d**). For efficient delivery into target tissues, the pre-miRNA construct was mixed with a liposomal transfection reagent (Roche Biomedicals, Indianapolis, IN). A 10% (v/v) fast green solution was concurrently added during the injection as a dye indicator. The mixtures were injected into the ventral side, near the liver primordia below the heart, using capillary needles. After injection, the embryonic eggs were sealed with sterilized scotch tapes and incubated in a humidified incubator at 39–40°C until day 12 when the embryos were examined and photographed under a dissection microscope. Several malformations were observed. There was no visible overt toxicity or overall perturbation of embryo development. The liver was the closest organ to the injection site and hence was most dramatically affected in phenotypes. Other regions, particularly the skin close to the injection site, were also affected by the diffused miRNA effects. As shown in **Fig. 14.4e**, Northern blot analysis detecting the target β-catenin mRNA expression in the dissected livers showed that β-catenin expression in the wild-type livers remained normal (lanes 1–3). However, the miRNA-treated samples decreased dramatically in β-catenin expression (lanes 4–6). The miRNA silencing effect degraded greater than 98% of β-catenin mRNA expression in the embryonic chicken, but has no influence in the housekeeping gene GAPDH expression, indicating its high target specificity and very limited interferon-related cytotoxicity in vivo.

After ten days of primordial injection with the anti-β-catenin pre-miRNA template, the embryonic chicken livers showed an enlarged and engorged first lobe, but the size of the second and third lobes of the livers were dramatically decreased (**Fig. 14.4f**). Histological sections of normal livers showed hepatic cords and sinusoidal space with few blood cells. In the anti-β-catenin miRNA-treated embryos, the general architecture of the hepatic cells in lobes 2 and 3 remained unchanged; however, there were islands of abnormal regions in lobe 1. The endothelium development appeared to be defective, and blood leaked outside of the blood vessels. Abnormal types of hematopoietic cells were also observed between the space of hepatocytes, principally dominated by a population of small cells with round nuclei and scanty cytoplasm. In severely affected regions, hepatocytes were disrupted (**Fig. 14.4f**, small windows) and the diffused miRNA effect further inhibited the feather growth in the skin area near the injection site. These findings showed that the anti-β-catenin miRNA was effective over a long period of time (≥10 days) and at a very low dose (25 nM) in knocking out the targeted gene expression. Further, the miRNA gene silencing effect appeared to be very specific as off-targeted organs appear to be normal, indicating that the small single-strand composition of miRNA herein possessed no overt toxicity. In another attempt to silencing noggin expression in the mandible beak area using the same approach (**Fig. 14.4g**), it was observed that an enlarged lower beak morphology is reminiscent of those of BMP4-overexpressing chicken embryos reported previously (69, 70). Skeleton staining showed the outgrowth of bone and cartilage tissues in the injected mandible area (**Fig. 14.4g**, right panels) and Northern blot analysis further confirmed that about 60% of noggin mRNA expression was knocked out in this region (insets). Since bone morphogenetic protein 4 (BMP4), a member of transforming growth factor-β (TGF-β) superfamily, is known to promote bone development and the noggin is an antagonist of BMP2/4/7 genes, it is not surprising to find out that our miRNA-mediated noggin knock outs created a morphological change resembling the BMP4-overexpressing results shown in previously mentioned chicken and other avian models. Thus, the gene silencing in chicken by the pre-miRNA transfection presents a great potential of localized transgene-like approach in creating animal models for developmental biology research.

7. Localized RNA Interference Effects on Mouse Skin

To evaluate the efficacy and safety of intronic miRNA in animals, we have tested the vector-based intronic miRNA transfection in mice as previously described (61). As shown in **Fig. 14.5**, patched albino (white) skins of melanin-knockdown mice were created by a succession of intra-cutaneous (i.c.) injection of an anti-*tyrosinase* (*Tyr*) pre-miRNA agent (50 μg) for 4 days (total 200 μg). The Tyr, a type-I membrane protein and copper-containing enzyme, catalyzes the critical and rate-limiting step of tyrosine hydroxylation in the biosynthesis of melanin (black pigment) in skins and hairs. After 13-day incubation, the expression of melanin has been blocked in the miRNA transfections due to a significant loss of its intermediates resulted from the anti-*Tyr* miRNA-triggered gene silencing effect. Contrarily, the blank control and

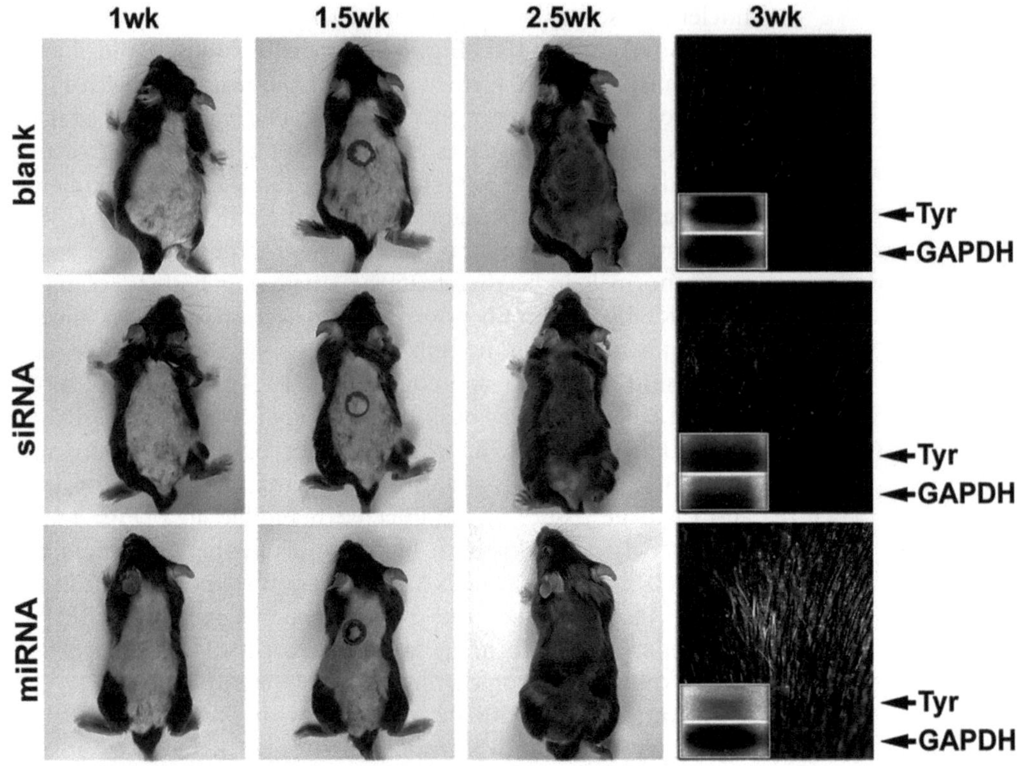

Fig. 14.5. In vivo effects of anti-tyrosinase miRNA on the mouse pigment production of local skins. Transfection of the miRNA induced strong gene silencing of *tyrosinase* (*Tyr*) mRNA expression but not housekeeping *GAPDH*, whereas that of U6-directed siRNA triggered mild non-specific RNA degradation of both *Tyr* and *GAPDH* gene transcripts. Since Tyr is an essential enzyme for black pigment melanin production, the success of gene silencing can be observed by a significant loss of the *black color* in mouse hairs. The *circles* indicate the location of injections. Northern blot analysis of *Tyr* mRNA expression in local hair follicles confirmed the effectiveness and specificity of the miRNA-mediated gene silencing effect (*small windows*).

the Pol-III (U6)-directed siRNA transfections presented normal black skin color under the same dosage. Northern blot analysis, using RNA-PCR-amplified mRNAs from hair follicles, showed a 76.1 ± 5.3% reduction of *Tyr* expression 2-day after the miRNA transfection. This is in consistent with the immunohistochemical staining results from the same skin area, while mild and non-specific degradation of common gene transcripts was detected in the siRNA-transfected skins (seen from smearing patterns of both housekeeping control *GAPDH* and targeted *Tyr* mRNAs). Given that Grimm et al. (49) have reported that high siRNA/shRNA concentrations generated from the Pol-III-directed RNAi systems may over-saturate the cellular miRNA pathway and cause global miRNA dysregulation, the siRNA pathway is likely incompatible with the native miRNA pathway in skin tissues. Therefore, these results demonstrate that the intronic miRNA expression system is a powerful new tool for transgenic miRNA expression in animals. It is noted that non-targeted skin hairs appear to be normal after miRNA transfection. This underscores the fact that the intronic miRNA is safe and effective in vivo. The results also specify that the miRNA-mediated gene silencing effect is stable and efficient in knocking down targeted gene expression over a relatively long period of time since the hair re-growth takes over ten days. Taken together, the intronic miRNA-mediated transgene approach may offer relatively safe, effective, and long-term gene manipulation in animals and prevents the non-specific lethal effects of the conventional transgenic methods.

More recent advances of the applications of intronic miRNA expression systems have been reported in mice. Chung et al. (53) have succeeded in expression of a cluster of polycistronic miRNAs using the Pol-II-mediated intronic miRNA expression system. A polycistronic miRNA cluster can be processed into multiple miRNAs via the cellular miRNA pathway. This new RNAi approach has a few advantages over the conventional Pol-III-mediated shRNA expression systems. First, Pol-II expression is tissue specific, whereas Pol-III expression is not. Second, Pol-II expression is compatible with the native microRNA pathway, while Grimm et al. (49) have reported some incompatibility issues between the Pol-III-mediated shRNA and Pol-II-mediated native miRNA pathways. Third, excessive RNA accumulation and toxicity can be prevented by the NMD mechanism of a cellular Pol-II-mediated intron expression system, but not a Pol-III exon-like expression system (71). Lastly, one Pol-II is able to express a large size cluster (>10 kb) of polycistronic shRNAs, which can be further excised into multiple shRNAs via the native miRNA pathway, preventing the promoter conflict that often occurs in a multiple promoter vector system. For example, in many commercial U6-mediated shRNA expression systems, a self-inactivated vector promoter is often used to increase the U6 promoter activity.

8. Development of MicroRNA/piRNA-Based Gene Therapy

The following experimentations have shown the preliminary success of silencing exogenous retrovirus replication in an ex vivo cell model of patient-extracted $CD4^+$ T lymphocytes. The specific anti-HIV *SpRNAi-rGFP* vectors were designed to target against the *gag-pol* region of about nts +2113 to +2450 of HIV-1 genome. This region is relatively conserved and can serve as a good target for anti-HIV treatment (72). The viral genes located in this target region include 3′-proximal $Pr55^{gag}$ polyprotein (i.e., matrix p17 + capsid p24 + nucleocapsid p7) and 5′-proximal $p66/p51^{pol}$ polyprotein (i.e., protease p10 + reverse transcriptase). All of these components have critical roles in viral replication and infectivity. During the early infection phase, the viral reverse transcriptase transcribes the HIV RNA genome into a double-stranded cDNA sequence, which forms a pre-integration complex with the matrix, integrase, and viral protein R (Vpr). This complex is then transferred to the cell nucleus and integrated into the host chromosome, consequently establishing the HIV provirus. We hypothesized that, although HIV carries few reverse transcriptase and matrix proteins during its first entry into host cells, the co-suppression of $Pr55^{gag}$ and $p66/p51^{pol}$ gene expressions by miRNAs is expected to eliminate the production of infectious viral particles in the late infection phase. Silencing $Pr55^{gag}$ may prevent the assembly of intact viral particles due to the lack of matrix and capsid proteins, while suppression of protease in $p66/p51^{pol}$ can inhibit the maturation of several viral proteins. HIV expresses about nine viral gene transcripts, which encode at least 15 various proteins. Thus, the separation of a polyprotein into individual functional proteins requires the viral protease activity. As shown in **Fig. 14.6**, this therapeutic approach has been reported to be feasible (22, 47).

The anti-HIV *SpRNAi-rGFP* vectors were tested in the $CD4^+$ T-lymphocyte cells from HAART-treated, HIV-seropositive patients. Because only partial complementarity between miRNA and its target RNA is necessary to trigger the gene silencing effect, it would be very advantageous to overcome the daunting challenge of high HIV mutations, which frequently generate new drug resistance to current small molecule drugs. Northern blotting (**Fig. 14.6a**) demonstrated that the ex vivo gene silencing effect of anti-HIV miRNA transfections ($n = 3$ for each set) on HIV-1 replication in $CD4^+$ T lymphocytes from both acute and chronic phase AIDS patients. In the acute phase ($\leq$1 month), the 50 nM miRNA vector transfection degraded on average of 99.8% viral RNA genome (lane 4), whereas the same treatment knocked down only on average of 71.4 $\pm$ 12.8% viral genome replication in the chronic phase (about 2-year infection). Immunocytochemical staining of HIV p24 marker protein confirmed the results of Northern blot analysis (**Fig. 14.6b**). Sequencing analysis has revealed that at least two HIV-1b mutants in the acute phase and seven HIV-1b mutants in the chronic phase were found within the targeted HIV genome domain. It is likely that the higher genome complexity of HIV mutations in chronic infections is able to counteract the miRNA-mediated silencing efficacy. Transfection of 50 nM miRNA* vector homologous to the HIV-1 genome failed to induce any RNAi effect on viral genome, indicating the specificity of the miRNA effect (**Fig. 14.6b**, lane 5). Expression of cellular housekeeping gene, β-actin, was at a normal level and showed no interferon-induced non-specific RNA degradation. These results suggest that the designed anti-HIV *SpRNAi-rGFP* vector is highly specific and

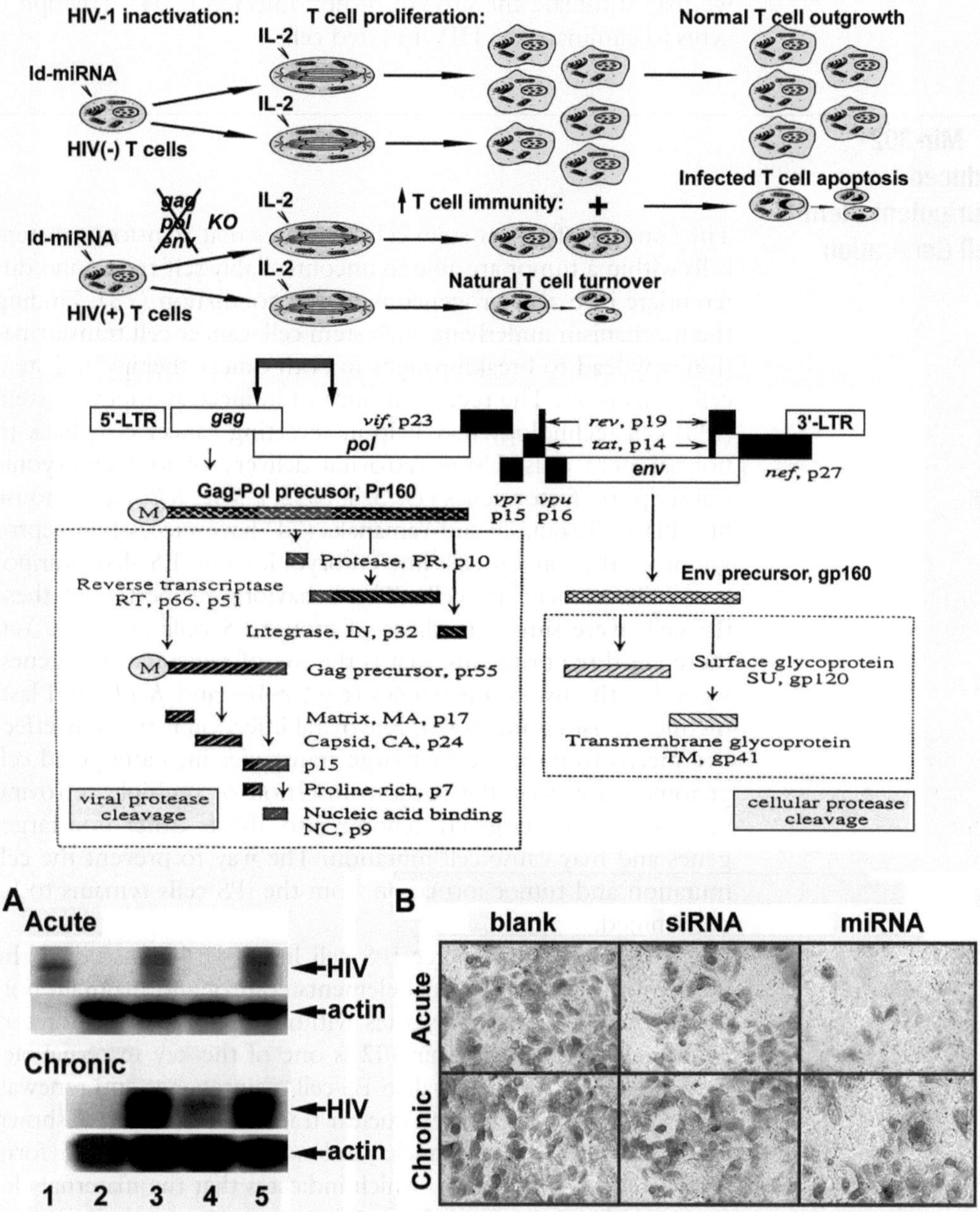

Fig. 14.6. Silencing of HIV-1 genome replication using anti-*gag/pro/pol* miRNA transfections into $CD4^+$ T lymphocytes isolated from the acute and chronic phases of AIDS infections. (**a**) Northern blot analysis showed about 98 and 70% decreases of HIV genome in the acute and chronic infections after miRNA treatments (*lanes* 4), respectively. No effect was detected in the T cells transfected by miRNA* targeting the same *gag/pro/pol* region of the viral genome (*lanes* 5). The size of pure HIV-1 provirus was measured to be about 9,700 nucleotide bases (*lanes* 1). RNA extracts from normal non-infected $CD4^+$ T lymphocytes were used as a negative control (*lanes* 2), and those from HIV-infected T cells were used as a positive control (*lanes* 3). (**b**) Immunostaining of HIV p24 marker confirmed the results of (**a**). Since the ex vivo HIV-silenced T lymphocytes were resistant to any further infection by the same strains of HIV, they may be transfused back to the donor patient for eliminating HIV-infected cells.

efficient in suppressing HIV-1 replication in the early infections. In conjunction with an intermittent interleukin-2 therapy (47), we may stimulate the growth of non-infected CD4+ T lymphocytes to eliminate the HIV-infected cells.

9. Mir-302-Induced Pluripotent Stem Cell Generation

The concept of cancer stem cells indicates that transformed stem cells within a tumor are able to uncontrollably self-renew and differentiate into a heterogeneous tumor population (73). Finding the mechanism underlying such stem cell–cancer cell transformation may lead to breakthroughs in both cancer therapy and stem cell generation. The recent advance of induced pluripotent stem (iPS) cell technology may help in reverting cancer cells back to normal stem cells. Using retroviral delivery of four embryonic transcription factor genes (*Oct3/4*, *Sox2*, *c-Myc*, *Klf4*) into mouse fibroblasts, Takahashi and Yamanaka (74) have successfully reprogrammed the somatic cells into embryonic stem (ES)-like pluripotent cells, namely iPS cells. The behavioral properties of these iPS cells were similar to those of mouse ES cells (75, 76). Yet, there are three problems: first is the use of retroviral transgenes, second is the use of oncogenes (e.g., *c-Myc* and *Klf4*), and last, the mechanism is unknown. Retroviral infection is the only effective means to integrate four large transgenes into a targeted cell genome. However, the random insertion of multiple retroviral vectors into the targeted genome also affects other non-target genes and may cause cell mutation. The way to prevent the cell mutation and tumor formation from the iPS cells remains to be determined.

To generate tumor-free iPS cell lines, we have studied the mechanism of how maternal elements control and maintain normal ES cell renewal in zygotes without the risk of tumor formation. We found that mir-302 is one of the key maternal elements essential for mammalian ES cell maintenance and renewal. Experiments of somatic cell nuclear transfer (SCNT) have shown that hybrids of somatic nuclei and oocyte cytoplasm can form pluripotent stem-like cells, which indicates that the maternal elements in the oocyte cytoplasm play an important role in nuclear reprogramming (77). Also, mouse oocytes lack Dicer, a conserved ribonuclease required for miRNA biogenesis, arrest in the division phase of meiosis I, indicating that miRNAs play a critical role in oogenesis (78). It is noted that the mir-302 family (mir-302's) is expressed at extremely high levels in mouse oocytes and human ES cells compared to other differentiated tissue cells (79, 80). The mir-302's are expressed together as a cluster in a long non-coding RNA transcript containing mir-302b, mir-302c, mir-302a, mir-302d, and mir-367 in a 5′- to 3′-sequential order (80). All of the mir-302 members share an identical sequence in their 5′ first 17 nucleotides, including the whole-seed motif, and an average of 87–89% homology in their 23-nucleotide mature miRNA sequences. Based on the databases of the miR-Base::Sequences program (http://microrna.sanger.ac.uk/), they are highly conserved in mammals and target almost the same cellular genes. In fact, many of these target genes are developmental signals and transcriptional factors involved in initiation and/or facilitation of lineage-specific cell differentiation during early embryogenesis (81).

Using ectopic expression of a mir-302's-encoding transgene, we have successfully generated several novel miRNA-induced iPS (mirPS) cell lines derived from human somatic cells (81). The previously established Pol-II-driven intronic miRNA expression system was used to generate these new iPS cell lines (**Figs. 14.3** and **14.7**). The intronic miRNA expression system transcribes a recombinant transgene of red-shifted fluorescent protein (RGFP), mainly *SpRNAi-RGFP*, which contains an artificial splicing-competent intron (*SpRNAi*) capable of producing intronic miRNA and/or small hairpin RNA (shRNA) precursors (2, 60, 61). The mature intronic mir-302s can be released by intracellular RNA splicing and processing machineries, such as components of spliceosomes, exosomes, and NMD systems. Then, they trigger a specific silencing effect on targeted genes (82). As previously reported (81), we found several evidences of somatic cell reprogramming. First, various mir-302-induced iPS (mirPS) cell lines can be generated, derived from primary cultures of normal human skin epidermal cells (hHFC) and cancerous skin melanoma Colo cells, both of which can form embryoid bodies in vitro (**Fig. 14.8**). Second, the elevated expression of mir-302s has been confirmed using microRNA (miRNA) microarray and Northern blot analyses of the mirPS cell transcriptomes (**Fig. 14.9**). Third, the elevated expression of standard embryonic stem (ES) cell markers has been detected, including Oct3/4, SSEA-3, SSEA-4, Sox2, and Nanog (**Fig. 14.9b**). Fourth, global genomic DNA demethylation has been observed, similar to the status of a zygotic genome undergoing the reprogramming process. Fifth, genome-wide gene expression patterns of these mirPS cells have been shown to share over 86% high similarity to those of human ES WA01 (H1) and WA09 (H9) cells (**Fig. 14.9c**). Last, xenotransplantation of the mirPS cell-derived embryoid bodies into pseudopregnant immunocompromised SCID-beige mice can form teratoma-like tissue cysts, containing all three embryonic germ layers (ectoderm, mesoderm, and definitive endoderm) (**Fig. 14.9d**). However, unlike teratomas, these cysts grow slowly and form a very clear boundary to the surrounding mouse tissues. Since the transgenic delivery of

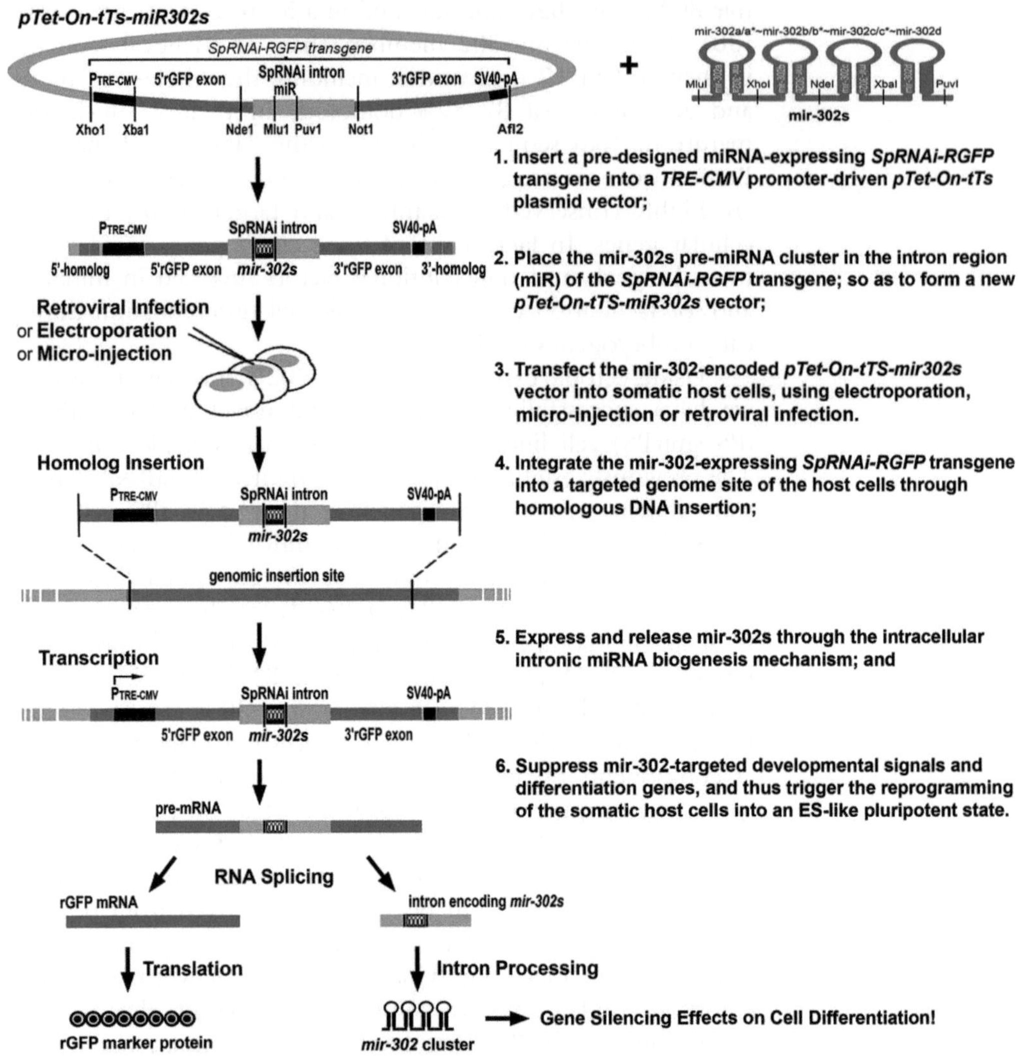

Fig. 14.7. Strategy for generating transgenic mir-302-induced pluripotent stem (mirPS) cells, using vector-based transgene delivery. The transgene vector can be transfected into the tested somatic/cancerous cells using electroporation, micro-injection, and/or retroviral infection. In the transfected cells, the vector induces the integration of a Pol-II promoter-driven *SpRNAi-RGFP* transgene into the cell genomes for steadily expressing a pre-designed mir-302 pre-miRNA cluster (mir-302s). In our design, the mir-302s is placed in the intron region of the *SpRNAi-RGFP* transgene and generated as a part of the transgene transcript (pre-mRNA), containing a non-coding intron flanking with two red-shifted fluorescent protein (*RGFP*)-coding exons. After transgenic expression, the intron is spliced out of the pre-mRNA and further excised into small miRNA-like mir-302 molecules capable of triggering targeted gene silencing, while the *RGFP* exons are ligated together to form a mature mRNA for synthesis of a red fluorescent RGFP-marker protein. The presence of RGFP served as an indicator for intracellular mir-302s expression and effectiveness.

mir-302s into somatic cells can be performed with electroporation, the risks of random retroviral insertion and cell mutation are prevented. These advantages provide a safe solution for the problems of previous iPS technology, preventing the risks of retroviral infection, oncogenic mutation, and uncertain tumorigenecity.

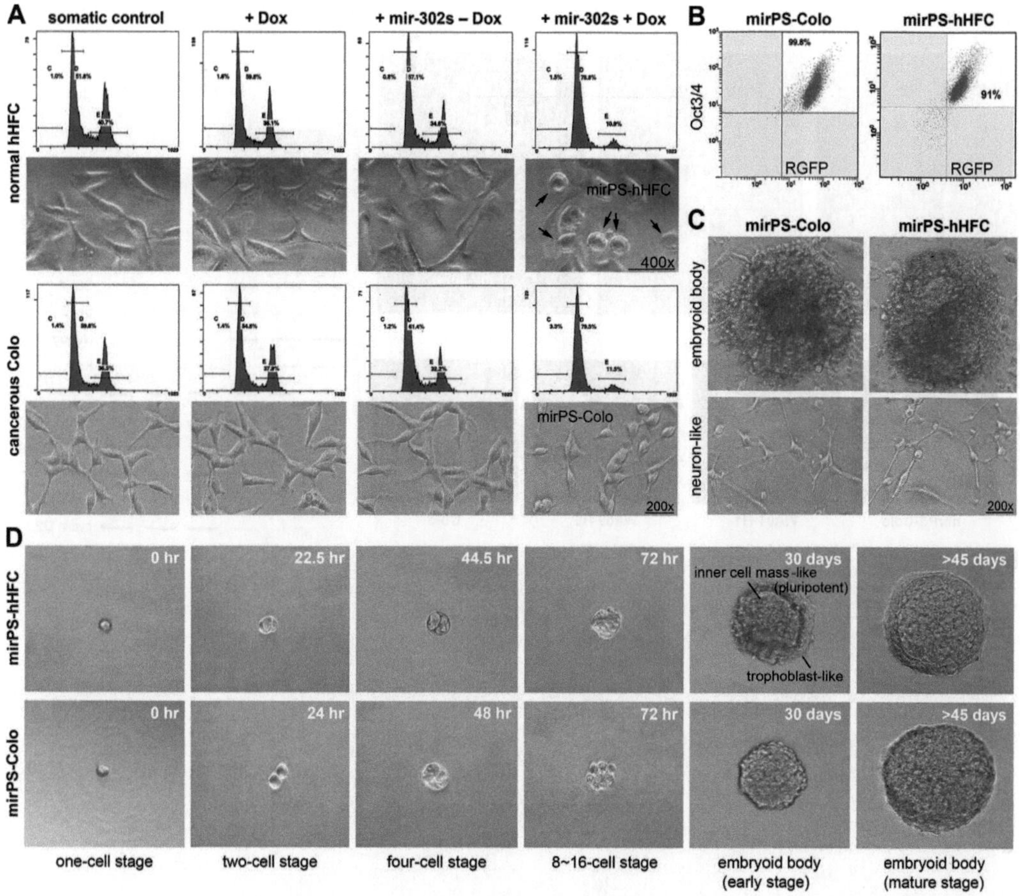

Fig. 14.8. Reprogramming human normal skin cells (hHFC) and cancerous melanoma Colo cells into embryonic stem (ES)-like pluripotent state. (**a**) Changes of morphologies and cell proliferation rates in the mir-302-induced pluripotent stem (mirPS) cells, namely mirPS-hHFC and mirPS-Colo, respectively. Both mirPS cell lines show ES-like round resting cell morphology and very slow cell renewal cycle in the presence of doxycyclin (Dox) induction. (**b**) Selection of positive mirPS cells using FACS flow cytometry with both Oct3/4 and RGFP markers. The transfection efficiency is measured to be over 91%. (**c**) Formation of embryoid bodies (EB) derived from these mirPS cells and the guided differentiation into neuron-like cells. (**d**) Time-course EB formation from a single mirPS cell after limiting dilution. The cell cycles are estimated to be approximately 20–24 h.

Given that the function of mir-302s can reprogram both normal and cancerous tissue cells into an ES-like pluripotent cell state, the findings of this somatic cell reprogramming method may lead to novel therapeutical interventions for both stem cell and cancer therapies.

Based on the microarray results, we further noted that cell cycle checkpoint genes, i.e., CDK2, cyclin D1 and D2, and DNA methylation facilitator, i.e., MECP2 and MECP1-p66, are all strong targets of mir-302s (**Fig. 14.9**). It is known that cyclin E-dependent kinase CDK2 is required for the entry of S-phase cell cycle and inhibition of CDK2 results in G1-phase checkpoint arrest, whereas cyclin D1 can override G1-phase arrest in response to DNA damage (83). Based on this principle, the suppression of

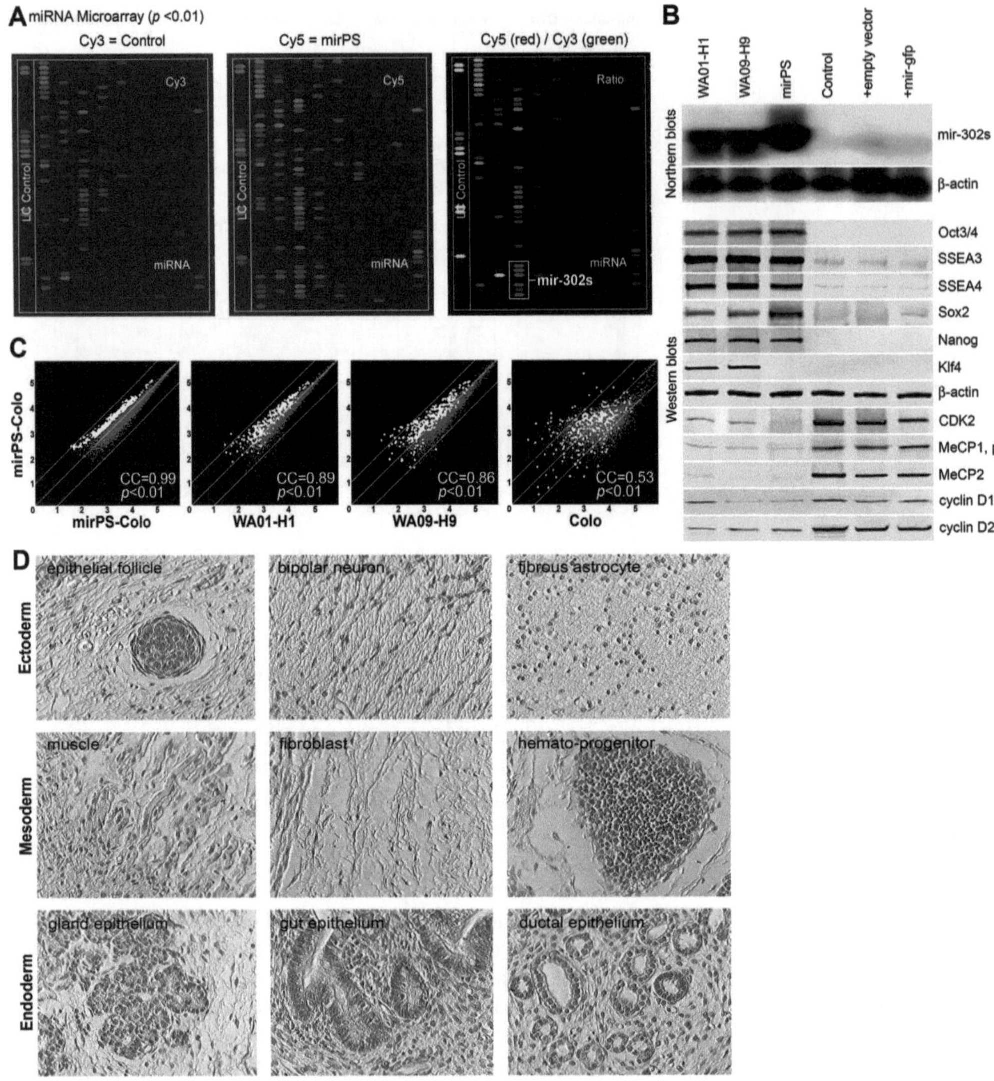

Fig. 14.9. Correlation between mir-302s expression and ES cell marker activation. (**a**) Microarray analyses of global miRNA expression, revealing that all mir-302 familial members (mir-302s), are highly expressed in the mirPS cells, but not the original somatic cells ($n = 3$, $p < 0.01$). (**b**) Northern and Western blot analyses, showing that the mirPS cells express abundant ES cell markers, including Oct3/4 (Oct4), SSEA-3, SSEA-4, Sox2, and Nanog but less oncogenic Klf4 ($n = 4$, $p < 0.01$), highly similar to those observed in human ES WA01-H1 and WA09-H9 cells. (**c**) Comparison of altered gene expression patterns using Human genome GeneChip U133A&B and plus 2.0 arrays (Affymetrix), showing high similarity between mirPS-Colo and H1 (89%) as well as H9 (86%), but not original Colo (53%) cells. *White dots* referred the highly variable genes as compared to the stably expressed genes (*dark dots*). (**d**) Teratoma-like tissues cysts derived from the mirPS-derived EB implants in the uterus or peritoneal cavity of female pseudopregnant immunocompromised SCID-beige mice (n/total = 6/6).

both CDK2 and cyclin D1 in mirPS cells reveals a fact that the cell cycle of mirPS cells can reach a very slow cell division rate. This result of such somatic stem cell cycle transition indicates that mir-302 is a strong tumor suppressor which may help to generate tumor-free iPS cells. In addition, the suppression of MECP2 and MECP1-p66 activities causes global genomic DNA demethylation and thus may trigger epigenetic reprogramming of somatic cell genomes into an ES-like status, which in turn activates the expression of many ES cell marker genes, such as *Oct3/4*, *SSEA-3*, *SSEA-4*, *Sox2*, and *Nanog*. Therefore, our intronic mir-302 transgene approach is able to not only reprogram somatic and cancerous cells to ES-like pluripotent stem cells but also maintain this ES-like state under a tumor-fee and feeder-free cell cultural condition.

10. Conclusions

The consistent evidence of miRNA-induced gene silencing effects in zebrafish, chicken embryos, mouse stem cells, and human diseases demonstrates the preservation of an ancient intron-mediated gene regulation system in eukaryotes. In these animal models, the intron-derived miRNAs determine the activation of RNAi-like gene silencing pathways. We herein provide the first evidence for the biogenesis and function of intronic miRNAs in vivo. Given that natural evolution gives raise to more complexity and variety of introns in higher animal and plant species for coordinating their vast gene expression volumes and interactions, dysregulation of these miRNAs due to intronic expansion or deletion will likely to cause genetic diseases, such as myotonic dystrophy and fragile X mental retardation. Thus, gene expression produces not only gene transcripts for its own protein synthesis, but also intronic miRNAs, capable of interfering with other gene expressions. Thence, the expression of a gene results in gain-of-function of the gene and also loss-of-function of some other genes, which contain complementarity to the mature intronic miRNAs. An array of genes can swiftly and accurately coordinate their expression patterns with each other through the mediation of their intronic miRNAs, bypassing the time-consuming translation processes under quickly changing environments. Conceivably, intron-mediated gene regulation may be as important as the mechanisms by which transcription factors regulate the gene expression. It is likely that intronic miRNA is able to trigger cell transitions rapidly in response to external stimuli without the tedious protein synthesis. Undesired gene products are reduced by both transcriptional inhibition and/or translational suppression via miRNA regulation. This could enable a rapid switch to a new gene expression pattern without the need to produce various transcription factors. This regulatory property of miRNAs may serve as one of the most ancient gene modulation systems before the emergence of proteins. According to the variety of miRNAs and the complexity of genomic introns, a thorough investigation of miRNA variants in the human genome will markedly improve the understanding of genetic diseases and the design of miRNA-based drugs. Learning how to exploit such a novel gene regulation system in the future therapy will be a forthcoming challenge.

References

1. Ambros, V., Lee, R.C., Lavanway, A., Williams, P.T., and Jewell, D. (2003) MicroRNAs and other tiny endogenous RNAs in *C. elegans*. *Curr Biol*, **13**, 807–818.
2. Lin, S.L., Chang, D., Wu, D.Y., and Ying, S.Y. (2003) A novel RNA splicing-mediated gene silencing mechanism potential for genome evolution. *Biochem Biophys Res Commun*, **310**, 754–760.
3. Rodriguez, A., Griffiths-Jones, S., Ashurst, J.L., and Bradley, A. (2004) Identification of mammalian microRNA host genes and transcription units. *Genome Res*, **14**, 1902–1910.
4. Lin, S.L., Chuong, C.M., and Ying, S.Y. (2001) A novel mRNA-cDNA interference phenomenon for silencing bcl-2 expression in human LNCaP cells. *Biochem Biophys Res Commun*, **281**, 639–644.
5. Ying, S.Y. and Lin, S.L. (2004) Intron-derived microRNAs – fine tuning of gene functions. *Gene*, **342**, 25–28.
6. Clement, J.Q., Qian, L., Kaplinsky, N., and Wilkinson, M.F. (1999) The stability and fate of a spliced intron from vertebrate cells. *RNA*, **5**, 206–220.
7. Parrish, S., Fleenor, J., Xu, S., Mello, C., and Fire, A. (2000) Functional anatomy of a dsRNA trigger: differential requirement for the two trigger strands in RNA interference. *Mol Cell*, **6**, 1077–1087.
8. Holen, T., Amarzguioui, M., Wiiger, M.T., Babaie, E., and Prydz, H. (2002) Positional effects of short interfering RNAs targeting the human coagulation trigger Tissue Factor. *Nucleic Acids Res*, **30**, 1757–1766.
9. Hutvagner, G. and Zamore, P.D. (2002) A microRNA in a multiple-turnover RNAi enzyme complex. *Science*, **297**, 2056–2060.
10. Zeng, Y., Yi, R., and Cullen, B.R. (2003) MicroRNAs and small interfering RNAs can inhibit mRNA expression by similar mechanisms. *Proc Natl Acad Sci U S A*, **100**, 9779–9784.
11. Hall, I.M., Shankaranarayana, G.D., Noma, K., Ayoub, N., Cohen, A., and Grewal, S.I. (2002) Establishment and maintenance of a heterochromatin domain. *Science*, **297**, 2232–2237.
12. Llave, C., Xie, Z., Kasschau, K.D., and Carrington, J.C. (2002) Cleavage of scarecrow-like mRNA targets directed by a class of arabidopsis miRNA. *Science*, **297**, 2053–2056.
13. Rhoades, M.W., Reinhart, B.J., Lim, L.P., Burge, C.B., Bartel, B., and Bartel, D.P. (2002) Prediction of plant microRNA targets. *Cell*, **110**, 513–520.
14. Lee, R.C., Feibaum, R.L., and Ambros, V. (1993) The *C. elegans* heterochromic gene *lin-4* encodes small RNAs with antisense complementarity to *lin-14*. *Cell*, **75**, 843–854.
15. Reinhart, B.J., Slack, F.J., Basson, M., Pasquinelli, A.E., Bettinger, J.C., Rougvie, A.E., Horvitz, H.R., and Ruvkun, G. (2000) The 21-nucleotide let-7 RNA regulates developmental timing in *Caenorhabditis elegans*. *Nature*, **403**, 901–906.
16. Lau, N.C., Lim, L.P., Weinstein, E.G., and Bartel, D.P. (2001) An abundant class of tiny RNAs with probable regulatory roles in *Caenorhabditis* elegans. *Science*, **294**, 858–862.
17. Brennecke, J., Hipfner, D.R., Stark, A., Russell, R.B., and Cohen, S.M. (2003) Bantam encodes a developmentally regulated microRNA that controls cell proliferation and regulates the proapoptotic gene hid in *Drosophila*. *Cell*, **113**, 25–36.
18. Xu, P., Vernooy, S.Y., Guo, M., and Hay, B.A. (2003) The Drosophila microRNA Mir-14 suppresses cell death and is required for normal fat metabolism. *Curr Biol*, **13**, 790–795.
19. Lagos-Quintana, M., Rauhut, R., Meyer, J., Borkhardt, A., and Tuschl, T. (2003) New microRNAs from mouse and human. *RNA*, **9**, 175–179.
20. Mourelatos, Z., Dostie, J., Paushkin, S., Sharma, A., Charroux, B., Abel, L., Rappsilber, J., Mann, M., and Dreyfuss, G. (2002) miRNPs: a novel class of ribonucleoproteins containing numerous microRNAs. *Genes Dev*, **16**, 720–728.
21. Zeng, Y., Wagner, E.J., and Cullen, B.R. (2002) Both natural and designed micro RNAs can inhibit the expression of cognate mRNAs when expressed in human cells. *Mol Cell*, **9**, 1327–1333.

22. Lin, S.L., Chuong, C.M., and Ying, S.Y. (2001) D-RNAi (messenger RNA-antisense DNA interference) as a novel defense system against cancer and viral infections. *Curr Cancer Drug Targets*, **1**, 241–247.
23. Lin, S.L. and Ying, S.Y. (2005) D-RNAi-based therapeutics. In: Los, M. and Gibson, S. (Eds.) *Apoptotic Pathways as Targets for Novel Therapies in Cancer & Other Diseases.* Kluwer Press, New York, pp. 275–296.
24. O'Donnell, K.A. and Boeke, J.D. (2007) Mighty Piwis defend the germline against genome intruders. *Cell*, **129**, 37–44.
25. Handa, V., Saha, T., and Usdin, K. (2003) The fragile X syndrome repeats form RNA hairpins that do not activate the interferon-inducible protein kinase, PKR, but are cut by Dicer. *Nucleic Acids Res*, **31**, 6243–6248.
26. Krol, J., Fiszer, A., Mykowska, A., Sobczak, K., de Mezer, M., and Krzyzosiak, W.J. (2007) Ribonuclease dicer cleaves triplet repeat hairpins into shorter repeats that silence specific targets. *Mol Cell*, **25**, 575–586.
27. Lin, S.L., Chang, D.C., and Ying, S.Y. (2005) Asymmetry of intronic pre-microRNA structures in functional RISC assembly. *Gene*, **356**, 32–38.
28. Danin-Kreiselman, M., Lee, C.Y., and Chanfreau, G. (2003) RNAse III-mediated degradation of unspliced pre-mRNA and lariat introns. *Mol Cell*, **11**, 1279–1289.
29. Yi, R., Qin, Y., Macara, I.G., and Cullen, B.R. (2003) Exportin-5 mediates the nuclear export of pre-microRNAs and short hairpin RNAs. *Genes Dev*, **17**, 3011–3016.
30. Lin, S.L., Chang, S.J.E., and Ying, S.Y. (2006) First in vivo evidence of microRNA-induced fragile X mental retardation syndrome. *Mol Psychiatry*, **11**, 616–617.
31. Lee, Y.S., Nakahara, K., Pham, J.W., Kim, K., He, Z., Sontheimer, E.J., and Carthew, R.W. (2004) Distinct roles for Drosophila Dicer-1 and Dicer-2 in the siRNA/miRNA silencing pathways. *Cell*, **117**, 69–81.
32. Tang, G. (2005) siRNA and miRNA: an insight into RISCs. *Trends Biochem Sci*, **30**, 106–114.
33. Liquori, C.L., Ricker, K., Moseley, M.L., Jacobsen, J.F., Kress, W., Naylor, S.L., Day, J.W., and Ranum, L.P.W. (2001) Myotinic dystrophy type 2 caused by a CCTG expansion in intron 1 of ZNF9. *Science*, **293**, 864–867.
34. Fu, Y.H., Kuhl, D.P.A., Pizzuti, A., Pieretti, M., Sutcliffe, J.S., Richards, S. et al. (1991) Variation of the CGG repeat X site results in genetic instability. Resolution of Sherman paradox. *Cell*, **67**, 1047–1058.
35. Hagerman, R.J., Staley, L.W., O'Conner, R., Lugenbeel, K., Nelson, D., McLean, S.D., and Taylor, A. (1996) Learning-disabled males with a fragile X CGG expansion in the upper premutation size range. *Pediatrics*, **97**, 122–126.
36. Jin, P., Zarnescu, D.C., Zhang, F., Pearson, C.E., Lucchesi, J.C., Moses, K., and Warren, S.T. (2003) RNA-mediated neurodegeneration caused by the fragile X premutation rCGG repeats in *Drosophila*. *Neuron*, **39**, 739–747.
37. Genc, B., Muller-Hartmann, H., Zeschnigk, M., Deissler, H., Schmitz, B., Majewski, F., von Gontard, A., and Doerfler, W. (2000) Methylation mosaicism of 5′-(CGG)(n)-3′ repeats in fragile X, premutation and normal individuals. *Nucleic Acids Res*, **28**, 2141–2152.
38. Jin, P., Alisch, R.S., and Warren, S.T. (2004) RNA and microRNAs in fragile X mental retardation. *Nat Cell Biol*, **6**, 1048–1053.
39. Miyagishi, M. and Taira, K. (2002) U6 promoter-driven siRNAs with four uridine 3′ overhangs efficiently suppress targeted gene expression in mammalian cells. *Nat Biotechnol*, **20**, 497–500.
40. Lee, N.S., Dohjima, T., Bauer, G., Li, H., Li, M.J., Ehsani, A., Salvaterra, P., and Rossi, J. (2002) Expression of small interfering RNAs targeted against HIV-1 rev transcripts in human cells. *Nat Biotechnol*, **20**, 500–505.
41. Paul, C.P., Good, P.D., Winer, I., and Engelke, D.R. (2002) Effective expression of small interfering RNA in human cells. *Nat Biotechnol*, **20**, 505–508.
42. Xia, H., Mao, Q., Paulson, H.L., and Davidson, B.L. (2002) siRNA-mediated gene silencing in vitro and in vivo. *Nat Biotechnol*, **20**, 1006–1010.
43. McCaffrey, A.P., Meuse, L., Pham, T.T., Conklin, D.S., Hannon, G.J., and Kay, M.A. (2002) RNA interference in adult mice. *Nature*, **418**, 38–39.
44. Gunnery, S., Ma, Y., and Mathews, M.B. (1999) Termination sequence requirements vary among genes transcribed by RNA polymerase III. *J Mol Biol*, **286**, 745–757.
45. Schramm, L. and Hernandez, N. (2002) Recruitment of RNA polymerase III to its target promoters. *Genes Dev*, **16**, 2593–2620.
46. Sledz, C.A., Holko, M., de Veer, M.J., Silverman, R.H., and Williams, B.R. (2003) Activation of the interferon system by short-interfering RNAs. *Nat Cell Biol*, **5**, 834–839.
47. Lin, S.L. and Ying, S.Y. (2004) Combinational therapy for HIV-1 eradication and vaccination. *Int J Oncol*, **24**, 81–88.

48. Stark, G.R., Kerr, I.M., Williams, B.R., Silverman, R.H., and Schreiber, R.D. (1998) How cells respond to interferons. *Annu Rev Biochem*, **67**, 227–264.
49. Grimm, D., Streetz, K.L., Jopling, C.L., Storm, T.A., Pandey, K., Davis, C.R., Marion, P., Salazar, F., and Kay, M.A. (2006) Fatality in mice due to oversaturation of cellular microRNA/short hairpin RNA pathways. *Nature*, **441**, 537–541.
50. Lin, S.L. and Ying, S.Y. (2004) New drug design for gene therapy – taking advantage of introns. *Lett Drug Des Discov*, **1**, 256–262.
51. Lin, S.L. and Ying, S.Y. (2004) Novel RNAi therapy – intron-derived microRNA drugs. *Drug Des Rev*, **1**, 247–255.
52. Zhou, H.S.L., Xia, X.G., and Xu, Z. (2005) An RNA polymerase II construct synthesizes short hairpin RNA with a quantitative indicator and mediates high efficient RNAi. *Nucleic Acid Res*, **33**, e62.
53. Chung, K.H., Hart, C.C., Al-Bassam, S., Avery, A., Taylor, J., Patel, P.D., Vojtek, A.B., and Turner, D.L. (2006) Polycistronic RNA polymerase II expression vectors for RNA interference based on BIC/miR-155. *Nucleic Acid Res*, **34**, e53.
54. Xia, X.G., Zhou, H., Samper, E., Melov, S., and Xu, Z. (2006) Pol II-expressed shRNA knocks down *Sod2* gene expression and causes phenotypes of the gene knockout in mice. *PLoS Genet*, **2**, e10.
55. Nott, A., Meislin, S.H., and Moore, M.J. (2003) A quantitative analysis of intron effects on mammalian gene expression. *RNA*, **9**, 607–617.
56. Zhang, G., Taneja, K.L., Singer, R.H., and Green, M.R. (1994) Localization of pre-mRNA splicing in mammalian nuclei. *Nature*, **372**, 809–812.
57. Ghosh, S. and Garcia-Blanco, M.A. (2000) Coupled in vitro synthesis and splicing of RNA polymerase II transcripts. *RNA*, **6**, 1325–1334.
58. Bateman, J.R. and Wu, C.T. (2007) DNA replication and models for the origin of piRNAs. *Bioessays*, **29**, 382–385.
59. Lin, S.L., Chang, D.C., and Ying, S.Y. (2007) Hyaluronan stimulates transformation of androgen-independent prostate cancer. *Carcinogenesis*, **28**, 310–320.
60. Lin, S.L. and Ying, S.Y. (2006) Gene silencing in vitro and in vivo using intronic microRNAs. *Methods Mol Biol*, **342**, 295–312.
61. Lin, S.L., Chang, S.J.E., and Ying, S.Y. (2006) Transgene-like animal model using intronic microRNAs. *Methods Mol Biol*, **342**, 321–334.
62. Lee, Y., Ahn, C., Han, J., Choi, H., Kim, J., Yim, J., Lee, J., Provost, P., Radmark, O., Kim, S., and Kim, V.N. (2003) The nuclear RNase III Drosha initiates microRNA processing. *Nature*, **425**, 415–419.
63. Butz, S. and Larue, L. (1995) Expression of catenins during mouse embryonic development and in adult tissues. *Cell Adhes Commun*, **3**, 337–352.
64. Lin, S.L., Suksaweang, S., Chuong, C.M., and Ying, S.Y. (2001) D-RNAi (Messenger RNA-antisense DNA interference) as a novel defense system against cancer and viral infections. *Curr Cancer Drug Targets*, **1**, 241–247.
65. Filipovska, J. and Konarska, M.M. (2000) Specific HDV RNA-templated transcription by pol II in vitro. *RNA*, **6**, 41–54.
66. Modahl, L.E., Macnaughton, T.B., Zhu, N., Johnson, D.L., and Lai, M.M. (2000) RNA-dependent replication and transcription of hepatitis delta virus RNA involve distinct cellular RNA polymerases. *Mol Cell Biol*, **20**, 6030–6039.
67. Vagin, V.V., Sigova, A., Li, C., Seitz, H., Gvozdev, V., and Zamore, P.D. (2006) A distinct small RNA pathway silences selfish genetic elements in the germline. *Science*, **313**, 320–324.
68. Carmell, M.A., Girard, A., van de Kant, H.J., Bourc'his, D., Bestor, T.H., de Rooij, D.G., and Hamnon, G.J. (2007) MIWI2 is essential for spermatogenesis and repression of transposons in the mouse male germline. *Dev Cell*, **12**, 503–514.
69. Abzhanov, A., Protas, M., Grant, B.R., Grant, P.R., and Tabin, C.J. (2004) Bmp4 and morphological variation of beaks in Darwin's finches. *Science*, **305**, 1462–1465.
70. Wu, P., Jiang, T.X., Suksaweang, S., Widelitz, R.B., and Chuong, C.M. (2004) Molecular shaping of the beak. *Science*, **305**, 1465–1466.
71. Sheth, U. and Parker, R. (2006) Targeting of aberrant mRNAs to cytoplasmic processing bodies. *Cell*, **125**, 1095–1109.
72. Kovacs, J.A., Vogel, S., Albert, J.M., Falloon, J., Davey, R.T., Jr., Walker, R.E., Polis, M.A., Spooner, K., Metcalf, J.A., Baseler, M., Fyfe, G., and Lane, H.C. (1996) Controlled trial of interleukin-2 infusions in patients infected with the human immunodeficiency virus. *N Engl J Med*, **335**, 1350–1356.
73. Reya, T., Morrison, S.J., Clarke, M.F., and Weissman, I.L. (2001) Stem cells, cancer, and cancer stem cells. *Nature*, **414**, 105–111.
74. Takahashi, K. and Yamanaka, S. (2006) Induction of pluripotent stem cells from

mouse embryonic and adult fibroblast cultures by defined factors. *Cell*, **126**, 663–676.
75. Okita, K., Ichisaka, T., and Yamanaka, S. (2007) Generation of germline-competent induced pluripotent stem cells. *Nature*, **448**, 313–317.
76. Wernig, M., Meissner, A., Foreman, R., Brambrink, T., Ku, M., Hochedlinger, K., Bernstein, B.E., and Jaenisch, R. (2007) In vitro reprogramming of fibroblasts into a pluripotent ES-cell-like state. *Nature*, **448**, 318–324.
77. Simonsson, S. and Gurdon, J.B. (2004) DNA demethylation is necessary for the epigenetic reprogramming of somatic cell nuclei. *Nat Cell Biol*, **6**, 984–990.
78. Murchison, E.P., Stein, P., Xuan, Z., Pan, H., Zhang, M.Q., Schultz, R.M., and Hannon, G.J. (2007) Critical roles for Dicer in the female germline. *Genes Dev*, **21**, 682–693.
79. Tang, F., Kaneda, M., O'Carroll, D., Hajkova, P., Barton, S.C., Sun, Y.A., Lee, C., Tarakhovsky, A., Lao, K., and Surani, M.A. (2007) Maternal microRNAs are essential for mouse zygotic development. *Genes Dev*, **21**, 644–648.
80. Suh, M.R., Lee, Y., Kim, J.Y., Kim, S.K., Moon, S.H., Lee, J.Y., Cha, K.Y., Chung, H.M., Yoon, H.S., Moon, S.Y., Kim, V.N., and Kim, K.S. (2004) Human embryonic stem cells express a unique set of microRNAs. *Dev Biol*, **270**, 488–498.
81. Lin, S.L., Chang, D.C., Chang-Lin, S., Lin, C.H., Wu, D.T., Chen, D.T., and Ying, S.Y. (2008) Mir-302 reprograms human skin cancer cells into a pluripotent ES-cell-like state. *RNA*, **14**, 2115–2124.
82. Lin, S.L., Kim, H., and Ying, S.Y. (2008) Intron-mediated RNA interference and microRNA (miRNA). *Front Biosci*, **13**, 2216–2230.
83. Sherr, C.J. and Roberts, J.M. (2008) Living with or without cyclins and cyclin-dependent kinases. *Genes Dev*, **18**, 2699–2711.

Chapter 15

Inhibition of the microRNA Pathway in Zebrafish by siRNA

Anders Fjose and Xiao-Feng Zhao

Abstract

The microRNA (miRNA) pathway and the phenomenon of RNA interference (RNAi), which have both been shown to involve targeting of mRNAs by small RNA molecules, are interconnected and depend partly on the same cellular machinery. RNAi in vertebrates was first reported in zebrafish (*Danio rerio*) 10 years ago. However, reliable RNAi-based gene silencing techniques, based on injection of small interfering RNAs (siRNAs) into zygotes, have not been established for this important vertebrate model because of unspecific developmental defects. We have recently shown that these side effects can be attributed to inhibition of the miRNA pathway by siRNAs at early embryonic stages. This review highlights these findings and the function of microRNAs in zebrafish development.

Key words: RNAi, siRNA, microRNA, miR-430, zebrafish.

1. Introduction: Connections Between the RNAi and miRNA Pathways

The discovery of RNA interference (RNAi) demonstrated that exogenously introduced double-stranded RNA (dsRNA) molecules could trigger degradation of cognate messenger RNA (mRNA) sequences in the nematode *Caenorhabditis elegans* (1). This RNA silencing process was subsequently shown to involve the generation of small interfering RNA (siRNA) molecules of ~22 nucleotides (nt), which guide the RISC complex to target and cleave mRNAs (2). Research on developmental regulatory genes in *C. elegans* also led to the discovery of a new category of small endogenous RNA molecules, microRNAs (miRNAs), generated by processing of transcripts containing certain stem-loop structures (3–6). These miRNAs have approximately the same size as siRNAs, and their generation and function depend to a

M. Sioud (ed.), *RNA Therapeutics*, Methods in Molecular Biology 629,
DOI 10.1007/978-1-60761-657-3_15, © Springer Science+Business Media, LLC 2010

large extent on the same cellular machinery (7–9, *see* **Chapters 14** and **16**).

During the past 10 years, rapid progress in the research on these two RNA-mediated silencing mechanisms has revealed that they are widespread in eukaryotic organisms, where they are involved in cellular control of gene expression or protection against parasitic and pathogenic invaders such as transposons and viruses (8, 10). In addition, various studies have uncovered that they have essential roles in developmental processes and can be associated with diseases (11, 12). Importantly, these small-RNA-mediated pathways can also be utilized experimentally or therapeutically to knock down expression of specific genes (13, 14).

Generation of siRNAs and miRNAs from dsRNA precursors that are produced in or introduced into cells depend on the ribonuclease III (RNase III) enzyme known as Dicer (8, 15). In the case of miRNAs, the dsRNA precursors (pre-miRNAs) are ~65 nt stem-loop structures, which are excised from longer primary transcripts (pri-miRNAs) in the nucleus by the RNAse III enzyme Drosha and exported to the cytoplasm by nucleo-cytoplasmic transport factors (8, 9, 16). In the cytoplasm, one of the strands from the duplexes of miRNAs or siRNAs, which has the weakest base pairing at its 5′ terminus, is preferably incorporated into the RNA-induced silencing complexes (RISCs) miRISC or siRISC, respectively (8). The loaded single-stranded RNA (ssRNA), sometimes called the guide strand, targets RISCs to mRNAs that contain homologous sequences.

Whereas siRNAs, in general, and the miRNAs in plants are perfectly complementary to their targets, animal miRNAs usually base pair with imperfect complementarity to target sequences located within the 3'UTRs of mRNAs. The first 2–8 nucleotides at the 5′ end of miRNAs, commonly referred to as the "seed" region, is most important for target recognition and silencing (17, 18). However, the presence of secondary structures and other features of the 3'UTR sequence surrounding the target site may also influence how miRNAs and mRNAs interact.

Each RISC contains an Argonaute (AGO) protein as its core component and several additional protein factors that may vary in different types of complexes (8). Importantly, different members of the AGO family, which contain the highly conserved RNA-binding domains PAZ and PIWI, are core components in siRISCs and miRISCs. AGO proteins in siRISCs silence targeted mRNAs by catalyzing their endonucleolytic cleavage. In the animal miRNA pathway, AGO proteins are known to inhibit the expression of mRNAs by various mechanisms, including translational inhibition, mRNA degradation, and rapid degradation of nascent peptides (19). In some cases, animal miRNAs may also have sufficient complementarity to sequences in the targeted mRNAs to cause endonucleolytic cleavage, indicating that there is no clear distinction between miRNAs and siRNAs (20).

The miRNA gene number in vertebrate species is currently estimated to exceed 500 genes, which can be grouped into >40 families based on seed sequence identity (21–24). Furthermore, various computational predictions and experimental studies indicate that single miRNAs can target >100 different mRNAs (25, 26). Hence, at least 30% of human genes are probably regulated by miRNAs (17). Considering these estimates, it can be assumed that miRNAs have global impact on gene-expression profiles, and may in some cases function as developmental switches. Several recent studies in mice and zebrafish have provided experimental evidence in support of these ideas (11, 25). In particular, the family of miR-430 miRNAs in zebrafish has been shown to be important for the removal of a large group of maternal mRNAs after the initiation of zygotic transcription at the mid-blastula stage, which is critical for normal development (25, 27). Furthermore, establishment of new techniques to manipulate and assay the activity of specific miRNAs in zebrafish, which are based on microinjection of reporter mRNAs and inhibitors of processing and target binding, have led to important contributions to our understanding of miRNA functions in vertebrates (25, 28–30).

RNAi is a powerful tool to knock down expression of specific genes, and it has been employed as a research tool in a variety of species, including animal models such as *C. elegans*, *Drosophila melanogaster*, and mice (31–33). In addition, RNAi methods have been used extensively to study gene functions in mammalian cell culture systems (34). Furthermore, strong efforts have been made to develop RNAi-based therapeutics in humans (35, 36).

Although the first observation of RNAi in vertebrates were reported for zebrafish (37), RNAi has not been established as a reliable tool for analysis of gene functions in this important animal model due to frequent occurrence of unspecific developmental aberrations (28, 38). The factors causing this problem have remained unknown until recently, when it was demonstrated that siRNA treatment led to inhibition of the miRNA pathway in zebrafish embryos (39). In particular, siRNA injections were shown to reduce the level of mature miR-430 miRNAs, and in this way affect the maternal to zygotic transition (25, 39). This finding may help us to develop new RNAi techniques for zebrafish that can circumvent the problem leading to unspecific developmental defects. The observations in zebrafish also suggest a need for investigating whether similar inhibition of the miRNA pathway may occur in mammals under some experimental or therapeutic conditions.

2. Functions of miRNAs in Zebrafish Development

2.1. Zebrafish miRNAs and Analytical Tools

Identification and characterization of miRNAs in zebrafish have provided important information regarding the conservation of miRNA genes in vertebrates. Zebrafish have also proven to be a particularly valuable system to study miRNA functions in developmental processes. In the initial investigations, small-RNA cDNA libraries from different developmental stages of zebrafish were used for large-scale sequencing to obtain a comprehensive picture of the total number of miRNAs (40–42). An additional approach has been to use prediction algorithms to search in the zebrafish genome database to identify conserved hairpin structures of 60–100 nucleotides (22). These analyses have led to identification of 415 zebrafish miRNAs, and based on seed sequence identity it has been possible to group these into 44 families (22). However, since zebrafish genome sequencing and annotation is not yet complete, the exact number of miRNAs remains unknown.

Synchronous populations of several hundred fertilized zebrafish eggs (zygotes) are easily available for microinjection of reporter constructs, RNA molecules or chemically modified nucleic acids, which can facilitate investigations of miRNA functions in developmental processes. For gain-of-function studies, mature miRNA duplexes, which have been chemically synthesized, can be injected into zygotes to analyze how overexpression of specific miRNAs may affect development (28). Mature miRNA duplexes can also be coinjected with reporter mRNAs to test whether predicted miRNA-binding sites can mediate silencing of the presumptive target mRNA. This type of assay is based on coinjections of specific miRNA duplexes with reporter mRNAs containing a green fluorescent protein (GFP) coding sequence and a 3'UTR region with the predicted miRNA-binding site (25).

Antisense morpholino phosphorodiamidate oligonucleotides (MOs) have been used extensively to knock down gene functions in zebrafish embryos and larvae by inhibition of splicing and/or translation of specific mRNAs (43, 44). Due to the chemical modifications, these oligonucleotides are not susceptible to enzymatic degradation and can be applied to inhibit gene functions in zebrafish larvae several days after microinjection at the zygote stage. MOs have also proved to be very useful as tools to inhibit specific miRNAs, and can be designed to block processing of the pre-miRNA or the activity of the mature miRNA (29, 45). In addition, MOs designed to be complementary to specific miRNA-binding sites in target mRNAs have been shown to efficiently protect the target mRNA from translational inhibition or degradation (30). The efficiency of "target protectors" to block the interaction of miRNAs with a particular binding site in a target mRNA can be investigated by coinjections with reporter mRNAs containing a GFP coding sequence and a 3'UTR region with the miRNA-binding site (30). Subsequently, a tested target protector can be used to study whether protection of a specific target mRNA from silencing has any biological effect.

A key step toward understanding the function of specific miRNAs is to determine the spatiotemporal expression pattern during development. However, detection of miRNAs by in situ hybridization has been technically difficult because of their small size. This problem was solved when it was discovered that locked nucleic acid (LNA) modified DNA oligonucleotide probes can provide increased sensitivity for detection of miRNAs by Northern blots (46), and can be applied for whole-mount in situ detection of specific miRNAs in zebrafish embryos and larvae (40, 47).

The temporal and spatial expression patterns of many zebrafish miRNAs have been investigated by in situ hybridization, microarray, and/or Northern analyses of different developmental stages (40, 41, 47). Studies of the temporal profiles revealed that only a few are expressed before the end of gastrulation. The majority of the zebrafish miRNAs also display tissue-specific expression patterns, suggesting functional roles in differentiation or maintenance of tissue identity (47). Similar spatiotemporal features have been observed in other vertebrates, and generally miRNA homologues are expressed in the same type of tissue (12, 48).

2.2. Functional Roles of Specific miRNAs

Studies of abnormalities caused by *dicer* mutations in mice and zebrafish provided the first evidence that generation of mature miRNAs is essential for normal development, but alternative explanations involving failure to produce endogenous siRNAs could not be excluded (49, 50). In zebrafish, homozygous *dicer* mutants were shown to produce mature miRNAs during the first few days, due to the presence of maternally contributed *dicer* (49). As a consequence these zygotic *dicer* (Z*dicer*) mutants have no visible defects other than a developmental arrest around day 10 postfertilization (49).

The problem of maternal *dicer* activity could be circumvented when a germ line replacement technique was used to generate embryos that were maternal–zygotic mutant for *dicer* (MZ*dicer*). These MZ*dicer* embryos did not generate mature miRNAs and displayed severe morphogenesis defects during gastrulation, brain formation, somitogenesis, and heart development (51). Further analyses revealed a causal connection between some of these abnormalities and members of the miR-430 family of miRNAs, which are the only abundant miRNAs at blastula and gastrula stages (41, 51). These miRNAs are expressed ubiquitously from multiple gene copies, mainly located in two large clusters on chromosome 4 (22, 41).

Experimental evidence that normal development depends on the miR-430 family was obtained from microinjections of preprocessed duplexes of these miRNAs into MZ*dicer* zygotes, which resulted in almost complete rescue of the brain and gastrulation defects (51). In particular, these injections were shown to rescue the formation of normal brain ventricles and brain constrictions, including the characteristic fold of the midbrain–hindbrain boundary (MHB), revealing essential roles for the miR-430 miRNAs during brain morphogenesis. Hence, it could also be concluded that this particular phenotype depended on the loss of a specific mature miRNA and was not due to the inability of MZ*dicer* mutants to generate endogenous siRNAs. However, since miR-430 injections failed to rescue some of the abnormalities (e.g., heart defects) in MZ*dicer* embryos, it remains a possibility that endogenous siRNAs are required for normal development. Alternatively, these additional defects may be associated with the function of other miRNAs.

To elucidate how the lack of mature miR-430 miRNAs could lead to morphogenesis defects, it was important to consider the temporal expression profile and how it related to processes known to occur at the same embryonic stages. Interestingly, this showed that expression of the miR-430 family members begins at the moment of transition from maternal to zygotic expression, and their mature miRNAs rapidly accumulate to high levels (41, 51). This maternal–zygotic transition, which is a universal process in animal development, is also known to involve elimination of many maternal mRNAs (52). In zebrafish, this transition occurs at the mid-blastula stage (~2.5 hpf [hours postfertilization]) around 2 h before initiation of gastrulation (25).

Comparison of mRNA levels showed that >700 genes were expressed at significantly higher levels in MZ*dicer* mutant embryos than in wild-type embryos (25). Interestingly, target predictions indicated an enrichment of putative miR-430 target sites in the upregulated mRNAs, and several of these 3'UTR targets were validated in injection experiments with reporter mRNAs (25). On the basis of these results, it was estimated that the miR-430 family of miRNAs directly regulate several hundred maternal mRNAs (25). Further studies investigated the mechanism by which miR-430 promoted removal of the maternal target mRNAs. These investigations provided evidence that binding of miR-430 to the 3'UTR of maternal mRNAs leads to deadenylation, thereby triggering degradation (25). Hence, miR-430 seems to have a major role in facilitating clearance of a large group of maternal mRNAs at the transition to zygotic expression. In MZ*dicer* mutants, the absence of mature miR-430 leads to delayed degradation of the maternal target mRNAs, and consequently an extended period of mixed maternal and zygotic gene expression. This mixed state probably interferes with processes

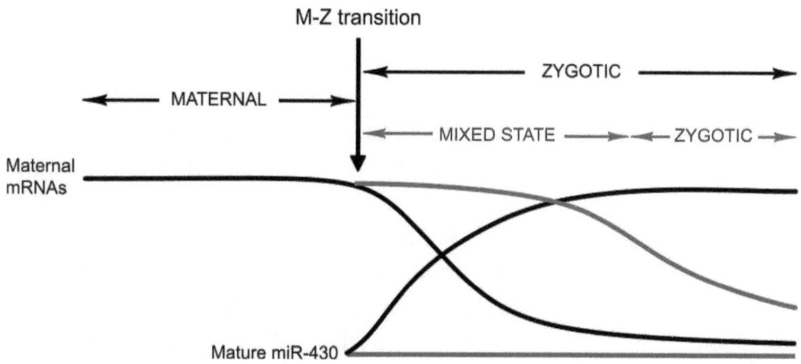

Fig. 15.1. Removal of maternal mRNAs at the maternal–zygotic (M–Z) transition in early zebrafish embryos is mediated by miRNAs from the miR-430 family. In MZ*dicer* embryos (*greyish lines* and characters), the mature forms of miR-430 miRNAs are not generated and as a result the removal of maternal mRNAs is delayed. The delayed removal leads to a mixed state of zygotic and maternal expression.

responsible for gastrulation and other types of morphogenesis (**Fig. 15.1**).

Similar miRNA-based mechanisms are likely to operate at the maternal–zygotic transition in other species of fish. In support of this assumption, conserved and clustered miR-430 miRNAs have been found in other fish genomes, including *Fugu rubripes* and *Tetradon nigroviridis* (51). This miRNA family is also conserved in amphibians, and studies on *Xenopus* have detected high levels of miR-430 during the blastula period (53). However, investigations on mammals and *Drosophila* have shown that unrelated types of miRNAs and/or endogenous siRNAs are involved in targeting of maternal mRNAs for degradation during the transition to zygotic expression (54, 55).

Although the MZ*dicer* mutants displayed morphogenesis defects, such as lack of brain ventricles and brain constrictions, detailed morphological analysis revealed that the body axis and all major subregions and cell types were not affected by the lack of mature miRNAs (51). Hence, mature miRNAs apparently do not have widespread essential roles in cell fate specification or signaling during early zebrafish development. However, several recent studies have demonstrated important roles of particular miRNAs in specific processes during zebrafish development, and similar results have been reported for various miRNAs mammals (11, 12, 21, 28). One important example of conservation of miRNA functions in vertebrates has been revealed for the Hox clusters, which encode transcription factors that are required for correct development of different tissues along the anterior–posterior axis (56). In the Hox clusters of different vertebrates, two conserved miRNAs, miR-10 and miR-196, are located in the same positions (56–58). Furthermore, studies in both zebrafish and mice have shown that these miRNAs target the mRNAs of nearby Hox genes and contribute to define their expression domains along the body axis (56, 57, 59).

In addition to an involvement in patterning of the anterior–posterior axis, several zebrafish miRNAs have been shown to participate in regulating the differentiation of highly specialized cell types. By using MOs to block the maturation of miR-375, it was revealed that this miRNA is essential for the formation of the insulin-secreting pancreatic islet (29). Similarly it was demonstrated that zebrafish miR-214 contributes to precise specification of muscle cell types by modulating a component of the hedgehog signal transduction pathway (45). Inhibition of miRNA maturation has also been applied in several other investigations where specific zebrafish miRNAs have been shown to have roles in erythroid maturation, fin regeneration, and cardiac patterning (60–62).

3. RNAi in Zebrafish

Following the discovery of RNAi in *C. elegans* (1), the first study of RNAi in vertebrates was reported for zebrafish by our laboratory (37). Using dsRNA microinjection in zygotes to target three genetically characterized genes, *ntl*, *flh*, and *pax2a*, we demonstrated that the mRNA levels were reduced and the embryonic defects were similar to the known mutant phenotypes of these loci (37). However, the efficiency to induce phenocopies of the mutants was quite low (<30%) and treatment with increasing concentrations of dsRNA was in some cases shown to cause a higher frequency of unspecific defects (37). Despite these limitations, it was assumed that RNAi-based techniques could be established as important tools for loss-of-function studies of specific zebrafish genes. Further attempts were made by several laboratories to explore the potential of dsRNA to knock down expression of specific genes, but these investigations showed conflicting results. Whereas some research groups claimed to generate specific effects with high efficiency (63, 64), other scientists reported that dsRNA injections resulted only in unspecific effects (65, 66).

At the same time as these investigations were made, it was discovered that antisense MOs, designed to inhibit translation and/or splicing, could be used as efficient tools to knock down expression of specific genes in zebrafish (67–69). However, the phenotypes induced by MO injections were also shown to include variable side effects that made interpretations difficult (67). An additional disadvantage of MOs was the high costs required for synthesis of these chemically modified oligonucleotides. Hence, it was still of interest to explore the possibilities to establish efficient

RNAi-based techniques in zebrafish, and in this connection it was important to consider the potential causes for the high frequencies of unspecific abnormalities in dsRNA-injected embryos. One likely explanation was that these side effects could reflect a global antiviral/interferon response to long dsRNA (70). This issue was also discussed in relation to the application of similar methods in mammals, and using siRNAs to trigger RNAi in mammalian cells, the global responses associated with longer dsRNAs could be avoided (71).

Following the first demonstration of siRNA-mediated gene silencing in mammalian cells, several laboratories reported that siRNA injections could be used to induce gene-specific effects in zebrafish (72–74). In studies on three different zebrafish cell lines, highly efficient gene-specific knock down has also been demonstrated in experiments where several siRNAs were introduced by microinjection and transfection (38). However, when the same siRNAs were used for microinjection into zebrafish zygotes, most of the embryos at later stages displayed severe unspecific defects, including truncated tails, loss of brain structures, and enlarged heart cavities (38). The conflicting results reported by different laboratories are difficult to explain, but they may partially reflect differences in the injection procedures (e.g., amounts of siRNA injected) and variations in target-binding efficiencies of the siRNAs. It has also been important to find an explanation to the observation that siRNAs, which mediated specific gene silencing in zebrafish cell lines, were only able to cause unspecific abnormalities in embryos (38).

4. Inhibition of the miRNA Pathway by siRNA

Remarkably, the MZ*dicer* mutant embryos displayed several defects similar to those caused by siRNAs, and the defects in mutants were partially rescued by injection of preprocessed miR-430 miRNAs (38, 51). Based on these observations, and considering the common factors in the RNAi and miRNA pathways, we have investigated whether the function of miR-430 could be associated with the unspecific defects caused by siRNA treatment (39). Following injection of siRNAs corresponding to sequences in two endogenous genes (*eri1* and *six3a*) and the exogenous reporter gene GFP, we found by Northern analysis of RNA extracted from early embryos (12 hpf) that the levels of processed miRNAs were significantly reduced for the four miRNAs tested, including the miR-430 family member miR-430b (39) (**Fig. 15.2**). Importantly, these siRNA injections also caused brain morphogenesis defects and tail truncations similar to the MZ*dicer*

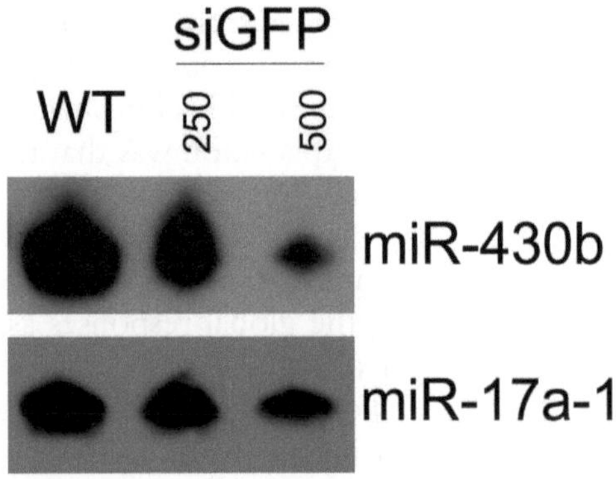

Fig. 15.2. Injection of siRNA affects the endogenous level of miRNAs. Following injection of different amounts (250 and 500 ng) of siRNA containing a GFP sequence (siGFP), Northern analysis of the endogenous levels of the processed forms of miR-430b and miR-17a-1 showed significant reductions as compared to noninjected embryos (WT) at the same developmental stage (12 hpf). Reproduced from (39) with permission from Blackwell Publishing Ltd.

phenotype (39) (**Fig. 15.3**). By contrast, injections of other types of RNA molecules such as single-stranded and double-stranded RNAs without 3′ overhangs did not affect embryonic development or the levels of mature miRNAs (39). Hence, these results suggested that the characteristic features of siRNAs were critical for reducing the levels of processed miRNAs and causing the unspecific defects in zebrafish embryos. Interestingly, we also observed strongest reduction (~70%) in the level of mature miR-430b (39) (**Fig. 15.2**).

To investigate a possible involvement of miR-430, we checked whether coinjection of preprocessed miR-430b duplexes could rescue the effects of siRNA treatment (39) (**Fig. 15.3**). Consistent with the previously reported rescue of MZ*dicer* mutants by miR-430 (51), we observed that a control injection of miR-430b alone did not induce any embryonic defects. However, in additional control experiments with the mature forms of miR-206, and a mutated version of miR-430b (miR-430b-mis), which lacks the power to rescue MZ*dicer* embryos (51), the embryos were affected in a similar way as when injected with siRNA (39). These results indicated that treatment with miRNA duplexes generally leads to the same kind of unspecific abnormalities as in the case of siRNAs, and this assumption was supported by observations of reduced levels of mature miR-430b following miR-206 injection (39). By contrast, coinjections of the three different siRNAs with preprocessed miR-430b duplexes resulted in efficient rescue of the unspecific defects, which were observed when these siRNAs were injected alone (39) (**Fig. 15.3**). Such rescue was not observed in coinjections with miR-430b-mis, which cannot bind target sites because of two-point substitutions in the 5′

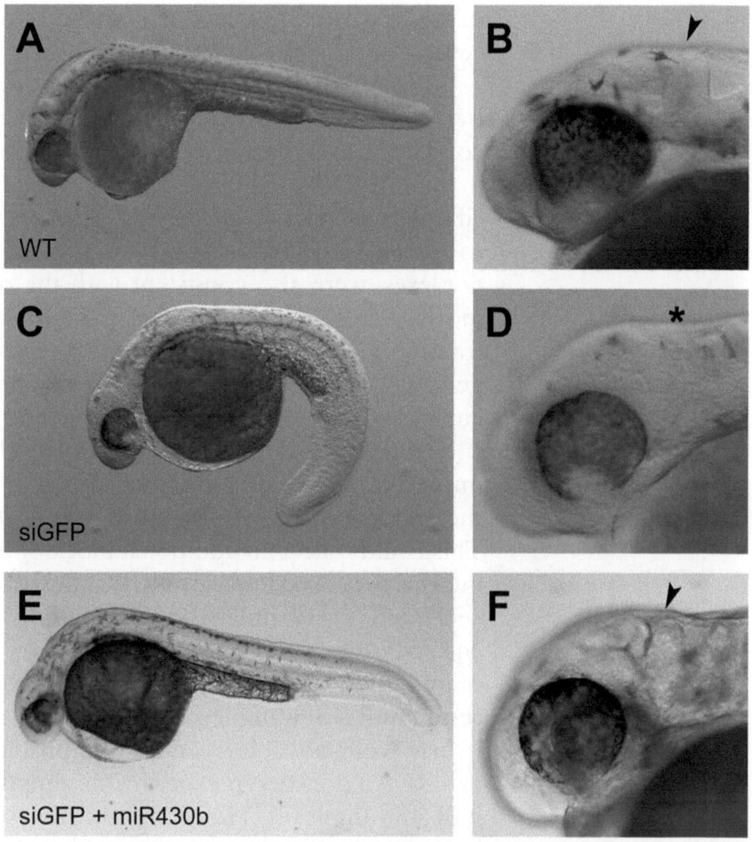

Fig. 15.3. Coinjection of mature miRNAs from the miR-430 family can rescue siRNA-induced abnormalities in zebrafish embryos. (**a**, **b**) Normal embryos have a straight tail and a characteristic fold at the MHB (*arrowhead*). (**c**, **d**) Injection of siRNA containing a GFP sequence (siGFP) caused tail defects and abnormalities at the MHB (*star* in **d**). The unspecific brain and tail defects caused by siRNA injection were rescued by coinjection of preprocessed miR-430b duplexes (**e, f**). Images in **c–f** are reproduced from (39) with permission from Blackwell Publishing Ltd.

seed region (51). These results clearly showed that the unspecific brain and tail defects induced by siRNAs, which correlated with a strong reduction of the endogenous level of mature miR-430b, could be prevented by coinjection of this particular miRNA (39).

Taken together, these observations strongly suggest a direct or indirect inhibitory effect of injected siRNA (and miRNA) duplexes on the processing of endogenous miRNAs, particularly miR-430, which is the most abundant miRNA at early embryonic stages. Such inhibition probably results from a competition between pre-miRNAs and injected siRNAs for binding to essential factors. This idea is based on the knowledge that siRNAs can associate with Dicer and other factors required for the generation of siRNAs and mature miRNAs from dsRNAs and pre-miRNAs, respectively (75–77). Accordingly, the injection of large amounts of siRNAs would affect the availability of these factors for processing of pre-miRNAs.

The relatively strong reduction of processed miR-430 caused by siRNA injection would be at least partially equivalent to the conditions in MZ*dicer* embryos where there is no production of mature miRNAs (51). Furthermore, the similarities of the unspecific defects relative to the MZ*dicer* phenotype, and the ability to rescue some of these by injection of preprocessed miR-430 duplexes, were also consistent with the role of miR-430 in promoting clearance of maternal mRNAs at the maternal–zygotic transition (25). Hence, our results suggest that siRNA treatment generally will induce these kinds of abnormalities in zebrafish embryos.

Notably, we also observed relatively high frequencies of unspecific defects after injecting smaller amounts of siRNAs, when the endogenous level of mature miR-430 was less affected (39). This could reflect an additional block of miRNA function involving the effector complex miRISC, due to competition with miR-430 (and other miRNAs) for binding to Argonaute proteins and/or other factors (**Fig. 15.4**). It seems likely that this type of inhibition was partially prevented by the coinjection of preprocessed miR-430b duplexes as well. Based on these considerations, it may seem possible to avoid the problem of unspecific effects in siRNA knockdown experiments simply by coinjecting mature miR-430 duplexes. However, with this strategy the efficiency of inducing specific phenotypes would probably be reduced because

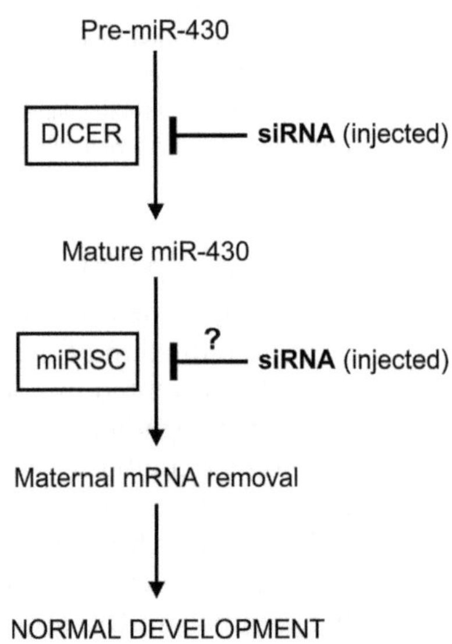

Fig. 15.4. A model proposed to explain how injection of siRNAs may inhibit the function of miR-430, which is essential for normal development. Injection of siRNA has been shown to reduce the level of mature miR-430 (39), but competition for critical factors may also affect miRISC function.

of the competition between siRNAs and miR-430 for binding to essential components. Furthermore, the inhibition of other miRNAs expressed at early stages may cause some additional defects that cannot be rescued by coinjected miR-430.

Another approach to avoid inhibition of the miRNA pathway in early embryos may be to use chemically modified siRNAs ("caged" siRNAs) that can be activated by photoexposure at later stages during zebrafish development (78). Alternatively, it may be possible to establish transgenic RNAi techniques in zebrafish that are similar to the techniques currently being used in *Drosophila* and mice (79, 80). Vectors developed for inducible or tissue-specific expression of short-hairpin RNAs (shRNAs) should make it possible to circumvent the problems associated with maternal mRNA clearance during the maternal–zygotic transition (81–83). However, it remains possible that the general inhibitory effect of siRNAs (or shRNAs) on the miRNA pathway at later developmental stages could lead to some unspecific defects as well. This general inhibition may also be a problem in studies on mammals, and it will be particularly important to investigate these aspects in relation to the therapeutic use of RNAi in humans.

It cannot be excluded that siRNA treatment can cause some additional side effects in zebrafish. In the case of MO knockdown experiments, about 15–20% of the MOs used in zebrafish show off-targeting effects, and these have been shown to result from induction of a p53-dependent cell death pathway (84). Although the mechanism of MO-induced p53 activation is not known, it appears to depend on the MO sequences used (84). Studies on mammalian cells have also revealed that some siRNA sequences can cause a significant increase in the protein level of p53 (85). However, it is still unclear whether siRNA treatment can result in similar side effects in zebrafish.

5. Conclusions

The discovery of RNAi has provided us with powerful new tools to investigate specific gene functions in several animal models. Similarly, the identification of miRNAs has revealed novel cellular mechanisms that control gene expression. However, the rapid progress in our understanding of these two pathways has shown that they are quite strongly interconnected, and use of RNAi to knock down gene expression may in some cases influence the miRNA pathway and cause side effects. Our work on zebrafish has uncovered a general inhibitory effect of siRNAs on the miRNA pathway, which particularly affects the global role of miR-430 in the removal of maternal mRNAs at the point of transition from maternal to zygotic expression. This finding may help us to establish reliable siRNA-mediated gene silencing techniques in the zebrafish model. It also emphasizes the need to investigate how experimental and therapeutic use of RNAi may affect miRNA functions and cause unwanted side effects.

Acknowledgments

This work was supported by the *Research Council of Norway* project 174979/I30, and the Faculty of Mathematics and Natural Sciences at the University of Bergen.

References

1. Fire, A., Xu, S., Montgomery, M.K. et al. (1998) Potent and specific genetic interference by double-stranded RNA in *Caenorhabditis elegans*. *Nature*, **391**, 806–811.
2. Bernstein, E., Caudy, A.A., Hammond, S.M. et al. (2001) Role for a bidentate ribonuclease in the initiation step of RNA interference. *Nature*, **409**, 363–366.
3. Lau, N.C., Lim, L.P., Weinstein, E.G. et al. (2001) An abundant class of tiny RNAs with probable regulatory roles in *Caenorhabditis elegans*. *Science*, **294**, 858–862.
4. Reinhart, B.J., Slack, F.J., Basson, M. et al. (2000) The 21-nucleotide let-7 RNA regulates developmental timing in *Caenorhabditis elegans*. *Nature*, **403**, 901–906.
5. Wightman, B., Ha, I., and Ruvkun, G. (1993) Posttranscriptional regulation of the heterochronic gene lin-14 by lin-4 mediates temporal pattern formation in *C. elegans*. *Cell*, **75**, 855–862.
6. Lee, R.C., Feinbaum, R.L., and Ambros, V. (1993) The *C. elegans* heterochronic gene lin-4 encodes small RNAs with antisense complementarity to lin-14. *Cell*, **75**, 843–854.
7. Zeng, Y., Yi, R., and Cullen, B.R. (2003) MicroRNAs and small interfering RNAs can inhibit mRNA expression by similar mechanisms. *Proc Natl Acad Sci USA*, **100**, 9779–9784.
8. Rana, T.M. (2007) Illuminating the silence: understanding the structure and function of small RNAs. *Nat Rev Mol Cell Biol*, **8**, 23–36.
9. Kim, D.H., Behlke, M.A., Rose, S.D. et al. (2005) Synthetic dsRNA Dicer substrates enhance RNAi potency and efficacy. *Nat Biotechnol*, **23**, 222–226.
10. Baulcombe, D. (2004) RNA silencing in plants. *Nature*, **431**, 356–363.
11. Stefani, G. and Slack, F.J. (2008) Small non-coding RNAs in animal development. *Nat Rev Mol Cell Biol*, **9**, 219–230.
12. Kloosterman, W.P. and Plasterk, R.H. (2006) The diverse functions of microRNAs in animal development and disease. *Dev Cell*, **11**, 441–450.
13. Dillon, C.P., Sandy, P., Nencioni, A. et al. (2005) RNAi as an experimental and therapeutic tool to study and regulate physiological and disease processes. *Annu Rev Physiol*, **67**, 147–173.
14. Beal, J. (2005) Silence is golden: can RNA interference therapeutics deliver? *Drug Discov Today*, **10**, 169–172.
15. Hammond, S.M. (2005) Dicing and slicing: the core machinery of the RNA interference pathway. *FEBS Lett*, **579**, 5822–5829.
16. Yi, R., Qin, Y., Macara, I.G. et al. (2003) Exportin-5 mediates the nuclear export of pre-microRNAs and short hairpin RNAs. *Genes Dev*, **17**, 3011–3016.
17. Lewis, B.P., Burge, C.B., and Bartel, D.P. (2005) Conserved seed pairing, often flanked by adenosines, indicates that thousands of human genes are microRNA targets. *Cell*, **120**, 15–20.
18. Doench, J.G. and Sharp, P.A. (2004) Specificity of microRNA target selection in translational repression. *Genes Dev*, **18**, 504–511.
19. Pillai, R.S., Bhattacharyya, S.N., and Filipowicz, W. (2007) Repression of protein

synthesis by miRNAs: how many mechanisms? *Trends Cell Biol*, **17**, 118–126.
20. Yekta, S., Shih, I.H., and Bartel, D.P. (2004) MicroRNA-directed cleavage of HOXB8 mRNA. *Science*, **304**, 594–596.
21. Williams, A.E. (2008) Functional aspects of animal microRNAs. *Cell Mol Life Sci*, **65**, 545–562.
22. Thatcher, E.J., Bond, J., Paydar, I. et al. (2008) Genomic organization of zebrafish microRNAs. *BMC Genomics*, **9**, 253.
23. Landgraf, P., Rusu, M., Sheridan, R. et al. (2007) A mammalian microRNA expression atlas based on small RNA library sequencing. *Cell*, **129**, 1401–1414.
24. Berezikov, E., Guryev, V., van de Belt, J. et al. (2005) Phylogenetic shadowing and computational identification of human microRNA genes. *Cell*, **120**, 21–24.
25. Giraldez, A.J., Mishima, Y., Rihel, J. et al. (2006) Zebrafish MiR-430 promotes deadenylation and clearance of maternal mRNAs. *Science*, **312**, 75–79.
26. Lim, L.P., Lau, N.C., Garrett-Engele, P. et al. (2005) Microarray analysis shows that some microRNAs downregulate large numbers of target mRNAs. *Nature*, **433**, 769–773.
27. Cohen, S.M. and Brennecke, J. (2006) Developmental biology. Mixed messages in early development. *Science*, **312**, 65–66.
28. Begemann, G. (2008) MicroRNAs and RNA interference in zebrafish development. *Zebrafish*, **5**, 111–119.
29. Kloosterman, W.P., Lagendijk, A.K., Ketting, R.F. et al. (2007) Targeted inhibition of miRNA maturation with morpholinos reveals a role for miR-375 in pancreatic islet development. *PLoS Biol*, **5**, e203.
30. Choi, W.Y., Giraldez, A.J., and Schier, A.F. (2007) Target protectors reveal dampening and balancing of Nodal agonist and antagonist by miR-430. *Science*, **318**, 271–274.
31. Lents, N.H. and Baldassare, J.J. (2006) RNA interference takes flight: a new RNAi screen reveals cell cycle regulators in *Drosophila* cells. *Trends Endocrinol Metab*, **17**, 173–174.
32. Coumoul, X. and Deng, C.X. (2006) RNAi in mice: a promising approach to decipher gene functions in vivo. *Biochimie*, **88**, 637–643.
33. Cullen, L.M. and Arndt, G.M. (2005) Genome-wide screening for gene function using RNAi in mammalian cells. *Immunol Cell Biol*, **83**, 217–223.
34. Paddison, P.J. (2008) RNA interference in mammalian cell systems. *Curr Top Microbiol Immunol*, **320**, 1–19.
35. Nguyen, T., Menocal, E.M., Harborth, J. et al. (2008) RNAi therapeutics: an update on delivery. *Curr Opin Mol Ther*, **10**, 158–167.
36. Durcan, N., Murphy, C., and Cryan, S.A. (2008) Inhalable siRNA: potential as a therapeutic agent in the lungs. *Mol Pharm*, **5**, 559–566.
37. Wargelius, A., Ellingsen, S., and Fjose, A. (1999) Double-stranded RNA induces specific developmental defects in zebrafish embryos. *Biochem Biophys Res Commun*, **263**, 156–161.
38. Gruber, J., Manninga, H., Tuschl, T. et al. (2005) Specific RNAi mediated gene knockdown in zebrafish cell lines. *RNA Biol*, **2**, 101–105.
39. Zhao, X.F., Fjose, A., Larsen, N. et al. (2008) Treatment with small interfering RNA affects the microRNA pathway and causes unspecific defects in zebrafish embryos. *FEBS J*, **275**, 2177–2184.
40. Kloosterman, W.P., Steiner, F.A., Berezikov, E. et al. (2006) Cloning and expression of new microRNAs from zebrafish. *Nucleic Acids Res*, **34**, 2558–2569.
41. Chen, P.Y., Manninga, H., Slanchev, K. et al. (2005) The developmental miRNA profiles of zebrafish as determined by small RNA cloning. *Genes Dev*, **19**, 1288–1293.
42. Lim, L.P., Lau, N.C., Weinstein, E.G. et al. (2003) The microRNAs of *Caenorhabditis elegans*. *Genes Dev*, **17**, 991–1008.
43. Eisen, J.S. and Smith, J.C. (2008) Controlling morpholino experiments: don't stop making antisense. *Development*, **135**, 1735–1743.
44. Chen, E. and Ekker, S.C. (2004) Zebrafish as a genomics research model. *Curr Pharm Biotechnol*, **5**, 409–413.
45. Flynt, A.S., Li, N., Thatcher, E.J. et al. (2007) Zebrafish miR-214 modulates Hedgehog signaling to specify muscle cell fate. *Nat Genet*, **39**, 259–263.
46. Valoczi, A., Hornyik, C., Varga, N. et al. (2004) Sensitive and specific detection of microRNAs by northern blot analysis using LNA-modified oligonucleotide probes. *Nucleic Acids Res*, **32**, e175.
47. Wienholds, E., Kloosterman, W.P., Miska, E. et al. (2005) MicroRNA expression in zebrafish embryonic development. *Science*, **309**, 310–311.
48. Ason, B., Darnell, D.K., Wittbrodt, B. et al. (2006) Differences in vertebrate microRNA expression. *Proc Natl Acad Sci USA*, **103**, 14385–14389.
49. Wienholds, E., Koudijs, M.J., van Eeden, F.J. et al. (2003) The microRNA-producing enzyme Dicer1 is essential for zebrafish development. *Nat Genet*, **35**, 217–218.

50. Bernstein, E., Kim, S.Y., Carmell, M.A. et al. (2003) Dicer is essential for mouse development. *Nat Genet*, **35**, 215–217.
51. Giraldez, A.J., Cinalli, R.M., Glasner, M.E. et al. (2005) MicroRNAs regulate brain morphogenesis in zebrafish. *Science*, **308**, 833–838.
52. Newport, J. and Kirschner, M. (1982) A major developmental transition in early *Xenopus* embryos: II. Control of the onset of transcription. *Cell*, **30**, 687–696.
53. Watanabe, T., Takeda, A., Mise, K. et al. (2005) Stage-specific expression of microRNAs during *Xenopus* development. *FEBS Lett*, **579**, 318–324.
54. Lykke-Andersen, K., Gilchrist, M.J., Grabarek, J.B. et al. (2008) Maternal Argonaute 2 is essential for early mouse development at the maternal–zygotic transition. *Mol Biol Cell*, **19**, 4383–4392.
55. Bushati, N., Stark, A., Brennecke, J. et al. (2008) Temporal reciprocity of miRNAs and their targets during the maternal-to-zygotic transition in *Drosophila*. *Curr Biol*, **18**, 501–506.
56. Pearson, J.C., Lemons, D., and McGinnis, W. (2005) Modulating Hox gene functions during animal body patterning. *Nat Rev Genet*, **6**, 893–904.
57. Woltering, J.M. and Durston, A.J. (2008) MiR-10 represses HoxB1a and HoxB3a in zebrafish. *PLoS ONE*, **3**, e1396.
58. Tanzer, A., Amemiya, C.T., Kim, C.B. et al. (2005) Evolution of microRNAs located within Hox gene clusters. *J Exp Zoolog B Mol Dev Evol*, **304**, 75–85.
59. Mansfield, J.H., Harfe, B.D., Nissen, R. et al. (2004) MicroRNA-responsive 'sensor' transgenes uncover Hox-like and other developmentally regulated patterns of vertebrate microRNA expression. *Nat Genet*, **36**, 1079–1083.
60. Pase, L., Layton, J.E., Kloosterman, W.P. et al. (2009) miR-451 regulates zebrafish erythroid maturation in vivo via its target gata2. *Blood*, **113**, 1794–1804.
61. Yin, V.P., Thomson, J.M., Thummel, R. et al. (2008) Fgf-dependent depletion of microRNA-133 promotes appendage regeneration in zebrafish. *Genes Dev*, **22**, 728–733.
62. Morton, S.U., Scherz, P.J., Cordes, K.R. et al. (2008) microRNA-138 modulates cardiac patterning during embryonic development. *Proc Natl Acad Sci USA*, **105**, 17830–17835.
63. Li, Y.X., Farrell, M.J., Liu, R. et al. (2000) Double-stranded RNA injection produces null phenotypes in zebrafish. *Dev Biol*, **217**, 394–405.
64. Hsieh, D.J. and Liao, C.F. (2002) Zebrafish M2 muscarinic acetylcholine receptor: cloning, pharmacological characterization, expression patterns and roles in embryonic bradycardia. *Br J Pharmacol*, **137**, 782–792.
65. Zhao, Z., Cao, Y., Li, M. et al. (2001) Double-stranded RNA injection produces nonspecific defects in zebrafish. *Dev Biol*, **229**, 215–223.
66. Oates, A.C., Bruce, A.E., and Ho, R.K. (2000) Too much interference: injection of double-stranded RNA has nonspecific effects in the zebrafish embryo. *Dev Biol*, **224**, 20–28.
67. Heasman, J. (2002) Morpholino oligos: making sense of antisense? *Dev Biol*, **243**, 209–214.
68. Ekker, S.C. and Larson, J.D. (2001) Morphant technology in model developmental systems. *Genesis*, **30**, 89–93.
69. Nasevicius, A. and Ekker, S.C. (2000) Effective targeted gene 'knockdown' in zebrafish. *Nat Genet*, **26**, 216–220.
70. Stark, G.R., Kerr, I.M., Williams, B.R. et al. (1998) How cells respond to interferons. *Annu Rev Biochem*, **67**, 227–264.
71. Elbashir, S.M., Harborth, J., Lendeckel, W. et al. (2001) Duplexes of 21-nucleotide RNAs mediate RNA interference in cultured mammalian cells. *Nature*, **411**, 494–498.
72. Kanungo, J., Li, B.S., Zheng, Y. et al. (2006) Cyclin-dependent kinase 5 influences Rohon–Beard neuron survival in zebrafish. *J Neurochem*, **99**, 251–259.
73. Liu, W.Y., Wang, Y., Sun, Y.H. et al. (2005) Efficient RNA interference in zebrafish embryos using siRNA synthesized with SP6 RNA polymerase. *Dev Growth Differ*, **47**, 323–331.
74. Dodd, A., Chambers, S.P., and Love, D.R. (2004) Short interfering RNA-mediated gene targeting in the zebrafish. *FEBS Lett*, **561**, 89–93.
75. Kok, K.H., Ng, M.H., Ching, Y.P. et al. (2007) Human TRBP and PACT directly interact with each other and associate with dicer to facilitate the production of small interfering RNA. *J Biol Chem*, **282**, 17649–17657.
76. Haase, A.D., Jaskiewicz, L., Zhang, H. et al. (2005) TRBP, a regulator of cellular PKR and HIV-1 virus expression, interacts with Dicer and functions in RNA silencing. *EMBO Rep*, **6**, 961–967.
77. Chendrimada, T.P., Gregory, R.I., Kumaraswamy, E. et al. (2005) TRBP recruits the Dicer complex to Ago2 for microRNA processing and gene silencing. *Nature*, **436**, 740–744.

78. Blidner, R.A., Svoboda, K.R., Hammer, R.P. et al. (2008) Photoinduced RNA interference using DMNPE-caged 2′-deoxy-2′-fluoro substituted nucleic acids in vitro and in vivo. *Mol Biosyst*, **4**, 431–440.
79. Hitz, C., Steuber-Buchberger, P., Delic, S. et al. (2009) Generation of shRNA transgenic mice. *Methods Mol Biol*, **530**, 1–29.
80. Dietzl, G., Chen, D., Schnorrer, F. et al. (2007) A genome-wide transgenic RNAi library for conditional gene inactivation in *Drosophila*. *Nature*, **448**, 151–156.
81. Emelyanov, A. and Parinov, S. (2008) Mifepristone-inducible LexPR system to drive and control gene expression in transgenic zebrafish. *Dev Biol*, **320**, 113–121.
82. Asakawa, K., Suster, M.L., Mizusawa, K. et al. (2008) Genetic dissection of neural circuits by Tol2 transposon-mediated Gal4 gene and enhancer trapping in zebrafish. *Proc Natl Acad Sci USA*, **105**, 1255–1260.
83. Davison, J.M., Akitake, C.M., Goll, M.G. et al. (2007) Transactivation from Gal4-VP16 transgenic insertions for tissue-specific cell labeling and ablation in zebrafish. *Dev Biol*, **304**, 811–824.
84. Robu, M.E., Larson, J.D., Nasevicius, A. et al. (2007) p53 activation by knockdown technologies. *PLoS Genet*, **3**, e78.
85. Scacheri, P.C., Rozenblatt-Rosen, O., Caplen, N.J. et al. (2004) Short interfering RNAs can induce unexpected and divergent changes in the levels of untargeted proteins in mammalian cells. *Proc Natl Acad Sci USA*, **101**, 1892–1897.

Chapter 16

Profiling of miRNA Expression and Prediction of Target Genes

Mouldy Sioud and Lina Cekaite

Abstract

MicroRNAs (miRNAs) represent a new class of small noncoding RNAs expressed in plants and animals. They are responsible for the regulation of up to 30% of human genes, thus underlying their influence on almost all cellular pathways. Recent studies have shown miRNAs to be differentially expressed in cancers. Furthermore, several miRNAs are associated with fragile sites, preferential sites of translation, deletion and amplification that are often altered in cancers. However, despite the progress in identifying miRNA genes, knowledge about their functions and specific target genes is still limited. This chapter highlights the technical advances in miRNA gene profiling and discusses the limitations of target prediction programs and the necessity of defining the roles that govern miRNA specificity and target discrimination in vivo.

Key words: miRNAs, siRNAs, RNA interference, microarrays.

1. Introduction

MicroRNAs (miRNAs) are evolutionarily conserved subpopulations of small noncoding RNAs of about 19–25 nucleotides in length and are derived from introns and exons of both protein- and noncoding genes (*see* **Chapter 14**). They regulate gene expression by binding to mRNA and blocking translation or triggering mRNA degradation (1). These small RNAs are initially transcribed in the nucleus as long primary miRNAs (pri-miRNAs) that are then processed by the double-strand-specific ribonuclease III Drosha into hairpin-like miRNA precursors (pre-miRNAs) that are $\approx$ 70 nt in length. These precursors are then transported

M. Sioud (ed.), *RNA Therapeutics*, Methods in Molecular Biology 629,
DOI 10.1007/978-1-60761-657-3_16, © Springer Science+Business Media, LLC 2010

to the cytoplasm by Exportin 5 and further processed by the enzyme Dicer into mature miRNAs with a final length of ≈ 22 nt. Subsequently, mature miRNAs are incorporated into the RNA-induced silencing complex (RISC), leading to the inhibition of gene expression (**Fig. 16.1**).

Currently, nearly 600 human miRNAs have been identified (http://microrna.sanger.ac.uk/) and it is postulated that up to 1,000 miRNAs exist (2, 3). Present estimates suggest that about one-third of human mRNAs appear to be miRNA targets (4). This makes the miRNA genes one of the most abundant classes of regulatory genes in mammals. The picture gets even more complicated when one miRNA may have several mRNA targets and several miRNAs may regulate a single target mRNA (5).

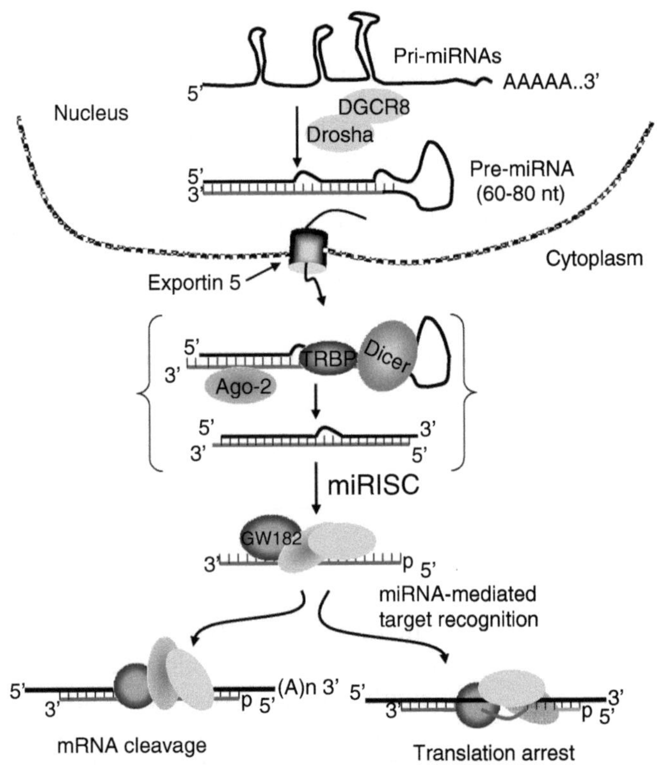

Fig. 16.1. Schematic representation of miRNA silencing pathway. miRNA are encoded in the genome and are transcribed to yield primary transcripts, which are processed by Drosha with the aid of DGCR8 to generate short precursor miRNAs (pri-miRNAs). PremiRNAs are exported from the nucleus to the cytoplasm, where they are further processed by Dicer to generate short RNA duplexes containing two strands. Subsequent to processing, miRNAs are assembled into miRISC. Only one strand of the complex is stably associated with an miRISC complex. Unwinding of the complex seems to occur so rapidly after duplex formation due to the Ago2 presence in the complex with Dicer and TRBP. The loaded strand RNA guides the RISC complex to recognize target mRNA sequences. A perfect complementarity allows Ago-catalyzed cleavage of mRNA, whereas central mismatches promote repression of mRNA translation.

Furthermore, bioinformatic analyses have predicted that each miRNA recognizes on average about 100 different mRNA targets (6).

Despite the success in identification of miRNA genes, knowledge about their functions and specific target genes is limited, and only a handful miRNAs have been fully characterized. In mammals, the function for specific miRNAs has been described in the processes of B-cell differentiation (7), adipocyte differentiation (8), insulin secretion (9), embryonic brain development (10), and embryonic cardiogenesis (11). Thus, one major function is to control processes that underlie cell proliferation and differentiation in diverse organisms during normal development (12). The essential roles of miRNAs in vertebrate development have been strongly demonstrated by removing the Dicer enzyme in zebrafish (10, 13) and mice (14). In both cases, removal of Dicer resulted in severe developmental defects and embryonic death in mice. Further, it has been demonstrated that an individual miRNA can trigger large-scale changes in development (15). Moreover, some miRNAs are differentially expressed in stem cells (16, 17), and recently it was found that miRNAs are required for stem cell division (18). Collectively, these studies indicate that the regulatory potential of miRNAs can be widespread.

2. Arrays

The need to analyze global gene-expression profiles from multiple cell types using microarray is critical not only to miRNA research, but also more generally to studies of the flow of genetic information. Microarray technology is based on basic principles of nucleic acid hybridization (19), whereby fragmented DNA is attached to a substrate and then hybridized with a known gene or fragment. Typically, spotted oligo or cDNA microarray is hybridized with fluorescent cDNA probes prepared from polyA+ RNA using standard methods for making cDNA libraries. Subsequent to hybridization, the captured images are then analyzed using software that quantifies the signal of each spot corresponding to individual capture probes. The intensity of each spot is expected to be proportional to the amount of mRNA present within the analyzed samples. Since the first expression microarray report was published in 1995 (20), the technology has seen huge applications in basic and applied research. In contrast to mRNAs, the analysis of miRNAs by microarrays is challenging because they are small, lack poly(A) tails, display a wide range of predicted melting temperature versus their DNA capture probes, and exhibit a strong sequence homology between family members. These features complicate the design of appropriate capture probes across the complete miRnome.

To take full advantage of the cDNA arrays, one would like to be able to prepare sensitive probes from miRNAs. Using standard procedures to prepare internally labeled cDNA probes from RNAs, we have described a method that allows the preparation of cDNA probes from gel-purified miRNAs. In this method, small RNA recovered from 15% denaturing acrylamide gel slices were reversed transcribed using hexamer random primers to produce a labeled probe for analyzing miRNA gene expression between two cell populations (21, 22). During cDNA synthesis miRNA can be also labeled with fluorescent molecules. In contrast to end labeling using T4 polynucleotide kinase (23), cDNA labeling incorporate multiple radioactive or fluorescent bases, and thus the specific activity of such probes is expected to be high. Since most miRNAs comprise less than 0.001% of a total RNA sample, this type of labeling is important for obtaining sensitive and accurate miRNA microarray data. In addition to internal labeling, purified miRNAs can be 3′ end labeled by polyA polymerase, which adds several radioactive or fluorescence-labeled ATP molecules (22). This simple method also provides high sensitivity comparable to that obtained with the cDNA-labeling methods. In the case of the cDNA labeling, the microarray should contain the processed miRNA sequences, whereas in the case of the 3′ end labeling the microarrays should contain the miRNA complementary strands. To reduce the nonspecific cross hybridization of longer mRNA or ribosomal RNA, we have purified miRNAs (21). However, Liu and colleagues demonstrated that total RNA can be used directly without the need for miRNA purification (24). In this method, total RNA was reverse-transcribed with biotinylated random primers and hybridized to oligonucleotide spotted arrays. Subsequently, miRNA expression was detected using streptavidin-bound fluorophores. This method of labeling is crucial for small biological materials. However, the data obtained with total RNA must be interpreted with caution and validated with additional strategies such as Northern and PCR techniques.

Since the first description of dot blot arrays using a radioactive label (21, 23), several labeling techniques were described and direct labeling of miRNAs can be accomplished chemically or enzymatically. Ambion's miRNA-labeling kit relies on polyA polymerase and incorporates amine-modified adenosines, which are then treated with reactive fluorescent dye. In contrast, Invitrogen NCcode-labeling kit involves the ligation of DNA polymer with either Alexa Fluor 3 or Alexa Fluor 5 dye molecules to the polyA-tailed RNA via T4 DNA ligase, and this in turn is labeled with fluorescent DNA dendrimers. A label IT kit from Mirus relies on alkylation chemistry for covalent linkage of reactive Cy3 or Cy5 dye molecules to virtually any nucleic acid target

(25). An universal linkage system (ULS) labeling method was also used to label miRNAs (26). ULS labeling is based on an ability to label directly guanine nucleotides without any size discrimination and without the need of enzymes. Notably, the labeling strategies indicated above label miRNAs prior to hybridization to the arrays. An RNA-primed array-based Klenow enzyme (RAKE) assay labels miRNA subsequent to hybridization. A major advantage of RAKE is that it requires no sample RNA manipulation prior to hybridization (27). Also, LNA-modified capture probes that enable the discrimination between miRNA family members were designed and used for expression profiling (28).

3. Probe Amplification

Generally, 20–100 μg of total RNAs is required to prepare miRNA probe for each hybridization. This requirement for large amounts of starting materials severely limit the use of microarrays to profile miRNA gene expression in small specimens of human tissue, such as obtained by core needle or fine needle aspiration biopsies. To overcome this problem, we have developed a PCR-based amplification method that allows the analysis of miRNAs gene expression in small biopsies (22). The amplification method we used is shown in **Fig. 16.2**. In this procedure, purified miRNAs were polyA tailed using *E. coli* polyA polymerase and the RNA is primed by oligo-dT carrying on its 5′ end a T7 RNA polymerase promoter sequence. The use of the reverse transcriptase powerscript during cDNA synthesis allows the addition of few additional nucleotides, primarily deoxycytidines, to the 3′ end of the cDNA. The 3′-tailed first-strand cDNA is then primed with SP6 RNA polymerase promoter sequence carrying GGG at its 3′ end, leading to the generation of double-stranded cDNA. The resulting double-stranded DNA can either be PCR amplified for a desired number of cycles or in vitro transcribed using T7 (or SP6) RNA polymerase, resulting in linear amplification of miRNAs. The protocol is expected to generate large amounts of miRNA starting from few nanograms of miRNAs. **Figure 16.3** shows the hybridization signals using an internally labeled 32p PCR probe, which was prepared from miRNA that had undergone 25 rounds of amplification. In these experiments, the probe was hybridized to dot blots containing 160 human miRNA sequences. We have obtained reproducible results between different samples, and there is no indication of a bias due to the amplification step. Moreover, a hybridization of fluorescently labeled PCR products to a glass slide upon which specifically designed miRNA-specific probes have been arrayed in-house, revealed satisfying results,

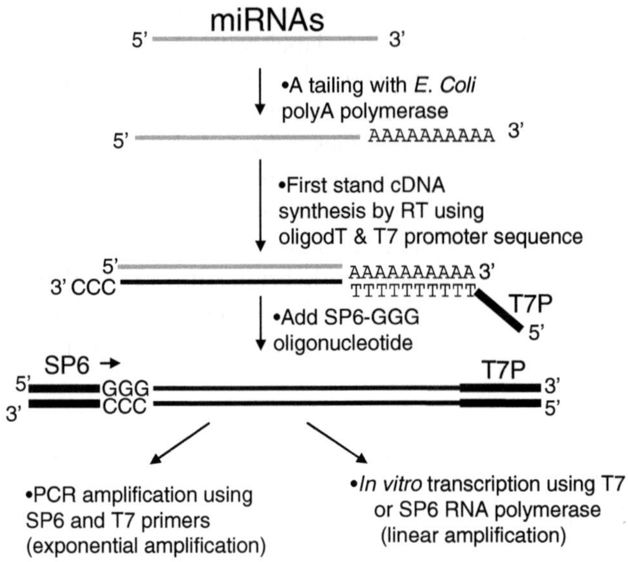

Fig. 16.2. Schematic illustration of miRNA amplification (for details *see* text).

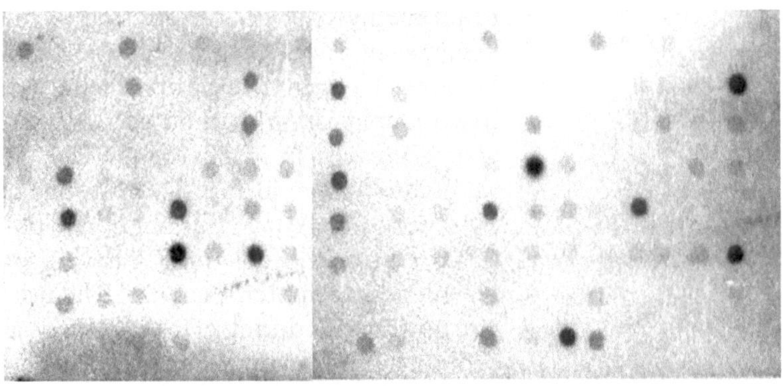

Fig. 16.3. Amplified miRNAs produce significant hybridization signals. Oligonucleotides representing 160 miRNA sequences were spotted onto Hybond-N+ nylon membranes and then hybridized with a cDNA probe prepared from PCR-amplified miRNAs prepared from the breast cancer cell line MCF-7.

thus, enabling the use of microarray technology for qualitative and quantitative studies of miRNAs expression in small number of cells.

4. miRNAs in Cancer

It is speculated that abnormalities in miRNA expression might similarly contribute to the generation or maintenance of cancer stem cells, a tumor-initiating cells (29). One of the arguments connecting several miRNAs to cancer was the finding that more than half of the mapped miRNA genes are localized to fragile sites and genomic regions involved in cancer (30). And the first evidence for direct involvement of miRNAs in cancer was the finding that miRNA-15 and miRNA-16 were deleted or underexpressed in most patients with chronic lymphocytic leukemia (31). Reduced accumulation of specific miRNAs has further been reported for let-7 in lung cancer (32, 33), miRNA26a and miRNA-99a in lung cancer cell lines (30), miRNA-143 and miRNA-145 in colon cancer (34) and miRNA-125b, miRNA-145, miRNA-21 and miRNA-155 in human breast cancer (35), miRNA23b, miRNA145, miRNA 100, and mRNA-205 in prostate cancer (36). An accumulation of specific miRNAs in cancer has also been observed: miRNA-155 in lymphomas (37) and overexpression of seven miRNAs encoded by the gene *c13orf25* amplified in certain lymphomas, nominates this miRNA cluster the first noncoding oncogene (24). Another example of an miRNA-oncogene overexpressed in brain tumor is the anti-apoptotic factor miR-21. On the other hand, downregulation of the Ras oncogene by let-7 implicates let-7 as a tumor suppressor in lung tissue (33). Thus, dependent on what genes the miRNAs are regulating, the miRNAs may act either as oncogenes or tumor suppressors. The loss of miRNA function by mutation, deletion, promoter methylation, or defect in processing might result in an abnormal expression of the target oncogene, which subsequently contributes to tumor formation. Also, the amplification or overexpression of a miRNA that downregulates a tumor suppressor or other important genes involved in differentiation may contribute to tumor formation by stimulating proliferation, angiogenesis, and invasion. For example, amplification of the oncogenic miRNA, miRNA-17-92 cluster, miRNA-21, and miRNA-155 have been associated with tumor initiation and progression (38). MiRNA-155 is implicated as a collaborator of Myc protein.

One important finding of miRNA profiling in cancer is that specific miRNAs have altered expression in tumors compared to normal cells. Another is that miRNA reveals distinct signatures in specific cancer forms (39), and that miRNA expression profiles enabled successful classification of poorly differentiated tumors that were inaccurately classified using mRNA profiles (29). For diagnostic purposes, miRNAs could potentially be more directly associated with gene function than mRNA, since they do not have to be first translated into proteins to have a biological effect. Furthermore, miRNAs may offer greater diagnostic sensitivity compared to proteins since they can be detected using quantitative amplification methods such as real-time PCR. A recent study demonstrated that serum levels of miR-141 can distinguish patients with prostate cancer from healthy donors (40).

5. Target Predictions

Several groups have developed computational algorithms designed to predict target genes of miRNA sequences. In general, the basis for these prediction programs is the degree of sequence complementarity between a miRNA and a target mRNA gene. In animal cells, miRNAs bind to specific target mRNAs through partly complementary sequences that are predominantly in the 3′ untranslated region (3′UTR). This recognition relies mainly on the sequences at the miRNA 5′ end, known as the seed. Due to this imperfect complementarity, computational approaches to find target transcripts have been hindered. In addition, many known target genes of miRNAs contain several miRNA-binding sites and several genes are predicted to be regulated by a single miRNAs (41). As the sequences that miRNA seeds recognize are found in many genes, their flexibility makes it difficult to identify functional relevant target genes from sequence analysis alone. Thus, experimental validation is required for miRNA target genes.

To examine whether specific miRNAs contribute to monocyte differentiation into dendritic cells (DCs), we have profiled the expression of miRNAs in monocytes in response to GM-CSF and IL-4. Among the differentially expressed miRNA, miRNA-21, miRNA-24, and miRNA-223 were found to be upregulated in most experiments (42). **Figure 16.4** shows relative expression levels of miRNAs as measured by quantitative real-time PCR.

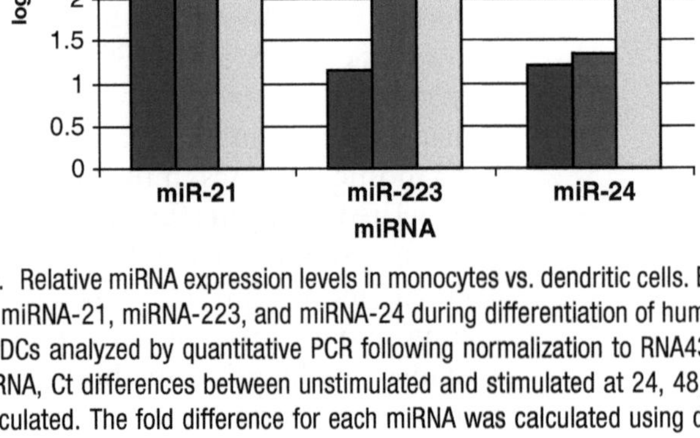

Fig. 16.4. Relative miRNA expression levels in monocytes vs. dendritic cells. Expression levels of miRNA-21, miRNA-223, and miRNA-24 during differentiation of human monocytes to DCs analyzed by quantitative PCR following normalization to RNA43 (42). For each miRNA, Ct differences between unstimulated and stimulated at 24, 48, and 72 h were calculated. The fold difference for each miRNA was calculated using delta–delta Ct method.

To examine miRNA-21 target genes, the miRNA target prediction programs were used. The numbers of computationally predicted target transcripts using miRBASE Targets (43), TargetScan (44), and Pictar (45) programs are presented in Venn diagrams (**Fig. 16.5**). The number in the intersection indicates commonly predicted target genes by the used databases. A substantial number of target genes were identified by one program but not the other. This observation underlies the complexity of target prediction using bioinformatics.

As indicated above, prediction of miRNA target genes provides an alternative approach to assign biological functions. Therefore, miRNA-21 target genes were used for Gene Ontology (GO) analysis. The enriched biological processes of the predicted targets for miR-21 are presented in **Table 16.1**. As shown several pathways seem to be affected by miRNA-21 (42). Given the potential of miRNAs to recognize many targets through imperfect base pairing, these pleiotropic effects of miRNA-21 may simply reflect the different spectrum of target mRNAs available in a given system, and this may represent a recurring theme within miRNA-mediated regulatory pathways. Further efforts toward experimental validation should provide new insight into the prediction of miRNA-21 target sites. Given the large number of predicted target genes, however, it is important to determine the roles that govern miRNA specificity in vivo and how to best identify a target gene. Therefore, it will be instructive to develop bioinformatics programs that take into account the miRNA whole

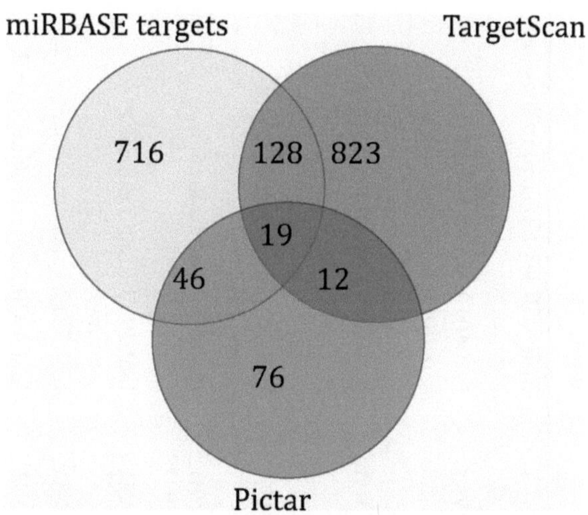

Fig. 16.5. Computationally predicted target transcripts of miR-21. miRanda Targets, TargetScan, and Pictar miRNA target prediction databases were used to identify a potential binding site for miR-21 in genomic sequences. The numbers of computationally predicted target transcripts are shown in the Venn diagram. The numbers in the intersection indicate the target genes that were predicted commonly using the indicated databases (adapted from ref. 42).

Table 16.1
Numbers of the miRNA-regulated target genes predicted by miRanda algorithm

Category	Term	Count	%	*P*-value	Genes	Total
KEGG_PATHWAY	hsa04630:Jak-STAT signaling pathway	7	3.45	0.007	STAT3, CNTFR, IL9, SPRY1, SPRY2, SOS2, IL12A,	50
KEGG_PATHWAY	hsa04010:MAPK signaling pathway	8	3.94	0.028	MAPK10, MAP3K8, CACNG1, NTF3, FGF14, RASA1, FASLG, SOS2,	50
KEGG_PATHWAY	hsa04060:Cytokine–cytokine receptor interaction	7	3.45	0.074	ACVR2A, CNTFR, IL9, FASLG, CCL1, IL12A, CCR7,	50
KEGG_PATHWAY	hsa04940:Type I diabetes mellitus	3	1.48	0.074	GLCCI1, FASLG, IL12A,	50
KEGG_PATHWAY	hsa04510:Focal adhesion	5	2.46	0.199	PPP1CC, MAPK10, SOS2, THBS1, PPP1CB,	50
KEGG_PATHWAY	hsa04910:Insulin signaling pathway	4	1.97	0.203	PPP1CC, MAPK10, SOS2, PPP1CB,	50
KEGG_PATHWAY	hsa05212:Pancreatic cancer	3	1.48	0.208	STAT3, MAPK10, E2F3,	50
KEGG_PATHWAY	hsa04920:Adipocytokine signaling pathway	3	1.48	0.208	STAT3, MAPK10, PPARA,	50
KEGG_PATHWAY	hsa04664:Fc epsilon RI signaling pathway	3	1.48	0.212	MAPK10, BTK, SOS2,	50
KEGG_PATHWAY	hsa04810:Regulation of actin cytoskeleton	5	2.46	0.236	PPP1CC, FGF14, SOS2, ARHGEF7, PPP1CB,	50
KEGG_PATHWAY	hsa04350:TGF-beta signaling pathway	3	1.48	0.281	ACVR2A, PITX2, THBS1,	50
KEGG_PATHWAY	hsa04660:T-cell receptor signaling pathway	3	1.48	0.298	MAP3K8, MALT1, SOS2,	50
KEGG_PATHWAY	hsa04620:Toll-like receptor signaling pathway	3	1.48	0.333	MAPK10, MAP3K8, IL12A,	50

sequence and the sequence/structure that flank the target site that might influence the target site functionally in vivo.

6. Target Validation

To date, different strategies have been used to validate miRNA target genes. An indirect approach to determining a miRNA expression pattern has been developed, in which a reporter transgene (e.g., GFP, luciferase) under the control of a ubiquitous promoter carrying a 3'-UTR with a miRNA complementary sequence has been used (46). In cotransfection experiments, active miRNAs are expected to reduce luciferase activity or GFP expression. However, placing a potential miRNA target site within the 3'UTR of a reporter gene will not predict whether the site in question will be available when it is part of the natural transcripts. It should be noted that RNA folding and function are tightly linked.

The development of antisense oligonucleotides represents a powerful tool for dissecting the complexity of miRNA biological pathways (47). These anti-miRNA oligonucleotides are useful in specifically inhibiting individual miRNAs, thereby helping to uncover the function of miRNAs and their target genes. In this assay, the inhibition of miRNA is expected to lead to the upregulation of its target gene. By transfecting cells with candidate miRNAs, it will be possible to see whether the overexpression of specific miRNAs will affect the expression of the identified target genes in vitro and in vivo. However, a more global approach to find target genes is to use proteomics to uncover target proteins that are affected as a consequence of miRNA expression in target cells. This strategy will work only when the cell populations that are compared are identical except for artificial increase of miRNA in one. Proteins that are altered as a probable consequence of higher miRNA expression may be purified from gels and subsequently identified using MALDI technology.

To experimentally identify transcripts whose levels might be affected by miRNA-21, we first transfected mature miRNAs miRNA-21 into freshly isolated human monocytes and used oligonucleotide microarrays (Illumina Human-6 v3 Expression BeadChips containing probes covering the complete human transcriptome (48,000 transcripts) to determine changes in the mRNA levels. Totally, 1534 transcripts were changed by one-fold or more with 854 transcripts downregulated and 680 were induced (42). The data indicated that only 31 (3.0%), 51 (5.6%), and 12 (6.9%) of the repressed transcripts after overexpression of miRNA-21 were computationally predicted to have a binding site by miRBASE targets, TargetScan, and Pictar miRNA target prediction databases, respectively (42). A number of genes involved in mRNA transcription regulation, including PDCD4, TGFβ, ING3, MMP7, and CYP27B1 were downregulated subsequent to miRNA-21 transfection. Of note, previous studies identified the programmed cell death 4 (PDCD4) gene as miRNA-21 target gene (48–50). To gain additional information about miRNA-21 target genes, it is necessary to analyze the expression data obtained by knocking down the miRNA-21 in monocytes. As indicated above, anti-miR oligonucleotides that allow functional characterization in model organisms or cell lines without the need for genomic manipulations are now available commercially.

Although steady progress in understanding miRNA biogenesis has been achieved in recent years, there is still a long road ahead in this emerging field before we decipher many of the miRNA functions. There is a growing number of reports that link miRNA to the regulation of genes/pathways associated with human cancer. However, there is no conclusive evidence that causatively links the malfunction of a miRNA or miRNA target gene to the initiation of cancer. Therefore, we need to understand whether changes in miRNA expression are the initiating cause or the result of the disease. Also, further knowledge of the complete microRNome is necessary for experimental and computational dissection of miRNA regulatory pathways.

Acknowledgments

We thank Dr. Anne Dybwad for critical reading of the manuscript. This work is supported in part by Helse Sør to M. Sioud.

References

1. Bartel, D.P. (2009) MicroRNAs: target recognition and regulatory functions. *Cell*, **136**, 215–233.
2. Bentwich, I., Avniel, A., Karov, Y., Aharonov, R., Gilad, S., Barad, O., Barzilai, A., Einat, P., Einav, U., Meiri, E. et al. (2005) Identification of hundreds of conserved and nonconserved human microRNAs. *Nat Genet*, **37**, 766–770.
3. Berezikov, E., Guryev, V., van de Belt, J., Wienholds, E., Plasterk, R.H., and Cuppen, E. (2005) Phylogenetic shadowing and computational identification of human microRNA genes. *Cell*, **120**, 21–24.
4. Lewis, B.P., Burge, C.B., and Bartel, D.P. (2005) Conserved seed pairing, often flanked by adenosines, indicates that thousands of human genes are microRNA targets. *Cell*, **120**, 15–20.
5. Krek, A., Grun, D., Poy, M.N., Wolf, R., Rosenberg, L., Epstein, E.J., MacMenamin, P., da Piedade, I., Gunsalus, K.C., Stoffel,

M., and Rajewsky, N. (2005) Combinatorial microRNA target predictions. *Nat Genet*, **37**, 495–500.
6. Bushati, N. (2007) microRNA functions. *Ann Rev Cell Dev Biol*, **23**, 175–205.
7. Chen, C.Z., Li, L., Lodish, H.F., and Bartel, D.P. (2004) MicroRNAs modulate hematopoietic lineage differentiation. *Science*, **303**, 83–86.
8. Esau, C., Kang, X., Peralta, E., Hanson, E., Marcusson, E.G., Ravichandran, L.V., Sun, Y., Koo, S., Perera, R.J., Jain, R., Dean, N.M., Freier, S.M., Bennett, C.F., Lollo, B., and Griffey, R. (2004) MicroRNA-143 regulates adipocyte differentiation. *J Biol Chem*, **279**, 52361–52365.
9. Poy, M.N., Eliasson, L., Krutzfeldt, J., Kuwajima, S., Ma, X., Macdonald, P.E., Pfeffer, S., Tuschl, T., Rajewsky, N., Rorsman, P., and Stoffel, M. (2004) A pancreatic islet-specific microRNA regulates insulin secretion. *Nature*, **432**, 226–230.
10. Giraldez, A.J., Cinalli, R.M., Glasner, M.E., Enright, A.J., Thomson, J.M., Baskerville, S., Hammond, S.M., Bartel, D.P., and Schier, A.F. (2005) MicroRNAs regulate brain morphogenesis in zebrafish. *Science*, **308**, 833–838.
11. Zhao, Y., Samal, E., and Srivastava, D. (2005) Serum response factor regulates a muscle-specific microRNA that targets Hand2 during cardiogenesis. *Nature*, **436**, 214–220.
12. Ambros, V. (2004) The functions of animal microRNAs. *Nature*, **431**, 350–355.
13. Wienholds, E., Koudijs, M.J., van Eeden, F.J., Cuppen, E., and Plasterk, R.H. (2003) The microRNA-producing enzyme Dicer1 is essential for zebrafish development. *Nat Genet*, **35**, 217–218.
14. Bernstein, E., Kim, S.Y., Carmell, M.A., Murchison, E.P., Alcorn, H., Li, M.Z., Mills, A.A., Elledge, S.J., Anderson, K.V., and Hannon, G.J. (2003) Dicer is essential for mouse development. *Nat Genet*, **35**, 215–217.
15. Lim, L.P., Lau, N.C., Garrett-Engele, P., Grimson, A., Schelter, J.M., Castle, J., Bartel, D.P., Linsley, P.S., and Johnson, J.M. (2005) Microarray analysis shows that some microRNAs downregulate large numbers of target mRNAs. *Nature*, **433**, 769–773.
16. Suh, M.R., Lee, Y., Kim, J.Y., Kim, S.K., Moon, S.H., Lee, J.Y., Cha, K.Y., Chung, H.M., Yoon, H.S., Moon, S.Y., Kim, V.N., and Kim, K.S. (2004) Human embryonic stem cells express a unique set of microRNAs. *Dev Biol*, **270**, 488–498.
17. Houbaviy, H.B., Murray, M.F., and Sharp, P.A. (2003) Embryonic stem cell-specific microRNAs. *Dev Cell*, **5**, 351–358.
18. Hatfield, S.D., Shcherbata, H.R., Fischer, K.A., Nakahara, K., Carthew, R.W., and Ruohola-Baker, H. (2005) Stem cell division is regulated by the microRNA pathway. *Nature*, **435**, 974–978.
19. Southern, E.M. (1975) Detection of specific sequences among DNA fragments separated by gel electrophoresis. *J Mol Biol*, **98**, 503–517.
20. Schena, M., Shalon, D., Davis, R.W., and Brown, P.O. (1995) Quantitative monitoring of gene expression patterns with a complementary DNA microarray. *Science*, **270**, 467–470.
21. Sioud, M. and Rosok, O. (2004) Profiling microRNA expression using sensitive cDNA probes and filter arrays. *Biotechniques*, **37**(574–576), 578–580.
22. Sioud, M. and Rosok, O. (2005) High-throughput analysis of microRNA gene expression using sensitive probes. *Methods Mol Biol*, **309**, 311–320.
23. Babak, T., Zhang, W., Morris, Q., Blencowe, B.J., and Hughes, T.R. (2004) Probing microRNAs with microarrays: tissue specificity and functional inference. *RNA*, **10**, 1813–1819.
24. Liu, C.G., Calin, G.A., Meloon, B., Gamliel, N., Sevignani, C., Ferracin, M., Dumitru, C.D., Shimizu, M., Zupo, S., Dono, M., Alder, H., Bullrich, F., Negrini, M., and Croce, C.M. (2004) An oligonucleotide microchip for genome-wide microRNA profiling in human and mouse tissues. *Proc Natl Acad Sci USA*, **101**, 9740–9744.
25. Goff, L.A., Yang, M., Bowers, J., Getts, R.C., Padgett, R.W., and Hart, R.P. (2005) Rational probe optimization and enhanced detection strategy for microRNAs using microarrays. *RNA Biol*, **2**, 93–100.
26. Wiegant, J.C., van Gijlswijk, R.P., Heetebrij, R.J., Bezrookove, V., Raap, A.K., and Tanke, H.J. (1999) ULS: a versatile method of labeling nucleic acids for FISH based on a monofunctional reaction of cisplatin derivatives with guanine moieties. *Cytogenet Cell Genet*, **87**, 47–52.
27. Nelson, P.T., Baldwin, D.A., Scearce, L.M., Oberholtzer, J.C., Tobias, J.W., and Mourelatos, Z. (2004) Microarray-based, high-throughput gene expression profling of microRNAs. *Nat Methods*, **1**, 155–161.
28. Castoldi, M., Schmidt, S., Benes, V., Noerholm, M., Kulozik, A.E., Hentze, M.W., and Muckenthaler, M.U. (2006) A sensitive array

for microRNA expression profling (miChip) based on locked nucleic acids (LNA). *RNA*, **12**, 913–920.

29. Lu, J., Getz, G., Miska, E.A., Alvarez-Saavedra, E., Lamb, J., Peck, D., Sweet-Cordero, A., Ebert, B.L., Mak, R.H., Ferrando, A.A., Downing, J.R., Jacks, T., Horvitz, H.R., and Golub, T.R. (2005) MicroRNA expression profiles classify human cancers. *Nature*, **435**, 834–838.
30. Calin, G.A., Liu, C.G., Sevignani, C., Ferracin, M., Felli, N., Dumitru, C.D., Shimizu, M., Cimmino, A., Zupo, S., Dono, M., Dell'Aquila, M.L., Alder, H., Rassenti, L., Kipps, T.J., Bullrich, F., Negrini, M.,, and Croce, C.M. (2004) MicroRNA profiling reveals distinct signatures in B cell chronic lymphocytic leukemias. *Proc Natl Acad Sci USA*, **101**, 11755–11760.
31. Calin, G.A., Dumitru, C.D., Shimizu, M., Bichi, R., Zupo, S., Noch, E., Aldler, H., Rattan, S., Keating, M., Rai, K., Rassenti, L., Kipps, T., Negrini, M., Bullrich, F., and Croce, C.M. (2002) Frequent deletions and down-regulation of micro-RNA genes miR15 and miR16 at 13q14 in chronic lymphocytic leukemia. *Proc Natl Acad Sci USA*, **99**, 15524–15529.
32. Takamizawa, J., Konishi, H., Yanagisawa, K., Tomida, S., Osada, H., Endoh, H., Harano, T., Yatabe, Y., Nagino, M., Nimura, Y., Mitsudomi, T., and Takahashi, T. (2004) Reduced expression of the let-7 microRNAs in human lung cancers in association with shortened postoperative survival. *Cancer Res*, **64**, 3753–3756.
33. Johnson, S.M., Grosshans, H., Shingara, J., Byrom, M., Jarvis, R., Cheng, A., Labourier, E., Reinert, K.L., Brown, D., and Slack, F.J. (2005) RAS is regulated by the let-7 microRNA family. *Cell*, **120**, 635–647.
34. Michael, M.Z., O'Connor, S.M., van Holst Pellekaan, N.G., Young, G.P., and James, R.J. (2003) Reduced accumulation of specific microRNAs in colorectal neoplasia. *Mol Cancer Res*, **1**, 882–891.
35. Iorio, M.V., Ferracin, M., Liu, C.G., Veronese, A., Spizzo, R., Sabbioni, S., Magri, E., Pedriali, M., Fabbri, M., Campiglio, M., Ménard, S., Palazzo, J.P., Rosenberg, A., Musiani, P., Volinia, S., Nenci, I., Calin, G.A., Querzoli, P., Negrini, M., and Croce, C.M. (2005) MicroRNA gene expression deregulation in human breast cancer. *Cancer Res*, **65**, 7065–7070.
36. Porkka, K.P., Pfeiffer, M.J., Waltering, K.K., Vessella, R.L., Tammela, T.L., and Visakorpi, T. (2007) MicroRNA expression profiling in prostate cancer. *Cancer Res*, **67**, 6130–6135.
37. Eis, P.S., Tam, W., Sun, L., Chadburn, A., Li, Z., Gomez, M.F., Lund, E., and Dahlberg, J.E. (2005) Accumulation of miR-155 and BIC RNA in human B cell lymphomas. *Proc Natl Acad Sci USA*, **102**, 3627–3632.
38. He, L., Thomson, J.M., Hemann, M.T., Hernando-Monge, E., Mu, D., Goodson, S., Powers, S., Cordon-Cardo, C., Lowe, S.W., Hannon, G.J., and Hammond, S.M. (2005) A microRNA polycistron as a potential human oncogene. *Nature*, **435**, 828–833.
39. Calin, G.A., Sevignani, C., Dumitru, C.D., Hyslop, T., Noch, E., Yendamuri, S., Shimizu, M., Rattan, S., Bullrich, F., Negrini, M., and Croce, C.M. (2004) Human microRNA genes are frequently located at fragile sites and genomic regions involved in cancers. *Proc Natl Acad Sci USA*, **101**, 2999–3004.
40. Mitchell, P.S., Parkin, R.K., Kroh, E.M., Fritz, B.R., Wyman, S.K., Pogosova-Agadjanyan, E.L., Peterson, A., Noteboom, J., O'Briant, K.C., Allen, A., Lin, D.W., Urban, N., Drescher, C.W., Knudsen, B.S., Stirewalt, D.L., Gentleman, R., Vessella, R.L., Nelson, P.S., Martin, D.B., and Tewari, M. (2008) Circulating microRNAs as stable blood-based markers for cancer detection. *Proc Natl Acad Sci USA*, **105**, 10513–10518.
41. Kiriakidou, M., Nelson, P.T., Kouranov, A., Fitziev, P., Bouyioukos, C., Mourelatos, Z., and Hatzigeorgiou, A. (2004) A combined computational–experimental approach predicts human microRNA targets. *Genes Dev*, **18**, 1165–1178.
42. Cekaite, L., Clancy, T., Sioud, M. (2010) MicroRNA 21 is upregulated during monocytes differentiation into dendritic cells: potential target genes. *Front Biosci* (in press).
43. Griffiths-Jones, S., Saini, H.K., van Dongen, S., and Enright, A.J. (2008) miRBase: tools for microRNA genomics. *Nucleic Acids Res*, (Database issue), D154–D158.
44. Friedman, R.C., Farh, K.K., Burge, C.B., and Bartel, D.P. (2009) Most mammalian mRNAs are conserved targets of microRNAs. *Genome Res*, **19**, 92–105.
45. Krek, A., Grün, D., Poy, M.N., Wolf, R., Rosenberg, L., Epstein, E.J., MacMenamin, P., da Piedade, I., Gunsalus, K.C., Stoffel, M., and Rajewsky, N. (2005) Combinatorial microRNA target predictions. *Nat Genet*, **37**, 495–500.
46. Aravin, A. and Tuschl, T. (2005) Identification and characterization of small RNAs

involved in RNA silencing. *FEBS Lett*, **579**, 5830–5840.
47. Weiler, J., Hunziker, J., and Hall, J. (2006) Anti-miRNA oligonucleotides (AMOs): ammunition to target miRNAs implicated in human disease? *Gene Ther*, **13**, 496–502.
48. Lu, Z., Liu, M., Stribinskis, V., Klinge, C.M., Ramos, K.S., Colburn, N.H., and Li, Y. (2008) MicroRNA-21 promotes cell transformation by targeting the programmed cell death 4 gene. *Oncogene*, **27**, 4373–4379.
49. Frankel, L.B., Christoffersen, N.R., Jacobsen, A., Lindow, M., Krogh, A., and Lund, A.H. (2008) Programmed cell death 4 (PDCD4) is an important functional target of the microRNA miR-21 in breast cancer cells. *J Biol Chem*, **283**, 1026–1033.
50. Asangani, I.A., Rasheed, S.A., Nikolova, D.A., Leupold, J.H., Colburn, N.H., Post, S., and Allgayer, H. (2008) MicroRNA-21 (miR-21) post-transcriptionally downregulates tumor suppressor Pdcd4 and stimulates invasion, intravasation and metastasis in colorectal cancer. *Oncogene*, **27**, 2128–2136.

Chapter 17

Small RNA Cloning

Seungil Ro and Wei Yan

Abstract

The next generation sequencing technologies have made the exhaustive sequencing of all small RNA species within a small RNA library possible. Thus, comparative studies on the changes in small-RNA transcriptomes between two biological samples represent a powerful means of revealing the functions of small RNAs in physiological and pathological processes. Here, we describe a small RNA cloning method that can be used to generate small RNA cDNA libraries for both conventional and next generation (deep) sequencing. It can also be used for detection and quantitative analyses of small RNAs using PCR.

Key words: Small RNAs, miRNAs, piRNAs, cloning, deep sequencing.

1. Introduction

In addition to mRNA/protein-encoding genes, the genomes of almost all species encode abundant small non-coding RNAs including microRNAs (miRNAs) (1–4), Piwi-interacting RNAs (piRNAs) (5–8), repeat-associated siRNAs (rasiRNAs) (9–12), 21U endo-siRNAs (13), and small nucleolar RNAs (snoRNAs) (14). Although functional studies on these small RNAs are just beginning, findings to date have already suggested that these small RNAs are involved in regulating a wide variety of functions including: genome stability, epigenetic programming, transcription, translation, and individual protein functions (15–18). The state-of-the-art next generation sequencing technologies have made the exhaustive sequencing of all small RNA species within a small RNA library possible (19). Thus, comparative studies by analyzing small-RNA transcriptomes between two samples

M. Sioud (ed.), *RNA Therapeutics*, Methods in Molecular Biology 629,
DOI 10.1007/978-1-60761-657-3_17, © Springer Science+Business Media, LLC 2010

would be a powerful way of investigating the relations between small RNA expression and physiological/pathophysiological conditions. The generation of a small RNA library is an essential step for either the identification of novel small RNA species, or comparative studies using next generation sequencing technologies. Here we described a method for generating small RNA cDNA libraries. In this method (**Fig. 17.1**), small RNA fractions isolated from tissues or cells are first polyadenylated using RNA polymerase. A RNA oligo linker (**Table 17.1**) is then added to the 5′ ends of the polyadenylated small RNAs followed by reverse transcription using an oligo with a linked sequence in its 5′ end and polyT in its 3′ end. The resultant small RNA cDNAs are then amplified with a pair of the 5′ linker and the 3′ linker primers (**Table 17.1**). The cDNAs can be cloned in a TA cloning vector for manual sequencing (a conventional sequencing). Alternatively, the adaptor sequences for 454 sequencing (a large-scale

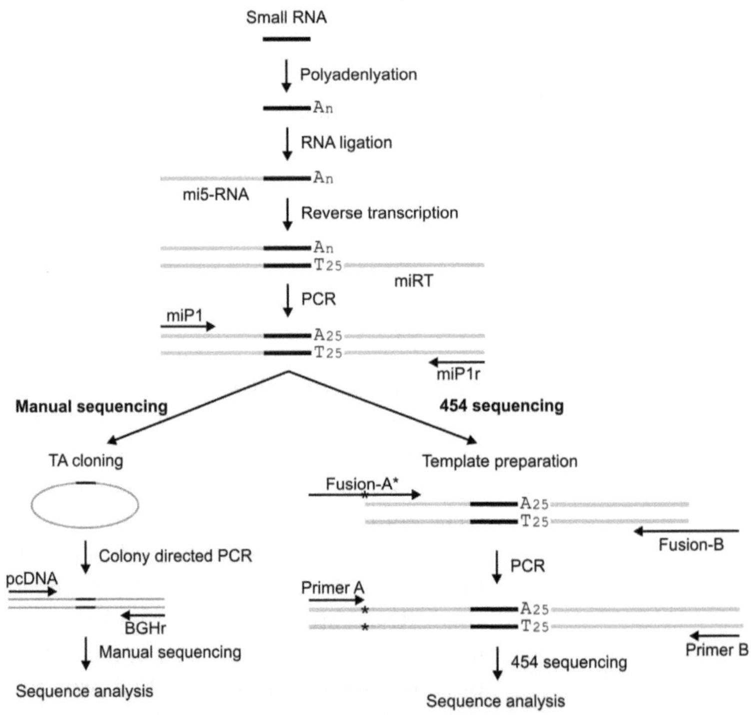

Fig. 17.1. Overview of cloning strategy. Small RNAs isolated from tissues or cells are polyadenylated and ligated with a 5′ RNA adapter mi5-RNA. Reverse transcription is performed using a RT primer miRT. Small RNAs are amplified by PCR using a pair of primers miP1 and miP1r from the small RNA cDNA library. For conventional (manual) sequencing, the PCR products are cloned into a TA cloning vector, transformed into bacteria, and amplified with a pair of primers, pcDNA and BGHr, by colony-directed PCR. The amplicons are sequenced with either pcDNA or BGHr, and analyzed. For 454 sequencing, a sequencing template is amplified with a pair of primers Fusion-A* and Fusion-B by a second PCR. The amplicons are sequenced with Primer A and then analyzed.

Table 17.1
Oligonucleotides used

Name	Sequence (5′ to 3′)	Quality	Usage
mi5-RNA	GACUGGAGCACGAGGACACUGACAUGGACUGAAGGAGUAGAAA	RNase-free HPLC	RNA ligation
miRT	CGAATTCTAGAGCTCGAGGCAGGCGACATGTTTTTTTTTTTTTTTTTTTTTTTTTVN[a]	RNase-free HPLC	Reverse transcription
miP1	CGAATTCTAGAGCTCGAGGCAGG	Regular desalting	PCR
miP1r	GACTGGAGCACGAGGACACTGA	Regular desalting	PCR
pcDNA	GCAGAGCTCTCTGGCTAAC	Regular desalting	Colony PCR/ sequencing
BGHr	CTAGAAGGCACAGTCGAGG	Regular desalting	Colony PCR/ sequencing
Fusion-A*	GCCTCCCTCGCGCCATCAGATGACTGGAGCACGAGGACACTGA	PAGE purification	454 sequencing template preparation
Fusion-B	GCCTTGCCAGCCCGCTCAGCGAATTCTAGAGCTCGAGGCAGG	PAGE purification	454 sequencing template preparation
Primer A	GCCTCCCTCGCGCCATCAG	Regular desalting	454 sequencing
Primer B	GCCTTGCCAGCCCGCTCAG	Regular desalting	454 sequencing

[a]V = A, C, or G; N = A, C, G, or T.
[b]**AT** are two tag nucleotides, which can be used for sample tags in sequencing multiple samples.

parallel pyrosequencing) can be incorporated at the ends of the small RNA cDNAs by a second round of PCR. The small RNA cDNA library generated using this method can be used for both manual sequencing and 454 sequencing analyses.

2. Materials

2.1. Isolation of Small RNAs and Polyadenylation

1. mirVana miRNA Isolation Kit (Ambion).
2. Phosphate buffered saline (PBS) buffer.
3. Poly(A) polymerase.

2.2. Purification, RNA Ligation, and RT-PCR

1. mirVana Probe and Marker Kit (Ambion).
2. RNA T4 ligase and mi5-RNA (*see* **Table 17.1**).
3. Superscript III First-Strand Synthesis System for RT-PCR and miRT (*see* **Table 17.1**).
4. AmpliTaq Gold PCR Master Mix.
5. miP1 and miP1r primers (*see* **Table 17.1**).
6. Agarose.
7. 100 bp DNA ladder.

2.3. Cloning and Transformation

1. QIAquick Gel Extraction Kit.
2. pcDNA3.1/V5-His TOPO TA Cloning Kit.
3. One Short Kit (Invitrogen).
4. Pre-poured LB agar ampicillin plates.
5. Parafilm.

2.4. Colony-Directed PCR, PCR Amplification, and Manual Sequencing

1. TE buffer.
2. GoTaq Green Master Mix.
3. pcDNA and BGHr (*see* **Table 17.1**).
4. Agarose.
5. 100 bp DNA ladder.
6. QIAquick PCR Purification Kit.

2.5. PCR Purification and Sequencing

1. Advantage-GC 2 PCR Kit.
2. Fusion-A* and Fusion-B primers (*see* **Table 17.1**).
3. QIAquick PCR Purification Kit.
4. Agarose.
5. 100 bp DNA ladder.

3. Methods

3.1. Isolation of Small RNAs

1. Harvest tissue (≤250 mg) or cells in a 1.7 mL tube with 500 μL of cold PBS.
2. Centrifuge at ~5,000 rpm for 2 min at room temperature (RT).
3. Remove PBS as much as possible. For cells, remove PBS carefully without breaking the pellet, leave ~100 μL of PBS, and resuspend cells by tapping gently.
4. Add 300–600 μL of lysis/binding buffer (10 volumes per tissue mass) on ice. When you start with frozen tissue or cells, immediately add lysis/binding buffer (10 volumes per tissue mass) on ice.
5. Cut tissue into small pieces using scissors, and grind it using a homogenizer. For cells, skip this step.
6. Vortex for 40 s to mix.
7. Add 1/10 volume of miRNA homogenate additive on ice, mix well by vortexing.
8. Leave the mixture on ice for 10 min. For tissue, mix it every 2 min.
9. Add an equal volume (330–660 μL) of acid-phenol: chloroform. Be sure to withdraw from the bottom phase (the upper phase is an aqueous buffer).
10. Mix thoroughly by inverting the tubes several times.
11. Centrifuge at 10,000 rpm for 5 min at RT.
12. Recover the aqueous phase carefully without disrupting the lower phase, and transfer it to a fresh tube.
13. Measure the volume using a scale (1g = ~1 mL) and note it.
14. Add 1/3 volume of 100% ethanol at RT to the recovered aqueous phase.
15. Mix thoroughly by inverting the tubes several times.
16. Transfer up to 700 μL of the mixture into a filter cartridge within a collection tube. Label the filter as "total RNA." When you have >700 μL of the mixture, apply it in successive application to the same filter.
17. Centrifuge at 10,000 rpm for 15 s at RT.
18. Collect the filtrate (the flow-through). Save the cartridge for total RNA isolation (go to **Step 24**).
19. Add 2/3 volume of 100% ethanol at RT to the flow-through.

20. Mix thoroughly by inverting the tubes several times.
21. Transfer up to 700 μL of the mixture into a new filter cartridge.
22. Centrifuge at 10,000 rpm for 15 s at RT.
23. Discard the flow-through, and repeat until all of the filtrate mixture is passed through the filter. Reuse the collection tube for the following washing steps.
24. Apply 700 μL of miRNA wash solution 1 (working solution mixed with ethanol) to the filter.
25. Centrifuge at 10,000 rpm for 15 s at RT.
26. Discard the flow-through.
27. Apply 500 μL of miRNA wash solution 2/3 (working solution mixed with ethanol) to the filter.
28. Centrifuge at 10,000 rpm for 15 s at RT.
29. Discard the flow-through and repeat step 27.
30. Centrifuge at 12,000 rpm for 1 min at RT.
31. Transfer the filter cartridge to a new collection tube.
32. Apply 100 μL of pre-heated (95°C) elution solution or RNase-free water to the center of the filter, and close the cap. Aliquot a desired amount of elution solution into a 1.7 mL tube, heat it on a heat block at 95°C for ~15 min. Open the cap carefully because it might be splash due to pressure buildup.
33. Let the filter tube stand for 1 min at RT.
34. Centrifuge at 12,000 rpm for 1 min at RT.
35. Measure total RNA and small RNA concentration using NanoDrop or another spectrophotometer.
36. Store it at –80°C until used.

3.2. Polyadenylation

1. Set up a reaction mixture with a total volume of 50 μL in a 0.5 mL tube containing 0.1–2 μg of small RNAs, 10 μL of 5× E-PAP buffer, 5 μL of 25 mM $MnCl_2$, 5 μL of 10 mM ATP, 1 μL (2U) of *Escherichia. coli* poly(A) polymerase I, and RNase-free water (up to 50 μL). When you have a low concentration of small RNAs, increase the total volume. 5× E-PAP buffer, 25 mM $MnCl_2$, and 10 mM ATP should be increased accordingly.
2. Mix well and spin the tube briefly.
3. Incubate for 1 h at 37°C.

3.3. Purification

1. Add an equal volume (50 μL) of acid-phenol: chloroform to the polyadenylation reaction mixture. When you have

>50 μL of the mixture, increase acid-phenol: chloroform accordingly.

2. Mix thoroughly by tapping the tube.
3. Centrifuge at 10,000 rpm for 5 min at RT.
4. Recover the aqueous phase carefully without disrupting the lower phase, and transfer it to a fresh tube.
5. Add 12 volumes (600 μL) of binding/washing buffer to the aqueous phase. When you have >50 μL of the aqueous phase, increase binding/washing buffer accordingly.
6. Transfer up to 460 μL of the mixture into a purification cartridge within a collection tube.
7. Centrifuge at 10,000 rpm for 15 s at RT.
8. Discard the filtrate (the flow-through), and repeat until all of the mixture is passed through the cartridge. Reuse the collection tube.
9. Apply 300 μL of binding/washing buffer to the cartridge.
10. Centrifuge at 12,000 rpm for 1 min at RT.
11. Transfer the cartridge to a new collection tube.
12. Apply 25 μL of pre-heated (95°C) elution solution to the center of the filter, and close the cap. Aliquot a desired amount of elution solution into a 1.7 mL tube, heat it on a heat block at 95°C for ~15 min. Open the cap carefully because it might be splash due to pressure buildup.
13. Let the filter tube stand for 1 min at RT.
14. Centrifuge at 12,000 rpm for 1 min at RT.
15. Repeat **Steps 12–14** with a second aliquot of 25 μL of pre-heated (95°C) elution solution.
16. Measure polyadenylated (tailed) RNA concentration using NanoDrop or another spectrophotometer.
17. Store it at –80°C until used. After polyadenylation, RNA concentration should increase up to 5–10 times of the starting concentration.

3.4. RNA Ligation

1. Set up a reaction mixture with a total volume of 50 μL in a 0.5 mL tube containing 1–2 μg of tailed RNAs, 3 μL (3 μg) of mi5-RNA, 5 μL of 1× T4 RNA ligase 1 reaction buffer, 1 μL (20U) of T4 RNA ligase (ssRNA ligase), and RNase-free water (up to 50 μL). When you have a low concentration of tailed RNAs, increase the total volume. 1× T4 RNA ligase 1 reaction buffer should be accordingly increased.
2. Mix well and spin the tube briefly.

3. Incubate for 30 min at 37°C and then for 15 min at 65°C to inactivate the ligase.

The ligated RNAs should be purified as described in **Section 3.3.**

3.5. Reverse Transcription

1. Mix 2 μg of ligated RNAs, 1 μL (1 μg) of miRT, and RNase-free water (up to 21 μL) in a PCR tube.
2. Incubate for 10 min at 65°C and for 5 min at 4°C.
3. Add 1 μL of 10 mM dNTP mix, 1 μL of RNaseOUT, 4 μL of 10× RT buffer, 4 μL of 0.1 MDTT, 8 μL of 25 mM $MgCl_2$, and 1 μL of SuperScript III reverse transcriptase to the mixture. When you have a low concentration of ligated RNAs, increase the total volume. 10× RT buffer, 0.1M DTT, and 25 mM $MgCl_2$ should be increased accordingly.
4. Mix well and spin the tube briefly.
5. Incubate for 60 min at 50°C and for 5 min at 85°C to inactivate the reaction.
6. Add 1 μL of RNase H to the mixture.
7. Incubate for 20 min at 37°C.
8. Add 60 μL of nuclease-free water.

3.6. PCR

1. Set up a reaction mixture with a total volume of 25 μL in a PCR tube containing 1 μL of small RNA cDNAs, 1 μL (10 pmol) of miP1, 1 μL (10 pmol) of miP1r, 12.5 μL of AmpliTaq Gold PCR Master Mix, and 9.5 μL of nuclease-free water.
2. Mix well and spin the tube briefly.
3. Start PCR with the following conditions: 95°C for 10 min, followed by 40 cycles at 95°C for 15 s, and at 60°C for 1 min.
4. Run 2 μL of PCR products along with a 100 bp DNA ladder on a 2% agarose gel. PCR products should be ~120–200 bp depending on the small RNA species (e.g., ~120–130 bp for miRNAs and piRNAs).

3.7. Cloning

3.7.1. Gel Elution

1. Excise the DNA fragments of ~120–130 bp on a 2% gel.
2. Add six volumes of buffer QG.
3. Incubate at 50°C for 10 min. Mix the tube every 2 min during incubation.
4. Add one volume of isopropanol and mix throughly.
5. Pass the sample through a QIAquick spin column in a collection tube.
6. Discard flow-through.

7. Add 0.5 mL of buffer QG.
8. Centrifuge at 12,000 rpm for 1 min.
9. Discard flow-through.
10. Add 0.75 mL of buffer PE.
11. Centrifuge at 12,000 rpm for 1 min.
12. Discard flow-through and centrifuge at 13,000 rpm for an additional 1 min.
13. Place the column into a new collection tube.
14. Add 30–50 μL of EB buffer to the center of the column and let it stand for 1 min.
15. Centrifuge at 12,000 rpm for 1 min.
16. Measure eluted DNA concentration using NanoDrop or another spectrophotometer.
17. Store it at –20°C until used.

3.7.2. TA Cloning (Optional for Manual Sequencing)

1. Mix 0.5–4 μL (50–100 ng) of eluted DNA, 1 μL of diluted salt, and sterile water to yield a final volume of 5 μL, and add 1 μL of TOPO vector. When you clone several samples, you can use 0.5 μL of TOPO vector with half the final volume in order to save costs.
2. Place the reaction mixture on ice for 5 min.
3. Proceed to transformation or store the TOPO cloning reaction at –20°C overnight.

3.7.3. Transformation (Optional for Manual Sequencing)

1. Add 1.5 μL of the TOPO cloning reaction into a 0.1 cm cuvette containing 25 μL of electrocompetent (EC) cells (One Shot TOP 10 or other EC cells).
2. Electroporate the sample using your own protocol and electroporator.
3. Immediately add 600 μL of RT SOC medium.
4. Transfer the solution to 15 mL Falcon tube.
5. Shake at 200 rpm for 1 h at 37°C.
6. Spread 10–100 μL of transformed cells on pre-warmed LB agar plates containing ampicillin (100 μg/mL) (LB-amp plates). Plate at least two different volumes to ensure that at least one plate will have well-spaced colonies. Store remaining cells at 4°C for additional plating.
7. Incubate plates at 37°C overnight until colonies are grown.
8. Spread an optimal amount of stored transformed cells (500–1,000 colonies/plate) on LB-amp plates.
9. Incubate plates at 37°C overnight until colonies are grown.

10. Proceed to colony-directed PCR or save plates sealed with Parafilm at 4°C.

3.7.4. Colony-Directed PCR (Optional for Manual Sequencing)

1. Pick up a single colony using a 10 μL tip on a 10 μL pipette and resuspend it in 10 μL TE buffer in a PCR tube. Take 5 μL of TE buffer with the pipette, pipette the buffer out gently onto a colony, and suction it back up into the tip in order to pick it up. Do not pick up too many cells.
2. Add 1 μL (5 pmol) of pcDNA, 1 μL (5 pmol) of BGHr, 12.5 μL of GoTaq Green Master Mix (2×) to the cell PCR tube as well as nuclease-free water to yield a final volume of 25 μL.
3. Mix well and spin the tube briefly.
4. Start PCR with the following conditions: [3 min at 95°C, then 40 cycles (for 30 s at 95°C, 30 s at 55°C, and 1 min at 72°C)].
5. Run 2 μL of the PCR products along with a 100 bp DNA ladder on a 2% agarose gel. PCR products with no insert and with no small RNA are 319 and 419 bp, respectively. Therefore, a PCR product with small RNA insert should be >439 bp.
6. Proceed to PCR purification with PCR product samples >439 bp.

3.7.5. PCR Purification (Optional for Manual Sequencing)

1. Add five volumes (115 μL) of buffer PBI to the PCR sample and mix well.
2. Transfer the sample to a QIAquick spin column in a collection tube.
3. Centrifuge at 12,000 rpm for 1 min.
4. Discard flow-through.
5. Add 0.75 mL of Buffer PE.
6. Centrifuge at 12,000 rpm for 1 min.
7. Discard flow-through and centrifuge at 13,000 rpm for an additional 1 min.
8. Place the column into a new collection tube.
9. Add 30–50 μL of EB buffer to the center of the column and let it stand for 1 min.
10. Centrifuge at 12,000 rpm for 1 min.
11. Measure the eluted DNA concentration using NanoDrop or another spectrophotometer.
12. Run 2 μL of the PCR products along with a 100 bp DNA ladder on a 2% agarose gel to double check the size and amount.
13. Proceed to sequencing or store it at –20°C until used.

3.7.6. Manual Sequencing

1. Set up a sequence sample with a total volume of 5–10 μL containing 40 ng of a PCR product and 2 pmol of pcDNA or BGHr primers
2. Submit samples for sequencing.

3.7.7. Sequencing Template Preparation (Optional for 454 Sequencing)

1. Set up a reaction mixture with a total volume of 25 μL in a PCR tube containing 5 μL of small RNA cDNAs (50–100 ng), 2 μL (20 pmol) of Fusion-A*, 2 μL (20 pmol) of Fusion-B, 5 μL of GC Melt (5M), 0.5 μL of dNTP mix (10 mM each), 5 μL of GC 2 PCR buffer, 0.5 μL of Advantage-GC 2 Polymerase Mix (50×), and 5 μL of PCR-Grade H_2O. In order to sequence multiple samples, each sample should be amplified with Fusion-A* containing two different nucleotide tags (*see* **Table 17.1** and **Notes 4**).
2. Mix well and spin the tube briefly.
3. Start PCR with the conditions: [3 min at 94°C, two cycles (at 94°C for 30 s, at 62°C for 30 s, and at 68°C for 1 min), three cycles (at 94°C for 30 s, at 59°C for 30 s, and at 68°C for 1 min), 20 cycles (at 94°C for 30 s, at 56°C for 30 s, and at 68°C for 1 min), and at 68°C for 3 min].
4. Run 2 μL of the PCR products along with a 100 bp DNA ladder on a 2% agarose gel.

   Major PCR products should be ~160–220 bp.

3.7.8. PCR Purification (Optional for 454 Sequencing)

1. Sequencing template PCR products should be purified as described above.
2. Measure the eluted DNA concentration using NanoDrop or another spectrophotometer.
3. Run 2 μL of the PCR products along with a 100 bp DNA ladder on a 2% agarose gel to double check the size and amount (save the gel image).
4. Proceed to 454 sequencing or store it at –20°C until used.

3.7.9. Sequencing

1. Set up a sequence sample with a total volume of 5–10 μL containing 10–50 ng of a purified PCR product and 2 pmol of Primer A (the sequencing company will add this primer).
2. Submit samples and the gel image for sequencing.

4. Notes

1. The cloning method can be used for manual sequencing and 454 sequencing.
2. This cloning method identifies small RNAs such as miRNAs (20–22), snoRNAs (12, 20, 22), tRNAs (12, 20, 22),

and rRNAs (12, 21, 22) by manual sequencing and by 454 sequencing (unpublished data).

3. Manual sequencing can be also used to check if small RNA cDNA libraries are properly constructed, or for prescreening if particular small RNA molecules are present in your cDNA library before engaging massive, costly, and time-consuming 454 sequencing.
4. Since 454 sequencing is massive (capable for 400–600 million base pairs with 400–500 bp read lengths) and expensive, multiple samples can be mixed and sequenced together. In this case, you should design Fusion-A* with a different combination of the two nucleotide tags for each sample (*see* **Table 17.1**) to amplify each sequencing template. After sequencing, reads should be sorted by each nucleotide tag to trace the source of the samples.
5. For 454 sequencing, either Primer A or Primer B can be used. However, Primer A is recommended because there is a poly "A" tail (25 As) at the end of a small RNA on the Primer B side (*see* **Fig. 17.1**).
6. A pair of Primer A and a small RNA specific reverse primer, or a small RNA-specific forward primer and Primer B can be used to check if sequencing templates are properly constructed by PCR.
7. Since our cloning method uses two degenerate nucleotides (VN) at the 3′ end to make small RNA cDNA libraries (*see* miRT in **Table 17.1**), the two nucleotides in sequencing may not be reliable.
8. In cloning miRNAs, our cloning method preferentially amplifies mature miRNAs over precursor miRNAs (pre-miRNAs). pre-miRNAs may be eliminated during cDNA synthesis due to a hairpin-loop structure. To clone pre-miRNAs, RT reactions should be performed at high temperature using a high-temperature stable reverse transcriptase. In addition, the sequencing reaction should be optimized with DMSO (5%) and 1.0M of betaine at an increased denaturation and annealing temperature.

Acknowledgments

The authors would like to thank Jonathan Cho for reading and editing the text. This work was supported by grants from the National Institute of Health (HD048855 and HD050281) to W.Y.

References

1. Aravin, A.A., Lagos-Quintana, M., Yalcin, A., Zavolan, M., Marks, D., Snyder, B., Gaasterland, T., Meyer, J., and Tuschl, T. (2003) The small RNA profile during *Drosophila melanogaster* development. *Dev Cell*, **5**, 337–350.
2. Lee, R.C. and Ambros, V. (2001) An extensive class of small RNAs in *Caenorhabditis elegans. Science*, **294**, 862–864.
3. Lau, N.C., Lim, L.P., Weinstein, E.G., and Bartel, D.P. (2001) An abundant class of tiny RNAs with probable regulatory roles in *Caenorhabditis elegans. Science*, **294**, 858–862.
4. Lagos-Quintana, M., Rauhut, R., Lendeckel, W., and Tuschl, T. (2001) Identification of novel genes coding for small expressed RNAs. *Science*, **294**, 853–858.
5. Lau, N.C., Seto, A.G., Kim, J., Kuramochi-Miyagawa, S., Nakano, T., Bartel, D.P., and Kingston, R.E. (2006) Characterization of the piRNA complex from rat testes. *Science*, **313**, 363–367.
6. Grivna, S.T., Beyret, E., Wang, Z., and Lin, H. (2006) A novel class of small RNAs in mouse spermatogenic cells. *Genes Dev*, **20**, 1709–1714.
7. Girard, A., Sachidanandam, R., Hannon, G.J., and Carmell, M.A. (2006) A germline-specific class of small RNAs binds mammalian Piwi proteins. *Nature*, **442**, 199–202.
8. Aravin, A., Gaidatzis, D., Pfeffer, S., Lagos-Quintana, M., Landgraf, P., Iovino, N., Morris, P., Brownstein, M.J., Kuramochi-Miyagawa, S., Nakano, T., Chien, M., Russo, J.J., Ju, J., Sheridan, R., Sander, C., Zavolan, M., and Tuschl, T. (2006) A novel class of small RNAs bind to MILI protein in mouse testes. *Nature*, **442**, 203–207.
9. Watanabe, T., Takeda, A., Tsukiyama, T., Mise, K., Okuno, T., Sasaki, H., Minami, N., and Imai, H. (2006) Identification and characterization of two novel classes of small RNAs in the mouse germline: retrotransposon-derived siRNAs in oocytes and germline small RNAs in testes. *Genes Dev*, **20**, 1732–1743.
10. Vagin, V.V., Sigova, A., Li, C., Seitz, H., Gvozdev, V., and Zamore, P.D. (2006) A distinct small RNA pathway silences selfish genetic elements in the germline. *Science*, **313**, 320–324.
11. Saito, K., Nishida, K.M., Mori, T., Kawamura, Y., Miyoshi, K., Nagami, T., Siomi, H., and Siomi, M.C. (2006) Specific association of Piwi with rasiRNAs derived from retrotransposon and heterochromatic regions in the Drosophila genome. *Genes Dev*, **20**, 2214–2222.
12. Ro, S., Song, R., Park, C., Zheng, H., Sanders, K.M., and Yan, W. (2007) Cloning and expression profiling of small RNAs expressed in the mouse ovary. *RNA*, **13**, 2366–2380.
13. Ruby, J.G., Jan, C., Player, C., Axtell, M.J., Lee, W., Nusbaum, C., Ge, H., and Bartel, D.P. (2006) Large-scale sequencing reveals 21U-RNAs and additional microRNAs and endogenous siRNAs in *C. elegans. Cell*, **127**, 1193–1207.
14. Terns, M.P. and Terns, R.M. (2002) Small nucleolar RNAs: versatile trans-acting molecules of ancient evolutionary origin. *Gene Expr*, **10**, 17–39.
15. Ouellet, D.L., Perron, M.P., Gobeil, L.A., Plante, P., and Provost, P. (2006) MicroRNAs in gene regulation: when the smallest governs it all. *J Biomed Biotechnol*, **2006**, 69616.
16. Maatouk, D. and Harfe, B. (2006) MicroRNAs in development. *ScientificWorldJournal*, **6**, 1828–1840.
17. Kim, V.N. and Nam, J.W. (2006) Genomics of microRNA. *Trends Genet*, **22**, 165–173.
18. Kim, V.N. (2006) Small RNAs just got bigger: Piwi-interacting RNAs (piRNAs) in mammalian testes. *Genes Dev*, **20**, 1993–1997.
19. Guryev, V. and Cuppen, E. (2009 Next-generation sequencing approaches in genetic rodent model systems to study functional effects of human genetic variation. *FEBS Lett*, **583,** 1668–1673.
20. Ro, S., Park, C., Jin, J., Sanders, K.M., and Yan, W. (2006) A PCR-based method for detection and quantification of small RNAs. *Biochem Biophys Res Commun*, **351**, 756–763.
21. Ro, S., Park, C., Sanders, K.M., McCarrey, J.R., and Yan, W. (2007) Cloning and expression profiling of testis-expressed microRNAs. *Dev Biol*, **311**, 592–602.
22. Ro, S., Park, C., Song, R., Nguyen, D., Jin, J., Sanders, K.M., McCarrey, J.R., and Yan, W. (2007) Cloning and expression profiling of testis-expressed piRNA-like RNAs. *RNA*, **13**, 1693–1702.

Chapter 18

In Situ Hybridization Detection of microRNAs

Rui Song, Seungil Ro, and Wei Yan

Abstract

MicroRNAs (miRNAs) are endogenous ~22 nucleotide RNAs that play critical roles in many cellular processes including cell differentiation, proliferation, and apoptosis. The analysis of spatiotemporal expression of miRNAs is important for dissecting their roles in development and during physiological/pathophysiological processes. In situ hybridization is a powerful technology that allows cellular localization. However, the detection of miRNAs by in situ hybridization has been challenging because of the low affinity of conventional RNA or DNA probes due to the small sizes of miRNAs. Here, we describe a protocol for miRNA in situ hybridization on mouse tissue cryosections using locked nucleic acid (LNA) probes. LNA probes demonstrate a much higher affinity to their complimentary RNAs compared to conventional RNA and DNA oligo probes, which allow detection of miRNAs in tissue sections with excellent specificity.

Key words: microRNAs, in situ hybridization, locked nucleic acid, cryosections.

1. Introduction

As posttranscriptional regulators, miRNAs play important roles in the regulation of gene expression and thus are important in many biological processes (1, 2). Spatiotemporal expression of specific miRNAs in specific organs can provide important clues on their potential functions. Therefore, it is usually the first essential step to determine the expression profiles and cellular localization of specific miRNAs. Since all organs contain multiple cell types with different proportions, expression profiling assays, e.g., miRNA Northern blots or miRNA PCR, cannot reveal cellular sources or localizations of miRNAs expression. In situ hybridization represents an ideal method to reveal cellular localization of mRNAs and long noncoding RNAs (3). miRNA in situ hybridization has

M. Sioud (ed.), *RNA Therapeutics*, Methods in Molecular Biology 629,
DOI 10.1007/978-1-60761-657-3_18,

been challenging because conventional RNA or DNA probes display low affinity to their targets due to the small sizes of miRNAs (~22 nt) and thus low specificity. The development of locked nucleic acid (LNA) appears to have solved this problem. LNA is a class of bicyclic high-affinity RNA analogs in which the furanose ring of the ribose sugar is chemically locked in an RNA mimicking N-type (C3′-endo) conformation by the introduction of an O2′, C4′-methylene bridge. Therefore, compared to traditional DNA probes, the LNA-modified DNA probes display a much higher hybridization affinity toward the complementary RNA molecules (4, 5). Moreover, the increased thermal stability of the LNA–DNA/RNA duplex make it possible to generate probes as short as ~20 nt with high melting temperatures (≥70°C), providing the basis for high stringency required by miRNA in situ hybridization (6, 7). Because of these properties, LNA-modified DNA probes have become the first choice in miRNA in situ hybridization assays. Here, we describe a step-by-step protocol for miRNA in situ hybridization using LNA probes. This protocol is designed for digoxigenin (DIG)-labeled LNA probes, which can be visualized using nitro blue tetrazolium/5-bromo-4-chloro-3-indolyl phosphate (NBT/BCIP) in order to enhance the signal strength. The NBT/BCIP visualization generates a blue precipitate that can be detected with light microscopy.

2. Materials

2.1. Labeling miRCURY™ LNA miRNA Detection Probes and Purification

1. miRCURY™ LNA miRNA detection probes (Exiqon).
2. DIG Oligonucleotide Tailing Kit, second generation (Roche).
3. 0.1 M EDTA, pH 8.0.
4. Illustra MicroSpin G-25 Columns (GE Healthcare).

2.2. Determination of Labeling Efficiency

1. Control DIG-dUTP/dATP tailed oligonucleotide (2.5 pmol/μL, provided in DIG Oligonucleotide Tailing Kit).
2. 1× PBS buffer.
3. Blocking solution: 1% bovine serum albumin (BSA) in 1×PBS.
4. Antibody solution: Dilute 10 μL of anti-DIG alkaline phosphatase-conjugated antibody (Roche) in 10 mL of blocking solution.
5. NBT/BCIP solution: prepared from NBT/BCIP stock solution as manufacturer's instruction (Roche).
6. TE buffer: 10 mM Tris-Cl, pH 7.5, 1 mM EDTA.

2.3. Preparation of Frozen Sections

1. 1× PBS.
2. Acetone.
3. 5 and 20% Sucrose solutions prepared in 1× PBS.
4. 20% Sucrose/OCT solution (4 mL): Mix 2 mL of 20% sucrose and 2 mL of O.C.T. compound.
5. 2-Methylbutane.

2.4. In Situ Hybridization

1. 4% PFA: add 16 g of paraformaldehyde to 40 mL of 10× PBS, and then add water to 400 mL. Adjust pH to 7.4–7.6. Use preheated water (60°C) to dissolve paraformaldehyde.
2. IsHyb in situ hybridization (ISH) kit (Biochain). This kit contains prehybridization solution, hybridization solution, 20× SSC buffer, 100× blocking solution, anti-DIG alkaline phosphatase-conjugated antibody, 10× alkaline phosphatase buffer, and NBT/BCIP reagents.

2.5. Washing and Immuno Detection

1. Washing buffers: 2× SSC, 1× SSC, and 0.2× SSC prepared from the 20× SSC.
2. 1× blocking solution prepared from 100× blocking solution included in the ISH kit.
3. Anti-DIG alkaline phosphatase-conjugated antibody solution (50 μL) containing 0.1 μL of the antibody and 50 μL of 1× blocking solution.
4. 1× alkaline phosphatase buffer prepared from 10× alkaline phosphatase buffer.
5. NBT/BCIP solution containing 6.6 μL of NBT and 3.3 μL of BCIP in 1 mL 1× alkaline phosphatase buffer.
6. Aqua-Mount solution.

3. Methods

3.1. Labeling miRCURY™ LNA miRNA Detection Probes

1. miRCURY™ LNA miRNA detection probes (25 pmol/μL) can be purchased from Exiqon.
2. Set a reaction mixture on ice with a final volume of 20 μL in a PCR tube containing 4 μL (100 pmol) of a LNA miRNA detection probe, 4 μL of reaction buffer, 4 μL of $CoCl_2$ solution, 1 μL of DIG–dUTP solution, 1 μL of dATP solution, and 1 μL of 400 U terminal transferase, and 5 μL of nuclease-free water.
3. Mix them well and centrifuge briefly.

4. Incubate the mixture at 37°C for 30 min and then place on ice.
5. Stop reaction by adding 5 μL of 0.1 M EDTA, pH 8.0.

3.2. Purification of DIG-Labeled LNA Probes

1. For preparation of a column, resuspend the resin in the column by vortexing. Loosen the cap by a one-quarter turn and snap off the bottom closure. Place the column in the supplied collection tube. Prespin the column at 1,000×*g* for 1 min.
2. Transfer the column to a clean 1.7 mL microcentrifuge tube. Slowly apply the DIG-labeled LNA probe sample to the center of the resin. Do not disturb the resin bed.
3. Centrifuge at 1,000×*g* for 4 min. The purified probes are collected at the bottom of the tube.
4. Measure purified probe concentrations using NanoDrop or other spectrophotometers. Probe concentration should be around 30 ng/μL.
5. Aliquot labeled probes into multiple tubes and store at –20°C or –80°C.

3.3. Determination of Labeling Efficiency

1. Dilute DIG-labeled LNA probes (30 ng/μL) and the control DIG-dUTP/dATP tailed oligonucleotide (2.5 pmol/μL) in nuclease-free water at serial dilutions 10^{-1}, 10^{-2}, 10^{-3}, and 10^{-4} in a volume of 50 μL.
2. Spot 1 μL of undiluted probe and 1 μL of each dilution onto nitrocellulose.
3. Air dry membrane and UV-crosslink both sides to fix the nucleic acid.
4. Equilibrate the membrane with 1× PBS for 3 min and incubate it in 10 mL of 1× PBS in a container containing 1% BSA at room temperature (RT) for 30 min.
5. Prepare 10 mL of antibody solution freshly (*see* **Section 2**) and incubate the membrane at RT for 30 min.
6. Wash the membrane with 1× PBS twice (10 min each).
7. Add 10 mL of NBT/BCIP solution to the membrane and incubate for 10 min.
8. Stop the reaction by rinsing the membrane with TE buffer several times and then expose the membrane (*see* **Fig. 18.1**).

3.4. Preparation of Frozen Sections

1. Dissect a mouse testis tissue.
2. Wash tissue with 1× PBS.
3. Fix tissue in 4 mL of acetone in a tube at 4°C for 30 min.
4. Wash with 5% sucrose.

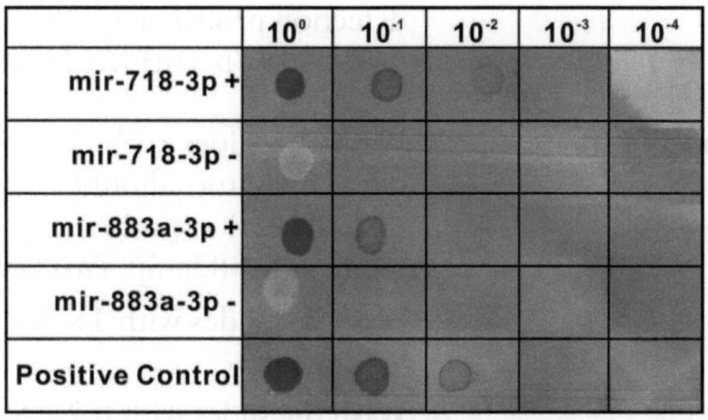

Fig. 18.1. Labeling efficiency test. "+" stands for DIG-labeled probe, and "–" for unlabeled probe. DIG-dUTP/dATP tailed oligonucleotide provided in Roche DIG Oligonucleotide Tailing Kit was used as a positive control. 10^0, undiluted probes. 10^{-1}, 10^{-2}, 10^{-3}, and 10^{-4}, serial diluted probes.

5. Incubate with 10% sucrose, 12.5% sucrose, and 15% sucrose by gentle shaking at 4°C for 30 min, respectively.
6. Finally transfer the tissue to 20% sucrose, shake at 4°C overnight.
7. Put the tissue in 4 mL of 20% sucrose/OCT solution, shake at 4°C for 30 min.
8. Freeze the tissue in 20% sucrose/OCT mixture in liquid nitrogen cooled 2-methylbutane and store the block at –80°C until use.
9. Cut sections of 10 μm on a cryostat and mount onto slides.
10. Air dry sections at least 20 min, but no longer than 3 h.
11. Proceed to in situ hybridization or store them at –20°C (frost-free) (*see* **Note 1**).

3.5. In Situ Hybridization

1. Warm the frozen slides at RT. Dry them at 50°C for 15 min.
2. Fix the slides with 4% PFA at RT for 20 min.
3. Wash the slides three times (5 min each) with 1× PBS at RT.
4. Prehybridize the slides with Prehybridization Solution at 60°C for 3 h. Prehybridization temperature should be 15°C below T_m of miRCURY™ LNA detection probes (e.g., 60°C calculated from 75°C of mmu-miR-883a-3p Tm) (*see* **Note 2**).
5. Prepare hybridization mixture by adding 0.2 μL of DIG-labeled LNA probe to 20 μL of hybridization solution.
6. Heat hybridization mixture at 65°C for 5 min to linearize probes and then chill on ice.
7. Apply the prepared hybridization mixture onto the slides. Hybridize the slides at 55°C for 12 h. Hybridization temperature should be 20°C below T_m of miRCURY™ LNA

detection probes (e.g., 55°C calculated from 75°C of mmu-miR-883a-3p Tm) (*see* **Note 2**).

3.6. Washing

1. Wash the slides with 2× SSC at 55°C for 10 min. Washing temperature should be 20°C below T_m of miRCURY™ LNA detection probes (e.g., 55°C calculated from 75°C of mmu-miR-883a-3p Tm) (*see* **Note 2**).
2. Wash the slides with 1× SSC at the same temperature 55°C for 10 min.
3. Wash the slides with 0.2× SSC twice (20 min each) at 37°C.
4. Incubate the slides with 1× blocking solution at RT for 1 h (*see* **Note 3**).

3.7. Immunological Detection

1. Incubate the slides in 50 μL of the anti-DIG alkaline phosphatase-conjugated antibody solution at 4°C overnight.
2. Wash the slides with 1× PBS three times (10 min each) at RT.
3. Wash the slides with 1× alkaline phosphatase buffer twice (5 min each) at RT.

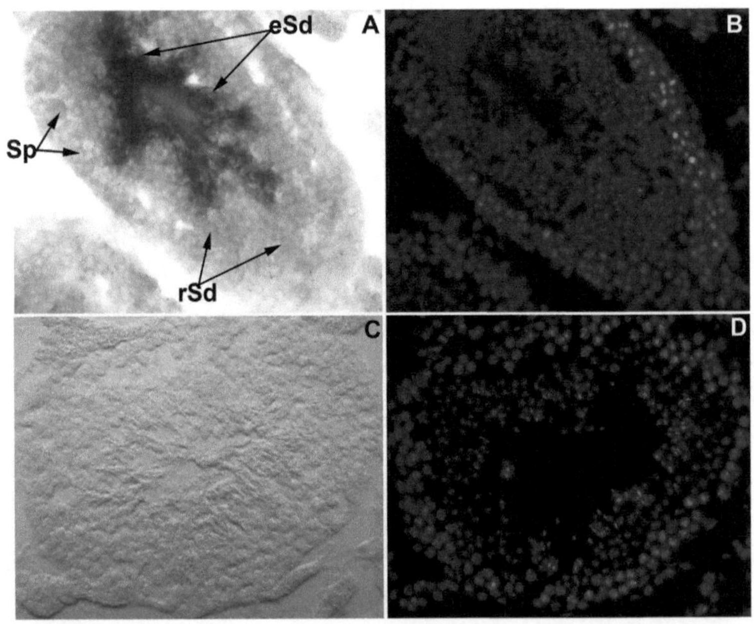

Fig. 18.2. In situ hybridization analyses of mir-92a-3p in adult mouse testes. (**a**) mir-92a-3p is detected in round spermatids (rSd) at lower levels and in elongated spermatids (eSd) at higher levels, whereas early spermatogenic cell types including spermatogonia and spermatocytes (Sp) show no detectable hybridization signals. (**b**) DAPI counterstaining of cell nuclei in the section shown in **panel a**. (**c**) Control in situ hybridization using DIG-labeled mir-92a-3p probe competed with 100× excessive amounts of unlabeled mir-92a-3p probe. (**d**) DAPI counterstaining of cell nuclei in the section shown in (**c**). Magnification: 400×.

4. Incubate the slides with NBT/BCIP solution in a dark room until a desired signal appears.
5. Discard the NBT/BCIP solution. Rinse the slides with distilled water.
6. Counterstain the slides with nuclear fluorescent dye 4″,6-diamidino-2-phenylindole (DAPI) or propidium iodide.
7. Mount the slides with aqua mount.
8. Analyze on a microscope (**Fig. 18.2**).

4. Notes

1. Fresh sections are recommended for best results although several months old slides can still be usable.
2. It is important to determine optimal temperatures of prehybridization, hybridization, and washing for each probe. Optimal temperatures can be calculated from the T_m of each probe, which is provided by the company.
3. Blocking step can also be performed at 4°C for overnight because longer incubation will generally give lower background.
4. We have used this protocol and examined localization of 3 X-linked miRNAs in the mouse testes (8).
5. To better demonstrate the cellular sources of miRNA in situ hybridization signals, conventional immunofluorescent staining of marker proteins can be performed after the in situ hybridization procedures, as described in reference (8). The fluorescent images from the red and green channels can then be merged with NBT/BCIP images (designated as the blue channel) using Adobe Photoshop software.
6. Control experiments: Since both the 5′ and the 3′ strands of a stem-loop structure of a precursor miRNAs can be expressed in certain tissues (9), the traditional way of using the sense probe as a control probe is not applicable for miRNA in situ hybridization. We use either antisense LNA oligos with several mutations in their sequences, or DIG-labeled ("hot") LNA probes competed with 100× excessive amounts of unlabeled ("cold") LNA probe during hybridization. The later method is more economic (no need to order mutant LNAs as controls) and convenient. Briefly, the concentrations of the DIG-labeled and unlabeled LNA probes are adjusted to 5 and 25 pmol/μL, respectively. Prepare the control hybridization mixture by adding 0.2 μL

of DIG-labeled “hot” LNA probe and 4.0 μL of unlabeled “cold” LNA probe to 20 μL of hybridization solution. This will result in 100× excessive numbers of “cold” probe compared to “hot” probe in the hybridization solution. Proceed to hybridization and the rest of the steps as described in **Sections 3.5–3.7**.

Acknowledgments

The authors would like to thank Jonathan Cho for reading and editing the manuscript. This work was supported by grants from the National Institute of Health (HD048855 and HD050281) to W. Y.

References

1. Bartel, D.P. (2004) MicroRNAs: genomics, biogenesis, mechanism, and function. *Cell*, **116**, 281–297.
2. Kim, V.N. and Nam, J.W. (2006) Genomics of microRNA. *Trends Genet*, **22**, 165–173.
3. Mitchell, B.S., Dhami, D., and Schumacher, U. (1992) In situ hybridisation: a review of methodologies and applications in the biomedical sciences. *Med Lab Sci*, **49**, 107–118.
4. Kloosterman, W.P., Wienholds, E., de Bruijn, E., Kauppinen, S., and Plasterk, R.H. (2006) In situ detection of miRNAs in animal embryos using LNA-modified oligonucleotide probes. *Nat Methods*, **3**, 27–29.
5. Wheeler, G., Valoczi, A., Havelda, Z., and Dalmay, T. (2007) In situ detection of animal and plant microRNAs. *DNA Cell Biol*, **26**, 251–255.
6. Obernosterer, G., Martinez, J., and Alenius, M. (2007) Locked nucleic acid-based in situ detection of microRNAs in mouse tissue sections. *Nat Protoc*, **2**, 1508–1514.
7. Silahtaroglu, A.N., Nolting, D., Dyrskjot, L., Berezikov, E., Moller, M., Tommerup, N., and Kauppinen, S. (2007) Detection of microRNAs in frozen tissue sections by fluorescence in situ hybridization using locked nucleic acid probes and tyramide signal amplification. *Nat Protoc*, **2**, 2520–2528.
8. Song, R., Ro, S., Michaels, J.D., Park, C., McCarrey, J.R., and Yan, W. (2009) Many X-linked microRNAs escape meiotic sex chromosome inactivation. *Nat Genet*, **41**, 488–493.
9. Ro, S., Park, C., Young, D., Sanders, K.M., and Yan, W. (2007) Tissue-dependent paired expression of miRNAs. *Nucleic Acids Res*, **35**, 5944–5953.

Chapter 19

Detection and Quantitative Analysis of Small RNAs by PCR

Seungil Ro and Wei Yan

Abstract

Increasing lines of evidence indicate that small non-coding RNAs including miRNAs, piRNAs, rasiRNAs, 21U endo-siRNAs, and snoRNAs are involved in many critical biological processes. Functional studies of these small RNAs require a simple, sensitive, and reliable method for detecting and quantifying levels of small RNAs. Here, we describe such a method that has been widely used for the validation of cloned small RNAs and also for quantitative analyses of small RNAs in both tissues and cells.

Key words: Small RNAs, miRNAs, piRNAs, expression, PCR.

1. Introduction

The past several years have witnessed the surprising discovery of numerous non-coding small RNAs species encoded by genomes of virtually all species (1–6), which include microRNAs (miRNAs) (7–10), piwi-interacting RNAs (piRNAs) (11–14), repeat-associated siRNAs (rasiRNAs) (15–18), 21U endo-siRNAs (19), and small nucleolar RNAs (snoRNAs) (20). These small RNAs are involved in all aspects of cellular functions through direct or indirect interactions with genomic DNAs, RNAs, and proteins. Functional studies on these small RNAs are just beginning, and some preliminary findings have suggested that they are involved in regulating genome stability, epigenetic marking, transcription, translation, and protein functions (5, 21–23). An easy and sensitive method to detect and quantify levels of these small RNAs in organs or cells during developmental courses, or under different

M. Sioud (ed.), *RNA Therapeutics*, Methods in Molecular Biology 629,
DOI 10.1007/978-1-60761-657-3_19, © Springer Science+Business Media, LLC 2010

physiological and pathophysiological conditions, is essential for functional studies. Quantitative analyses of small RNAs appear to be challenging because of their small sizes [~20 nucleotides (nt) for miRNAs, ~30 nt for piRNAs, and 60–200 nt for snoRNAs]. Northern blot analysis has been the standard method for detection and quantitative analyses of RNAs. But it requires a relatively large amount of starting material (10–20 μg of total RNA or >5 μg of small RNA fraction). It is also a labor-intensive procedure involving the use of polyacrylamide gel electrophoresis, electrotransfer, radioisotope-labeled probes, and autoradiography. We have developed a simple and reliable PCR-based method for detection and quantification of all types of small non-coding RNAs. In this method, small RNA fractions are isolated and polyA tails are added to the 3′ ends by polyadenylation (**Fig. 19.1**). Small RNA cDNAs (srcDNAs) are then generated by reverse

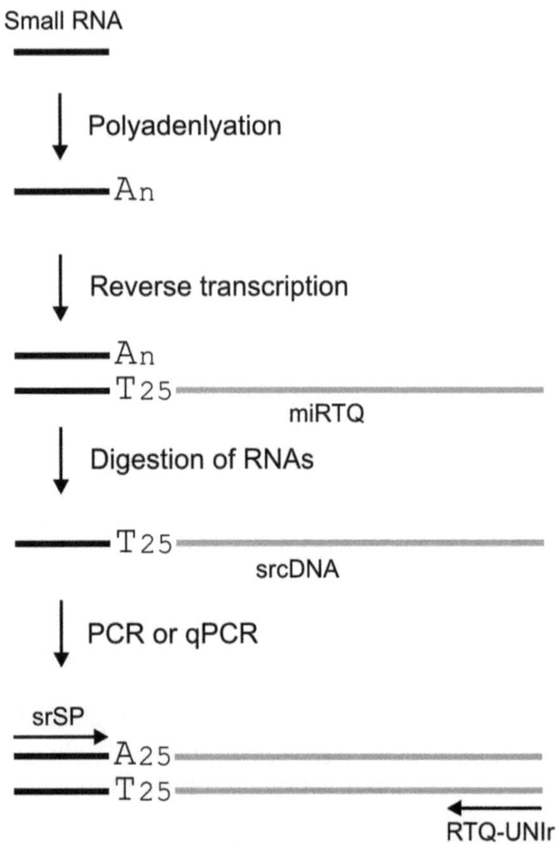

Fig. 19.1. Overview of small RNA complementary DNA (srcDNA) library construction for PCR or qPCR analysis. Small RNAs are polyadenylated using a polyA polymerase. The polyA-tailed RNAs are reverse-transcribed using a primer miRTQ containing oligo dTs flanked by an adaptor sequence. RNAs are removed by RNase H from the srcDNA. The srcDNA is ready for PCR or qPCR to be carried out using a small RNA-specific primer (srSP) and a universal reverse primer, RTQ-UNIr.

transcription using a primer consisting of adaptor sequences at the 5′ end and polyT at the 3′ end (miRTQ). Using the srcDNAs, non-quantitative or quantitative PCR can then be performed using a small RNA-specific primer and the RTQ-UNIr primer. This method has been utilized by investigators in numerous studies (18, 24–38). Two recent technologies, 454 sequencing and microarray (39, 40) for high-throughput analyses of miRNAs and other small RNAs, also need an independent method for validation. 454 sequencing, the next-generation sequencing technology, allows virtually exhaustive sequencing of all small RNA species within a small RNA library. However, each of the cloned novel small RNAs needs to be validated by examining its expression in organs or in cells. Microarray assays of miRNAs have been available but only known or bioinformatically predicted miRNAs are covered. Similar to mRNA microarray analyses, the up- or down-regulation of miRNA levels under different conditions needs to be further validated using conventional Northern blot analyses or PCR-based methods like the one that we are describing here.

2. Materials

2.1. Isolation of Small RNAs, Polyadenylation, and Purification

1. mirVana miRNA Isolation Kit (Ambion).
2. Phosphate-buffered saline (PBS) buffer.
3. Poly(A) polymerase.
4. mirVana Probe and Marker Kit (Ambion).

2.2. Reverse Transcription, PCR, and Quantitative PCR

1. Superscript III First-Strand Synthesis System for RT-PCR (Invitrogen).
2. miRTQ primers (**Table 19.1**).
3. AmpliTaq Gold PCR Master Mix for PCR.
4. SYBR Green PCR Master Mix for qPCR.
5. A miRNA-specific primer (e.g., let-7a) and RTQ-UNIr (**Table 19.1**).
6. Agarose and 100 bp DNA ladder.

3. Methods

3.1. Isolation of Small RNAs

1. Harvest tissue (≤250 mg) or cells in a 1.7-mL tube with 500 μL of cold PBS.

Table 19.1
Oligonucleotides used

Name	Sequence (5′–3′)	Note	Usage
miRTQ	CGAATTCTAGAGCTCGAGGCAGGCGACATGGCTGGCTAGTTAAGCTTGGTACCGAGCTAGTCCTTTTTTTTTTTTTTTTTTTTTTTTTVN*	RNase free, HPLC	Reverse transcription
RTQ-UNIr	CGAATTCTAGAGCTCGAGGCAGG	Regular desalting	PCR/qPCR
let-7a	TGAGGTAGTAGGTTGTATAG	Regular desalting	PCR/qPCR

*V = A, C, or G; N = A, C, G, or T

2. Centrifuge at ~5,000 rpm for 2 min at room temperature (RT).
3. Remove PBS as much as possible. For cells, remove PBS carefully without breaking the pellet, leave ~100 μL of PBS, and resuspend cells by tapping gently.
4. Add 300–600 μL of lysis/binding buffer (10 volumes per tissue mass) on ice. When you start with frozen tissue or cells, immediately add lysis/binding buffer (10 volumes per tissue mass) on ice.
5. Cut tissue into small pieces using scissors and grind it using a homogenizer. For cells, skip this step.
6. Vortex for 40 s to mix.
7. Add one-tenth volume of miRNA homogenate additive on ice and mix well by vortexing.
8. Leave the mixture on ice for 10 min. For tissue, mix it every 2 min.
9. Add an equal volume (330–660 μL) of acid-phenol:chloroform. Be sure to withdraw from the bottom phase (the upper phase is an aqueous buffer).
10. Mix thoroughly by inverting the tubes several times.
11. Centrifuge at 10,000 rpm for 5 min at RT.
12. Recover the aqueous phase carefully without disrupting the lower phase and transfer it to a fresh tube.
13. Measure the volume using a scale (1g = ~1 mL) and note it.
14. Add one-third volume of 100% ethanol at RT to the recovered aqueous phase.
15. Mix thoroughly by inverting the tubes several times.
16. Transfer up to 700 μL of the mixture into a filter cartridge within a collection tube. Label the filter as total RNA. When you have >700 μL of the mixture, apply it in successive application to the same filter.
17. Centrifuge at 10,000 rpm for 15 s at RT.
18. Collect the filtrate (the flow-through). Save the cartridge for total RNA isolation (go to **Step 24**).
19. Add two-third volume of 100% ethanol at RT to the flow-through.
20. Mix thoroughly by inverting the tubes several times.
21. Transfer up to 700 μL of the mixture into a new filter cartridge. Label the filter as small RNA. When you have >700 μL of the filtrate mixture, apply it in successive application to the same filter.

22. Centrifuge at 10,000 rpm for 15 s at RT.
23. Discard the flow-through and repeat until all of the filtrate mixture is passed through the filter. Reuse the collection tube for the following washing steps.
24. Apply 700 μL of miRNA wash solution 1 (working solution mixed with ethanol) to the filter.
25. Centrifuge at 10,000 rpm for 15 s at RT.
26. Discard the flow-through.
27. Apply 500 μL of miRNA wash solution 2/3 (working solution mixed with ethanol) to the filter.
28. Centrifuge at 10,000 rpm for 15 s at RT.
29. Discard the flow-through and repeat **Step 27**.
30. Centrifuge at 12,000 rpm for 1 min at RT.
31. Transfer the filter cartridge to a new collection tube.
32. Apply 100 μL of pre-heated (95°C) elution solution or RNase-free water to the center of the filter and close the cap. Aliquot a desired amount of elution solution into a 1.7-mL tube and heat it on a heat block at 95°C for ~15 min. Open the cap carefully because it might splash due to pressure buildup.
33. Leave the filter tube alone for 1 min at RT.
34. Centrifuge at 12,000 rpm for 1 min at RT.
35. Measure total RNA and small RNA concentrations using NanoDrop or another spectrophotometer.
36. Store it at –80°C until used.

3.2. Polyadenylation

1. Set up a reaction mixture with a total volume of 50 μL in a 0.5-mL tube containing 0.1–2 μg of small RNAs, 10 μL of 5× E-PAP buffer, 5 μL of 25 mM $MnCl_2$, 5 μL of 10 mM ATP, 1 μL (2U) of *Escherichia coli* poly(A) polymerase I, and RNase-free water (up to 50 μL). When you have a low concentration of small RNAs, increase the total volume; 5× E-PAP buffer, 25 mM $MnCl_2$, and 10 mM ATP should be increased accordingly.
2. Mix well and spin the tube briefly.
3. Incubate for 1 h at 37°C.

3.3. Purification

1. Add an equal volume (50 μL) of acid-phenol:chloroform to the polyadenylation reaction mixture. When you have >50 μL of the mixture, increase acid-phenol:chloroform accordingly.
2. Mix thoroughly by tapping the tube.

3. Centrifuge at 10,000 rpm for 5 min at RT.
4. Recover the aqueous phase carefully without disrupting the lower phase and transfer it to a fresh tube.
5. Add 12 volumes (600 μL) of binding/washing buffer to the aqueous phase. When you have >50 μL of the aqueous phase, increase binding/washing buffer accordingly.
6. Transfer up to 460 μL of the mixture into a purification cartridge within a collection tube.
7. Centrifuge at 10,000 rpm for 15 s at RT.
8. Discard the filtrate (the flow-through) and repeat until all of the mixture is passed through the cartridge. Reuse the collection tube.
9. Apply 300 μL of binding/washing buffer to the cartridge.
10. Centrifuge at 12,000 rpm for 1 min at RT.
11. Transfer the cartridge to a new collection tube.
12. Apply 25 μL of pre-heated (95°C) elution solution to the center of the filter and close the cap. Aliquot a desired amount of elution solution into a 1.7-mL tube and heat it on a heat block at 95°C for ~15 min. Open the cap carefully because it might be splash due to pressure buildup.
13. Let the filter tube stand for 1 min at RT.
14. Centrifuge at 12,000 rpm for 1 min at RT.
15. Repeat **Steps 12–14** with a second aliquot of 25 μL of pre-heated (95°C) elution solution.
16. Measure polyadenylated (tailed) RNA concentration using NanoDrop or another spectrophotometer.
17. Store it at –80°C until used. After polyadenylation, RNA concentration should increase up to 5–10 times of the starting concentration.

3.4. Reverse Transcription

1. Mix 2 μg of tailed RNAs, 1 μL (1 μg) of miRTQ, and RNase-free water (up to 21 μL) in a PCR tube.
2. Incubate for 10 min at 65°C and for 5 min at 4°C.
3. Add 1 μL of 10 mM dNTP mix, 1 μL of RNaseOUT, 4 μL of 10× RT buffer, 4 μL of 0.1M DTT, 8 μL of 25 mM $MgCl_2$, and 1 μL of SuperScript III reverse transcriptase to the mixture. When you have a low concentration of ligated RNAs, increase the total volume; 10× RT buffer, 0.1M DTT, and 25 mM $MgCl_2$ should be increased accordingly.
4. Mix well and spin the tube briefly.
5. Incubate for 60 min at 50°C and for 5 min at 85°C to inactivate the reaction.

6. Add 1 μL of RNase H to the mixture.
7. Incubate for 20 min at 37°C.
8. Add 60 μL of nuclease-free water.

3.5. PCR and qPCR

1. Set up a reaction mixture with a total volume of 25 μL in a PCR tube containing 1 μL of small RNA cDNAs (srcDNAs), 1 μL (5 pmol of a miRNA-specific primer (srSP), 1 μL (5 pmol) of RTQ-UNIr, 12.5 μL of AmpliTaq Gold PCR Master Mix, and 9.5 μL of nuclease-free water. For qPCR, use SYBR Green PCR Master Mix instead of AmpliTaq Gold PCR Master Mix.
2. Mix well and spin the tube briefly.
3. Start PCR or qPCR with the conditions: 95°C for 10 min and then 40 cycles at 95°C for 15 s, at 48°C for 30 s and at 60°C for 1 min.
4. Adjust annealing Tm according to the Tm of your primer
5. Run 2 μL of the PCR or qPCR products along with a 100 bp DNA ladder on a 2% agarose gel. ~PCR products should be ~120–200 bp depending on the small RNA species (e.g., ~120–130 bp for miRNAs and piRNAs).

4. Notes

1. This PCR method can be used for quantitative PCR (qPCR) or semi-quantitative PCR (semi-qPCR) on small RNAs such as miRNAs, piRNAs, snoRNAs, small interfering RNAs (siRNAs), transfer RNAs (tRNAs), and ribosomal RNAs (rRNAs) (18, 24–38).
2. Design miRNA-specific primers to contain only the "core sequence" since our cloning method uses two degenerate nucleotides (VN) at the 3′ end to make small RNA cDNAs (srcDNAs) (*see* let-7a, **Table 19.1**).
3. For qPCR analysis, two miRNAs and a piRNA were quantitated using the SYBR Green PCR Master Mix (41). Cycle threshold (Ct) is the cycle number at which the fluorescence signal reaches the threshold level above the background. A Ct value for each miRNA tested was automatically calculated by setting the threshold level to be 0.1–0.3 with auto baseline. All Ct values depend on the abundance of target miRNAs. For example, average Ct values for let-7 isoforms range from 17 to 20 when 25 ng of each srcDNA sample from the multiple tissues was used (*see* (41).

4. This method amplifies over a broad dynamic range up to 10 orders of magnitude and has excellent sensitivity capable of detecting as little as 0.001 ng of the srcDNA in qPCR assays.
5. For qPCR, each small RNA-specific primer should be tested along with a known control primer (e.g., let-7a) for PCR efficiency. Good efficiencies range from 90% to 110% calculated from slopes between –3.1 and –3.6.
6. On an agarose gel, mature miRNAs and precursor miRNAs (pre-miRNAs) can be differentiated by their size. PCR products containing miRNAs will be ~120 bp long in size while products containing pre-miRNAs will be ~170 bp long. However, our PCR method preferentially amplifies mature miRNAs (*see* Results and Discussion in (41)). We tested our PCR method to quantify over 100 miRNAs, but never detected pre-miRNAs (18, 29–31, 38).

Acknowledgments

The authors would like to thank Jonathan Cho for reading and editing the text. This work was supported by grants from the National Institute of Health (HD048855 and HD050281) to W.Y.

References

1. Ambros, V. (2004) The functions of animal microRNAs. *Nature*, **431**, 350–355.
2. Bartel, D.P. (2004) MicroRNAs: genomics, biogenesis, mechanism, and function. *Cell*, **116**, 281–297.
3. Chang, T.C. and Mendell, J.T. (2007) The roles of microRNAs in vertebrate physiology and human disease. *Annu Rev Genomics Hum Genet.*
4. Kim, V.N. (2005) MicroRNA biogenesis: coordinated cropping and dicing. *Nat Rev Mol Cell Biol*, **6**, 376–385.
5. Kim, V.N. (2006) Small RNAs just got bigger: Piwi-interacting RNAs (piRNAs) in mammalian testes. *Genes Dev*, **20**, 1993–1997.
6. Kotaja, N., Bhattacharyya, S.N., Jaskiewicz, L., Kimmins, S., Parvinen, M., Filipowicz, W., and Sassone-Corsi, P. (2006) The chromatoid body of male germ cells: similarity with processing bodies and presence of Dicer and microRNA pathway components. *Proc Natl Acad Sci U S A*, **103**, 2647–2652.
7. Aravin, A.A., Lagos-Quintana, M., Yalcin, A., Zavolan, M., Marks, D., Snyder, B., Gaasterland, T., Meyer, J., and Tuschl, T. (2003) The small RNA profile during *Drosophila melanogaster* development. *Dev Cell*, **5**, 337–350.
8. Lee, R.C. and Ambros, V. (2001) An extensive class of small RNAs in *Caenorhabditis elegans. Science*, **294**, 862–864.
9. Lau, N.C., Lim, L.P., Weinstein, E.G., and Bartel, D.P. (2001) An abundant class of tiny RNAs with probable regulatory roles in *Caenorhabditis elegans. Science*, **294**, 858–862.
10. Lagos-Quintana, M., Rauhut, R., Lendeckel, W., and Tuschl, T. (2001) Identification of novel genes coding for small expressed RNAs. *Science*, **294**, 853–858.
11. Lau, N.C., Seto, A.G., Kim, J., Kuramochi-Miyagawa, S., Nakano, T., Bartel, D.P., and Kingston, R.E. (2006) Characterization of the piRNA complex from rat testes. *Science*, **313**, 363–367.
12. Grivna, S.T., Beyret, E., Wang, Z., and Lin, H. (2006) A novel class of small RNAs in mouse spermatogenic cells. *Genes Dev*, **20**, 1709–1714.
13. Girard, A., Sachidanandam, R., Hannon, G.J., and Carmell, M.A. (2006) A germline-specific class of small RNAs binds mammalian Piwi proteins. *Nature*, **442**, 199–202.

14. Aravin, A., Gaidatzis, D., Pfeffer, S., Lagos-Quintana, M., Landgraf, P., Iovino, N., Morris, P., Brownstein, M.J., Kuramochi-Miyagawa, S., Nakano, T., Chien, M., Russo, J.J., Ju, J., Sheridan, R., Sander, C., Zavolan, M., and Tuschl, T. (2006) A novel class of small RNAs bind to MILI protein in mouse testes. *Nature*, **442**, 203–207.
15. Watanabe, T., Takeda, A., Tsukiyama, T., Mise, K., Okuno, T., Sasaki, H., Minami, N., and Imai, H. (2006) Identification and characterization of two novel classes of small RNAs in the mouse germline: retrotransposon-derived siRNAs in oocytes and germline small RNAs in testes. *Genes Dev*, **20**, 1732–1743.
16. Vagin, V.V., Sigova, A., Li, C., Seitz, H., Gvozdev, V., and Zamore, P.D. (2006) A distinct small RNA pathway silences selfish genetic elements in the germline. *Science*, **313**, 320–324.
17. Saito, K., Nishida, K.M., Mori, T., Kawamura, Y., Miyoshi, K., Nagami, T., Siomi, H., and Siomi, M.C. (2006) Specific association of Piwi with rasiRNAs derived from retrotransposon and heterochromatic regions in the Drosophila genome. *Genes Dev*, **20**, 2214–2222.
18. Ro, S., Song, R., Park, C., Zheng, H., Sanders, K.M., and Yan, W. (2007) Cloning and expression profiling of small RNAs expressed in the mouse ovary. *RNA*, **13**, 2366–2380.
19. Ruby, J.G., Jan, C., Player, C., Axtell, M.J., Lee, W., Nusbaum, C., Ge, H., and Bartel, D.P. (2006) Large-scale sequencing reveals 21U-RNAs and additional microRNAs and endogenous siRNAs in *C. elegans*. *Cell*, **127**, 1193–1207.
20. Terns, M.P. and Terns, R.M. (2002) Small nucleolar RNAs: versatile trans-acting molecules of ancient evolutionary origin. *Gene Expr*, **10**, 17–39.
21. Ouellet, D.L., Perron, M.P., Gobeil, L.A., Plante, P., and Provost, P. (2006) MicroRNAs in gene regulation: when the smallest governs it all. *J Biomed Biotechnol*, **2006**, 69616.
22. Maatouk, D. and Harfe, B. (2006) MicroRNAs in development. *ScientificWorldJournal*, **6**, 1828–1840.
23. Kim, V.N. and Nam, J.W. (2006) Genomics of microRNA. *Trends Genet*, **22**, 165–173.
24. Bohnsack, M.T., Kos, M., and Tollervey, D. (2008) Quantitative analysis of snoRNA association with pre-ribosomes and release of snR30 by Rok1 helicase. *EMBO Rep*, **9**, 1230–1236.
25. Hertel, J., de Jong, D., Marz, M., Rose, D., Tafer, H., Tanzer, A., Schierwater, B., and Stadler, P.F. (2009) Non-coding RNA annotation of the genome of *Trichoplax adhaerens*. *Nucleic Acids Res*, **37**, 1602–1615.
26. Kim, M., Patel, B., Schroeder, K.E., Raza, A., and Dejong, J. (2008) Organization and transcriptional output of a novel mRNA-like piRNA gene (mpiR) located on mouse chromosome 10. *RNA*, **14**, 1005–1011.
27. Mishima, T., Takizawa, T., Luo, S.S., Ishibashi, O., Kawahigashi, Y., Mizuguchi, Y., Ishikawa, T., Mori, M., Kanda, T., and Goto, T. (2008) MicroRNA (miRNA) cloning analysis reveals sex differences in miRNA expression profiles between adult mouse testis and ovary. *Reproduction*, **136**, 811–822.
28. Papaioannou, M.D., Pitetti, J.L., Ro, S., Park, C., Aubry, F., Schaad, O., Vejnar, C.E., Kuhne, F., Descombes, P., Zdobnov, E.M., McManus, M.T., Guillou, F., Harfe, B.D., Yan, W., Jegou, B., and Nef, S. (2009) Sertoli cell Dicer is essential for spermatogenesis in mice. *Dev Biol*, **326**, 250–259.
29. Ro, S., Park, C., Sanders, K.M., McCarrey, J.R., and Yan, W. (2007) Cloning and expression profiling of testis-expressed microRNAs. *Dev Biol*, **311**, 592–602.
30. Ro, S., Park, C., Song, R., Nguyen, D., Jin, J., Sanders, K.M., McCarrey, J.R., and Yan, W. (2007) Cloning and expression profiling of testis-expressed piRNA-like RNAs. *RNA*, **13**, 1693–1702.
31. Ro, S., Park, C., Young, D., Sanders, K.M., and Yan, W. (2007) Tissue-dependent paired expression of miRNAs. *Nucleic Acids Res*, **35**, 5944–5953.
32. Siebolts, U., Varnholt, H., Drebber, U., Dienes, H.P., Wickenhauser, C., and Odenthal, M. (2009) Tissues from routine pathology archives are suitable for microRNA analyses by quantitative PCR. *J Clin Pathol*, **62**, 84–88.
33. Smits, G., Mungall, A.J., Griffiths-Jones, S., Smith, P., Beury, D., Matthews, L., Rogers, J., Pask, A.J., Shaw, G., VandeBerg, J.L., McCarrey, J.R., Renfree, M.B., Reik, W., and Dunham, I. (2008) Conservation of the H19 noncoding RNA and H19-IGF2 imprinting mechanism in therians. *Nat Genet*, **40**, 971–976.
34. Song, R., Ro, S., Michaels, J.D., Park, C., McCarrey, J.R., and Yan, W. (2009) Many X-linked microRNAs escape meiotic sex chromosome inactivation. *Nat Genet*, **41**, 488–493.

35. Wang, W.X., Wilfred, B.R., Baldwin, D.A., Isett, R.B., Ren, N., Stromberg, A., and Nelson, P.T. (2008) Focus on RNA isolation: obtaining RNA for microRNA (miRNA) expression profiling analyses of neural tissue. *Biochim Biophys Acta*, **1779**, 749–757.
36. Wu, F., Zikusoka, M., Trindade, A., Dassopoulos, T., Harris, M.L., Bayless, T.M., Brant, S.R., Chakravarti, S., and Kwon, J.H. (2008) MicroRNAs are differentially expressed in ulcerative colitis and alter expression of macrophage inflammatory peptide-2 alpha. *Gastroenterology*, **135**(1624–1635), e24.
37. Wu, H., Neilson, J.R., Kumar, P., Manocha, M., Shankar, P., Sharp, P.A., and Manjunath, N. (2007) miRNA profiling of naive, effector and memory CD8 T cells. *PLoS ONE*, **2**, e1020.
38. Yan, W., Morozumi, K., Zhang, J., Ro, S., Park, C., and Yanagimachi, R. (2008) Birth of mice after intracytoplasmic injection of single purified sperm nuclei and detection of messenger RNAs and microRNAs in the sperm nuclei. *Biol Reprod*, **78**, 896–902.
39. Guryev, V. and Cuppen, E. (2009) Next-generation sequencing approaches in genetic rodent model systems to study functional effects of human genetic variation. *FEBS Lett.*
40. Li, W. and Ruan, K. (2009) MicroRNA detection by microarray. *Anal Bioanal Chem.*
41. Ro, S., Park, C., Jin, JL., Sanders, KM., and Yan, W. (2006) A PCR-based method for detection and quantification of small RNAs. *Biochem and Biophys Res Commun*, **351**, 756–763.

Chapter 20

In Vivo Reprogramming of Human Telomerase Reverse Transcriptase (hTERT) by *Trans*-Splicing Ribozyme to Target Tumor Cells

Sang-Jin Lee, Seong-Wook Lee, Jin-Sook Jeong, and In-Hoo Kim

Abstract

Our understanding of RNA has evolved over the last 20 years from the initial concept that RNA is simply an intermediate in protein synthesis or a structural component maintaining and expressing genetic information. Subsequently, the non-coding RNAs have attracted huge interest and have been developed as therapeutic reagents as well as research tools. An example of RNA-based therapeutic application is the *Tetrahymena* group I intron-based *trans*-splicing ribozyme, which cleaves target RNA and *trans*-ligates an exon tagged at its 3′ end onto the downstream U nucleotide of the targeted RNA. Here, we describe the specific *trans*-splicing ribozyme that can sense and reprogram human telomerase reverse transcriptase (hTERT)-encoding RNA. This ribozyme converts hTERT RNA to therapeutic transgene herpes simplex virus (HSV) thymidine kinase (tk) and exhibits cytotoxicity to various hTERT-expressing cancer cells. For use in cancer therapy, CMV promoter-driven hTERTRibozyme.HSVtk expression cassette is inserted into adenovirus genome and delivered into either subcutaneous or intraspleenic liver-metastasized xenograft. We present here an evaluation of the inhibitory effects of CMV.hTERTRibozyme.HSVtk on tumor growth.

Key words: Ribozyme, *trans*-splicing, adenovirus, cancer gene therapy.

1. Introduction

RNA was first identified as an intermediate for protein synthesis and subsequently as a structural component involved in the maintenance and expression of genetic information. Beyond its distinct coding and structural functions variously expressed as mRNA, tRNA, or rRNA, the past 20 years have witnessed an amazing

M. Sioud (ed.), *RNA Therapeutics*, Methods in Molecular Biology 629,
DOI 10.1007/978-1-60761-657-3_20, © Springer Science+Business Media, LLC 2010

revelation that a much broader class of non-coding RNAs are endowed with a variety of hitherto unknown critical functions in gene expression and delineation of the roles of non-coding RNAs in a variety of cellular mechanisms became a whole new research field. It soon became clear that RNA, itself a target for regulation because of its ability to form unique secondary structures and/or interact with other RNA molecules, can be utilized as a powerful therapeutic alternative to conventional gene replacement therapies for various diseases. In this regard, there are three RNA-based strategies as a drug: (1) use of *trans*-splicing ribozyme for reprogramming pre-mRNA transcript; (2) use of small RNA for targeted silencing of gene transcripts; and (3) use as aptamers, like antibodies, for neutralizing the protein functions.

Ribozymes (ribonucleic acid + enzyme), which are RNAs with catalytic activity, are involved in gene expression and/or regulation, e.g., in RNA processing, RNA splicing, RNA genome duplication among other functions. A ribozyme recognizes the specific sequence of target RNAs and forms the specific three-dimensional feature (based on Watson–Crick base pairs), resulting in the cleavage or ligation of phosphodiester linkage in a RNA molecule. This unique feature of ribozymes permits the design of specific ribozymes that can serve as gene therapeutic reagents by linking different recognition sequences adjacent to the defined cleavage sites (1). On the basis of size, structure, and catalytic reaction, ribozymes can be divided into two groups: (1) the so-called small endonucleolytic ribozymes that include hammerhead, hairpin, hepatitis delta virus, and *Neurospora Varkud* satellite small ribozyme and (2) large ribozymes that include RNase P, group I intron, and group II intron (2). Unlike small ribozymes and RNase P for cleaving target sequences, group I and II introns are utilized for self-splicing (3, 4). Among them, *trans*-splicing ribozyme derived from self-splicing group I intron has been highlighted for its potential as a riboswitch.

Self-splicing group I intron from *Tetrahymena thermophila* has catalytic *trans*-splicing activity and can mediate cleavage of a target RNA and *trans*-ligation of an exon tagged at the 3′ end of the intron onto the downstream U nucleotide (nt) of the cleaved target RNA not only in vitro but also in bacteria and mammalian cells (5–7). Therefore, the *trans*-splicing ribozyme can target and reprogram a specific disease-associated or causative RNA with transgene transcript harboring a therapeutic effect, selectively in cells that express the target RNA, thus providing an attractive gene therapy tool against diverse human diseases, genetic or infectious (8–12). Moreover, we suggested that the *trans*-splicing ribozyme could be a potent anti-cancer agent via cancer-specific transcript targeting and RNA replacement with new RNA exerting anti-cancer activity (13). We validated the potential of this approach by developing *trans*-splicing ribozyme specifically targeting and replacing human telomerase reverse transcriptase (hTERT) RNA to trigger transgene activity selectively in hTERT-positive cancer cells. This hTERT-specific ribozyme selectively marks hTERT-positive tumor cells or specifically and efficiently retards tumor growth in tumor xenografts as well as in tissue cultures (13–16).

2. Materials

2.1. hTERT Ribozyme Construction

1. RNA mapping library generation: *Trans*-splicing ribozyme library is constructed by PCR amplification of plasmid pT7L-21, which encodes a slightly shortened version of the *Tetrahymena* group I intron (6), with a 5′ primer containing a randomized sequence at positions corresponding to the ribozyme's internal guide sequence (IGS) (5′-GGG GGG ATC CTA ATA CGA CTC ACT ATA GNN NNN AAA AGT TAT CAG GCA TGC ACC-3′) and a 3′ primer specific for 3′ exon tag sequences present in the pT7L-21 plasmid (5′-AGT AGT CTT ACT GCA GGG GCC TCT TCG CTA TTA CG-3′). The resulting cDNA library is in vitro transcribed using T7 RNA polymerase to generate the RNA mapping library (*see* **Note 1**).
2. hTERT RNA generation: hTERT substrate RNA is constructed by in vitro transcription using T7 RNA polymerase from hTERT cDNA amplified from pCl-neo vector using 5′ primer (5′-GGG GAA TTC TAA TAC GAC TCA CTA TAG GGC AGG CAG CGC TGC GTC CT-3′) and 3′ primer (5′-CGG GAT CCC TGG CGG AAG GAG GGG GCG GCG GG-3′) (*see* **Note 2**). The in vitro transcription was performed using the following conditions: 40 mM Tris-HCl, pH 7.5; 1.5 mM $MgCl_2$; 10 mM DTT; 4 mM spermidine; 80 U RNase inhibitor; and 0.5 mM NTP.

2.2. Adenovirus Vector Construction and Adenovirus Production

1. Vector construction: CMV.hTERTRibozyme.HSVtk expression cassette in pCDNA3.1(+) is digested with *Sac*II and *Spe*I restriction enzymes (20 mM Tris-acetate, pH 7.9; 50 mM potassium acetate; 10 mM magnesium acetate; 1 mM dithiothreitol). The fragment containing CMV.hTERTRibozyme.HSVtk is ligated with the linearized pZAP1.2 shuttle vector with T4 DNA ligase (50 mM Tris-HCl, pH 7.5; 10 mM $MgCl_2$; 1 mM ATP; 10 mM dithiothreitol). The shuttle vector containing CMV.hTERTRibozyme.HSVtk expression cassette is digested with *Pac*I and *PflM*I (10 mM

Bis-Tris-Propane-HCl, pH 7.0; 10 mM $MgCl_2$; 1 mM dithiothreitol). The DNA fragment containing the CMV. hTERTRibozyme.HSVtk expression cassette and the adenovirus left ITR, which is called the left arm for adenovirus generation, is purified on agarose gel (*see* **Note 3**).

2. Adenovirus recovery by clone isolation: The left arm is ligated with the linearized RightZAP1.2 with T4 DNA ligase (50 mM Tris-HCl, pH 7.5; 10 mM $MgCl_2$; 1 mM ATP; 10 mM dithiothreitol). The resulting ligation mixture is transfected into HEK 293 helper cell, which is maintained in DMEM (high glucose with L-glutamine), 10% fetal bovine serum (FBS), 1% A/A, trypsin/EDTA. The transfection into HEK 293 cell is performed using lipofectamine 2000 in Opti-MEM buffer (available from Invitrogen). After transfection, the cell monolayer is overlaid with a 1:1 mixture of 2% SeaPlaque agarose and 2× MEM mix (*see* **Note 4**).
3. Adenovirus amplification: The recovered adenoviruses from plaque assay are used to infect HEK 293 cells seeded on a 6-cm dish. When more than 90% cells are detached from the dish, the cells and the medium are harvested and adenovirus particles are released by three courses of freeze/thaw. Then, one half of crude viral extract is used for infecting HEK 293 cells seeded in a 175-cm flask. HEK 293 cells are maintained in DMEM media. Finally, adenoviruses obtained from cell pellet are purified by CsCl ultracentrifugation and saved in storage buffer after dialysis (*see* **Note 5**).

2.3. In Vitro Activity of CMV.hTERTRibozyme.HSVtk in Adenovirus

1. Uptake assay of isotope-labeled PCV: HT-29 maintained in DMEM media is seeded in a 12-well plate and exposed to various MOI (multiplicity of infectivity) of adenovirus harboring CMV.hTERTRibozyme.HSVtk. Two days post-infection, ^{3}H-PCV (1 μCi/ml) is added to the medium and incubated for 2 h. Then, cells are washed three times with phosphate buffered saline (PBS) and lysed with 0.1% SDS. The total amount of ^{3}H-PCV inside the cell is counted by scintillation counter and normalized with used protein concentration (*see* **Note 6**).
2. Cell proliferation assay: HT-29 are cultured in DMEM media supplemented with 10% fetal bovine serum (FBS) and 1% A/A and seeded in a 96-well plate. After cells on the 96-well plate are exposed to appropriate PFU of adenovirus and ganciclovir (GCV), the number of viable cells are quantitated by using tetrazolium compound MTS [3-(4,5-dimethylthiazol-2-yl)-5-(3-carboxymethoxyphenyl)-2-(4-sulfophenyl)-2*H*-tetrazolium, inner salt] (available from Promega). MTS is converted to colored formazan by

NADH and NADPH and thus the reduced products are directly proportional to the number of living cells. The colored formazan is quantitated by measuring absorbance at 490 nm (*see* **Note 7**).

3. PCR analysis of *trans*-spliced molecule (TSM): hTERT-positive HT-29 cells and hTERT-negative IMR-90 cells, both of which are maintained in DMEM media supplemented with 10% FBS and 1% A/A, are seeded in 100-cm plates. Next day, cells are infected with either AdCMV.hTERTRibozyme.HSVtk or AdCMV.HSVtk. Total RNAs extracted from cells infected by adenoviruses are isolated with Trizol (available from Invitrogen) supplemented with 20 mM EDTA and are reverse transcribed for cDNA synthesis with HSVtk-specific primer (5′-CGG GAT CCT CAG TTA GCC TCC CCC AT-3′) in the presence of 10 mmol/L L-argininamide. L-Argininamide is added to the reaction to quench the self-splicing during the reverse transcription reaction. To detect TSMs, the nested PCR is performed using Taq polymerase (*see* **Note 8**).

2.4. In Vivo Anti-Cancer Activity of CMV.hTERTRibozyme.HSVtk Delivered by Adenovirus

1. Four- to five-week-old male Balb/cAnNCrj-nu/nu nude mice are injected with HT-29 cells grown in DMEM medium supplemented with 10% FBS and 1% A/A. Appropriate doses of adenovirus diluted in PBS are injected directly into tumors and GCV (available from Roche) dissolved in PBS is administered the next day intraperitoneally. Tumor sizes are measured every third day using digital caliper (*see* **Note 9**).
2. Four- to five-week-old male Balb/cAnNCrj-nu/nu nude mice are also used as a mouse model of liver metastasis by intrasplеenic inoculation of HT-29 cell. Prior to injection, NK cells are depleted with intraperitoneal injection of anti-asialo GM1 (rabbit) antibody dissolved in PBS (available from Wako Pure Chemicals), which facilitates the xenograft formation of HT-29 cells. A small longitudinal left upper flank incision is made to visualize the spleen and then HT-29 cells prepared in PBS are injected under the spleen capsule with a 29-gauge needle. Adenoviruses prepared in PBS are injected into the mice through the tail vein with a 29-gauge needle and GCV dissolved in PBS is administered the next day intraperitoneally (*see* **Note 10**).

3. Methods

In this study, we describe the hTERT RNA-targeting specific *trans*-splicing ribozyme as a new cancer-targeting agent, which directs the expression of therapeutic genes selectively in telomerase-positive cancer cells via the targeted RNA replacement of the hTERT transcript. This *trans*-splicing ribozyme is expected to have the advantage of cumulative effects of hTERT level reduction and cytotoxin gene induction in the hTERT-positive cancer cells combined with its tissue specificity.

Although any uridine residue in a target RNA can be theoretically targeted by *Tetrahymena trans*-splicing ribozyme via alteration of nucleotide composition of 6-nt-long IGS on the ribozyme, only a limited number of uridines on the target RNA will be accessible to the ribozyme due to the complex tertiary structure of the substrate RNA (17). Therefore, an RNA mapping strategy should be initially employed that is based on RNA tagging and a *trans*-splicing ribozyme library to identify which regions of the hTERT transcript were most accessible to the ribozymes (7, 8). To construct ribozyme-expressing vectors, we modified a ribozyme, which was directed at the most targetable uridine on the hTERT RNA, to contain an extension of the P1 helix, an addition of the P10 helix, and an antisense sequence which was complementary to the downstream region of the targeted uridine of the hTERT RNA because group I *trans*-spicing ribozymes with only a 6-nt-long IGS were inefficient in their activity and specificity when expressed in mammalian cells (7).

The hTERTRibozyme, the hTERT-recognizing group I *trans*-splicing ribozyme derivative with antisense sequences at the downstream region of IGS, is further modified to specifically harbor the suicide herpes simplex thymidine kinase (*HSVtk*) gene as its 3′ exon under the control of the CMV promoter. As a result, when exposed to hTERTRibozyme.HSVtk and GCV prodrug, hTERT-abundant cells suffer cell death caused by phosphorylated GCV converted by HSVtk. Furthermore, CMV-driven hTERTRibozyme.HSVtk is incorporated into replication-defective adenovirus for effective delivery into tumor cells inside animal xenograft model.

3.1. hTERT Ribozyme Construction

1. To map the hTERT RNA, the ribozyme library with randomized IGS (50 nM) is incubated with the hTERT transcripts (500 nM) in a *trans*-splicing reaction buffer with guanosine (0.1 mM) at 37°C for 3 h. The resulting *trans*-splicing products are then reverse transcribed with primer specific for 3′ exon tagged at the 3′ end of the ribozyme (5′-ATG TGC TGC AAG GCG ATT-3′) and amplified with 5′ primer specific for hTERT (5′-GGG GAA TTC AGC GCT GCG TCC TGC T-3′) and 3′ primer specific for the 3′ exon (5′-TGT AAA ACG ACG GCC AGT G-3′). The amplified *trans*-spliced cDNA is cloned into pUC19 vector and sequenced to identify splicing junction. An example of

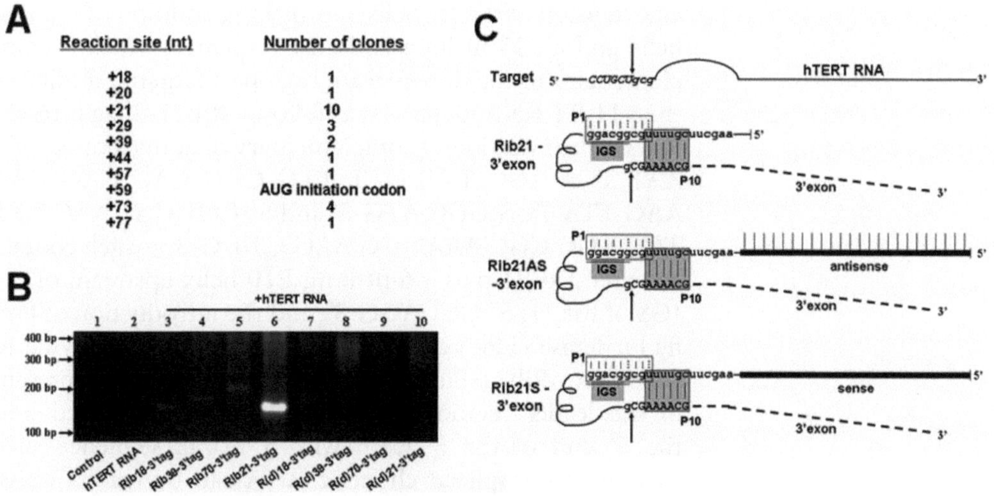

Fig. 20.1. *Trans*-splicing ribozyme. (**a**) Mapping results of the ribozyme-accessible sites in the target hTERT RNA. Nucleotide positions in the hTERT cDNA sequence are presented for the uridines that can be accessible to the group I ribozyme. The number of individual clones containing a given uridine at the splice site identified by the mapping analysis is indicated. (**b**) RT-PCR amplification of *trans*-spliced RNA products generated in vitro. Target hTERT RNA (10 nM) was incubated alone (lane 2) or with series of active (Rib-3′tag, 100 nM; lanes 3–6) or inactive (R(d)-3′tag, 100 nM; lanes 7–10) ribozymes. Reaction products were then amplified and subjected to electrophoresis. (**c**) Schematic diagram of the modified *trans*-splicing ribozymes. The target hTERT RNA is shown with sequences around the splice site *italicized*. Three *trans*-splicing ribozymes with extended IGS are represented with the 3′ exon sequences capitalized. Potential base parings between the ribozymes and the target transcript are indicated by *vertical lines*. The *arrows* indicate the 5′ and 3′ splicing sites.

the mapping results produced is shown in **Fig. 20.1a** (*see* **Note 2**).

2. To verify that the sites predicted by the mapping studies are indeed the most accessible, several ribozymes targeting different uridines are assessed with regard to their in vitro *trans*-splicing abilities. To this end, different ribozymes targeting uridines at positions 18, 39, 70, or 21 in the hTERT RNA (Rib18-, Rib39-, Rib70-, and Rib21-3′tag ribozymes, respectively) are constructed through exchange of original IGS of *Tetrahymena* group I intron. The deleted form in the catalytic site of each ribozyme is used as an inactive control. Each specific ribozyme is incubated with hTERT RNA in a splicing buffer with guanosine and the resulting *trans*-splicing products are analyzed as described above. As shown, in **Fig. 20.1b**, as an example of the analysis produced, most notable among these ribozymes was Rib21 recognizing uridine at position 21 (U21) on hTERT RNA, which exhibits the most pronounced ability to *trans*-splice a 3′ exon tag onto the target RNA. This suggests that the relative *trans*-splicing efficiency at the chosen sites is correlated with the predicted accessibility determined from mapping analysis.
3. We modified the ribozyme (designated Rib21AS-3′exon), which is directed at U21 on the hTERT RNA, to contain

an extension of the P1 helix, a 6-nt-long addition of the P10 helix and a 325-nt-long antisense sequence, which is complementary to the downstream region of targeted uridine of the hTERT RNA, described as follows. Rib21-3′tag is modified via the insertion of complementary oligonucleotides (5′-AAT TCA AGC TTC GTT TTG CGG CAG CAG GAA AAG TTA TCA GGC ATG-3′ and 5′-CCT GAT AAC TTT TCC TGC CGC AAA ACG AAG CTT G-3′) which contain an extended P1 plus a 6-nt-long P10 helix upstream of the IGS of Rib21, 5′-GGCAGG-3′, and the introduction of firefly luciferase (Fluc) cDNA as a 3′ exon of the ribozyme to construct pRib21-Fluc. Since the target sequence is present in the leader region, a new start codon is inserted into the 5′ end of the 3′ exon with a Kozak sequence such that the *trans*-spliced chimeric RNA can be readily translated in the mammalian cells. The DNA fragment consisting of the Rib21 sequence with the extended IGS plus Fluc cDNA is inserted between the *Hind*III and *Xba*I sites of pSEAP that encodes alkaline phosphatase under the control of the SV40 promoter (Clontech). pRib21-Fluc is further modified by insertion of a 325-nt-long antisense sequence against the downstream region of the targeted uridine of the hTERT RNA [+30 to +354 residues, amplified from pCl-neo vector using 5′ primer (5′-GGG AAG CTT GGG AAG CCC TGG CCC-3′) and 3′ primer (5′-GGG AAG CTT AAG GCC AGC ACG TTC TT-3′)] into the *Hind*III site of the vector to construct pRib21AS-Fluc. The 325-nt-long sequence sense to the hTERT RNA is inserted for pRib21S-Fluc. To construct pRib21AS-TK which encodes Rib21AS ribozyme with *HSVtk* gene as its 3′ exon under CMV promoter, Fluc cDNA in pRib21AS-Fluc is replaced with the TK cDNA amplified from pHR-CMV-GFP-TK vector (primers: 5′-CGA AAA CGC CCA CCA TGG CTT CGT ACC CCT GCC ATC AAC ACG-3′ and 5′-CCT TAA TTA ATC TAG ATC AGT TAG CCT CCC CCA TCT C-3′). From the replaced vector, sequences corresponding to Rib21AS-TK are amplified (primers: 5′-AAG GAA AAA AGC GGC CGC AAG GCC AGC ACG TTC TT-3′ and 5′-AAG GAA AAA AGC GGC CGC AAG CTT CAG TTA GCC TCC CCC AT-3′), digested with *Not*I, and cloned into pcDNA3.1(+) plasmid. Schematic diagram of the modified *trans*-splicing ribozymes is shown in **Fig. 20.1c**.

3.2. Generation of Recombinant Adenoviruses Expressing Trans-Splicing Ribozyme

1. The ribozyme cDNA obtained by Fluc (Rib21AS-3′exon-Fluc) or HSVtk (Rib21AS-3′exon-HSVtk) is subcloned into a pZAP1.2 shuttle vector that has the adenovirus left ITR, packaging signal, and the cytomegalovirus (CMV) promoter as shown in **Fig. 20.2**. The resulting

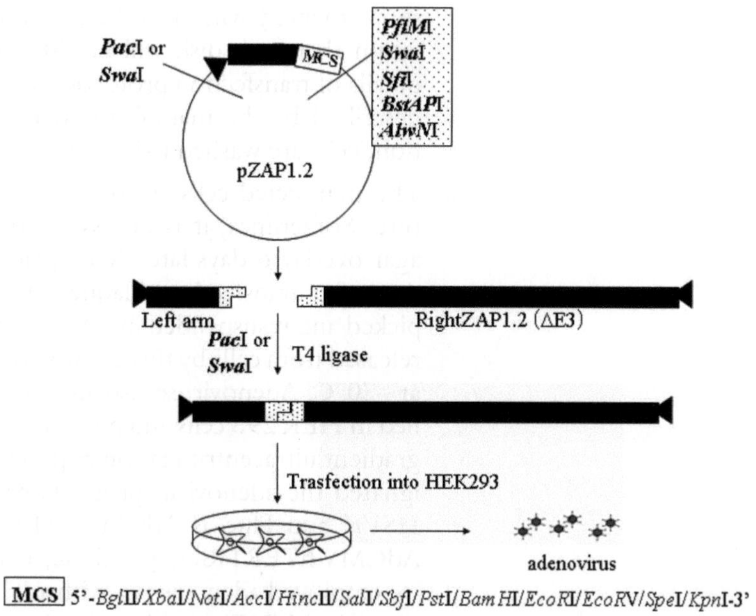

Fig. 20.2. Construction of adenovirus using AdenoZAP system (OD260). *Black bar* and *triangles* represent a schematic adenoviral genome and ITR, respectively. MCS in *white rectangle* indicates the multi-cloning site. *Dotted L-shaped boxes* are designed as non-symmetrical sticky ends to prevent the self-ligation. *Left* arm includes the left ITR and packaging signal. *Right* after transfection into HEK 293 cells, agarose overlay assay is ready for plaque isolation.

pZAP1.2.CMV.Rib21AS-3′exon-Fluc or pZAP1.2.CMV.-Rib21AS-3′exon-HSVtk is digested with two endonucleases *Pac*I/*Sfi*I and the DNA fragment containing ITR.CMV.Rib21AS-3′exon-Fluc or ITR.CMV.Rib21AS-3′exon-HSVtk is purified on agar gel. The pZAP1.2 shuttle vector harboring the CMV-driven expression cassette of Rib21AS-3′exon-Fluc or Rib21AS-3′exon-HSVtk is ligated in the presence of *PacI* restriction enzyme to the adenoviral backbone vector right arm with RightZAP1.2 (available from OD260) which has the most of adenovirus genome with the deletion of E1 and E3. The addition of *PacI* enzyme is to prevent the concatemers. After ligation with T4 ligase for 1 h at room temperature, enzymes are inactivated by incubating at 65°C for 10 min as illustrated in **Fig. 20.2**. For the integration of CMV-driven HSVtk expression cassette in adenovirus, first, CMV.HSVtk is subcloned into pZAP1.2 shuttle vector and then the following procedures are identical with ones for CMV.Ribozyme.HSVtk.

2. The ligation mixture prepared above is transfected into HEK 293 cells. One day before transfection, HEK 293 cells are transferred into six 6-cm dishes containing DMEM media

supplemented with 10% FBS and 1% AA with the expectation that each dish will be 80% confluent next day. The details of transfection protocol using Lipofectamin 2000 are described by the manufacturer. Eight hours after transfection, cells are washed twice with DMEM/10% FBS/1% AA.

3. The transfected cells are overlaid with 10 mL of agar mixture. Sometimes, it is necessary to add 5 mL of a second agar overlay 6 days later. Viral plaques usually appear 7–10 days after transfection. Plaques of 2–3 mm in diameter are picked and resuspended in 500 μL of DMEM. Adenoviruses released from cells by three courses of freeze/thaw are stored at –70°C. Adenoviruses isolated by plaque assays are amplified in HEK293 cells and purified by double cesium chloride gradient ultracentrifugation as previously described. We designated the adenovirus products harboring ribozymes with HSV*tk* and Fluc as AdCMV.hTERTRibozyme.HSVtk and AdCMV.hTERTRibozyme.Fluc, respectively. As a control, we employed adenoviruses with *HSVtk* (AdCMV.HSVtk) or Fluc (AdCMV.Fluc) driven by the CMV promoter and Ad-Mock, which harbors only the adenoviral backbone.

3.3. In Vitro Activity by CMV.hTERTRibozyme.HSVtk Delivered by Adenovirus

1. In vitro CMV.hTERTRibozyme.HSVtk activity is determined by uptake of PCV into cell interior. HT-29 cells (1 × 10^5) maintained in DMEM medium are seeded in 24-well plates. The next day, cells are infected with various MOI of adenovirus harboring CMV.hTETRibozyme.HSVtk. After incubation with adenovirus for 48 h, cells are exposed to media containing [^{3}H]PCV (1.0 μCi/ml) at 37°C for 2 h, and 50 μL of media are collected for the measurement of free [^{3}H]PCV. Cells are washed with cold PBS three times, dissolved by 500 μL of 0.1% SDS, and then radioactivities are counted by a scintillation counter.

2. HT-29 cells are cultured and seeded on a 96-well plate one day before the addition of adenovirus. On the following day, appropriate PFUs of adenovirus are added to cells in the 96-well plate. Cell proliferation assay, using [3-(4, 5-dimethylthiazol-2-yl)-5-(3-carboxymethoxyphenyl)-2-(4-sulfophenyl)-2*H*-tetrazolium, inner salt (MTS)], is done using CellTiter 96 AQueous One solution reagent (available from Promega). Two days post-infection, one solution reagent is pipetted into each well of 96-well plate containing 100 μL of culture media and incubated for 2 h at 37°C in 5% CO_2 incubator. Then, the absorbance at 490 nm is measured using a 96-well plate recorder.

3. The *trans*-spliced molecules (TSMs) produced by adenovirus harboring hTERTRibozyme.HSVtk is confirmed by nested RT-PCR as shown in **Fig. 20.3**. Total RNAs

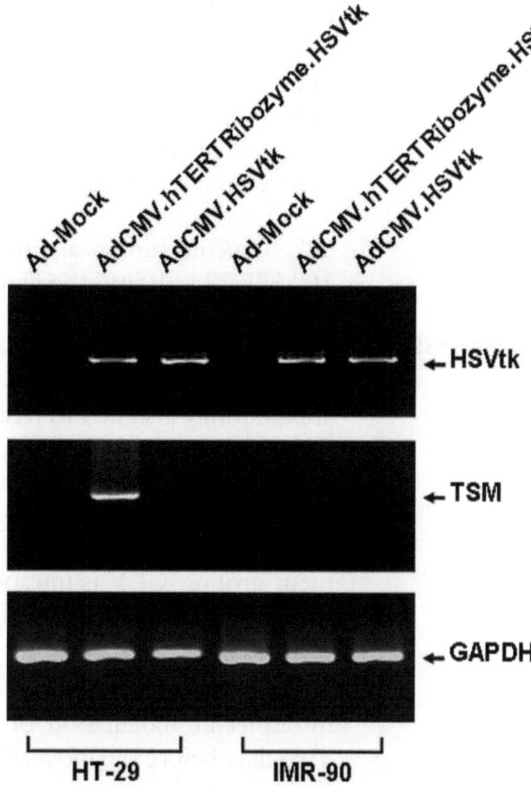

Fig. 20.3. Investigation of TSMs produced by CMV.hTERTRibozyme.HSVtk. Allele-specific RT-PCR is adopted to investigate *trans*-spliced molecules from HT-29 cells post-infection with AdCMV.hTERTRibozyme.HSVtk. The hTERT-positive HT-29 and hTERT-negative IMR-90 cells are transfected with AdCMV.hTERTRibozyme.HSVtk. The corresponding cells are treated with AdCMV.HSVtk as a negative *trans*-splicing control. The production of ribozyme or transgene RNA is represented by the amplification of HSVtk RNA. As an internal control, human glyceraldehyde-3-phosphate dehydrogenase (GAPDH) RNA is amplified.

extracted from cells infected by adenoviruses are templates for cDNA synthesis with primer for HSVtk (5′-CGG GAT CCT CAG TTA GCC TCC CCC AT-3′) in the presence of 10 mmol/L L-argininamide. To detect TSMs, the first PCR is performed with a 5′ primer specific for the *trans*-splicing junction (5′-GGG GAA TTC AGC GCT GCG TCC TGC T-3′) and with HSVtk (5′-GTT ATC TGG GCG CTT GTC AA-3′). Conditions for PCR reaction are as follows: preheating at 95°C for 3 min, followed by 30 cycles of PCR (95°C, 30 s; 57°C, 30 s; 72°C, 30 s), and final extension at 72°C for 5 min. Then, the second PCR is done with inward primers: 5′-GCT GCG TCC TGC TAA AAC-3′ and 5′-CAG TAG CGT GGG CAT TTT CT-3′. For this amplification, PCR mixtures are reacted for 30 cycles of denaturation for 1 min at 94°C, annealing for 1 min at

60°C, and polymerization for 1 min at 72°C. As shown in **Fig. 20.3**, TSMs obtained from cells infected specifically with AdCMV.hTERTRibozyme.HSVtk are visualized only inside HT-29 cells highly expressing hTERT. Human glyceraldehydes-3-phosphate dehydrogenase (GAPDH) is used as an internal control in PCR reaction (13).

3.4. In Vivo Anti-tumoral Activity of CMV.hTERTRibozyme.HSVtk in Adenovirus

1. Subcutaneous tumors are induced by injecting 200 μl of 5 × 10^6 HT-29 cells into the flank regions of nude mice. Tumor size is measured every 2–3 days with digital calipers and calculated according to the formula: $LW^2/2$ (length L and width W). When the tumors reach a size of 140 mm^3, they are randomly assigned to treatment groups. In order to evaluate the anti-tumoral effect by adenoviruses, 200 μL of 5 × 10^7 plaque forming units are intratumorally injected three times at 3-day intervals. One day after the initial injection of adenovirus injection, GCV treatment is initiated in the treatment groups. GCV is intraperitoneally injected (50 mg/kg twice a day) for 10 days, while PBS is injected into the control group.
2. A mouse model of liver metastasis is established by intraspleenic inoculation of HT-29 cell as described (18). Two days before the inoculation of cells, 50 μL of anti-asialo GM1 (rabbit) antibody is administered intraperitoneally in order to facilitate the xenograft formation of HT-29 cells by inhibiting natural killer cell activity. A small longitudinal left upper flank incision is made, and 50 μL of 2 × 10^6 HT-29 cells prepared in PBS is injected under the spleen capsule with a 29-gauge needle. After removal of the needle, the injection site is pressed with an aseptic cotton sponge for several minutes to prevent further leakage. Then, the spleen is returned into abdominal cavity and the peritoneum and abdominal wall are sutured with silk. The animals show multiple intrahepatic tumor nodules, easily detectable by gross inspection, within 8–10 days. Adenoviruses are diluted to 100 μL volume with PBS and then injected into the mice through the tail vein with a 29-gauge needle. One day later, the animals are treated with GCV (50 mg/kg) intraperitoneally twice daily for 10–15 days. In order to calculate tumor growth, the total mass of the liver lobes of each animal is measured after euthanizing them. Whole liver lobes are then fixed in 10% neutral buffered formalin, serially sliced in 2–3 mm thickness, and embedded in paraffin. The paraffin blocks are cut into 4-μm sections and stained with H&E. All sections of each lobe are examined, photographed, and the tumor mass is calculated. The tumor fraction in each mouse is calculated and the actual tumor weights estimated,

using the following formula: tumor weight = total liver mass (mg) × measured tumor fraction.

3. Statistical analyses are performed using Statistical Analysis System software (SAS Institute). Inter-group differences are assessed by analysis of variance. In the cases of highly skewed distribution of measurements and small sample sizes, non-parametric statistical tests (Kruskal–Wallis test for overall comparison and Wilcoxon rank-sum test for pairwise comparison) are employed. All data are expressed as the mean values ± SD. *P* values < 0.01 are determined to be statistically significant.

4. Notes

1. With the in vitro transcription condition, mainly two RNA bands can be generated (longer transcript, unspliced ribozyme-3′ exon chimeric transcript; shorter one, self-spliced ribozyme RNA). For the mapping study, transcribed RNAs are separated on a 5% acrylamide gel with urea. Longer transcripts are then eluted with elution buffer (0.6 M ammonium acetate, 1 mM EDTA, 0.2% SDS) at 37°C for 3 h, extracted with phenol, and concentrated with ethanol.
2. To confer proper conformation into substrate and ribozyme RNA before co-incubation of two transcripts, hTERT RNA is preheated at 37°C for 5 min in the *trans*-splicing reaction buffer with GTP. Ribozyme RNA is heated at 50°C for 5 min in the reaction buffer, incubated at 37°C for 2 min, and then added into the preheated substrate RNA.
3. Prior to the ligation, the DNA fragment CMV.hTERTRibozyme.HSVtk obtained from pCDNA3.1(+) is extracted with phenol and concentrated with ethanol. All of the silica-based protocols for DNA fragment purification from agarose gels are appropriate for the following procedure. We use QIAquick for all gel-purification works in this study (available from QIAGEN).
4. The ligation of both RightZAP1.2 and left arm is carried out in the presence of *Pac*I restriction enzyme to prevent the DNA concatemer formation. The resulting ligation efficiency can be verified on agarose gel by confirming the band shift after ligation reaction. Overlay agarose is made by mixing 1:1 of 2% SeaPlaque agarose (42°C) and 2× MEM (37°C) in a 50-ml conical tube and waiting until temperature 37°C. Sometimes, a second agar overlay

of about 5 mL is required to feed cells due to the delayed plaque formation.

5. HEK 293 cells are prepared in a 6-cm or a 175-cm^2 flask with 80% confluency before infection. Depending on the concentration of adenoviral vectors, the cells may detach 2–3 days post-infection. Sometimes, cell monolayers become so acidic without any evident cytopathic effect that it needs to be split and fed with fresh DMEM media. The bulk of produced adenovirus is dialyzed using the dialysis bag of 30 kDa MW cutoff (available from PIERCE).
6. Prior to cells washed with PBS, free ^{3}H-PCV in the medium is counted and used for the calculation of ^{3}H-PCV trapped in cell interior. The values of ^{3}H-PCV uptaken inside cells are presented by cpm of cells/cpm of media/μg of protein.
7. MTS cell proliferation assay can be replaced by numerous competitive reagents available from other commercial sources. The addition of 10% SDS to each well to stop reaction enables us to measure the absorbance as late as up to 18 h.
8. The enzymes for reverse transcription and PCR reaction are available from a number of manufacturers. The appropriate buffers for those enzymes are also supplied by manufacturers.
9. When cells are harvested from culture plate, minimum amount of trypsin–EDTA is used to avoid damage to cells. Cells for xenograft formation are counted after mixing cells 1:1 with 0.8 mM trypan blue solution in PBS. Viable cells exclude trypan blue, while dead cells stain blue due to trypan blue uptake. The inoculation area of the mice is sterilized with ethanol.
10. Mouse is operated on while anesthetized with halothane, since inhalation of anesthesia is superior to most injectable forms of anesthesia in safety and efficacy. After injection of cells with 29-gauge needle, the injection site is pressed with an aseptic cotton sponge for several minutes to prevent further leakage. The spleen is returned to abdominal cavity and the peritoneum and the abdominal wall are sutured with silk.

References

1. Puerta-Fernandez, E., Romero-Lopez, C., Barroso-delJesus, A., and Berzal-Herranz, A. (2003) HIV-1 TAR as anchoring site for optimized catalytic RNAs. *FEMS Microbiol Rev*, **27**, 75–97.
2. Pavco, P.A., Bouhana, K.S., Gallegos, A.M., Agrawal, A., Blanchard, K.S., Grimm, S.L., Jensen, K.L., Andrews, L.E., Wincott, F.E., Pitot, P.A., Tressler, R.J., Cushman, C., Reynolds, M.A., and Parry, T.J. (2000)

Antitumor and antimetastatic activity of ribozymes targeting the messenger RNA of vascular endothelial growth factor receptors. *Clin Cancer Res*, **6**, 2094–2103.
3. Bagheri, S. and Kashani-Sabet, M. (2004) Ribozymes in the age of molecular therapeutics. *Curr Mol Med*, **4**, 489–506.
4. Lewin, A.S. and Hauswirth, W.W. (2001) Ribozyme gene therapy: applications for molecular medicine. *Trends Mol Med*, **7**, 221–228.
5. Been, M.D. and Cech, T.R. (1986) One binding site determines sequence specificity of Tetrahymena pre-rRNA self-splicing, trans-splicing, and RNA enzyme activity. *Cell*, **47**, 207–216.
6. Sullenger, B.A. and Cech, T.R. (1994) Ribozyme-mediated repair of defective mRNA by targeted, trans-splicing. *Nature*, **371**, 619–622.
7. Jones, J.T., Lee, S.W., and Sullenger, B.A. (1996) Tagging ribozyme reaction sites to follow trans-splicing in mammalian cells. *Nat Med*, **2**, 643–648.
8. Lan, N., Howrey, R.P., Lee, S.W., Smith, C.A., and Sullenger, B.A. (1998) Ribozyme-mediated repair of sickle beta-globin mRNAs in erythrocyte precursors. *Science*, **280**, 1593–1596.
9. Phylactou, L.A., Darrah, C., and Wood, M.J. (1998) Ribozyme-mediated trans-splicing of a trinucleotide repeat. *Nat Genet*, **18**, 378–381.
10. Rogers, C.S., Vanoye, C.G., Sullenger, B.A., and George, A.L., Jr. (2002) Functional repair of a mutant chloride channel using a trans-splicing ribozyme. *J Clin Invest*, **110**, 1783–1789.
11. Ryu, K.J., Kim, J.H., and Lee, S.W. (2003) Ribozyme-mediated selective induction of new gene activity in hepatitis C virus internal ribosome entry site-expressing cells by targeted trans-splicing. *Mol Ther*, **7**, 386–395.
12. Shin, K.S., Sullenger, B.A., and Lee, S.W. (2004) Ribozyme-mediated induction of apoptosis in human cancer cells by targeted repair of mutant p53 RNA. *Mol Ther*, **10**, 365–372.
13. Hong, S.H., Jeong, J.S., Lee, Y.J., Jung, H.I., Cho, K.S., Kim, C.M., Kwon, B.S., Sullenger, B.A., Lee, S.W., and Kim, I.H. (2008) In vivo reprogramming of hTERT by trans-splicing ribozyme to target tumor cells. *Mol Ther*, **16**, 74–80.
14. Kwon, B.S., Jung, H.S., Song, M.S., Cho, K.S., Kim, S.C., Kimm, K., Jeong, J.S., Kim, I.H., and Lee, S.W. (2005) Specific regression of human cancer cells by ribozyme-mediated targeted replacement of tumor-specific transcript. *Mol Ther*, **12**, 824–834.
15. Jeong, J.S., Lee, S.W., Hong, S.H., Lee, Y.J., Jung, H.I., Cho, K.S., Seo, H.H., Lee, S.J., Park, S., Song, M.S., Kim, C.M., and Kim, I.H. (2008) Antitumor effects of systemically delivered adenovirus harboring trans-splicing ribozyme in intrahepatic colon cancer mouse model. *Clin Cancer Res*, **14**, 281–290.
16. Song, M.S., Jeong, J.S., Ban, G., Lee, J.H., Won, Y.S., Cho, K.S., Kim, I.H., and Lee, S.W. (2009) Validation of tissue-specific promoter-driven tumor-targeting trans-splicing ribozyme system as a multifunctional cancer gene therapy device in vivo. *Cancer Gene Ther*, **16**, 113–125.
17. Lan, N., Rooney, B.L., Lee, S.W., Howrey, R.P., Smith, C.A., and Sullenger, B.A. (2000) Enhancing RNA repair efficiency by combining trans-splicing ribozymes that recognize different accessible sites on a target RNA. *Mol Ther*, **2**, 245–255.
18. Giavazzi, R., Jessup, J.M., Campbell, D.E., Walker, S.M., and Fidler, I.J. (1986) Experimental nude mouse model of human colorectal cancer liver metastases. *J Natl Cancer Inst*, **77**, 1303–1308.

Chapter 21

Design and Function of Triplex Hairpin Ribozymes

Guillermo Aquino-Jarquin, Ramiro Rojas-Hernández, and Luis Marat Alvarez-Salas

Abstract

Triplex ribozymes allow for the individual activity of multiple *trans*-acting ribozymes producing higher target cleavage relative to tandem-expressed RZs. A triplex expression system based on a single hairpin ribozyme for the multiple expression (multiplex) vectors can be engineered to target RNAs with single or multiple antisense-accessible sites. System construction relies on triplex expression modules consisting of hairpin ribozyme cassettes flanked by ribozymes lacking catalytic domains. Multiplex vectors can be generated with single or multiple specificity by tandem cloning of triplex expression modules. Triplex ribozymes are initially tested in vitro using *cis*- and *trans*-cleavage assays against radioactive-labeled targets. In addition, triplex ribozymes are tested for *cis* and *trans* cleavage in vivo by transfection in cultured cells followed by ribonuclease protection assays (RPAs) and RT-PCR. The use of triplex configurations with multiplex ribozymes will provide the basis for the development of future RZ-based therapies and technologies.

Key words: Ribozyme, hairpin ribozyme, catalytic RNA, antisense, RNA structure, antisense RNA, papillomavirus, RNA, ribozyme expression, multiplex expression.

1. Introduction

The discovery of catalytic nucleic acids (i.e., ribozymes and DNAzymes) has been one of the most relevant advances in biochemistry in the last decades attracting considerable interest because of their potential use for gene therapy through gene silencing (1). The hairpin ribozyme is a small *cis*-cleaving

M. Sioud (ed.), *RNA Therapeutics*, Methods in Molecular Biology 629,
DOI 10.1007/978-1-60761-657-3_21, © Springer Science+Business Media, LLC 2010

ribozyme found in some plant viroids and satellite RNAs. This ribozyme contains guide sequences that allow hybridization and subsequent cleavage of specific substrate RNA. Furthermore, ribozymes display multiple turnover allowing processing of other target RNAs following initial cleavage (2).

The hairpin RZ is the 50-nt catalytic core of three plant satellite RNAs, including the negative strand of the satellite RNA from the tobacco ringspot virus (sTRSV), the chicory yellow mottle virus type I (sCYMV1), and the arabis mosaic virus (sArMV) (3, 4). The catalytic domain (domain B) of hairpin ribozymes contains two short intramolecular helices that flank an internal loop associated with the cleavage process. Ribozyme–substrate complex is accomplished through hybridization of the target RNA with domain A, forming two intermolecular helices flanking a symmetrical internal loop containing the substrate cleavage site. Minimal amounts of Mg^{2+} are required for correct positioning of domain B relative to the target (docking) and there is no apparent dependence on other co-factors for cleavage (5, 6). Cleavage occurs through a trans-esterification reaction pathway using the 2′-hydroxy group at the scissile linkage primary nucleophile generating cleavage products with 5′-hydroxy and 2′,3′-cyclophosphate termini (7). In addition, the hairpin ribozyme has a modular structure that allows the physical separation of domains A and B. Thus, the catalytic domain B can recognize single or multiple domains A, regardless of their position (8). Therefore, the same hairpin ribozyme can be used for both *cis*- and *trans*-cleavage activities.

Despite major in vivo issues, ribozymes designed to cleave targets on HIV-1, HBV, HPV, and several cellular genes have been successfully tested in vitro and in vivo (9, 10). Variables such as nuclease sensitivity, target co-localization, endogenous ion concentration, and ribozyme expression levels have hampered application of ribozymes as efficient therapeutic agents (11). The continuous search for better ribozyme in vivo performance has led to the engineering of multiple expression (multiplex) systems harboring several *trans*-acting ribozymes within a single transcript. A particular multiplex system uses a triplex unit consisting of two *cis*-cleaving ribozymes flanking a therapeutic ribozyme allowing for the independent activity of each ribozyme, thus increasing overall efficiency (12–14).

We have further developed the triplex concept by designing a multiplex expression system based on a single hairpin ribozyme that produces higher *trans*-cleavage activity than do canonical three-ribozyme-based triplex and tandem ribozymes (15). With simple modifications, the system can be adapted to target different genes, making triplex ribozymes excellent candidates as potential gene inhibitors.

2. Materials

2.1. Cell Culture and Transfection

1. C33-A (HTB-31) and SiHa (HTB-35) cervical cancer cell lines.
2. Dulbecco's modified minimum essential medium (DMEM) growth medium supplemented with 5% fetal bovine serum (FBS) (Invitrogen), 50 μg/mL gentamicin, 50 U/mL penicillin, and 50 μg/mL streptomycin.
3. Phosphate-buffered saline (PBS, 1×): 2.7 mM KCl, 1.8 mM KH_2PO_4, 136 mM NaCl, 10 mM Na_2HPO_4, pH 7.4. Filter through 0.22-μm filter before storing at room temperature.
4. Trypsin–versene solution: Mix 0.5 mL of 2.5% trypsin with 25 mL of 1× PBS and 25 mL of versene.
5. Lipofectin™ transfection reagent, stored at 4°C.
6. Tissue-culture 100-mm plastic dishes.
7. Cell scrapers and spatulas.

2.2. Plasmids and Oligonucleotides

1. The pBtV5-X plasmids are derivatives from a pBSKS vector containing a heavily modified multiple cloning site (MCS) with a hairpin ribozyme flanked by a mutant $tRNA^{Val}$ and a tetraloop sequence expressed from a T3 promoter (*see* **Note 1**).
2. The pCR3.1-GFP plasmid contains the humanized green fluorescent protein (GFP) gene cloned in the pCR3.1 vector (16).
3. The pSilencer3.0-H1MCS vector (15) contains a modified MCS compatible with pBtV5-X inserts cloned in the pSilencer3.0-H1 eukaryotic expression vector (Ambion Inc., Austin TX).
4. The pE6-GFP plasmid contains the HPV-16 *E6* gene cloned in the pGreenLantern 2 vector (Invitrogen). It is used to produce target transcripts for in vitro ribozyme cleavage assays from a T7 promoter (*see* **Note 2**).
5. Chemically synthesized single-stranded oligodeoxynucleotides (ssODNs) containing the coding and complementary sequences for TL and TR (TL, TR, TLC, and TRC, respectively) with linkers for *Eco*RI/*Hin*dIII and *Mlu*I/*Sac*I, respectively (**Table 21.1**) (*see* **Note 3**).

2.3. Oligonucleotide Purification

1. Denaturing gel-loading buffer (80% formamide, 0.25% bromophenol blue, 0.25% xylene cyanol FF, 1 mM EDTA).

Table 21.1
Single-stranded oligodeoxynucleotides (ssODNs) required for triplex ribozyme construction and in vivo analysis. Relevant restriction sites are in italics (*see* text)

Name	Sequence
TL	5′-*AATTC*AAACAGAGAAGTCAACCATGGTACCTCCTGACAGTCCTGTTTA-3′
TLC	5′-*AGCTT*AAACAAGACTGTCAGGAGGTACCATGGTTGACTTCTCTGTTTG-3′
TR	5′-*CGCGT*GACAGTCCTGTTTCCTCCAAACAGAGAAGTCAACCAGAGAA *GAGCT*-3′
TRC	5′-CTGGTTGACTTCTCTGTTTGGAGGAAACAGGACTGTCA-3′
Tp*Sac*I	5′-CCC*GAGCTC*TACCAGGTAATA-3′
Tp*Bss*HII	5′-CCC*GCGCGC*AAACAGAGAAGT-3′
DTpF	5′-CCC*GGATCCGAATTC*AAACAGAGAAGT-3′
DTpR	5′-CCC*TCTAGAGAGCTC*TTCTCTGGTTGA-3′

2. E buffer (1 mM EDTA, 0.5 M ammonium acetate, 0.1% SDS).
3. Thirty percent acrylamide/*N,N*'-methylene-*bis*-acrylamide solution (19:1) in deionized water (*see* **Note 4**).
4. Urea ultrapure.
5. TBE buffer (5×): 0.45 M Tris–HCl ,pH 8.3; 0.45 M boric acid, 10 mM EDTA. Store at room temperature.
6. *N,N,N,N*′-Tetramethylethylenediamine (TEMED). Store desiccated at 4°C.
7. Ammonium persulfate. Freshly prepare 10% solution in deionized water.
8. TLC plates with fluorescent indicator.
9. Syringe filters (0.45 μm).
10. Three-milliliter syringes (3 mL).

2.4. Cloning

1. *Escherichia coli* DH5α competent cells preserved at –70°C.
2. Luria–Bertani (LB) broth and agar stored at room temperature.
3. Ampicillin stock solution (100 mg/mL) stored in aliquots at –20°C.
4. *Bam*HI, *Eco*RI, *Hind*III, *Xho*I, *Mlu*I, *Sac*I, *Xba*I restriction enzymes (New England Biolabs, Inc., Ipswich, MA) and T4 DNA ligase stored at –20°C.
5. T4 ligase buffer (10×): 500 mM Tris–HCl, pH 7.8; 100 mM $MgCl_2$; 100 mM DTT; 10 mM ATP; and 250 mg/mL BSA.

6. Agarose powder stored at room temperature.
7. DNA gel-loading buffer (6×): 0.25% bromophenol blue, 0.25% xylene cyanol FF, 15% Ficoll type 400 stored at 4°C.
8. Plasmid mini and midi purification kits. Stored at room temperature.

2.5. Polymerase Chain Reaction

1. Appropriate chemically synthesized DNA primers stored in aliquots at –20°C. Primers may be purified by denaturing polyacrylamide gel electrophoresis (D-PAGE) or high-performance liquid chromatography (HPLC) to ensure length homogeneity.
2. Deoxynucleoside 5′-triphosphate (dNTP) mix (10×): 10 mM each deoxyguanosine 5′-triphosphate (dGTP), deoxyadenosine 5′-triphosphate (dATP), thymidine 5′ triphosphate (TTP), and deoxycytidine 5′-triphosphate (dCTP).
3. PCR buffer (10×): 100 mM Tris–HCl, pH 8.3; 500 mM KCl; 15 mM $MgCl_2$.
4. *Taq* DNA polymerase.
5. Reagents for agarose gel electrophoresis.
6. Fifty-microliter PCR microtubes.

2.6. In Vitro Transcription

1. Riboprobe™ T7 and T3 in vitro transcription systems.
2. Phenol–chloroform–isoamyl alcohol solution (25:24:1) stored at 4°C.
3. Ammonium acetate (7.5 M, pH 7.5) stored at room temperature.
4. Absolute and 70% (v/v) ethanol stored at –20°C and room temperature, respectively.
5. α-[^{32}P]-UTP (3,000 Ci/mmol).
6. D-PAGE reagents.

2.7. Ribozyme Cis- and Trans-Cleavage Assays

1. RZ buffer (10×): 400 mM Tris–HCl, pH 7.0; 120 mM $MgCl_2$; and 20 mM spermidine stored at –20°C.
2. D-PAGE reagents.

2.8. Ribozyme In Vivo Analysis

1. After transfection, ribozyme expression and cleavage can be analyzed through a method of your choice. Because of the short ribozyme length, we recommend the use of ribonuclease protection assays. In the present example we used the Direct Protect™ RPA kit as described by the manufacturer's instructions with minor changes.
2. Target transcript processing was estimated by quantitative RT-PCR (QRT-PCR) using SuperScriptII™ One-Step

RT-PCR kit and the Vista Green^{TM} fluorescent dye in a Rotor-Gene^{TM} RG-3000.

3. Methods

The effectiveness of an in vivo system for the expression of *trans*-acting hairpin ribozymes (taHR) relies on the amount of transcripts produced from the vector promoter, the longevity of the ribozymes under physiological conditions, and the independent activity of each ribozyme over the target RNA (15). This chapter examines the design and construction of a multiplex ribozyme expression vector based on a novel triplex-like configuration consisting of a single hairpin ribozyme that can produce *cis* and *trans* cleavage with intracellular activity.

The construction of a basic triplex ribozyme module starts with a ribozyme cassette with a taHR protected at the 5′ and 3′ ends by a mutant $tRNA^{Val}$ and a tetraloop sequence, respectively. This cassette is cloned in a heavily modified suitable for the easy exchange of ribozyme target and nuclease protection moieties (10). In addition to the ribozyme cassette, the basic triplex module includes two flanking trimming hairpin ribozymes (TL and TR) lacking the entire catalytic domain B. Thus, taHR remains as the only catalytic moiety of the construct being tasked with both trimming and *trans*-cleavage activities (**Fig. 21.1a**). This design can be used with differentially targeted taHR ribozymes to produce multiplex expression vectors (**Fig. 21.1b**).

3.1. Construction of the Basic Triplex Module Expression Vector

1. Prepare 18% polyacrylamide/7 M urea denaturing gels in 1× TBE buffer. In a clean and sterile beaker, weigh 21 g of urea and add 30 mL of the 30% acrylamide/*bis*-acrylamide solution and 5 mL of 5× TBE.
2. Dissolve by stirring and gentle warming for at least 1 h.
3. Let the mixture cool to room temperature and add 50 μL TEMED and 150 μL of 10% ammonium persulfate. Mix well and pour the mixture into the gel casting cassette and attach a comb. Leave to polymerize for at least 1 h at room temperature.
4. Resuspend the lyophilized ssODNs in 30 μL of denaturing gel-loading buffer and heat for 5 min at 65°C.
5. Remove the comb and set the electrophoresis apparatus with 1× TBE buffer. Rinse wells with 1× TBA buffer using a syringe and load the full sample volume.
6. Electrophorese at 250 V for 3 h; most ssODNs migrate close to the xylene cyanol band.

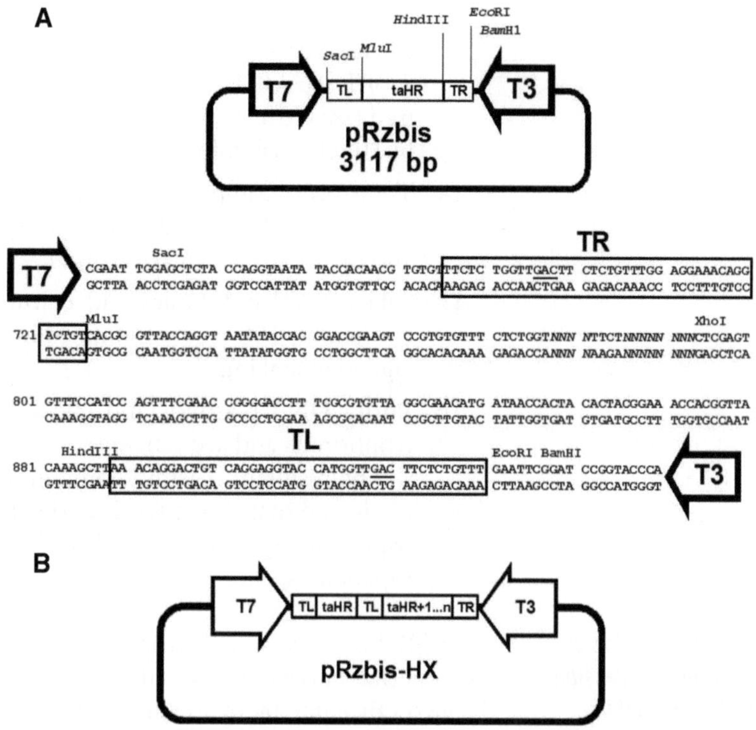

Fig. 21.1. (**a**) Map and sequence of pRzbis plasmid containing the basic triplex ribozyme module. TL and TR domains flanking *trans*-acting ribozyme (taHR) are shown as *open boxes*. Relevant transcription sites and T3 and T7 RNA polymerase promoters are indicated. (**b**) Map of multiplex expression vector based on ribozyme triplex modules. Modules (taHR) are cloned sequentially (from 1 to *n*) flanked by TL motifs and a single 3′ terminal TR motif.

7. Wrap the gel in plastic wrap and ODNs are visualized by DNA shadowing using an UV lamp and a fluorescent screen (*see* **Note 6**).
8. Excise the full-length ssODNs with a scalpel and elute in 350 μL of E buffer at 4°C overnight.
9. Filter the eluate using 0.45-μm syringe filters to exclude polyacrylamide debris.
10. Purify the eluted ssODNs through Sephadex G-50 with deionized water and quantitate by $UV_{260/280}$ absorbance.
11. Double-stranded oligodeoxynucleotides (dsODNs) containing TL and TR are prepared for cloning by hybridization of equimolar amounts of adequate coding and complementary ssODNs in 100 μL of 1× T4 DNA ligase reaction buffer.
12. Heat the mixture in boiling water for 3 min and then slowly cool to room temperature.

13. The dsODNs TL and TR are attached to flanking sites of the taHR cassette through successive cloning steps into the *Eco*RI/*Hind*III and *Mlu*I/*Sac*I sites of pBtV5-R434 plasmid (*see* **Note 1**), respectively, using 2 U of T4 DNA ligase in 10 μL final volume at 4°C overnight (*see* **Note 7**).
14. Transform competent bacteria with ligated DNA and spread on LB agar plates containing 100 μg/mL ampicillin and incubate at 37°C overnight.
15. Select positive colonies and culture in 1.5 mL LB broth containing 100 μg/mL ampicillin overnight at 37°C with vigorous shaking.
16. Purify DNA plasmids and check by 1% agarose gel electrophoresis and sequencing.

    The resultant plasmid containing the basic triplex module is designated as pRzbis and expresses a fully functional triplex ribozyme module with trimming and *trans*-cleavage capabilities.

3.2. Construction of Multiplex Expression Vectors Based in Triplex Modules with a Single Hairpin Ribozyme

The pRzbis expression vector can be used to produce multiplex vectors that encode various ribozyme triplex modules to improve in vivo efficiency by expressing large amounts of ribozymes and to prevent escape mutants by targeting multiple sites within the same transcript. This can be accomplished by simply concatenating triplex modules in tandem. However, this procedure causes the side-by-side cloning of TL and TR resulting in the derangement of their secondary and probably their tertiary structure, thus inhibiting the *cis*-cleaving activity on the particular cleavage site. To avoid this, cloning of multiplex vectors should be performed with partial triplex modules produced by PCR lacking TR, resulting in repeated TL-taHR units with a single terminal TR cloned in pRzbis (**Fig. 21.1b**).

1. The TL-taHR fragment of pRzbis is PCR amplified using 2 μM TpBssHI and TpSacI primers (**Table 21.1**) with 1 ng of plasmid DNA in 1× PCR buffer in 25 μL final volume. PCR conditions are set at 95°C denaturing for 2 min and 25 cycles of 95°C denaturing for 1 min, 55°C hybridization for 30 s, and 72°C polymerization for 1 min plus a final 10 min polymerization step at 72°C.
2. Amplified fragments can be visualized by native agarose gel electrophoresis, eluted and purified by phenol–chloroform–isoamyl alcohol extraction and ethanol–ammonium acetate precipitation.
3. The amplified fragments are cloned into the *Mlu*I/*Sac*I of pRzbis to replace TR moiety. The resulting plasmid is designated pRzbis-H2, which expresses two tandem copies of the triplex ribozyme modules.

4. Additional triplex ribozyme modules can be added by cloning the PCR-amplified module from pRzbis or pRzbis-H2 using primers DTp-F and DTp-R (**Table 21.1**), resulting in the pRzbis-HX plasmid series expressing up to four (X) triplex ribozyme modules (*see* **Note 1**) (**Fig. 21.1b**).
5. For in vivo testing, the PCR-amplified triplex ribozyme modules are cloned into the *Bam*HI/*Hind*III sites of p3.0-H1-MCS eukaryotic expression vector. Sequence all the constructs generated before testing.

3.3. In Vitro Transcription

1. Plasmids are linearized with a single-cut restriction enzyme and cleaned by phenol–chloroform–isoamyl alcohol extraction and ethanol/ammonium acetate precipitation.
2. Linearized plasmids are in vitro transcribed with T3 (ribozyme modules) or T7 RNA polymerases (target transcript and antisense probes) in the presence of 10 μCi α-[^{32}P]-UTP using Riboprobe™ T3 or T7 in vitro transcription kit. Labeled transcript production is optimized by the addition of 20- to 200-fold non-labeled UTP to the transcription reaction. DNA template is removed by incubation with 20 U of RNase-free DNase I for 30 min at 37°C. Labeled transcripts are purified through preparative denaturing PAGE, eluted in E buffer, and cleaned through phenol–chloroform–isoamyl alcohol extraction and ethanol/ammonium acetate precipitation as described above.
3. Resuspend labeled transcripts in 20 μL of deionized water and quantify in a scintillation counter. Adjust concentration to 5–20 × 10^3 cpm/μL and store at –70°C until use.

3.4. RZ Self-Digestion Assays

1. Labeled triplex ribozyme RNA (approximately 2 × 10^4 cpm) in 10 μL 1× RZ buffer is incubated at 37°C for 0–120 min.
2. Stop reactions by snap freezing in dry ice until used.
3. Samples are diluted in 40 μL denaturing gel-loading buffer and incubated at 65°C for 5 min before electrophoresis at 250 V for 50 min in 6% polyacrylamide/7 M urea gels.
4. After electrophoresis, gels are dried and quantified (*see* **Note 8**). An example of this is represented in **Fig. 21.2a** .

3.5. RZ Trans-Cleavage Assays

1. For multiple turnover conditions, a 30:1 molar ratio of ^{32}P-labeled target RNA/ribozyme RNA are incubated in 1× RZ buffer at 37°C for 0–120 min. Stop reactions by snap freezing in dry ice until used.
2. Samples are electrophoresed and analyzed as described above.

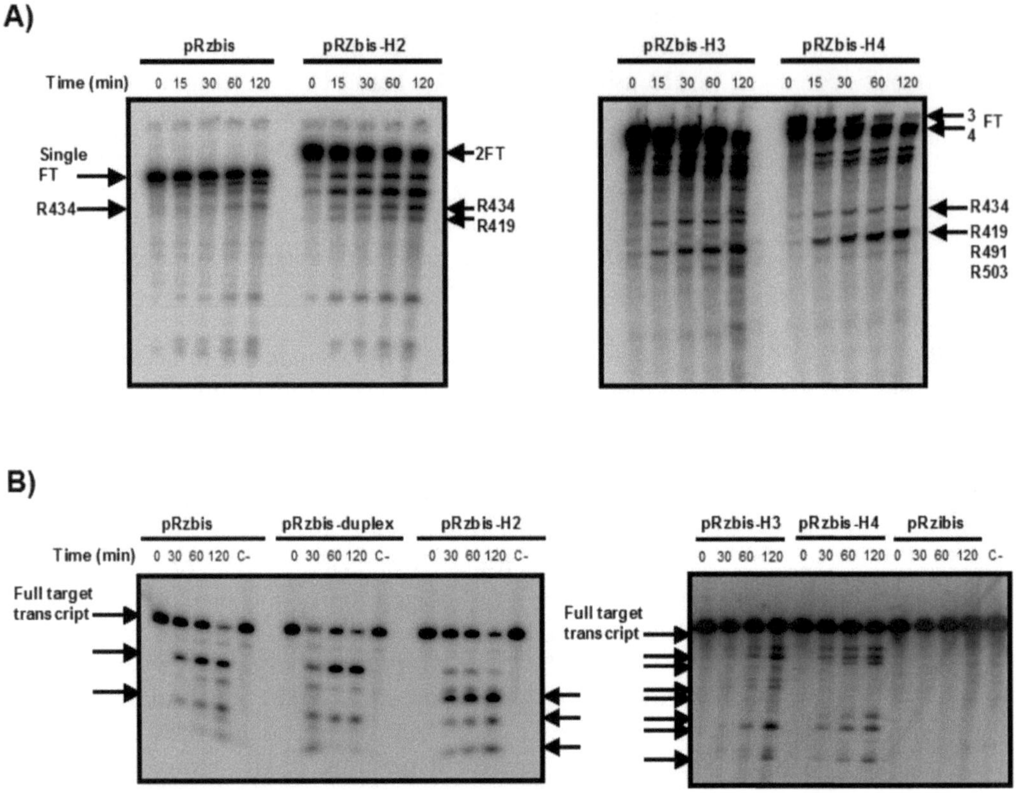

Fig. 21.2. Triplex ribozyme in vitro activity. (**a**) Trimming activity of triplex ribozymes from single and multiplex vectors. Labeled transcripts from one (pRzbis), two (pRzbis-H2), three (pRzbis-H3), and four (pRZbis-H4) ribozyme modules were incubated for up to 120 min in 1× RZ buffer at 37°C. Trimming fragments were separated by denaturing polyacrylamide gel electrophoresis (D-PAGE). *Arrows* indicate bands corresponding to full transcripts (FT) from one (single), two (2), three (3), and four (4) modules and free ribozymes (R419, R434, R491 and R503). (**b**) *Trans*-activity of triplex ribozymes from single and multiplex vectors. Full transcripts from pRzbis, Rzbis-H2, pRzbis-H3, and pRzbiz-H4 plasmids were incubated with a labeled target RNA in 1× RZ buffer at 37°C and digestion fragments were separated by D-PAGE. *Arrows* indicate the position of full transcripts and processed fragments.

3. Cleavage rates are calculated from cleavage percentage *vs.* time plots (*see* **Note 8**) as the average of at least three independent experiments (**Fig. 21.2b**).

3.6. In Vivo Testing of Ribozyme Cleavage

Transfection of plasmids expressing the different triplex ribozyme constructs is performed to test in vivo functionality. In the present example, we directed the ribozymes against the HPV-16 E6 mRNA and thus a cell line expressing such target (SiHa cells) must be used. Inactive ribozymes (i.e., R434i) with mutations within the catalytic domain will establish a possible antisense effect of the constructs and allow discrimination between self-processing and endogenous nuclease activity. Additionally, an active ribozyme with scrambled domain A and checked against GenBank database for negative cross-hybridization is used as control of specificity. As an extra control for self-processing and toxicity, we use a HPV-16-negative cervical cells (C33-A).

1. Confluent 100-mm dishes of C33-A and SiHa cells are split at a 1:10 ratio 24–48 h before transfection.
2. In a sterile 15-mL polystyrene conical tube, add 500 μL of basal DMEM, 9 μg of midiprep-purified (*see* **Note 9**) ribozyme expression vector, and 1 μg of pCR3.1-GFP plasmid (plasmid mixture).
3. In a second sterile conical tube, add 500 μL of basal DMEM (without FBS) and 25 μL of Lipofectin™ or transfection agent (lipofection mixture) (*see* **Note 10**).
4. Incubate tubes at room temperature for 5–10 min.
5. Add the total volume of lipofection mixture to the plasmid mixture tube and incubate at room temperature for 15–30 min (plasmid–lipofection mixture).
6. Cultured C33-A or SiHa cells (60–70% confluent) in 100-mm dishes are washed twice with PBS. Completely remove the liquid.
7. Add the plasmid–lipofection mixture to the dishes by gently dropping the mixture on the cell layer.
8. Incubate dishes at 37°C for 2 h in a CO_2 incubator tilting every 15–30 min to avoid drying.
9. Add 10 mL of warm DMEM supplemented with 5% FBS medium to the dishes and incubate at 37°C overnight in a CO_2 incubator.
10. Replace culture medium and incubate at 37°C overnight in a CO_2 incubator for further 48 h.
11. After transfection, cells are harvested and sorted using a FACScalibur fluorocytometer with a band-pass filter at 585/42 nm for GFP fluorescence. Excitation is performed with a 488-nm argon laser. This step allows removal of non-transfected cells for *trans*-activity analysis. It is therefore unnecessary for *cis*-cleavage analysis.
12. Extract total RNA using Trizol™ reagent following manufacturer's instructions and treat with DNase I. Clean RNA through phenol–chloroform–isoamyl alcohol extraction and ethanol–ammonium acetate precipitation. Store RNA at –70°C.
13. Confirm ribozyme expression and target transcript levels.

3.7. RNA Protection Assays (RPAs)

1. In a sterile 1.5-mL microcentrifuge tube, add 25 μg of total RNA and 5×10^4 cpm each of ^{32}P-labeled antisense RNA probe (*see* **Note 5**). There should be a 3- to 10-fold molar excess of probe over target mRNA.

2. Add 45 μL of lysis/denaturation solution to 10^4–10^5 cpm/μL and resuspend by vortex.
3. Incubate at 37°C overnight to permit full probe hybridization with complementary RNA in the sample lysate.
4. After hybridization, the mixture is treated with the provided ribonuclease cocktail (RNase A and RNase T1) to degrade non-hybridized probe. Hybridized fragments will be protected from ribonuclease digestion.
5. Include a minus and a plus RNase controls with non-hybridized probe to test probe integrity and ribonuclease performance.
6. Incubate tubes for 30 min at 37°C.
7. Add 20 μL of 10% sodium sarcosyl and 10 μL proteinase K (20 mg/mL) to each tube. Mix by flicking tube and incubate further for 30 min at 37°C.
8. Add 500 μL of isopropyl alcohol. Mix well by flipping the tube with your fingers.
9. Transfer tubes to –20°C freezer for 15–30 min. It is not necessary to add carrier during this precipitation.
10. Centrifuge for 15 min at 13,000 rpm, preferably at 4°C. Decant all supernatant from each tube. Dry briefly.
11. Resuspend each pellet in gel-loading buffer II, usually 4–10 μL.
12. Heat all tubes for 5 min at 65°C, vortex, and centrifuge tubes again briefly.
13. Prepare a denaturing polyacrylamide gel suitable for separation of protected fragments of expected size (typically a 6% polyacrylamide/7 M urea).
14. Electrophorese samples at 250 V for about 1 h. Dry gels.
15. Visualize and quantify the protected fragments (**Fig. 21.3a**).

3.8. Quantitative RT-PCR

1. Total RNA (1 μg) is amplified using target-specific primers and PCR conditions in the presence of Vista Green™ fluorescent dye using a real-time PCR apparatus in 50-μL PCR tubes (*see* **Note 11**). Fluorescence readings are set at 470 nm excitation and 510 nm emission.
2. The β-actin gene is used as internal control using the RT-PCR conditions as described (17).
3. In the present example, the E6/E7 mRNA levels are standardized against endogenous β-actin mRNA and plotted as a percentage of inhibition relative to untreated controls (**Fig. 21.3b**).

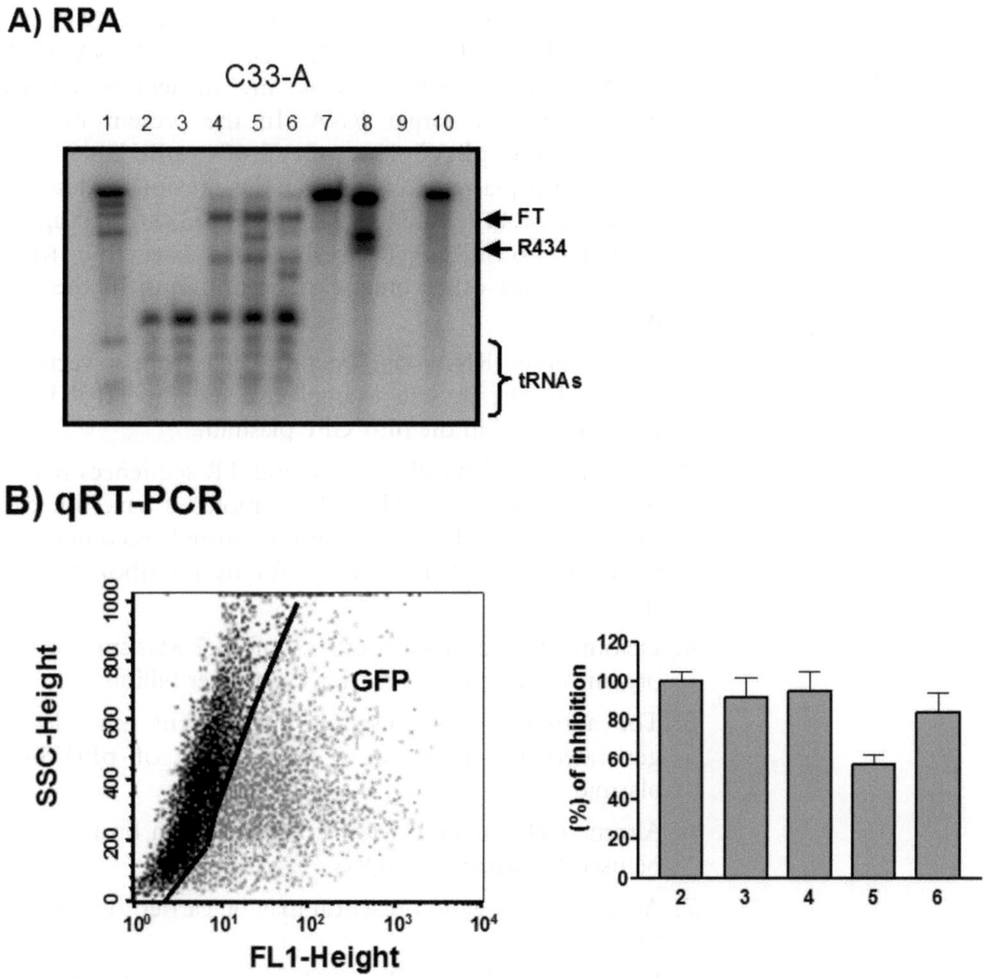

Fig. 21.3. Triplex ribozyme in vivo analysis. (**a**) RPA analysis of total RNA extracted from C33-A cells non-treated (2) and transfected with plasmids p3.0-H1-MCS (3), p3.0-Scrambled (4), p3.0-H1-R434 (5), and p3.0-H1-R434i (6). An in vitro produced triplex digestion was used as a molecular weight marker (1). Additional controls include the intact probe (7), a non-triplex R434 (8), the probe incubated plus (9), and minus (10) RNase cocktail in 1× RNase buffer. *Arrows* indicate the full triplex ribozyme transcript and the released ribozyme. A *bracket* indicates cross-hybridization with tRNAs. The remaining bands correspond to intracellular ribozyme transcript decay not associated with catalytic activity. (**b**) *Left panel:* flow cytometry of GFP-transfected cells (*gray*). *Right panel*: QRT-PCR using total RNA (1 μg) from GFP-sorted SiHa cells transfected with pCR3.1-GFP only (2) or co-transfected with p3.0-H1-MCS (3), p3.0-Scrambled (4), p3.0-H1-R434 (5), or p3.0-H1-R434i plasmids (6). The levels of target transcripts (HPV-16 E6/E7 mRNA) were corrected against β-actin mRNA levels before plotting.

4. Notes

1. Several pBtV5-X plasmids can be made with different specificity by cloning different hairpin ribozymes into the *Mlu*I/*Xho*I sites using dsODNs with the sequence

5′-CGCGTTACCAGGTAATATACCACGGACCGAAGT-CCGTGTGTT TCTCTGGTACTGTTCNNNNAAGAN-NNNNNNC-3′, where N is any nucleotide complementary to the target RNA. In the present example, we produced pBtV5-419, pBtV5-434, pBtV5-491, and pBtV5-503 plasmids containing the previously described ribozymes R419, R434, R91, and R503 directed against HPV-16 E6 mRNA (10). To clone over four triplex modules, add other unique restriction sites in the PCR primers.

2. Other target RNAs can be produced from a vector of choice. In the present example we use the HPV-16 E6 mRNA cloned in the pE6-GFP plasmid.
3. Although theoretically the TL and TR sequences may be modified, alternative TL and TR moieties have not been thoroughly tested yet and thus it is strongly recommended to use the provided sequences for triplex ribozyme construction.
4. Deionized water has a resistivity of 18.2 MΩ cm and total organic content of less than five parts per billion.
5. The antisense probe used in the present example was generated by using the T7 promoter of pBtV5-434 plasmid.
6. Alternatively, staining with 0.02% ethidium bromide may be used to visualize ssODNs.
7. Alternatively, ligation reaction may be carried out at 37°C for 1 h.
8. Radioactive gel quantification is greatly simplified by using a Typhoon 8600 fluorographic scanner (GE Healthcare Life Sciences, Piscataway NJ). However, quantification through scanning of radiographic films or scintillation counting can also be used. In such cases, cleavage percentage is calculated from the following formula:

$$\text{Cleavage percentage} = [\text{cleaved RNA dpm}/(\text{cleaved RNA dpm} + \text{non} - \text{cleaved RNA dpm})] \times 100$$

The dpm is calculated from cpm using the counting efficiency. For K_m and V_{max} determination, several *trans*-cleavage assays must be performed with different target RNA concentrations to produce cleavage rate *vs.* substrate molar concentration plots. Molar concentration of the ^{32}P-U-labeled RNA is obtained by converting dpm to μCi using the equivalency 1 μCi = 2.22×10^6 dpm.

The number of picomoles of U in the sample is calculated through the following formula (18):

$$\text{Picomoles of U in sample} = \text{picomoles of UTP added} \times [\mu\text{Ci in transcript}/(\mu\text{Ci UTP added} \times \text{decay factor})]$$

Decay factor is the correction for the half-life of ^{32}P at the moment of the experiment. The number of picomoles of U in the sample is then converted to picomoles of RNA including the number of U residues in the transcript sequence:

$$\text{Picomoles of transcript RNA} = \text{picomoles of U in sample}/(\text{number of U per transcript RNA})$$

9. High-quality plasmid DNA for transfection can be obtained through Qiagen columns or double CsCl isopycnic gradients.
10. Other lipofection reagents (i.e., Lipofectamine 2000™) may also be used.
11. In the present example, the primer set E6U (5′-CAGCAATACAACAAACCG-3′) and E7L (5′-TAGATTATGGTTTCTGAGAACA-3′) amplifies the HPV-16 E6 mRNA as described (19).

Acknowledgments

The authors would like to thank Dr. Joseph A. DiPaolo (NCI/NIH) for his unconditional support and encouragement. We also thank Dr. María Luisa Benítez-Hess for excellent technical assistance. This project was partially supported by CONACyT (Grant No. 45715Z).

References

1. Benitez-Hess, M.L. and Alvarez-Salas, L.M. (2006) Utilization of ribozymes as antiviral agents. *Lett Drug Des Discov*, **3**, 390–404.
2. Burke, J.M. (2002) Hairpin and hammerhead ribozymes: how different are they? *Biochem Soc Trans*, **30**, 1115–1118.

3. Haseloff, J. and Gerlach, W.L. (1989) Sequences required for self-catalysed cleavage of the satellite RNA of tobacco ringspot virus. *Gene*, **82**, 43–52.
4. Hampel, A. (1998) The hairpin ribozyme: discovery, two-dimensional model, and development for gene therapy. *Prog Nucleic Acid Res Mol Biol*, **58**, 1–39.
5. Berzal-Herranz, A., Joseph, S., Chowrira, B.M., Butcher, S.E., and Burke, J.M. (1993) Essential nucleotide sequences and secondary structure elements of the hairpin ribozyme. *EMBO J*, **12**, 2567–2573.
6. Walter, N.G. and Burke, J.M. (1998) The hairpin ribozyme: structure, assembly and catalysis. *Curr Opin Chem Biol*, **2**, 24–30.
7. Berzal-Herranz, A. and Burke, J.M. (1997) Ligation of RNA molecules by the hairpin ribozyme. *Methods Mol Biol*, **74**, 349–355.
8. Shin, C., Choi, J.N., Song, S.I., Song, J.T., Ahn, J.H., Lee, J.S., and Choi, Y.D. (1996) The loop B domain is physically separable from the loop A domain in the hairpin ribozyme. *Nucleic Acids Res*, **24**, 2685–2689.
9. Taylor, N.R. and Rossi, J.J. (1991) Ribozyme-mediated cleavage of an HIV-1 gag RNA: the effects of nontargeted sequences and secondary structure on ribozyme cleavage activity. *Antisense Res Dev*, **1**, 173–186.
10. Alvarez-Salas, L.M., Cullinan, A.E., Siwkowski, A., Hampel, A., and DiPaolo, J.A. (1998) Inhibition of HPV-16 E6/E7 immortalization of normal keratinocytes by hairpin ribozymes. *Proc Natl Acad Sci U S A*, **95**, 1189–1194.
11. Michienzi, A. and Rossi, J.J. (2001) Intracellular applications of ribozymes. *Methods Enzymol*, **341**, 581–596.
12. Taira, K., Oda, M., Shinshi, H., Maeda, H., and Furukawa, K. (1990) Construction of a novel artificial-ribozyme-releasing plasmid. *Protein Eng*, **3**, 733–737.
13. Altschuler, M., Tritz, R., and Hampel, A. (1992) A method for generating transcripts with defined 5′ and 3′ termini by autolytic processing. *Gene*, **122**, 85–90.
14. Ohkawa, J., Yuyama, N., Takebe, Y., Nishikawa, S., and Taira, K. (1993) Importance of independence in ribozyme reactions: kinetic behavior of trimmed and of simply connected multiple ribozymes with potential activity against human immunodeficiency virus. *Proc Natl Acad Sci U S A*, **90**, 11302–11306.
15. Aquino-Jarquin, G., Benitez-Hess, M.L., DiPaolo, J.A., and Alvarez-Salas, L.M. (2008) A triplex ribozyme expression system based on a single hairpin ribozyme. *Oligonucleotides*, **18**, 213–224.
16. Benitez-Hess, M.L., DiPaolo, J.A., and Alvarez-Salas, L.M. (2004) Antisense activity detection by inhibition of fluorescence resonance energy transfer. *Luminescence*, **19**, 85–93.
17. Guapillo, M.R., Marquez-Gutiérrez, M.A., Benitez-Hess, M.L., and Alvarez-Salas, L.M. (2006) A bacterial reporter system for the evaluation of antisense oligodeoxynucleotides directed against human papillomavirus type 16 (HPV-16). *Arch Med Res*, **37**, 584–592.
18. DeYoung, M.B., Siwkowski, A., and Hampel, A. (1997) Determination of catalytic parameters for hairpin ribozymes. *Methods Mol Biol*, **74**, 209–220.
19. Marquez-Gutierrez, M.A., Benitez-Hess, M.L., DiPaolo, J.A., and Alvarez-Salas, L.M. (2007) Effect of combined antisense oligodeoxynucleotides directed against the human papillomavirus type 16 on cervical carcinoma cells. *Arch Med Res*, **38**, 730–738.

Chapter 22

Catalytic M1GS RNA as an Antiviral Agent in Animals

Yong Bai, Paul Jay Fannin Rider, and Fenyong Liu

Abstract

The use of RNase P ribozyme (M1GS catalytic RNA) for inhibition of murine cytomegalovirus (MCMV) propagation in mice is described in this chapter. General information about RNase P based technology is included and followed by detailed protocols focused on (1) construction and in vitro cleavage assay of the customized M1GS ribozyme, (2) stable expression of the M1GS RNA and evaluation of its activity in inhibition of viral gene expression and growth in cultured cells, and (3) investigation of M1GS-mediated inhibition of viral infection and pathogenesis in animals. Using these methods, we have successfully constructed catalytic M1-1 RNA against the MCMV assembly protein (mAP) and M80 mRNA. Our recent study has demonstrated that an 80% reduction in the expression of mAP and M80 and a 2,000-fold reduction in viral growth were observed in cells expressing the ribozyme. Furthermore, after the functional ribozyme-expressing constructs were delivered into MCMV-infected SCID mice, a significant reduction of viral gene expression and infection was detected, and the survival of the infected animals was significantly improved. Collectively, our data demonstrate the feasibility of the use of RNase P ribozyme for inhibition of viral gene expression in animals and support the utility of RNase P ribozyme for gene-targeting applications in vivo.

Key words: RNase P, ribozyme, M1GS RNA, gene targeting, cytomegalovirus, herpesvirus, antiviral.

1. Introduction

RNase P is a ribonucleoprotein complex found in all examined organisms and is responsible for the 5′ maturation of tRNA. This enzyme catalyzes a hydrolysis reaction to remove the 5′ leader sequence of precursor tRNA (ptRNA), which is responsible for the maturation of 5′ termini of all tRNAs (1, 2). One unique feature of RNase P is that it recognizes the

M. Sioud (ed.), *RNA Therapeutics*, Methods in Molecular Biology 629,
DOI 10.1007/978-1-60761-657-3_22, © Springer Science+Business Media, LLC 2010

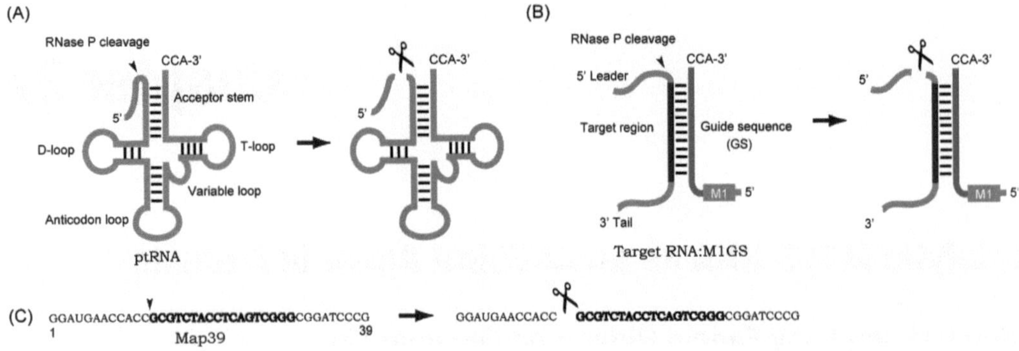

Fig. 22.1. Schematic representation of a natural substrate (ptRNA) (**a**) and a complex formed between a M1GS RNA and its mRNA substrate (**b**). The site of cleavage by RNase P or M1 RNA is marked with the *arrow*. (**c**) Schematic representation of the substrate used in the study. The targeted sequence that binds to the guide sequence of the ribozymes is *highlighted*.

structure, rather than the sequence, of the substrate, which allows the enzyme to hydrolyze different natural substrates in vivo or in vitro. Accordingly, any complex of two RNA molecules that resembles a tRNA molecule can be recognized and cleaved by RNase P (**Fig. 22.1a**) (3, 4). RNase P mediated inhibition of gene expression, an application based on this special feature, has been shown to be a novel and useful nucleic acid based gene interference strategy for specific inhibition of mRNA sequences of choice (5, 6).

1.1. M1GS Catalytic RNA as a Tool to Block Target mRNA Expression

RNase P of *Escherichia coli* contains a catalytic RNA subunit (M1 RNA) that can be engineered to cleave tRNA-like substrates and other target RNAs, including specific mRNAs (1–3, 7). A sequence-specific ribozyme, M1GS, constructed by attaching to M1 RNA an additional small RNA (guide sequence [GS]) which contains a sequence complementary to a target mRNA and a 3′ proximal CCA (**Fig. 22.1b**), is effective in blocking substrate mRNA expression in cultured cells (6, 8). Unlike other nucleic acid based interference approaches such as antisense oligonucleotides and RNA interference (RNAi) (9, 10), M1GS-based strategy is unique because of the use of M1 RNA which is one of the most efficient catalytic RNAs found in nature. Previous studies have shown that M1GS RNA and RNase P are effective in cleaving both viral and cellular mRNAs and blocking their expression in cultured cells, including inhibition of gene expression of human influenza viruses and herpesviruses (8, 11–13). Furthermore, our lab recently reported that engineered M1GS catalytic RNA also functioned to inhibit viral gene expression in animals (14). As such, M1GS catalytic RNA represents a promising tool for treatment of viral diseases in vivo.

1.2. M1GS Catalytic RNA as a Potential Therapeutic for CMV Infection

Human cytomegalovirus (CMV) is a common opportunistic pathogen and can cause serious clinical manifestations in immunocompromised or immunologically immature individuals, including neonates, AIDS patients, and transplant recipients (15). The emergence of drug-resistant strains of CMV has posed a need for the development of new drugs and treatment strategies. Murine cytomegalovirus (MCMV) infection of mice resembles in many ways its human counterpart with respect to pathogenesis, thus providing an animal model for studying CMV infection in vivo and for screening novel agents and developing new antiviral approaches (15–18). Immunodeficient animals such as SCID mice have been shown to be extremely susceptible to MCMV infection, primarily due to liver damage and failure associated with viral lytic replication in the organ (15, 16, 19, 20). In this chapter, the CB17 SCID mice, which lack functional T and B lymphocytes, were selected to analyze viral virulence, growth, and gene expression. This animal model can be used for studying whether new therapeutic approaches can block CMV opportunistic infection and prevent viral-associated diseases in immunocompromised hosts. Using this system, our recent study has shown that customized M1GS RNA is effective in inhibiting MCMV gene expression, blocking viral infection and pathogenesis, and leading to improved survival of animals.

In this chapter, detailed protocols of this system are described. To be an effective and efficient antiviral agent, the ribozyme M1-1 was constructed to target the region of the mRNA encoding MCMV capsid assembly protein (mAP) (21, 22). The mAP coding sequence completely overlaps with and is within the 3′ coding sequence of another viral capsid protein, M80 (23). Both mAP and M80 are essential for viral capsid formation and CMV replication (24, 25). We showed that the constructed ribozyme efficiently cleaved the target mRNA sequence in vitro and effectively inhibited mAP expression and viral growth in cultured NIH-3T3 cells. When a construct carrying M1-1 expression cassettes was introduced to MCMV-infected SCID mice via a modified "hydrodynamic transfection" procedure (26–28), viral gene expression and growth in various organs of these animals were reduced and animal survival was significantly improved. Our study represents the first report to investigate the activity of M1GS ribozyme in animals. Furthermore, these results demonstrate the feasibility of developing effective RNase P ribozyme as a novel class of antiviral agents for treatment of viral diseases in vivo.

2. Materials

2.1. Reagents and Solutions

1. 1 M Tris–HCl, pH 7.0.
2. 1 M Tris–HCl, pH 7.5.

3. 0.5 M ethylenediaminetetraacetic acid (EDTA).
4. 5 M NaCl.
5. 1 M $MgCl_2$.
6. 5 M NaOH.
7. 8 M urea.
8. Chloroform.
9. Isopropanol.
10. Formaldehyde.
11. Isoflurane.
12. Culture medium DMEM.
13. Nu-Serum.
14. Nonfat dry milk.
15. Phosphate-buffered saline (PBS).
16. Salmon Sperm DNA.
17. Neomycin.
18. [^{32}P]-labeled nucleotides.
19. Diethylpyrocarbonate (DEPC)-treated H_2O: double-distilled water is mixed with 0.1% DEPC and stirred overnight. The DEPC is inactivated by autoclaving for 20 min.
20. 20% sodium dodecyl sulfate (SDS).
21. 30% acrylamide/0.8% *N*,*N*′-methylene bis-acrylamide.
22. 1X Denhardt's solution: 0.1% bovine serum albumin (BSA), 0.1% polyvinylpyrrolidone, 0.1% Ficoll, 0.1% SDS, and 200 μg/mL denatured salmon sperm.
23. 20X standard saline citrate (SSC): 3 M NaCl, 0.3 M trisodium citrate.
24. 10X TBE: 0.89 M Tris-borate, 10 mM EDTA.
25. 10X buffer A (cleavage assay buffer): 500 mM Tris–HCl, pH 7.5, 1 M NH_4Cl, 1 M $MgCl_2$.
26. 2X RNA dye solution: 8 M urea, 20 mM EDTA, 0.25 mg/mL bromophenol blue, 0.25 mg/mL xylene cyanol FF.
27. Prehybridization buffer: 6X SSC, 0.05% sodium pyrophosphate, and 2X Denhardt's solution.
28. TNT: 10 mM Tris–HCl, pH 7.0, 150 mM NaCl, 0.05% Tween-20.

2.2. Enzymes and Reaction Buffers

1. 10X PCR buffer: 25 mM $MgCl_2$, four 10 mM deoxynucleotide 5′ triphosphate (dNTP).
2. *Taq* DNA polymerase.

3. T7 in vitro transcription kit.
4. Trizol reagent (total RNA isolation reagent).
5. Mammalian transfection kit.
6. ECL Western blotting detection kit.
7. Random Primed Labeling Kit.

2.3. Virus, Cells, Plasmids, and Animals

1. MCMV (Smith strain, ATCC).
2. PA317 cells (amphotropic retrovirus packaging cell line).
3. NIH-3T3 cells.
4. Plasmid FL117 (containing M1 DNA sequence driven by the T7 RNA polymerase promoter).
5. pLXSN (retroviral vector).
6. CB17 SCID mice.

3. Methods

3.1. Construction of M1-1 Catalytic RNA and In Vitro Cleavage Assay for Substrate Map39

Ribozyme M1-1 was constructed by covalently linking the 3′ terminus of M1 RNA with a guide sequence of 18 nucleotides that is complementary to the targeted mAP mRNA sequence. M1-1 was incubated with substrate Map39, which contained the targeted mAP mRNA sequence of 39 nucleotides (**Fig. 22.1c**). Efficient cleavage of Map39 by M1-1 yielded two hydrolyzed products of 12 and 27 nucleotides, respectively.

3.2. Construction of M1-1 Ribozyme and Substrate Map39

1. The DNA sequence of M1GS ribozyme (M1-1) targeting mAP mRNA is generated by PCR amplification from plasmid pFL117 (containing M1 DNA sequence driven by the T7 RNA polymerase promoter) with the following oligonucleotides as 5 and 3′ primers, respectively: AF25 (5′-GGAATTCTAATACGACTCACTATAG-3′) and M1mAP (5′-CCCGCTCGAGAAAAAATGGTGCGTCTACCTCAGTCGGGTGTGGAATTGTG-3′).
2. Add the following sequentially: 10 μL 10X PCR buffer (Mg^{2+} free), 2 μL 10 mM dATP, 2 μL 10 mM dTTP, 2 μL 10 mM dGTP, 2 μL 10 mM dCTP, 8 μL $MgCl_2$, 10 pmol 5′ primer (AF25), 100 pmol 3′ primer (M1mAP), 1 μg *Pvu*II-digested pFL117 plasmid, 1 μL *Taq* DNA polymerase. Add DEPC-treated water to adjust the final reaction volume to 100 μL.
3. PCR amplification of the DNA template using the following amplification program:
   (1) denaturing for 2 min at 94°C;

(2) 30 cycles of: denaturing for 2 min at 94°C, annealing for 1 min at 47°C, extension for 1 min at 72°C;

(3) final extension for 10 min at 72°C.

4. The PCR DNA products are separated in 5% polyacrylamide gels under nondenaturing conditions and are purified and used as the template for the in vitro transcription synthesis of the ribozymes.
5. Synthesize the M1-1 ribozyme by in vitro transcription using the following mixture: 4 μL (2 μg) M1-1 DNA from PCR reaction, 8 μL 5X transcription buffer, 4 μL 100 mM DTT, 4 μL 10 mM ATP, 4 μL 10 mM GTP, 4 μL 10 mM CTP, 4 μL 10 mM UTP, 4 μL H_2O, 1 μL RNasin (5 U/μL), 2 μL T7 RNA polymerase.
6. Incubate the reaction mixture at 37°C for 4 h to overnight.
7. Add an equal volume of 2X RNA dye solution and load onto an 8% polyacrylamide-8-M urea gels.
8. Place the gel on a TLC plate (silica gel UV256). Visualize RNA bands by briefly shadowing with a portable shortwave ultraviolet lamp. Extract RNA from the excised gel slice by the crush-soak method using DEPC-treated water.
9. The DNA sequence that encodes substrate Map39 was constructed by PCR using pGEM3zf (+) as a template and oligonucleotides AF25 (5′-GGAATTCTAATACGACTCACTATAG-3′) and smAP (5′-CGGGATCCGCCCGACTGAGGTAGACGCGG TGGTTCATCCTATAGTGAGTCGTATTA-3′) as 5′ and 3′ primers, respectively.
10. Synthesize labeled Map39 by in vitro transcription as described in **Section 3.2, steps 5–8**, except using 30 μCi α-[^{32}P]-GTP and 1 mM GTP instead of 10 mM GTP in the reaction.

3.3. In Vitro Cleavage Assay for Substrate Map39

1. Mix 10 nM of M1-1 RNA and trace amount of α-[^{32}P]-labeled Map39 mRNA (10,000 cpm).
2. Add 1 μL of 10X buffer A and use H_2O to adjust to a final volume of 10 μL.
3. Carry out the cleavage reaction at 37°C for 40 min.
4. Terminate the reaction by adding 10 μL of 2X RNA dye solution.
5. Separate the cleavage products by 8% polyacrylamide-7-M urea gel.
6. Expose the gel to a phosphorscreen and quantitate with a STORM840 PhosphorImager (Molecular Dynamics).

3.4. Expression and Functional Analysis of M1-1 Ribozyme in Cultured NIH-3T3 Cells

M1-1 DNA sequence was cloned into retroviral vector LXSN and placed under the control of the small nuclear U6 RNA promoter. To construct cell lines that express M1-1 ribozyme, amphotropic packaging cell line PA317 (29) was transfected with pLXSN-U6-M1-1 DNA, and retroviral vectors that contained the genes for M1-1 were produced. When NIH-3T3 cells were infected with these vectors, cells stably expressing the ribozyme were cloned. To determine the efficacy of the ribozymes, cells were challenged with MCMV Smith strain at a multiplicity of infection (MOI) of 0.1–1, and the expression pattern of mAP and M80 was studied by Northern blot and Western analysis. The growth of MCMV in cultured cells was determined by titering methods (plaque assay).

3.5. Stable Expression of M1-1 Ribozyme in Cultured Cells

1. One day before transfection, plate PA317 cells onto 6-well plates.
2. 12–16 h after cell plating, when the cells reach 80% confluence (*see* **Note 1**), transfect the cells with 10–20 μg of plasmid LXSN-U6-M1-1 DNA using a mammalian transfection kit (GIBCO/BRL) (*see* **Note 2**).
3. 48 h after transfection, collect culture supernatants (about 3 mL). These supernatants are retroviral stock used to infect NIH-3T3 cultured cells.
4. One day before infection, plate NIH-3T3 cells onto a 6-well plate. The cells should have 80% confluence the day of infection with the culture supernatant containing retroviruses.
5. Aspirate out the media and add 1.5 mL of the collected retroviral stock (from step 3) to the NIH-3T3 cell culture, and incubate for 4–12 h (*see* **Note 3**). Shake occasionally. Replace the inoculums with fresh DMEM supplemented with 10% Nu-Serum (Collaborative Research).
6. 48–72 h after infection, incubate the cells in cultured medium that contains 500 μg/mL neomycin.
7. Select the cells in the presence of neomycin for 2 weeks.
8. Cells infected with retrovirus are subsequently split sparsely over ten cultured flasks and placed under neomycin to select for cloned ribozyme-expressing cell lines (*see* **Note 4**).
9. Aliquot and freeze the cells for long-term storage in liquid nitrogen or use for further studies.

3.6. Northern Analysis of M1-1 Expression in Cultured Cells

3.6.1. Total RNA Extraction from Cultured Cells

T-75 flasks of cells (~10^7 infected cells) are used to isolate total RNA by Trizol reagent (*see* **Note 5**).

1. Wash cells twice with PBS and add 7.5 mL of Trizol to the cells. Pipette repeatedly and incubate at room temperature for 5 min.
2. Add 1.5 mL of chloroform, shake vigorously for 15 s, and incubate for 3 min at room temperature.

3. Centrifuge the mixture at no more than 12,000*g* at 4°C for 30 min.
4. Collect the aqueous layer, mix with 3.75 mL isopropanol, and incubate at room temperature for 10 min.
5. Centrifuge the mixture at no more than 12,000*g* at 4°C for 10 min.
6. Wash the pellet with 3.75 mL of ethanol and mix by briefly vortexing.
7. Centrifuge at no more than 7,500*g* at 4°C for 5 min.
8. Resuspend the RNA pellet in RNase-free water and test the concentration by Bio-spectrophotometer (BioRad).

3.6.2. Detection of M1-1 RNA in Cultured Cells by Northern Blots (*see* **Note 6**)

1. Load 10 μg of total RNA prepared as above onto a 2% formaldehyde agarose gel containing 45 mL of formaldehyde, 50 mL of 5X Northern buffer. Separate the RNAs by running the gel at a constant voltage (*see* **Note 7**).
2. Wash the gel three times (10 min each) in H_2O and transfer the RNAs from the agarose gel onto a nitrocellulose membrane as described in **steps 3–16**.
3. Set up a large glass dish filled with 1 L 20X SSC solution.
4. Place a long glass plate on the edge of dish so that it is perpendicular to the dish.
5. Place a small glass plate on top of the previous plate.
6. Wet a half sheet of Whatman 3MM paper in 20X SSC and place it on top of the glass plate. So its edges dip into the pool of 20X SSC.
7. Put three pieces of Whatman 3MM paper (larger than the gel) on top and wet with 20X SSC.
8. Place the gel on the filter paper and roll out air bubbles with plastic pipette.
9. Overlay the gel with saran warp, roll out bubbles, and cut a window into the saran warp to expose gel.
10. Prepare a piece of nitrocellulose membrane just large enough to cover the exposed surface of the gel. Wet the membrane with RNase-free water and place it on the surface of the gel. Remove any air bubbles under the membrane. Flood the surface of the membrane with 20X SSC.
11. Cut five sheets of Whatman 3MM paper to the same size as the membrane and place on top of the membrane.
12. Place a stack of paper towels on top of the Whatman 3MM paper to a height of about 4 cm.
13. Lay a glass plate on top of the structure and add a weight (1 kg) to hold everything in place.

14. Leave the transfer system overnight.
15. Rinse the membrane with deionized water 10 min three times.
16. Bake the nitrocellulose membrane at 80°C for 1.5 h in an oven.
17. Synthesize the radio-labeled DNA probe used to detect M1GS RNAs from plasmid pFL117 by using a random primed labeling kit (Boehringer Manheim).
18. The nitrocellulose membrane is prehybridized for 4 h and hybridized for 16 h with labeled probe at 65°C in hybridization buffer (6X SSC, 0.05% sodium pyrophosphate, 2X Denhardt's solution, 0.1% SDS, and 200 μg/mL of salmon sperm).
19. At 42°C, wash the membrane with 2X SSC, 1X SSC, and 0.5X SSC (containing 0.1% SDS). Each step lasts for about 15 min.
20. Expose the membrane to a phosphorscreen and scan with a STORM840 PhosphorImager.

3.7. Analysis of M1-1-Mediated Inhibition of Viral Gene Expression and Growth in Cultured Cells

3.7.1. Northern Blot Analysis

1. Isolate total RNA from T-25 flasks of cells (10^6 infected cells) by Trizol reagent as described in **Section 3.6.1**.
2. Determine the level of viral mAP and M80 mRNA inhibition by Northern blot as described in **Section 3.6.2**. The RNA fractions were separated in 1% agarose gels that contained formaldehyde, transferred to a nitrocellulose membrane, hybridized with the [^{32}P]-radio-labeled DNA probes that contained the MCMV DNA sequence, and analyzed with a STORM840 PhosphorImager. The DNA probes used to detect mouse M80 and mAP mRNA were synthesized from plasmids pM80.

3.7.2. Western Blot Analysis of mAP and M80 Proteins

1. Aspirate out the culture media and rinse the cells twice with 5 mL of PBS.
2. Collect the cells from flask and spin down with 3,000*g* at 4°C for 5 min.
3. Resuspend cell pellets into 50–100 μL of cold PBS, and then add the same volume of 2X disruption buffer and mix by vortexing for 1 min.
4. Sonicate the mixture three times on ice. Each lasts for 20–30 s.
5. Boil the sample for 1 min before loading on SDS-polyacrylamide gel electrophoresis (PAGE) gel. The level of viral mAP protein is determined by Western blot.
6. Load 50 μg of proteins onto 9% SDS-polyacrylamide gels cross-linked with *N*,*N*′-methylene bis-acrylamide with a

stacking layer of 4.5% acrylamide/bis-acrylamide. Separate the proteins by running the gels at a constant power setting.

7. Transfer the proteins from the polyacrylamide gel onto a nitrocellulose membrane, using an electrophoretic transfer apparatus with a constant current of 150 mA for 2 h.
8. Incubate the nitrocellulose membrane with TNT buffer plus 2.5% nonfat dry milk (NFDM) for 1 h in an orbital shaker (pre-blocking step).
9. After blocking, incubate for 1 h with primary antibody at a dilution of 1:500 in TNT supplemented with 1.25% of NFDM.
10. Wash the membrane with TNT buffer three times. Each wash is for 5 min.
11. Incubate the membrane with HRP-linked anti-species antibody (secondary antibody) diluted at 1:1,000 dilutions for 1 h with shaking.
12. Wash the membranes with TNT three times. Each wash is for 5 min.
13. Incubate with ECL substrates for 1 min at room temperature.
14. Expose membrane and develop film.

3.7.3. Growth of MCMV Analysis by Plaque Assay

The level of viral growth inhibition mediated by M1-1 ribozyme was determined by testing the viral titers in tissue culture cells.

1. Inoculate 5×10^5 ribozyme-expressing cells with MCMV Smith strain at a MOI of 1.
2. Incubate the cells with DMEM at 37°C with 5% CO_2 for 90 min.
3. Wash the cells with PBS, then add 0.5 mL fresh DMEM supplemented with 10% Nu-Serum for different periods of time.
4. Incubate for an interval of 1–5 days.
5. Harvest cells at 1-day intervals throughout 5 days after infection. Prepare viral stock by adding 0.5 mL of 10% Nu-Serum/10% skim milk solution. Scrape the cells and sonicate them three times.
6. Prepare tenfold serial dilution of the viral stock in 1 mL of DMEM for each dilution.
7. Infect NIH-3T3 cells in 6-well tissue culture plates (Corning Inc., Corning, NY) with 1 mL of viral dilutions.
8. Incubate at 37°C with 5% CO_2 for 1 h.
9. Wash the cells with DMEM medium.

10. Overlay the cells with fresh 1% agarose and DMEM containing 10% Nu-Serum in a 1:1 ratio.
11. Culture the cells for 4–6 days.
12. Count the number of viral plaques under an inverted microscope.
13. Plaque-forming unit (PFU) is determined by the highest viral dilution that yields plaque (*see* **Note 8**).

3.8. M1-1 Ribozyme Activity in SCID Mice

To determine the effect of M1GS ribozymes on the replication and infection of MCMV in vivo, we applied hydrodynamic transfection of plasmid LXSN-M1-1 DNAs to SCID mice that were intraperitoneally infected 24 h earlier with MCMV. To further allow sustained expression of M1-1 in these animals, similar hydrodynamic transfection procedure was repeated every 72 h. Transfection efficiency was assessed by detecting the expression of M1GS RNAs in the tissues. To further study the effect of M1-1 expression on MCMV infection in these immunodeficient mice, (1) the survival rate of the animals receiving the M1-1-expressing plasmid was measured; (2) viral replication in various organs of the animals was studied by titration method before the onset of mortality; and (3) viral gene expression in the tissues was examined by Northern blot methods.

3.9. Transfection of M1-1 to MCMV-Infected Animals

1. CB17 SCID mice were obtained from Jackson Laboratory (Bay Harbor, Maine); 4–6 weeks old male mice were selected for further study.
2. Infect the mice intraperitoneally with 1×10^4 PFU of MCMV (Smith strain).
3. 24 h after infection, restrain the animal in a rodent restraint and expose the tail completely.
4. Anesthetize the animal by short-term exposure to isoflurane.
5. Clean the tail with alcohol.
6. Warm the tail with heat source, such as gauze wet with warm autoclaved water, to make the veins visible.
7. Rapidly inject 1–2 mL PBS (*see* **Note 9**) containing 20 μg of pLXSN-U6-M1-1 DNA into one tail vein within 5–10 s (*see* **Note 10**).
8. Remove the needle and hold the injection site with gauze.
9. Return the animal back to the cage.
10. Repeat the hydrodynamic transfection procedures every 72 h (*see* **Note 11**).
11. The transfection efficiency was evaluated by detecting the expression of M1GS RNAs in the tissues using Northern analyses (*see* **Note 12**).

3.10. Investigation of M1-1-Mediated Inhibition of Viral Infection and Pathogenesis In Vivo

3.10.1. Animals' Survival Rates Analysis

1. Infect CB17 SCID mice with 1×10^4 PFU of MCMV Smith strain and transfect the M1-1 ribozyme to animals as described in **Section 3.9**.
2. Check the animals twice per day and keep monitoring the mortality of the infected animals for at least 41 days after infection.
3. Determine the survival rates.

3.10.2. Determination of Viral Growth in Various Organs

1. Infect CB17 SCID mice with 1×10^4 PFU of MCMV Smith strain and transfect the M1-1 ribozyme to animals as described in **Section 3.9**.
2. Kill the infected animals (at least three animals per group) at 1, 3, 7, 10, 14, and 21 days after inoculation.
3. Harvest the salivary glands, lungs, spleens, and livers.
4. Sonicate the tissues as a 10% (weight/volume) suspension in a 1:1 mixture of DMEM and 10% skim milk.
5. Use titration methods (plaque assay) to determine the viral titers of the samples as described in **Section 3.7.3**.

3.10.3. Determination of Viral Gene Expression in Various Organs

1. Infect the animals with 1×10^4 PFU of MCMV and transfect the M1-1 ribozyme to animals as described in **Section 3.9**.
2. Harvest the salivary glands, lungs, spleens, and livers.
3. RNA extraction is described in **Section 3.6.1** and the level of viral mAP and M80 mRNA inhibition is determined by Northern blot as described in **Section 3.6.2**.

4. Notes

1. In order to achieve optimal transfection efficiency, the cells should be evenly distributed throughout each well.
2. Alternative methods: The transfection works well when using the calcium phosphate transfection kit (Stratagene, San Diego, CA).
3. To increase the efficiency of retroviral infection to target cells, add polybrene to a final concentration of 8 μg/mL. The efficiency of infection can also be improved by carrying out multiple rounds of infection and using a high multiplicity of infection (MOI).
4. In our experiments, ten clones were selected. Most of the selected lines were found to express a high level of ribozymes.

5. According to the manual provided by Invitrogen, the amount of Trizol reagent added is based on the area of the culture dish (1 mL per 10 cm^2). An insufficient amount of Trizol reagent may result in contamination of the isolated RNA with DNA.
6. The M1GS RNAs expressed by the U6 promoter is primarily localized in the nucleus. This can be tested with Northern blot using the RNAs isolated from the nuclear and cytoplasmic fractions separately.
7. Since mRNA is extremely sensitive to degradation, this step should be completely RNase-free. DEPC-treated H_2O is recommended for the entire gel-running process.
8. The values obtained from plaque assay were the average of triplicate experiments.
9. The amount of injection volume is based on the animal blood volume and cardiac capacity. We showed that for 4–6 week old CB17 SCID mice, a 1–2 mL injected volume (20 μg plasmid) is sufficient to reach the hydrostatic pressure required for a high level of gene expression.
10. We used syringe with hypodermic needles of 24–27 ga size. If the needle is placed properly, the injection process will be very smooth. Otherwise, a "tuber" will appear in the tail and injection has to be repeated from another part of the vein.
11. It is recommended to check if there are significant differences in terms of growth and viability of the constructed ribozyme-expressing cell lines and the parental cells for up to 3 months. Our data suggest that M1-1 does not significantly interfere with cellular gene expression, and there is no significant cytotoxicity both in vitro and in vivo.
12. Besides using Northern blot method, the transfection efficiency can be evaluated by examining the GFP staining of the transfected cells in the tissues using fluorescence microscopy. The GFP-expressing cassette sequence was included in the LXSN vector sequence.

Acknowledgments

We are grateful to Dr. Phong Trang for helpful discussion. Yong Bai was partially supported by a predoctoral Block grant from University of California at Berkeley. This research has been supported by NIH (AI041927 and DE014842).

References

1. Frank, D.N. and Pace, N.R. (1998) Ribonuclease P: unity and diversity in a tRNA processing ribozyme. *Annu Rev Biochem*, **67**, 153–180.
2. Altman, S. and Kirsebom, L.A. (1999) Ribonuclease P. In: *The RNA World*, 2nd ed. Cold Spring Harbor Laboratory Press, Cold Spring Harbor, NY.
3. Forster, A.C. and Altman, S. (1990) External guide sequences for an RNA enzyme. *Science*, **249**, 783–786.
4. Yuan, Y., Hwang, E.S., and Altman, S. (1992) Targeted cleavage of mRNA by human RNase P. *Proc Natl Acad Sci U S A*, **89**, 8006–8010.
5. Gopalan, V., Vioque, A., and Altman, S. (2002) RNase P: variations and uses. *J Biol Chem*, **277**, 6759–6762.
6. Raj, S.M.L. and Liu, F. (2003) Engineering of RNase P ribozyme for gene-targeting applications. *Gene*, **313**, 59–69.
7. Guerriertakada, C., Li, Y., and Altman, S. (1995) Artificial regulation of gene-expression in *Escherichia-coli* by Rnase-P. *Proc Natl Acad Sci U S A*, **92**, 11115–11119.
8. Liu, F.Y. and Altman, S. (1995) Inhibition of viral gene-expression by the catalytic RNA subunit of Rnase-P from *Escherichia-coli*. *Genes Dev*, **9**, 471–480.
9. Scherer, L.J. and Rossi, J.J. (2003) Approaches for the sequence-specific knockdown of mRNA. *Nat Biotechnol*, **21**, 1457–1465.
10. Stein, C.A. and Cheng, Y.C. (1993) Antisense oligonucleotides as therapeutic agents – is the bullet really magical. *Science*, **261**, 1004–1012.
11. Plehn-Dujowich, D. and Altman, S. (1998) Effective inhibition of influenza virus production in cultured cells by external guide sequences and ribonuclease P. *Proc Natl Acad Sci U S A*, **95**, 7327–7332.
12. Cobaleda, C. and Sanchez-Garcia, I. (2000) In vivo inhibition by a site-specific catalytic RNA subunit of RNase P designed against the BCR-ABL oncogenic products: a novel approach for cancer treatment. *Blood*, **95**, 731–737.
13. Kilani, A.F., Trang, P., Jo, S., Hsu, A., Kim, J., Nepomuceno, E., Liou, K., and Liu, F.Y. (2000) RNase P ribozymes selected in vitro to cleave a viral mRNA effectively inhibit its expression in cell culture. *J Biol Chem*, **275**, 10611–10622.
14. Bai, Y., Trang, P., Li, H.J., Kim, K.H., Zhou, T.H., and Liu, F.Y. (2008) Effective inhibition in animals of viral pathogenesis by a ribozyme derived from RNase P catalytic RNA. *Proc Natl Acad Sci USA*, **105**, 10919–10924.
15. Mocarski, E.S., Shenk, T., and Pass, R.F. (2007) *Fields Virology*. Lippincott, Philadelphia, 2701–2772.
16. Collins, T.M., Quirk, M.R., and Jordan, M.C. (1994) Biphasic viremia and viral gene-expression in leukocytes during acute cytomegalovirus-infection of mice. *J Virol*, **68**, 6305–6311.
17. Hudson, J.B. (1979) Murine cytomegalovirus as a model for the study of viral pathogenesis and persistent infections. *Arch Virol*, **62**, 1–29.
18. Katzenstein, D.A., Yu, G.S.M., and Jordan, M.C. (1983) Lethal infection with murine cytomegalovirus after early viral replication in the spleen. *J Infect Dis*, **148**, 406–411.
19. Abenes, G., Chan, K.R., Lee, M., Haghjoo, E., Zhu, J.M., Zhou, T.H., Zhan, X.Y., and Liu, F.Y. (2004) Murine cytomegalovirus with a transposon insertional mutation at open reading frame m155 is deficient in growth and virulence in mice. *J Virol*, **78**, 6891–6899.
20. Pollock, J.L. and Virgin, H.W. (1995) Latency, without persistence, of murine cytomegalovirus in the spleen and kidney. *J Virol*, **69**, 1762–1768.
21. Liu, F.Y. and Altman, S. (1996) Requirements for cleavage by a modified RNase P of a small model substrate. *Nucleic Acids Res*, **24**, 2690–2696.
22. Rawlinson, W.D., Farrell, H.E., and Barrell, B.G. (1996) Analysis of the complete DNA sequence of murine cytomegalovirus. *J Virol*, **70**, 8833–8849.
23. Loutsch, J.M., Galvin, N.J., Bryant, M.L., and Holwerda, B.C. (1994) Cloning and sequence-analysis of murine cytomegalovirus protease and capsid assembly protein genes. *Biochem Biophys Res Comm*, **203**, 472–478.
24. Dunn, W., Chou, C., Li, H., Hai, R., Patterson, D., Stolc, V., Zhu, H., and Liu, F.Y. (2003) Functional profiling of a human cytomegalovirus genome. *Proc Natl Acad Sci U S A*, **100**, 14223–14228.
25. Welch, A.R., Woods, A.S., McNally, L.M., Cotter, R.J., and Gibson, W. (1991) A herpesvirus maturational proteinase, assemblin – identification of its gene, putative active-site domain, and cleavage site. *Proc Natl Acad Sci U S A*, **88**, 10792–10796.

26. Liu, F., Song, Y.K., and Liu, D. (1999) Hydrodynamics-based transfection in animals by systemic administration of plasmid DNA. *Gene Ther*, **6**, 1258–1266.
27. Song, E.W., Lee, S.K., Wang, J., Ince, N., Ouyang, N., Min, J., Chen, J.S., Shankar, P. et al. (2003) RNA interference targeting Fas protects mice from fulminant hepatitis. *Nat Med*, **9**, 347–351.
28. Zhang, G.F., Budker, V., and Wolff, J.A. (1999) High levels of foreign gene expression in hepatocytes after tail vein injections of naked plasmid DNA. *Hum Gene Ther*, **10**, 1735–1737.
29. Miller, A.D. and Rosman, G.J. (1989) Improved retroviral vectors for gene-transfer and expression. *Biotechniques*, 7, 980–990.

Chapter 23

Using Live Cells to Generate Aptamers for Cancer Study

Ling Meng, Kwame Sefah, Dalia Lopez Colon, Hui Chen, Meghan O'Donoghue, Xiangling Xiong, and Weihong Tan

Abstract

Aptamers are ssDNA, RNA, or modified nucleic acids, usually consisting of short strands of oligonucleotides. Aptamers have the ability to bind specifically to a range of targets, from small organic molecules to proteins. However, by using cell-based aptamer selection, we have developed a strategy to identify the molecular signatures on the surface of targeted cells by exploiting the differences at the molecular level between any two given cell types. By applying this method, we have generated a panel of aptamers for the specific recognition of leukemia cells, and we report the results in this study. The selected aptamers were found to bind to target cells with an equilibrium dissociation constant (K_d) in the nanomolar-to-picomolar range. Overall, the cell-based selection process is simple, fast, straightforward, and reproducible. Most importantly, since this strategy can be implemented without prior knowledge of a target's specific molecular signature, cell-based aptamer selection holds great promise for the development of specific molecular probes for cancer diagnosis and cancer biomarker discovery.

Key words: Cell-based SELEX, DNA aptamers, cell imaging, leukemia.

1. Introduction

Understanding human diseases at the molecular level has been extremely challenging because of the lack of effective probes to identify and recognize distinct molecular features of diseases. Cancers, as well as many other diseases, originate from the mutations of human genes. Such genetic alterations result in different behaviors of the diseased cells, as well as changes in cells at the morphological and molecular levels. Traditionally, cancer diagnosis has mostly been based on the morphology of tumor tissues or cells. However, these morphological features are difficult to

M. Sioud (ed.), *RNA Therapeutics*, Methods in Molecular Biology 629,
DOI 10.1007/978-1-60761-657-3_23, © Springer Science+Business Media, LLC 2010

identify by conventional methods, making the process ineffective for either early cancer diagnosis or the evaluation of complex molecular alterations that lead to cancer progression (1, 2). Therefore, the best means of classifying cancers would utilize molecular characteristics, especially at the proteomic level, because of the direct connection between genetic features and protein expression. At the molecular level, differences are present between any two given types of cells, such as normal vs. tumor cells or tumor cell type 1 vs. type 2. The ability to easily distinguish these differences would significantly advance our understanding of the biological processes and mechanisms of diseases. Especially, any method that would enable us to make such distinctions, e.g., comparing normal to cancer cells and identifying the molecular features of cancer cells, would be highly useful from a clinical perspective for disease diagnosis, prevention, and therapy. Here, we report a systematic approach that can conveniently circumvent the limitations of current technologies to produce a new class of molecular probes termed aptamers. Aptamers are single-stranded DNA, RNA, or modified nucleic acids. They have the ability to bind specifically to their targets, which range from small organic molecules to proteins (3, 4). Since nucleic acid aptamers are known to be excellent binders with high-binding affinity and selectivity, they can recognize molecular differences expressed on the surface membranes of two different types of cells. Target recognition is based on tertiary structures formed by the single-stranded oligonucleotides (5). These probes can therefore be used to specifically detect target cancer cells based on molecular characteristics in the presence of other cells, leading to effective disease studies and early diagnosis.

Aptamers have had many important applications in biomedicine, biotechnology, and bioanalysis (6–9). Most aptamers reported so far have been selected by using simple targets, such as a purified protein. Recently, aptamer selection against complex targets, such as red blood cell membranes and endothelial cells, was also demonstrated (10–13). Compared with molecular probes currently available for biomarker recognition, aptamers are emerging as candidates with ample potential because of their high specificity, low molecular weight, easy and reproducible production, versatility of application, and easy discovery and manipulation (14). Currently, however, the lack of aptamers for systems of medical relevance has limited their application in both clinical and laboratory settings. Thus, this study primarily aims to report the development of a group of effective aptamers for leukemia studies.

Aptamers are obtained through an in vitro selection process known as SELEX (Systematic Evolution of Ligands by EXponential enrichment) (15, 16), in which aptamers are selected from a library of random sequences of synthetic DNA or RNA by

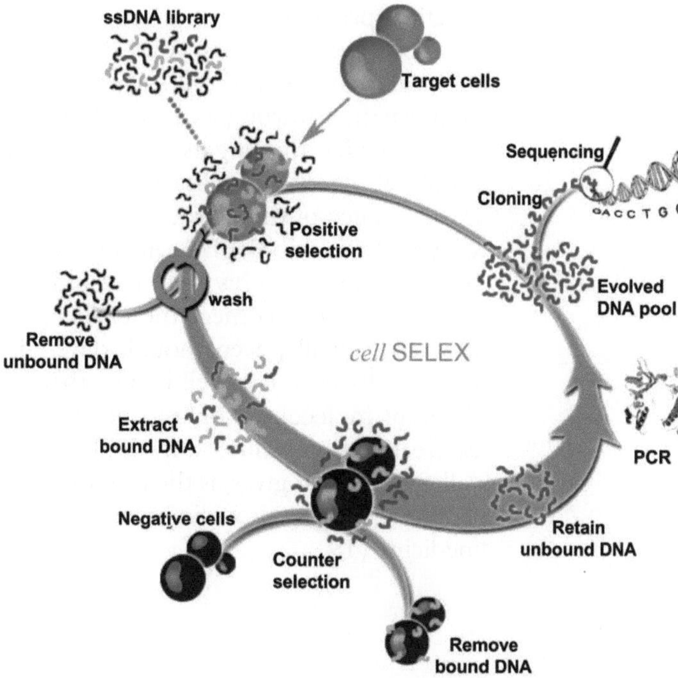

Fig. 23.1. Schematic representation of the cell-based aptamer selection. Briefly, the ssDNA pool was incubated with CCRF-CEM cells (target cells). After washing, the bound DNAs were eluted by heating to 95°C. The eluted DNAs were then incubated with Ramos cells (negative cells) for counterselection. After centrifugation, the supernatant was collected, and the selected DNA was amplified by PCR. The PCR products were separated into ssDNA for next round of selection or cloned and sequenced for aptamer identification in the last-round selection.

repetitive binding of the oligonucleotides to target molecules. This process is illustrated in **Fig. 23.1**. A key part of this method depends on counterselection. In counter-cell-SELEX, unbound probes are collected, while bound probes on negative cells are retained. In this way, nucleic acid sequences that interact with commonly expressed proteins are successfully removed so that only target-specific probes are selectively enriched. If negative selection is desired, the control cell type is incubated with the supernatant. By following these steps, the membrane protein targets of the selected aptamers represent the differences between the two cell lines used in this study (CEM, CCL-119 T cell line, human acute lymphoblastic leukemia and Ramos, CRL-1596, B-cell line, human Burkitt's lymphoma). Thus, the method allows us to discover the molecular signatures with ease, generating, at the same time, probes that can recognize such unique features with very high affinity and specificity. This gives clinicians real potential for early detection, diagnosis, and therapeutic intervention. More importantly, the use of a panel of probes has a clear advantage over the single-biomarker-based assays in clinical practice, providing much more information for accurate disease diagnosis and prognosis. Also, there is no need to purify specific molecular targets, and cell-surface proteins remain in their native conformations throughout the selection process. Since the probes can recognize the targets in their native state, a true molecular profile of the disease cells can be created, which adds to the clinical relevance of these probes. All of these factors make aptamers selected from cell-SELEX valuable tools capable of differentiating between the signatures of normal and disease cells (17–22), hence ideal candidates to meet the challenges of biomarker discovery (23, 24). Finally, even though different patients have the same types of diseases, it is well known that each patient can have very different molecular profiling and therefore can respond to medical treatment very differently. In this context, aptamer selection by cell-SELEX also gives us the tool we need to understand such differences, thereby providing new opportunities in "personalized" medicine (19).

2. Materials

2.1. Cell Culture

1. CCRF-CEM (CCL-119, T-cell line, human ALL) and Ramos (CRL-1596, B-cell line, human Burkitt's lymphoma).
2. RPMI-1640 Medium supplemented with 10% heat-inactivated fetal bovine serum and 100 U/mL penicillin–streptomycin.

2.2. Cell-SELEX Process

1. Wash buffer containing 4.5 g/L glucose and 5 mM $MgCl_2$ in Dulbecco's phosphate-buffered saline with $MgCl_2$ and $CaCl_2$ (Sigma, St. Louis, MO). Store at 4°C.
2. Binding buffer: Wash buffer supplemented with 0.1 mg/mL yeast tRNA and 1 mg/mL bovine serum albumin. Store at 4°C.
3. HPLC-purified library containing a central randomized sequence of 52 nucleotides flanked by two 18-nt primer hybridization sites (5′-ATA CCA GCT TAT TCA ATT- 52-nt -AGA TAG TAA GTG CAA TCT-3′). Store at –20°C (*see* **Note 1**).
4. Fluorescein isothiocyanate (FITC)-labeled 5′ forward primer (5′-FITC-ATA CCA GCT TAT TCA ATT-3′) or a tetramethyl-6-carboxyrhodamine (TAMRA)-labeled 5′ primer (5-TAMRA-ATA CCA GCT TAT TCA ATT-3′). Store at –20°C (*see* **Note 1**).

5. Triple-biotinylated (trB) 3′ reverse primer (5′-trB-AGA TTG CAC TTA CTA TCT 3′). Store at −20°C.
6. Sephadex G25, NAP 5 desalting column.
7. Streptavidin-coated Sepharose beads.
8. Commercially available *Taq* polymerase, dNTPs, and 10× PCR buffer.
9. Electrophoresis grade agarose.
10. Ethidium bromide, handle with gloves, as this is a potent carcinogen.
11. Gel running buffer: 10× Tris-borate-EDTA (TBE) buffer, pH 8.3.
12. 10 bp DNA ladder.
13. 10× Loading buffer: 50% glycerol (w/v), 100 mM Na_2EDTA, pH 8.0, 1% SDS, 0.1% bromophenol blue, 0.1% xylene cyanol. The glycerol is prepared in water.
14. 1× Phosphate-buffered saline buffer.
15. Fetal bovine serum (FBS).
16. 5 mL plastic syringe.
17. Empty DNA synthesis column (IDT Technologies).
18. 200 mM NaOH.
19. TA Cloning Kit.

3. Methods

3.1. Cell Culture and Preparation

1. CCRF-CEM and Ramos cell lines are cultured in RPMI medium 1640 supplemented with 10% FBS and 100 U/mL penicillin–streptomycin.
2. Cell concentration is determined using a hemocytometer. Also, cell viability is verified using Trypan blue, with a recommended 95% of the cells alive.
3. Cells are washed before and after incubation with wash buffer.

3.2. Designing the Primers and the Library

1. To design the primers, the Integrated DNA Technologies website can be accessed: http://www.idtdna.com/analyzer/Applications/OligoAnalyzer/.
2. The primers are between 18 and 20 bases long and should not form self-dimers or heterodimers with a GC content of 50% or more.

3. The primers must anneal into the library at a temperature between 55 and 60°C. To obtain such an annealing temperature, the primer GC content must be 50% or more. Since only one annealing temperature can be selected on the PCR machine, both primers must have the same annealing temperature (±1°C).
4. The library is designed using both primers with one as the primary primer. The sequence of one of the primers, followed by 30–45 N, and then the complementary sequence of the other primer are recorded. The library should have a melting temperature of 10°C or less (the lower the better).

3.3. Cell-SELEX Selection and Synthesis of Specific Aptamers for Cancer Cells

1. Before selection, 20 nmol ssDNA selection pool is dissolved in 500 μL of binding buffer and denatured by heating at 95°C for 5 min and then cooled immediately on ice for 10 min before binding.
2. For the first round of selection, the 500 μL ssDNA selection pool is incubated with 10×10^6 target CEM cells (*see* **Note 2**) at desired temperature for selection (e.g., on ice, room temperature, or 37°C) for 45 min with shaking on a platform at ~200 rpm.
3. Subsequently, the cells are washed for 1 min with 0.5 mL wash buffer. The cells are washed 3 times and then collected by centrifugation at 990 rpm for 4 min (*see* **Note 3**).
4. After washing, cells are reconstituted in 400 μL binding buffer. Bound DNA is eluted off the cells by heating the mixture at 95°C for 5 min. The mixture is then spun at either 4,000×*g* for 5 min or 14,000×*g* for 2 min.
5. The eluted ssDNA in the supernatant is collected and amplified by PCR for 10 cycles using the FITC- and biotin-labeled primers. Typical PCR conditions are 30 s at 94°C for denaturing, 30 s at 60°C for annealing, and 30 s for extension at 72°C, followed by 5 min extension at 72°C.
6. To perform PCR, the annealing temperature and number of cycles for the ssDNA selection pool are first optimized. A 10 μM stock solution for the primers (both primers in one solution) and a 1 nM stock solution of the ssDNA are prepared fresh each time.
7. To each PCR reaction, the following is added:

1× PCR buffer	= 5 μL
dNTPs	= 4 μL
Primer (10 μM)	= 2.5 μL
ssDNA pool (1 nM)	= 2.5 μL
Taq polymerase	= 0.15 μL

Complete with DNase-free H_2O to 50 μL. One negative control is required for every condition that is run. The negative control contains all the compounds, except the ssDNA pool.

8. The number of cycles and annealing temperature are optimized separately by choosing approximately six different conditions of each PCR. The PCR products are run on a 3% agarose gel in order to check the PCR conditions. The conditions that produce the brightest band with the least nonspecific amplification for the initial ssDNA library should be chosen.
9. Using the best conditions as determined by the gel, the entire enriched pool collected after the initial selection step is PCR-amplified in a reaction containing 80 μL dNTPs, 50 μL primer pool, 500 μL of the sample, 3.0 μL *Taq* polymerase, and DNase water to 1,000 μL.
10. This amplification will result in a dsDNA pool of aptamers having one strand labeled with biotin and the other with FITC. In order to separate the two strands, streptavidin-coated beads are used.
11. A 5 mL syringe is inserted into the top of the empty DNA synthesis column on the side without the filter. A 200 μL volume of streptavidin-coated bead solution is added into the syringe. The beads should coat the top of the filter inside the column.
12. The streptavidin bead solution is removed by gently pressing the plunger and then the beads are washed by adding 2.5 mL 1× PBS to the syringe, using the plunger to gently push through the liquid.
13. The PCR product is loaded onto the syringe, and the flow-through is collected. The dsDNA-enriched pool (one strand with biotin and one strand with FITC) will stay bound to the beads. The PCR sample is passed through the beads three times to make sure that the entire sample is bound.
14. The beads are then washed with 2–2.5 mL PBS to remove any forward FITC-labeled primer trapped on the beads.
15. Since we want the enriched pool of ssDNA labeled with FITC, we need to unzip the biotin-labeled strand from the dsDNA. This is done by adding 500 μL 200 mM NaOH to melt the DNA. Without putting much pressure on the plunger, the solution is allowed to drip slowly, and the enriched FITC-labeled ssDNA is then collected.
16. The ssDNA pool now needs to be desalted to remove the NaOH. A NAP5 desalting column is washed with 15 mL dH_2O. When the column is empty, the ssDNA pool is

added, followed by 1 mL dH_2O. The sample is in the flow-through. The column is washed with 20 mL dH_2O and reused for the next round.

17. Absorbance of the sample is measured at 260 nm using a UV-Vis Spectrophotometer. Subsequently, the number of moles of ssDNA in the sample is calculated.
18. The sample is dried using a speed-vac on low heat, and the ssDNA is resuspended in the binding buffer to a concentration of 250 nM.
19. For subsequent rounds of selection, the enriched sense FITC-labeled ssDNA is used as the pool for binding in the next round. Stringency of selection is gradually increased with each round. For example, the number of cells for selection is decreased (from 10×10^6 to 2×10^6 over several rounds), while the number of washes and amount of washing buffer used for each wash are increased (from 3 times with 0.5 mL washing buffer to 5 times with 5 mL washing buffer). Then, a counterselection step is performed.
20. Thus, for the second round, 8×10^6 target cells are added to 5–20 nmol of the enriched pool created in round one and incubated at the chosen temperature for 45 min shaking at 220 rpm. The cells are then washed four times with 2 min incubation time in 1 mL washing buffer and spun at 990 rpm for 4 min.
21. After washing, the cells are reconstituted in 400 μL binding buffer. Bound DNA is again eluted off the cells by heating the mixture at 95°C for 5 min. The mixture is spun at either 4,000×*g* for 5 min or 14,000×*g* for 2 min.
22. The supernatant is then incubated with 5 times the number of negative cells, here are Ramos cells (40×10^6), for counterselection for 1 h. Subsequent to centrifugation at 990 rpm for 4 min, the supernatant is collected. The supernatant contains the ssDNA molecules that did not bind to the negative cells.
23. PCR is performed on a portion of the enriched pool (supernatant) as described above, this time only optimizing the number of PCR cycles.
24. The sense FITC-labeled ssDNA is separated and collected using the streptavidin beads as described above.
25. Again, the concentration of DNA in the sample is determined by measuring the absorbance at 260 nm. The sample is dried with the speed-vac and reconstituted in 250 nM (*see* **Note 4**).

26. To 500 μL binding buffer, 5–20 nmol (denatured at 95°C for 5 min, followed by immediate cooling on ice) of the enriched ssDNA is added. Then, this solution is added to the cells to begin the next round of selection. After the second round of selection, 10% FBS is added to the incubation solution (50 μL in 500 μL) to reduce nonspecific adsorption of ssDNA to the cell surface.
27. After 10–20 rounds of selection, or when the observed shift of the selected pool bound to cells detected with flow cytometry is large enough and the flow data are reproducible (*see* **Section 3.4**), then the selected ssDNA pool is PCR-amplified using unmodified primers and cloned into the TA cloning kit following the manufacturer's instructions.
28. The positive clones are then sequenced and compared to determine if there is either sequence homology or conserved regions between the clones.
29. At least one good aptamer will have been produced by this series of steps, and it should be capable of specifically recognizing and binding to the target cells (*see* **Note 5**).

3.4. Flow Cytometric Analysis

1. To monitor the enrichment of aptamer candidates after selection, the FITC-labeled ssDNA pool is incubated with 2×10^5 CCRF-CEM cells or Ramos cells in 200 μL of binding buffer containing 20% FBS on ice for 30–50 min.
2. After incubation, 0.7 mL binding buffer is added to each sample. The cells are gently vortexed and then spun down at 990 rpm for 3 min. Following two washes, the cells are suspended in 0.4 mL of binding buffer and analyzed by flow cytometry.
3. The fluorescence is determined by counting 30,000 events. The FITC-labeled unselected ssDNA library is used as a negative control. An example of the results produced is shown in **Fig. 23.2**.

3.5. Analysis of Aptamer-Binding Affinity

1. Aptamers are incubated with CCRF-CEM cells (5×10^5) on ice for 30–50 min in the dark with varying concentrations (0.4, 1, 4, 10, 20, 50, and 200 nM) of FITC-labeled aptamer in a 500 μL volume of binding buffer containing 20% FBS.
2. Cells are then washed twice with 0.7 mL of the binding buffer, suspended in 0.4 mL of binding buffer, and subjected to flow cytometric analysis within 30 min.
3. The FITC-labeled unselected ssDNA library is used as a negative control to determine nonspecific binding.

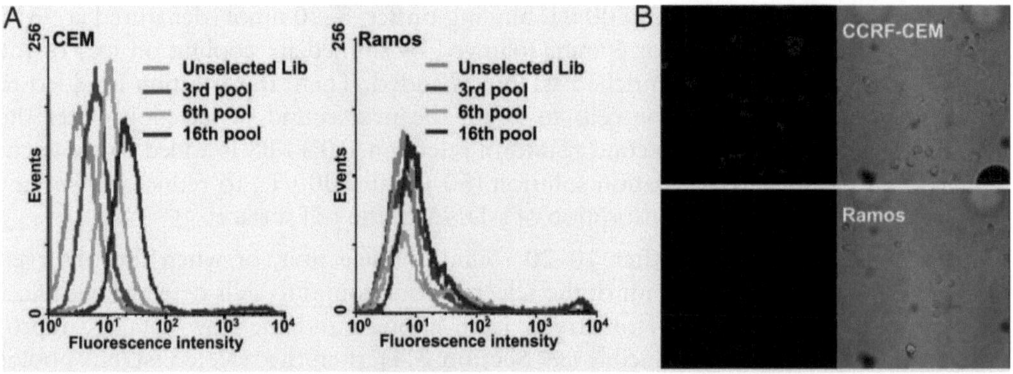

Fig. 23.2. Binding assay of selected pool with CCRF-CEM and Ramos cells. (**a**) Flow cytometry assay to monitor the binding of selected pool with CCRF-CEM cells (target cells) and Ramos cells (negative cells). For CEM cells, there was an increase in binding capacity of the pool as the selection was progressing, whereas there was little change for the control Ramos cells. (**b**) Confocal imaging of cells stained by the 20th-round selected pool labeled with tetramethylrhodamine dye molecules. (*Upper left*) Fluorescence image of CCRF-CEM cells. (*Upper right*) Optical image of CCRF-CEM cells. (*Lower left*) Fluorescence image of Ramos cells. (*Lower right*) Optical image of Ramos cells.

4. All of the experiments for binding assay are repeated two to four times. The mean fluorescence intensity of target cells labeled by aptamers is used to calculate for specific binding by subtracting the mean fluorescence intensity of nonspecific binding from unselected library DNAs (18–22). The equilibrium dissociation constants (K_d) of the aptamer–cell interaction are obtained by fitting the dependence of fluorescence intensity of specific binding on the concentration of the aptamers to the equation $Y = B \max X/(K_d + X)$, using SigmaPlot (Jandel, San Rafael, CA). An example of the results produced is shown in **Fig. 23.3**.

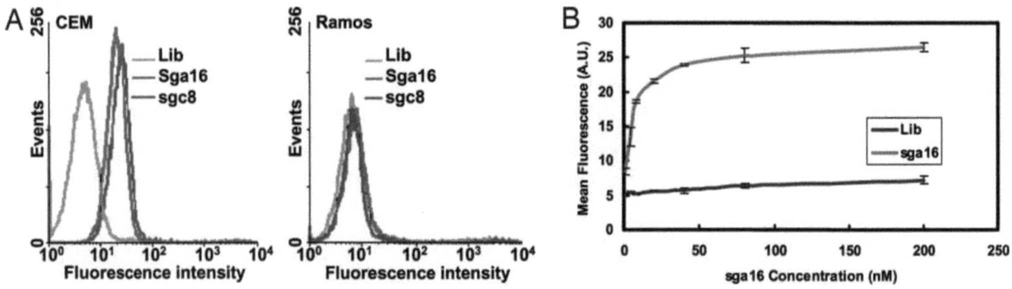

Fig. 23.3. Characterization of selected aptamers. (**a**) Flow cytometry assay for the binding of the FITC-labeled sequences sga16 and sgc8 with CCRF-CEM cells (target cells) and Ramos cells (negative cells). The concentration of the aptamers in the binding buffer was 250 nM. (**b**) Flow cytometry to determine the binding affinity of the FITC-labeled aptamer sequence sga16 to CCRF-CEM cells. The nonspecific binding was measured by using FITC-labeled unselected library DNA.

3.6. Confocal Imaging of Cell Bound with Aptamer

1. The selected ssDNA pools or aptamers are labeled with TAMRA.
2. Cells are incubated with 50 pmol of TAMRA-labeled ssDNA/aptamer in 100 μL of binding buffer containing 20% FBS on ice for 30–50 min. Other treatment steps are the same as described in **Section 3.3**.
3. A volume of 20 μL cell suspension bound with TAMRA-labeled ssDNA/aptamer is dropped on a thin glass slide.
4. The glass slide is placed above a ×60 objective on the confocal microscope and then covered with a cover slide. Imaging of the cells is performed on an Olympus FV500-IX81 confocal microscope. A 5-mW, 543-nm He–Ne laser is the excitation source for TAMRA throughout the experiments. The objective used for imaging was a PLAPO60XO3PH ×60 oil-immersion objective with a numerical aperture of 1.40 (Olympus). An example of the results produced is shown in **Fig. 23.2**.

4. Notes

1. To minimize the possibility of nonspecific amplification of random library sequences in PCR, the primers and library sequences need to be carefully optimized. In particular, the ssDNA pool cannot be too stable as it must easily denature at 95°C. Furthermore, the 5′ end of the pool DNA should not be G, as it can quench the fluorescence of the fluorophore. The primers should be between 18 and 20 nt long and designed such that they do not form self-dimers or heterodimers. Furthermore, the primers must anneal to the library at a temperature between 55 and 60°C. To obtain such an annealing temperature, the primer must have a GC content of 50% or more. Since PCR machines have one annealing temperature, both primers must have the same annealing temperature (±1°C). These design parameters can be easily optimized using free oligonucleotide prediction software such as Oligoanalyzer, Integrated DNA Technologies.
2. The cells used for selection should be healthy, with >95% cells alive. Dead cells nonspecifically allow ssDNA from the pool into their membranes, thereby affecting the efficiency of the selection. A full 8-h day for one person will be required to complete each round of selection.

3. For this selection, a suspension cell line is used as the target cell line. However, if an adherent cell line is desired, then the ssDNA pool can be incubated with a monolayer of the target cells in a T25 flask. After washing, the adhesive cells are gently scraped onto ice with a rubber policeman into a vial, and the eluted DNA is collected. Alternatively, a nonenzymatic cell disassociation solution can be used to disassociate a monolayer of cells. Counterselection is done by simply incubating the DNA from the target cells onto a monolayer of the negative cell line in a 60-cm^2 dish for 1 h, followed by a collection of the unbound supernatant and proceeding with PCR amplification as above.
4. If not enough PCR product is made, i.e., <5 nmol, the product of that round of selection is again subjected to PCR, until the required amount is made.
5. If all goes well, it is the experience of our lab that a complete selection from the first round until an aptamer is ready for screening takes 4–6 months.

Acknowledgments

This work was supported by NIH grants and a Center of Excellence grant from the State of Florida.

References

1. Luo, J., Isaacs, W.B., Trent, J.M., and Duggan, D.J. (2003) Looking beyond morphology: cancer gene expression profiling using DNA microarrays. *Cancer Invest*, **21**, 937–949.
2. Espina, V., Geho, D., Mehta, A.I., Petricoin, E.F., III, Liotta, L.A., and Rosenblatt, K.P. (2005) Pathology of the future: molecular profiling for targeted therapy. *Cancer Invest*, **23**, 36–46.
3. Osborne, S.E. and Ellington, A.D. (1997) Nucleic acid selection and the challenge of combinatorial chemistry. *Chem Rev*, **97**, 349–370.
4. Wilson, D.S. and Szostak, J.W. (1999) In vitro selection of functional nucleic acids. *Annu Rev Biochem*, **68**, 611–647.
5. Breaker, R.R. (2004) Natural and engineered nucleic acids as tools to explore biology. *Nature*, **432**, 838–845.
6. Fang, X., Sen, A., Vicens, M., and Tan, W. (2003) Synthetic DNA aptamers to detect protein molecular variants in a high-throughput fluorescence quenching assay. *Chem Bio Chem*, **4**, 829–834.
7. Guo, K., Wendel, H.P., Scheideler, L., Ziemer, G., and Scheule, A.M. (2005) Aptamer-based capture molecules as a novel coating strategy to promote cell adhesion. *J Cell Mol Med*, **9**, 731–736.
8. Yang, C.J., Jockusch, S., Vicens, M., Turro, N., and Tan, W. (2005) Light-switching excimer probes for rapid protein monitoring in complex biological fluids. *Proc Natl Acad Sci USA*, **102**, 17278–17283.
9. Liu, J.W. and Lu, Y. (2006) Fast colorimetric sensing of adenosine and cocaine based on a general sensor design involving aptamers and nanoparticles. *Angew Chem Int Ed*, **45**, 90–94.

10. Morris, K.N., Jensen, K.B., Julin, C.M., Weil, M., and Gold, L. (1998) High affinity ligands from in vitro selection: complex targets. *Proc Natl Acad Sci USA*, **95**, 2902–2907.
11. Blank, M., Weinschenk, T., Priemer, M., and Schluesener, H. (2001) Systematic evolution of a DNA aptamer binding to rat brain tumor microvessels. Selective targeting of endothelial regulatory protein pigpen. *J Biol Chem*, **276**, 16464–16468.
12. Daniels, D.A., Chen, H., Hicke, B.J., Swiderek, K.M., and Gold, L. (2003) A tenascin-C aptamer identified by tumor cell SELEX: systematic evolution of ligands by exponential enrichment. *Proc Natl Acad Sci USA*, **100**, 15416–15421.
13. Wang, C., Zhang, M., Yang, G., Zhang, D., Ding, H., Wang, H., Fan, M., Shen, B., and Shao, N. (2003) Single-stranded DNA aptamers that bind differentiated but not parental cells: subtractive systematic evolution of ligands by exponential enrichment. *J Biotechnol*, **102**, 15–22.
14. Jayasena, S.D. (1999) Aptamers: an emerging class of molecules that rival antibodies in diagnostics. *Clin Chem*, **45**, 1628–1650.
15. Ellington, A.D. and Szostak, J.W. (1990) In vitro selection of RNA molecules that bind specific ligands. *Nature*, **346**, 818–822.
16. Tuerk, C. and Gold, L. (1990) Systematic evolution of ligands by exponential enrichment: RNA ligands to bacteriophage T4 DNA polymerase. *Science*, **249**, 505–510.
17. Shangguan, D., Meng, L., Cao, Z.C., Xiao, Z., Fang, X., Li, Y., Cardona, D., Witek, R.P., Liu, C., and Tan, W. (2008) Identification of liver cancer-specific aptamers using whole live cells. *Anal Chem*, **80**, 721–728.
18. Tang, Z., Shangguan, D., Wang, K., Shi, H., Sefah, K., Mallikratchy, P., Chen, H.W., Li, Y., and Tan, W. (2007) Selection of aptamers for molecular recognition and characterization of cancer cells. *Anal Chem*, **79**(13), 4900–4907.
19. Shangguan, D., Li, Y., Tang, Z., Cao, Z.C., Chen, H.W., Mallikaratchy, P., Sefah, K., Yang, C.J., and Tan, W. (2006) Aptamers evolved from live cells as effective molecular probes for cancer study. *Proc Natl Acad Sci USA*, **103**, 11838–11843.
20. Sefah, K., Tang, Z., Shangguan, D., Chen, H., Lopez-Colon, D., Li, Y., Parekh, P., Martin, J., Meng, L., Phillips, J.A., Kim, Y., and Tan, W. (2009) Molecular recognition of acute myeloid leukemia using aptamers. *Leukemia*, **23**, 235–244.
21. Chen, H., Medley, C.D., Sefah, K., Shangguan, D., Tang, Z., Smith, J.E., Meng, L., and Tan, W. (2008) Molecular recognition of small cell lung cancer cells using aptamers. *ChemMedChem*, **3**, 991.
22. Tang, Z., Parekh, P., Turner, P., Moyer, R.W., and Tan, W. (2009) Generating aptamers for recognition of virus infected cells. *Clin Chem*, **55**, 813–822.
23. Shuangguan, D., Cao, Z., Meng, L., Mallikaratchy, P., Sefah, K., Wang, H., Li, Y., and Tan, W. (2008) Cell-specific aptamer probes for membrane protein elucidation in cancer cells. *J Proteome Res*, **7**, 2133–2139.
24. Mallikaratchy, P., Tang, Z., Meng, L., Shangguan, D., Sefah, K., and Tan, W. (2007) Aptamer directly evolved from live cells recognizes membrane bound immunoglobin heavy mu chain in Burkitt's lymphoma cells. *Mol Cell Proteomics*, **6**(12), 2230–2238.

Chapter 24

Primer-Free Aptamer Selection Using a Random DNA Library

Weihua Pan and Gary A. Clawson

Abstract

Aptamers are high-affinity oligonucleotides which can be selected from large random libraries by systematic evolution of ligands by exponential enrichment (SELEX) protocols, with affinities and specificities comparable or better than antibodies. The SELEX protocols comprise multiple rounds of selection, each of which require regeneration of bound ligands, which in turn require fixed primer sequences flanking the random library regions. These fixed primer sequences can interfere (with false positives and negatives) with the selection process. Here we present a primer-free protocol using a random DNA library.

Key words: Aptamer, primer-free selection, selex.

1. Introduction

Aptamers are highly structured oligonucleotides (DNA or RNA) that can bind to targets with affinities comparable to antibodies; for aptamers targeting proteins, K_ds usually lie within the 10^{-11}–10^{-9} M range (1). They are identified through an in vitro selection process called Systematic Evolution of Ligands by EXponential enrichment (SELEX) to recognize a wide variety of targets, from small molecules to proteins and other macromolecules (2–4). These oligonucleotides have properties that are well suited for in vivo diagnostic and/or therapeutic applications: Besides good specificity and affinity, they are easily synthesized, survive more rigorous processing conditions, they are poorly immunogenic, and their relatively small size can result in facile penetration of tissues. Less than two decades after the first applications of the technique, several lead aptamer compounds are currently in clinical trials, and the first aptamer drug Macugen was approved

M. Sioud (ed.), *RNA Therapeutics*, Methods in Molecular Biology 629,
DOI 10.1007/978-1-60761-657-3_24, © Springer Science+Business Media, LLC 2010

by the US FDA in 2004 for treatment of wet age-related macular degeneration.

Aptamers that are identified through the standard SELEX process usually comprise ~80 nt, since they are typically selected from nucleic acid libraries with ~40 nt long randomized regions plus fixed primer sites of ~20 nt on each side. The fixed primer sequences thus can comprise nearly ~50% of the library sequences, and therefore may positively or negatively compromise identification of aptamers in the selection process (3), although bioinformatics approaches suggest that the fixed sequences do not contribute significantly to aptamer structure after selection (5). To address these potential problems, primer sequences have been blocked by complementary oligonucleotides or switched to different sequences midway during the rounds of SELEX (6), or have been trimmed to only 6 and 4 nt (7), or to 9 and 7 nt (8), respectively. Wen and Gray (9) designed a primer-free genomic SELEX method, in which the primer sequences were completely removed from the library before selection and were then regenerated to allow amplification of the selected genomic fragments. However, to employ the technique, a unique genomic library has to be constructed (an elaborate procedure), which possesses limited diversity, and regeneration after rounds of selection relies on a linear reamplification step. Alternatively, efforts to circumvent problems caused by fixed primer sequences using high efficiency partitioning are met with problems regarding PCR ampli fication (10).

We have developed a primer-free (PF) selection method that significantly simplifies SELEX procedures and effectively eliminates primer-interference problems (11). The protocols work in a straightforward manner. The central random region of the library is purified without extraneous flanking sequences, and is bound to a suitable target (for example to a purified protein or complex mixtures such as cell lines). Then the bound sequences are obtained, reunited with flanking sequences, and reamplified to generate selected sublibraries. We have found that with purified protein target (specifically with S100B) after 10 rounds of

Fig. 24.1. **(A)** The PF_1 DNA-aptamer selection protocol. The library oligonucleotides are annealed together, and subjected to PCR amplification to yield a double-stranded DNA library (**a**). The 5′-end primer-free (PF_1) random region from the PF single-stranded DNA libraries is prepared by Nt.BstNBI /Nt.BbvCI digestions (cleavage sites are marked with lighter shaded *arrowheads*) and gel purifications, and the self-bridge (*black arrow*) is isolated by gel purification (**b**). After selection (**c**), the selected sequences (designated pS30-CC) are hybridized with their corresponding oligomers and the self-bridge and ligated to regenerate the previously removed primer regions. At this stage, primers are introduced that contain an Sp6 promoter at the 3′-end (**d**). This is then used to transcribe RNA containing the selected regions (**e**). Finally, the RNA is then exclusively reamplified (the template DNA is digested) using RT-PCR (**f**), and the selected sublibrary is ready for the next round of selection. **(B)** The PF_2 DNA-aptamer selection protocol. The library oligonucleotides are annealed together and subjected to PCR amplification to yield a double-stranded DNA library (**a**).

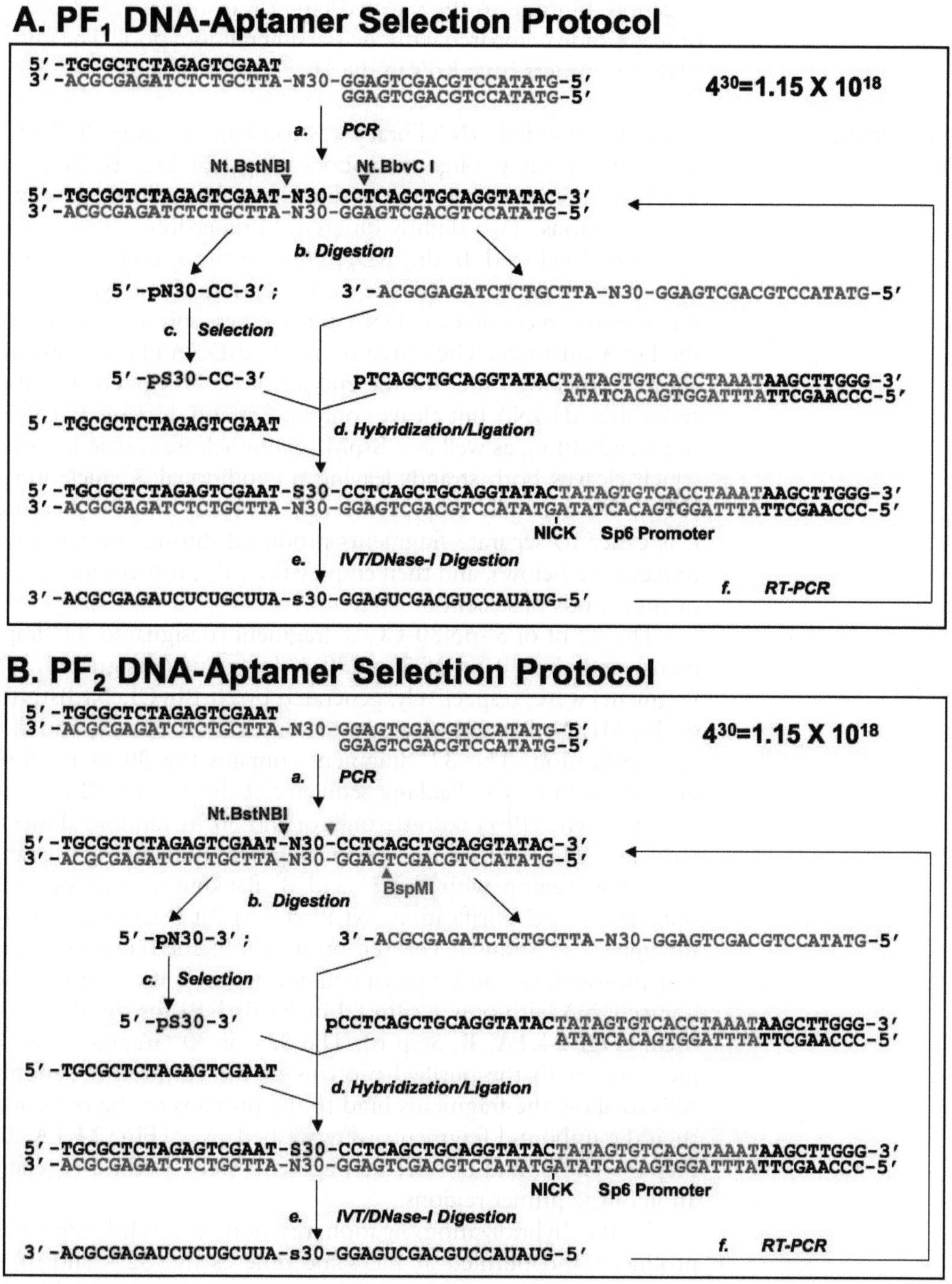

Fig. 24.1. (continued) The primer-free (PF_2) random region from the PF single-stranded DNA library is prepared by Nt.BstNBI/BspMI digestions (cut sites are denoted with *arrowheads*) and gel purification. The self-bridge is isolated either from the PF_1 protocol and/or by single digestion of the starting library with Nt.BstNBI (**b**), coupled with gel purification. After selection (**c**), the selected sequences (designated pS30) are hybridized with their corresponding oligomers and the self-bridge, and ligated to regenerate the previously removed primer regions. As in PF_1, primers are introduced that contain an Sp6 promoter at the 3′-end (**d**). This is then used to transcribe RNA containing the selected regions (**e**). The RNA is then exclusively reamplified using RT-PCR (**f**), and the selected sublibrary is ready for the next round of selection. (Based on figures in reference 11).

selection, binding to the selected target was increased ~3 orders of magnitude compared with the starting random pool. The initial S100B aptamers have Kd's in the 10^{-7}–10^{-8} M range.

1.1. Selection Protocols

A double-stranded DNA library was constructed using PCR with the corresponding oligonucleotides (**Fig. 24.1A, B**; Step a), which contain a central random domain of 30 nt, flanked by two primer regions. Two slightly different "Primer-free" (PF) protocols were developed. In the dsDNA library, the 5′-region contains an endonuclease "nicking" site for the endonuclease Nt.BstNBI; this enzyme recognizes dsDNA but cleaves only one strand of the DNA substrate. The 3′-region of the dsDNA library contains another "nicking" site, for the endonuclease Nt.BbvCI that also recognizes dsDNA but cleaves only one strand, leaving a –CC at the 3′end (PF_1), as well as a BspMI endonuclease restriction site, which cleaves both strands leaving no additional 3′ nucleotides (PF_2). We generally employ the PF_1 protocol initially (because it is easier to separate fragments produced during the selection process; see below), and then employ the PF_2 protocol for subsequent rounds of selection.

The 32 nt of 5′-pN30-CC-3′ fragment (designated 32^+ fragment) and the 30 nt of 5′-pN30-3′ fragment (designated 30^+ fragment) were, respectively, generated by Nt.BbvCI/Nt.BstNBI or BspMI/Nt.BstNBI cleavage of the dsDNA library, and the gel purification. The 32^+ fragment contains the 30 nt random domain, with a –CC flanking sequence at the 3′-end (PF_1). The 30^+ fragment (PF_2) consists only of the 30 nt random domain sequence. The "self-bridge" of 66^- fragment (containing the random N30 region with the 5′ and 3′ flanking sequences) was obtained by gel purification (Nt.BbvCI or Nt.BstNBI cuts only the upper "+" strand). This self-bridge is obtained directly in the PF_1 protocol, or can be generated and isolated after cutting the library DNA with only NtBbv.CI or Nt.BstNBI for the PF_2 protocol (**Fig. 24.1A, B**, Step b). The 32^+ or 30^+ fragments were incubated with the purified proteins or the cultured melanoma cells to allow the fragments bind to the proteins or the cells, and then the unbound fragments were washed away (**Fig. 24.1A, B**, Step c). The bound or selected fragments were used for regeneration of the primer regions.

In the hybridization/ligation reaction, the self-bridge was produced and purified at the same time as the 32^+- and 30^+-fragment purification; 5′-end primer was the same as the library construction and 3′-end primers were replaced with matching primers that also contained an additional Sp6 transcription promoter at the 3′-end (**Fig. 24.1A, B**, Step d). The products of the hybridization/ligation reaction were then used for in vitro RNA transcription (**Fig. 24.1A, B**, Step e). Following the RNA transcription, the ligated DNAs (including the unselected self-bridge DNA fragment) were digested by DNase I to remove interfering background sequences. The reverse transcription (RT)-PCR data have shown that the primer-regenerated products were efficiently reamplified for the next round of selection (**Fig. 24.1A, B**, Step f). The background amplification shown at high cycle numbers in the no-RT control PCR can be completely removed by an additional DNase I digestion, and is not detectable after lower cycle numbers, which we routinely employ.

2. Materials

2.1. Generation of PF DNA Library

2.1.1. Single-Stranded Semi-Random DNA Library

Single-stranded semi-random DNA library with the sequence shown below is synthesized and named N30. All of the oligos and primers used in the experiments are purchased from Integrated DNA Technologies, Inc. (IDT) with HPLC purification.

N30: GTATACCTGCAGCTGAGG$(N)_{30}$ATTCGTCTCTA GAGCGCA

2.1.2. Primers for Double-Stranded Semi-Random DNA Library Construction

P1: TGCGCTCTAGAGTCGAAT
P2: GTATACCTGCAGCTGAGG

2.1.3. Pilot and Large-Scale PCR

1. 20 mM Tris-HCl, pH 7.4.
2. HotMaster^{TM} TAQ DNA polymerase.
3. 10 mM dNTP mix.
4. 40% Acrylamide stock.
5. UltraPure^{TM} 10× TBE buffer.
6. Ammonium persulfate.
7. TEMED.
8. Phi×174 DNA/Hinf I DNA Marker.
9. 6× DNA loading buffer: 0.02% bromophenol blue, 0.02% xylene cyanol FF, 20% glycerol.
10. Ethidium bromide stock (10 mg/mL).
11. α-^{32}P-dCTP (3,000 Ci/mmol, 10 mCi/mL, Dopond).
12. Phenol:$ChCl_3$:IAA, pH 7.9.
13. 5 M NaCl.
14. 100% and 70% ethanol.

2.1.4. Enzymatic Digestion to Remove Primer Regions

All of the restriction endonucleic enzymes in the experiments are purchased from New England Biolabs.

1. First digestion: Nt.BbvC I; BspM I.
2. Second digestion: Nt.BstNB I.
3. Phenol:$ChCl_3$:IAA, pH 7.9.
4. 5 M NaCl.
5. 100% and 70% ethanol.

2.1.5. Gel Separation of the Digested Library DNAs

1. 20 mM Tris-HCl, pH 7.4.
2. 2× RNA loading buffer: 80% formamide, 100 mM EDTA, pH 8.0, 0.02% bromophenol blue, 0.02% xylene cyanol FF.
3. 40% Acrylamide stock.
4. UltraPure^{TM} 10× TBE buffer.
5. Urea.
6. Urea, ammonium persulfate, and TEMED.
7. Photographic film.

2.1.6. Extraction of the Isolated Library-DNA Fragments

1. Elution buffer: 250 mM NaCl, 20 mM Tris-HCl, pH 7.4.
2. 100% and 70% ethanol.

2.2. Purified Protein-Based Selection

1. 20 mM Tris-HCl, pH 7.4.
2. 32^+- and 30^+ fragments.
3. Selection buffer: 2.5 mM $CaCl_2$, 5 mM $MgCl_2$ in 1× PBS buffer, pH 7.4 (GIBCO).
4. Ni-NTA Agarose Beads (Qiagen).
5. Polypropylene column (Qiagen).
6. Binding buffer: 50 mM Na_2HPO_4–NaH_2PO_4, pH 7.2, 150 mM NaCl.
7. S100B (5 μg/μL) made in binding buffer.
8. Phenol:$ChCl_3$:IAA, pH 7.9.
9. 3 M NaAc, pH 5.2.
10. 100% and 70% ethanol.

2.3. Adherent Cell-Based Selection

1. 20 mM Tris-HCl, pH 7.4.
2. 32^+- and 30^+-fragments.
3. 50 mM $MgCl_2$.
4. Detroit-551 and SK-MEL-31 cells.
5. Serum-free MEM media.
6. TrypLE Express (Gibco)
7. Phenol:$ChCl_3$:IAA, pH 7.9.
8. 3 M NaAc, pH 5.2.
9. 100% and 70% ethanol.

2.4. Reamplification of the Bound Fragments

2.4.1. Primer Regeneration

1. 20 mM Tris-HCl, pH 7.4.
2. Primers and oligos used for primer regeneration:
   (a) PF_1:Selected $^+$32- and 66-fragments
   P1, TGCGCTCTAGAGTCGAAT
   Sp6TS-1, pTCAGCTGCAGGTATACTATAGTGTCA-CCTAAATAAGCTTGGG
   Sp6BS, CCCAAGCTTATTTAGGTGACACTATA
   (b) PF_2:Selected $^+$30- and $^-$66-fragments
   P1, TGCGCTCTAGAGTCGAAT
   Sp6TS-2, pCCTCAGCTGCAGGTATACTATAGTGT-CACCTAAATAAGCTTGGG
   Sp6BS, CCCAAGCTTATTTAGGTGACACTATA
3. T4 DNA ligase (2,000 U/μL).
4. Phenol:$ChCl_3$:IAA, pH 7.9.
5. 3 M NaAc, pH 5.2.
6. 100% and 70% ethanol.

2.4.2. In Vitro Transcription

1. 20 mM Tris-HCl, pH 7.4.
2. Dimethyl sulfoxide.
3. 10 mM NTP mix stock.
4. RNase inhibitor.
5. Sp6 RNA polymerase.
6. RNase-free DNase I.
7. Phenol:$ChCl_3$:IAA, pH 7.9.
8. 3 M NaAc, pH 5.2.
9. 100% and 70% ethanol.

2.4.3. Reverse Transcription

1. 20 mM Tris-HCl, pH 7.4.
2. P1, TGCGCTCTAGAGTCGAAT.
3. 5 mM dNTP mix.
4. RNase inhibitor.
5. SensiScript reverse transcriptase.

2.5. TOPO Cloning and Sequencing Analysis of the Selected dsDNA Libraries

1. 10th round selected PCR products
2. pCR®2.1-TOPO Vector (Invitrogen)
3. DH5α component cells
4. Plasmid purification Mini Kit
5. M13 forward primer
6. Informax's Vector NTI (Invitrogen)

2.6. Binding Test of the Aptamers

1. Aptamers and control oligos (N_{30}CC and N_{30}, IDT).
2. T4 polynucleotide kinase.
3. γ-^{32}P-ATP (3,000 Ci/mmol; 10 mCi/mL).
4. S100B in binding buffer (5 μg/μL).
5. Detroit-551 and SK-MEL-31 cells.

3. Methods

3.1. Generation of PF DNA Library

3.1.1. Pilot and Large-Scale PCR

1. To 100 pmol of PF-N30 add the following components on ice:

P1 (100 pmol/μL)	3 μL
P2 (100 pmol/μL)	2 μL
10× buffer	10 μL
dNTP (4 mM)	5 μL
HotMaster TAQ-DNA Polymerase	10 U
H_2O	up to 100 μL

2. Aliquot into seven tubes, one with 10 μL as a no-amplification control and another six containing 15 μL.
3. Perform PCR for 0, 3, 6, 9, 12, 15, and 18 cycles in an Eppendorf Mastercycler at the following conditions: Denaturing at

Denaturing at	
94°C	5 min
18 cycles at	
94°C	30 s
60°C	40 s
72°C	1 min
Additional extension at	
72°C	5 min

4. Place samples on ice after appropriate number of cycles. The 0 cycle tube is kept on ice, it has nothing done to it.
5. Run the samples on 8% polyacrylamide TBE gel with a DNA marker (Phi×174 DNA/*Hinf*I, Promega) at 280 V for 40 min.
6. Stain gel with EtBr (50 μL of 5 mg/mL stock EtBr in 200 mL H_2O) for 3 min, wash for a few seconds in H_2O, place on saran wrap and photograph.

7. Determine the optimal PCR cycle number for the reaction (optimum cycle number will be the cycle where the band of the correct size is the brightest and sharpest. Too many cycles will cause the band to shift up).
8. Perform a large-scale PCR reaction following the above recipe and the optimum cycle number, this time use 1,600 pmol of N30 in a total volume of 1,600 μL reaction (*see* **Note 1**).
9. Extract the PCR products twice with an equivalent volume of phenol:$ChCl_3$:IAA. Subsequently, adjust the NaCl concentration to 250 mM, add 2.5× volume of absolute and incubate at –70°C for 15 min.
10. Spin down at 21,000×*g* at 4°C for 15 min, and then discard the supernatant.
11. Rinse the pellet with 75% ethanol efficiently, spin down at 21,000×*g*, at 4°C for 5 min.
12. Discard the supernatant and dry the pellet in a speed vacuum for 3 min.

3.1.2. Enzymatic Digestion to Remove Primer Regions

1. First digestion.
   Resuspend the DNA library in 100 μL of 20 mM Tris-HCl, pH 7.4. Subsequently, add the following components on ice:
   (a) PF_1/Nt.BbvC I digestion

10× buffer 4	80 μL
Nt.BbvC I (10 U/μL)	400 U
H_2O	up to 800 μL

   (b) ½ PF_2/BspM I digestion

10× buffer 3	80 μL
*Bsp*M I (10 U/μL)	400 U
H_2O	up to 800 μL

   Incubate the reactions at 37°C overnight (about 16 h), and then purify the digested DNA library as described in **Section 3.1.1**, Steps 9–12.

2. Second digestion
   Resuspend the digested PF_1, ½ PF_2 and the undigested ½ PF_2 DNA library in 100 μL of 20 mM Tris-HCl, pH 7.4, respectively. Subsequently, add the following components on ice:

10× buffer 3	80 μL
Nt.BstNB I (10 U/μL)	400 U
H_2O	up to 800 μL

Incubate the reactions at 55°C for 4 h or overnight, and then purify the digested Library DNAs as described in **Section 3.1.1**, Steps 9–12.

3.1.3. Gel Separation of the Digested Library DNAs

1. Resuspend the double-digested library-DNAs in 400 μL of 20 mM Tris-HCl (pH 7.4), respectively.
2. Add a same volume of 2× RNA loading buffer (80% formamide, 100 mM EDTA, pH 8.0 and with 0.02% bromophenol blue and xylene cyanol FF) and heat at 85°C for 3 min, and then chill on ice.
3. Run the chilled samples on a pre-run 10% polyacrylamide, 7 M urea TBE gel until the blue dye to the bottom of the gel (*see* **Note 2**).
4. Open the gel plates and make three dots with the saved sample, and then seal the gel with plastic wrap.
5. Expose to a film and develop it, and then isolate the corresponding gel bands based on the exposed film:

PF_1 (*see* **Fig. 24.2**)	Band 32^+
	Band 66^-
½ PF_2 with 2-digestion (*see* **Fig. 24.2**)	Band 30^+
½ PF_2 with 1-digestion	Band 66^-

3.1.4. Extraction of the Isolated Library-DNA Fragments

1. Freeze the gel bands isolated above at –70°C for 15 min, and put on ice for 5 min.
2. Crush finely with a fire-sealed tip, add 600–800 μL elution buffer. Shake at RT for 30 min, heat to 85°C for 5 min, and then spin for 5 min at 21,000×*g*.
3. Transfer the supernatant to a fresh tube, and then precipitate the DNA fragments as described in **Section 3.1.3**, Steps 10–12.

3.2. Purified Protein-Based Selection

Human S100 calcium-binding protein B (S100B) was used as an example. The 6×His-tagged S100B protein (98 amino acids in length) was expressed and purified by using the QIAexpressionist™ system. During the selection, save 1 μL sample from each step for liquid scintillation counting to determine the binding rate.

3.2.1. Preparation of PF Library-DNA Fragments

1. Resuspend the 32^+- and 30^+ fragments in 40 μL of 20 mM Tris-HCl, pH 7.4.
2. Heat for 3 min at 85°C, and cool down to 37°C for 3 min in an incubator of 37°C.
3. Add 760 μL of selection buffer, incubate for 3 min at 37°C, and then keep at RT for 10 min.

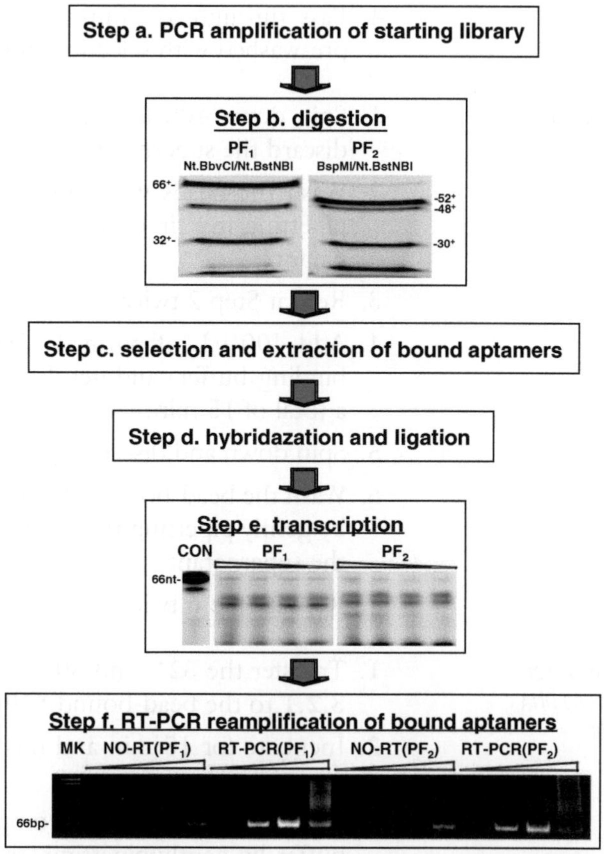

Fig. 24.2. Schematic representation and experimental results/reduction to practice of the PF selection protocol. Designated steps correspond to those depicted in **Fig. 24.1**. After digestion of the PF library constructed by PCR (**Step a**) with restriction endonucleases (as indicated), the digested products were separated by PAGE on a 10% gel under denaturing conditions. The corresponding fragments obtained with the PF_1 and PF_2 selection protocols are shown (**Step b**). After selection (**Step c**), the bound aptamers are hybridized with their corresponding oligomers and the self-bridge, and ligated to regenerate the previously removed primer regions. At 3′-end, primers are introduced that contain an Sp6 promoter at 3′-end (**Step d**). ^{32}P-labeled RNAs were transcribed using ligated products of 1, 0.5, 0.25, 0.125 μL from 5 μL reactions, and were separated by PAGE on 6% gels under denaturing conditions (**Step e**). The transcripts were then treated with DNase to remove the unselected random DNA regions, reverse transcribed into cDNA, and then reamplified for 7, 14, 21, or 28 PCR cycles. Products were separate by PAGE on 8% gels under nondenaturing conditions. The 66 nt fragment represents the full-length product containing the selected pS30-CC and pS30 sequences as re-embedded within the 5′ and 3′ flanking sequences. Migration of Phi×174/*Hinf*I markers is shown at the *left* (**Step f**). Controls included omission of the reverse transcription (RT) step. A small amount of product in such samples could be seen at high cycle numbers; this presumably can be completely eliminated with a second (or prolonged) DNase digestion step.

4. Pass through a column contained Ni-NTA agarose-beads pre-washed with selection buffer, then keep at RT until use.

3.2.2. Preparation of Ni-NTA Agarose-Bead-Bound S100B at RT

1. Spin down 400 μL of Ni-NTA agarose beads for 3 s and discard the supernatant.
2. Wash the beads with 400 μL of binding buffer by gently pipetting five times, then spin down and discard the supernatant.
3. Repeat Step 2 twice.
4. Add 400 μL of purified S100B (5 μg/μL, suspended in binding buffer) and gently pipette 5 times every 3 min for a total of 15 min.
5. Spin down and discard the supernatant.
6. Wash the bead-bound S100B with 400 μL of binding buffer by gently pipetting three times, then spin down, and discard the supernatant.
7. Repeat Step 6 twice.

3.2.3. Selection of S100B-Bound DNA Fragments

1. Transfer the 32^+- and 30^+ fragments mentioned in **Section 3.2.1** to the bead-bound S100B.
2. Incubate for 15 min and mix gently every 3 min by gently pipetting.
3. Wash the S100B-DNA complex with 800 μL of binding buffer by gently pipetting, then spin down and discard the supernatant.
4. Repeat Step 3 twice.

3.2.4. Recovery of the S100B-Selected DNA Fragments

1. Add 200 μL 20 mM Tris-HCl, pH 7.4, heat for 3 min at 85°C, vortex for 1 min, spin down, and then transfer the supernatant to a fresh tube.
2. Repeat Step 1 once and combine the supernatants.
3. Purify the selected DNA fragments as described in **Section 3.1.1** Steps 9–12.
4. Count the samples to determine the binding rate.

3.3. Adherent Cell-Based Selection

Here we use human skin-derived cell lines as an example. The cell lines are: SK-MEL-31, malignant melanoma, ATCC HTB-73; Detroit-551, normal, ATCC CCL-110. During the selection, save 1 μL sample from each step for liquid scintillation counting to determine the binding rate.

3.3.1. Preparation of PF Library-DNA Fragments

1. Resuspend the 32^+ and 30^+ fragments in 200 μL of 20 mM Tris-HCl, pH 7.4.
2. Heat for 3 min at 85°C, and cool down to 37°C for 3 min in an incubator of 37°C.

3. Add $MgCl_2$ to a final concentration of 5 mM and incubate for 3 min, then chill on ice until use.

3.3.2. Pre-wash the Cells

1. Plate Detroit-551 and SK-MEL-31 cells and grow for a day (2×10^6 cells in a 60 mm plate).
2. Wash the cells in 3 mL serum-free MEM media for 5 min in a rocker at 6 rpm, tilt at 6°, 37°C; repeat this step twice.

3.3.3. Counterselection with Detroit-551 Cells

1. Add 2.8 mL of serum-free MEM media to the fragments and warm for 3 min at 37°C, and then transfer to the pre-washed Detroit-551 cells.
2. Incubate for 30 min in the rocker with the same rocking conditions.

3.3.4. Real Selection with SK-MEL-31 Cells

1. Transfer the counterselected unbound fragments (media from Detroit-551 cell counterselection) to the pre-washed SK-MEL-31 cells.
2. Incubate for 30 min in the rocker with the same rocking condition.

3.3.5. Post-wash of the Fragment-Bound SK-MEL-31 Cells

1. Wash the cells in the same manner as in the pre-wash (**Section 3.3.2**).

3.3.6. Recovery of the SK-MEL-31 Cell-Bound Fragments

1. Add 800 μL TrypLE Express (Gibco) to the cells and incubate for 10 min in the rocker.
2. Transfer the cell suspension to a fresh tube and heat for 15 min at 85°C.
3. Extract and precipitate the SK-MEL-31 cell-bound fragments similar to that described in **Section 3.1.1**. Steps 9–12.
4. Count the samples to determine the binding rate.

3.4. Reamplification of the Bound Fragments

3.4.1. Primer Regeneration

1. Resuspend the selected 32^+ and 30^+ fragments in 100 μL of 20 mM Tris-HCl, pH 7.4, respectively.
2. Subsequently, add the following components on ice:

(a) PF_1:	P1 (100 pmol/μL)	20 μL
	66^-	40 μL
	Sp6TS-1 (100 pmol/μL)	20 μL
	Sp6BS (100 pmol/μL)	20 μL
(b) PF_2:	P1 (100 pmol/μL)	20 μL
	66^-	40 μL
	Sp6TS-2 (100 pmol/μL)	20 μL
	Sp6BS (100 pmol/μL)	20 μL

3. Heat for 3 min at 85°C, and cool down to RT dramatically in 1 h, and then add the following components on ice:

10× ligase buffer	40 μL
T4 DNA ligase (2,000 U/μL)	4 μL
H_2O	up to 400 μL

4. Incubate at 16°C overnight.
5. Extract and precipitate the ligated PF products as per Steps 9–12 of **Section 3.1.3**.

3.4.2. In Vitro Transcription

1. Resuspend the DNAs in 100 μL of 20 mM Tris-HCl, pH 7.4, and add the following components on ice:

DMSO	20 μL
5× buffer	80 μL
DTT (100 mM)	40 μL
NTP (10 mM)	40 μL
RNasin (40 U/mL)	10 μL
Sp6 RNA polymerase	200 U
H_2O	up to 400 μL

2. Incubate for 2 h at 37°C, and add another equivalent amount of Sp6 RNA polymerase.
3. Allow to incubate for an additional 2 h at 37°C, and then add the following components:

10× DNase I buffer	50 μL
RNase-free DNase I (1 U/μL)	50 μL

4. Incubate for 2 h at 37°C, and then extract and precipitate as per Steps 9–12 described in **Section 3.1.3** (*see* **Note 3**).

3.4.3. Reverse Transcription with SensiScript™ RT Kit

1. Resuspend the in vitro transcribed RNAs in 100 μL of 20 mM Tris-HCl, pH 7.4, and then add 1 nmol (10 μL of 100 μM working stock) of P1 and water to bring the volume up to 220 μL.
2. Heat both components for 3 min at 85°C, keep at RT for 10 min to form RNA–DNA complex and then add the following components on ice:

(a) NO-RT control:

RNA–DNA complex	20 μL
10× buffer	4 μL
dNTP (5 mM)	4 μL
RNasin (40 U/μL)	1 μL
H_2O	up to 40 μL

(b) Real RT reaction:

RNA–DNA complex	200 μL
10× buffer	40 μL
dNTP (5 mM)	40 μL
RNasin (40 U/μL)	10 μL
SensiScriptase (10 U/μL)	10 μL
H_2O	up to 400 μL

3. Incubate for 1 h at 37°C. The reverse transcribed DNAs are ready for PCR reamplification as described in **Section 3.1.3**.
4. Follow **Note 4** to do further rounds of selection.

3.5. TOPO Cloning and Sequencing Analysis of the Selected Double-Stranded DNA Libraries

1. Clone the 10th round selected PCR products to pCR®2.1-TOPO vector (Invitrogen). Additional cloning can be done as further selections are performed.
2. Sequence 40–50 single colonies with M13 forward primer (we use the Molecular Genetics Core Facility, Hershey Medical Center).
3. Align the selected sequences with Informax's Vector NTI (Invitrogen).
4. Determine the consensus sequences based on the alignment as protein or cell-specific aptamers.

3.6. Binding Test of the Aptamers

1. Take 50 pmol of each aptamer and control oligo, add the following components on ice:

10× PNK buffer	4 μL
γ-^{32}P-ATP (3,000 Ci/mmol, 10 mCi/mL)	1–5 μL
T4 PNK (10 U/mL)	2 μL
H_2O	up to 40 μL

2. Incubate for 30 min at 37°C, and add 40 μL of 2× RNA loading buffer.
3. Gel-purify as described in **Section 3.1.1**, Step 5–6.

4. Perform the binding test as for S100B (*see* **Section 3.2**) and SK-MEL-31 or Detroit-551 cells (*see* **Section 3.3**).
5. Count the recovered S100B–DNA complexes and the trypsinized cell samples in a liquid scintillation counter.

4. Notes

1. A trace amount of α-^{32}P-dCTP is incorporated in the reactions; and for PF_2 3,200 pmol of N30 should be used in a total volume of 3,200 μL reaction.
2. Save a few microliters of the sample for dotting after gel running, and keep the gel at around 45°C. For example, a 12 × 10 in. gel should be at 28–35 W.
3. The salt for precipitation here is replaced by 300 mM NaAc, pH 5.2, and the DNase digestion may be performed for an additional time in a 400 μL of reaction to completely remove the DNA background from 66^{-} fragments.
4. To amplify the reverse transcribed selected DNAs from **Section 3.4.3**, nonspecific products might be produced by PCR reaction. In this case, carry out an extra nonradioactive PCR reaction and gel-purify the expected products (66 bp), and then perform an advanced pilot PCR to determine the cycle number for radioactive PCR reaction. For further rounds of selection, follow **Table 24.1**.

Table 24.1
PF-selection plan (reagent/reaction volume, μL)

Round	1	2		3		4 to N	
PF	–1/–2	–1	–2	–1	–2	–1	–2
Pilot PCR	0.1 nm/100	10/100	10/100	10/100	10/100	10/100	10/100
Large PCR	1.6 nm/1,600 (3X)	800	2 × 800	400	2 × 400	200	2 × 200
Digestion 1	800	400	2 × 400	400	2 × 400	200	2 × 200
Digestion 2	800	400	2 × 400	400	2 × 400	200	2 × 200
Ligation	400	200	200	100	100	100	100
Transcription	400	200	200	100	100	100	100
NO-RT	40	20	20	10	10	10	10
RT	400	200	200	100	100	100	100

Acknowledgments

This work was supported by NIH/NCI grant #CA118591 from the IMAT Program.

References

1. Kulbachinskiy, A.V. (2007) Methods for selection of aptamers to protein targets. *Biochemistry (Mosc)*, **72**, 1505–1518.
2. Que-Gewirth, N.S. and Sullenger, B.A. (2007) Gene therapy progress and prospects: RNA aptamers. *Gene Ther*, **14**, 283–291.
3. Stoltenburg, R., Reinemann, C., and Strehlitz, B. (2007) SELEX – a (r)evolutionary method to generate high-affinity nucleic acid ligands. *Biomol Eng*, **24**, 381–403.
4. Mairal, T. et al. (2008) Aptamers: molecular tools for analytical applications. *Anal Bioanal Chem*, **390**, 989–1007.
5. Cowperthwaite, M.C. and Ellington, A.D. (2008) Bioinformatic analysis of the contribution of primer sequences to aptamer structures. *J Mol Evol*, **67**, 95–102.
6. Shtatland, T. et al. (2000) Interactions of *Escherichia coli* RNA with bacteriophage MS2 coat protein: genomic SELEX. *Nucleic Acids Res*, **28**, E93.
7. Vater, A., Jarosch, F., Buchner, K., and Klussmann, S. (2003) Short bioactive Spiegelmers to migraine-associated calcitonin gene-related peptide rapidly identified by a novel approach: tailored-SELEX. *Nucleic Acids Res*, **31**, e130.
8. Jarosch, F., Buchner, K., and Klussmann, S. (2006) In vitro selection using a dual RNA library that allows primerless selection. *Nucleic Acids Res*, **34**, e86.
9. Wen, J.D. and Gray, D.M. (2004) Selection of genomic sequences that bind tightly to Ff gene 5 protein: primer-free genomic SELEX. *Nucleic Acids Res*, **32**, e182.
10. Musheev, M.U. and Krylov, S.N. (2006) Selection of aptamers by systematic evolution of ligands by exponential enrichment: addressing the polymerase chain reaction issue. *Anal Chim Acta*, **564**, 91–96.
11. Pan, W., Xin, P., and Clawson, G.A. (2008) Minimal primer and primer-free SELEX protocols for selection of aptamers from random DNA libraries. *Biotechniques*, **44**, 351–360.

Chapter 25

Development of TLR7/8 Small RNA Antagonists

Mouldy Sioud

Abstract

Toll-like receptors (TLRs) expressed by innate immune cells recognize a spectrum of microbial structures. Recent studies have provided convincing evidence that TLR7 and TLR8 play a key role in sensing viral RNAs. Under certain conditions these receptors can also recognize self-RNA leading to the induction and/or perpetuation of autoimmune diseases. Recently, we have shown that the incorporation of 2′-modifications in RNAs not only evade immune activation but also suppress TLR7 and TLR8 signaling triggered by immunostimulatory RNAs. Comparable to immunosuppressive DNA oligonucleotides, 2′-modified RNAs may have utility as inhibitors of pathogenic inflammatory reactions mediated by TLR activation. This chapter describes the characterization of immunosuppressive RNA oligonucleotides carrying 2′-modified uridines.

Key words: Suppressive oligonucleotides, RNA interference, siRNA, chemical modification, Toll-like receptors, innate immunity.

1. Introduction

The recognition of pathogens by innate immunity is mediated by host pattern recognition receptors (PRRs) that recognize a set of conserved compounds, the so-called pathogen-associated molecular patterns (PAMPs) (1). Among the PRRs, Toll-like receptors (TLRs) represent a unique system that links innate and adaptive immunity (2, 3). At present several ligands for distinct TLRs have been identified. For example, viral double-stranded RNA is recognized by TLR3, lipopolysaccharide by TLR4, bacterial flagellin by TLR5, viral single-stranded RNAs by TLR7 and TLR8, and bacterial and viral DNA-containing unmethylated CpG motifs by TLR9 (1–4).

Although the activation of innate immunity via Toll-like receptors markedly enhances adaptive immunity, inappropriate

M. Sioud (ed.), *RNA Therapeutics*, Methods in Molecular Biology 629,
DOI 10.1007/978-1-60761-657-3_25, © Springer Science+Business Media, LLC 2010

control of this natural defense mechanism is expected to contribute to chronic disease progression ranging from autoimmunity to sepsis (5, 6). To regulate inflammation, immune cells use various counter regulative mechanisms including the production of immunosuppressive cytokines (7). One potential therapeutic strategy for limiting the activation of TLRs and the production of proinflammatory cytokines involves the use of suppressive RNA oligonucleotides. In this respect, we have shown that 2′-modified RNAs can block the production of proinflammatory cytokines triggered by the activation of TLR7/8 in human peripheral blood cells such as monocytes and dendritic cells (8, 9). The tested 2′-*O*-methyl uridine-modified RNA abolished cytokine response to immunostimulatory RNA in human monocytes that express TLR8 and peripheral mononuclear leukocytes that express TLR7 and TLR8. Interestingly, the RNA containing 2′-*O*-methyl uridines inhibited the activation of TLR7 by different RNA sequences, indicating that the observed antagonists effect is not sequence-dependent (8). This finding is also supported by the results showing a potent antagonism with 2′-*O*-methyl-modified RNA as short as ten bases in length (**Fig. 25.1**). In accordance with our data, Robbins and colleagues demonstrated that 2′-modified RNAs can also function as TLR7

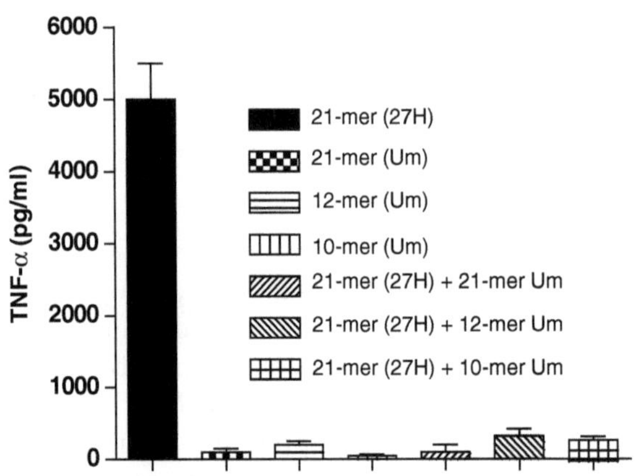

Fig. 25.1. A 10-mer 2′-*O*-methyl uridine-modified RNA inhibits TNF-α production by immunostimulatory RNAs. Human monocytes were transfected for 18 h with unmodified 21-mer single-stranded RNA (27H) or with its 2′-*O*-methyl uridine-modified versions (Um), namely 21-mer, 12-mer, and 10-mer. To test the immunosuppressive effects of the 2′-*O*-methyl-modified RNAs on the activation of TLR8 by the unmodified 21-mer, monocytes were cotransfected with the 21-mer (27H) in combination with either the 21-mer Um, 12-mer Um, or 10-mer Um. Subsequent to 18 h transfection time, TNF-α levels in culture supernatants were measured by ELISA. All molecules were tested at 500 ng/mL. Shorter 2′-modified RNAs were used in these experiments in order to define the part of the 27-mer oligonucleotide that binds and inactivates TLR signaling (*see* **Table 25.1**). Notably, the 10-mer sequence containing only five 2′-*O*-methyl-modified uridines exhibited a strong inhibition of TLR signaling induced by the unmodified 21-mer parental RNA oligonucleotide. The same results were obtained with PBMCs.

antagonists (10). In this chapter, we describe the inhibitory effects of 2′-modified RNA on TLR7/8 signaling in human monocytes. In addition, we further optimized the RNA sequence by reducing the length of the RNA oligonucleotides. Further studies are also planned to asses the therapeutic effects of the developed TLR7/8 antagonists in mouse model for SLE.

2. Materials

2.1. Isolation of Mononuclear Cells

1. Buffy coat from healthy adult donors.
2. RPMI 1640 medium.
3. Complete RPMI 1640 medium: RPMI supplemented with 10% fetal calf serum (FCS) and antibiotics (100 μg/mL penicillin and 100 μg/mL streptomycin).
4. Lymphoprep^{TM}.
5. Pasteur pipettes.
6. Z1 Coulter cell counter.

2.2. Delivery of siRNA into Human Monocytes Using Cationic Liposomes DOTAP

1. Unmodified and 2′-*O*-methyl uridine-modified RNA oligonucleotides. *See* **Table 25.1**.
2. *N*-[1-(2,3-Dioleoyloxy)propyl-*N*,*N*,*N*-trimethylammonium methylsulfate] (DOTAP).
3. Transfection buffer: 20 mM HEPES, 150 mM NaCl, pH 7.4.
4. R-848, a synthetic ligand for TLR7/8.
5. Human peripheral blood mononuclear cells (PBMC) (*see* **Note 1**).
6. Human monocytes.

Table 25.1
RNA sequences used in this study

RNA	Sequence
21-mer RNA (27H)	5′-GUCCGGGCAGGUCUACUUUTT-3′
21-mer RNA 2′-*O*-Me U	5′-GU$_m$CCGGGCAGGU$_m$CU$_m$ACU$_m$U$_m$U$_m$TT-3′
12-mer RNA 2′-*O*-Me U	5′-GGU$_m$CU$_m$ACU$_m$U$_m$U$_m$TT-3′
10-mer RNA 2′-*O*-Me U	5′-GGU$_m$CU$_m$ACU$_m$U$_m$U$_m$-3′

2.3. ELISA

1. Microplate reader.
2. Human TNF-α ELISA kit (*see* **Note 2**).
3. Coating buffer: 0.2 M sodium phosphate, pH 6.5. Dissolve 11.8 g Na_2HPO_4 and 16.1 g NaH_2PO_4 into 1 L distilled water, and then adjust pH to 6.5. Store at 2–8°C.
4. Phosphate-buffered saline (PBS): Dissolve 8 g NaCl, 1.16 g Na_2HPO_4, 0.2 g KH_2PO_4, and 0.2 g KCl into 1 L distilled water. Adjust the pH to 7.0.
5. Assay diluent: 10% fetal calf serum (FCS) in PBS buffer.
6. Washing solution: 0.1% tween in PBS.
7. OptEIA TMB substrate reagent (BD Biosciences).
8. Stop solution: 2M H_2SO_4.
9. 96-Well EIA/RIA plates.

3. Methods

3.1. Isolation of Mononuclear Cells

1. Transfer the buffy coat into a 75-cm^3 cell culture flask.
2. Add 50 mL RPMI and mix gently.
3. Carefully layer 20 mL of the diluted blood over 10 mL Lymphoprep™ in a 50-mL falcon tube. Alternatively, Lymphoprep™ can be underlayered (*see* **Note 3**).
4. Centrifuge at 800*g* for 30 min in swing-out rotor.
5. Subsequent to centrifugation, the mononuclear cells form a distinct band at the sample/medium interface.
6. Remove the cells from the interface using a Pasteur pipette.
7. Dilute the harvested cells with medium and pellet the cells by centrifugation for 5 min at 800*g*.
8. Wash the cells twice with serum-free RPMI 1640 medium or PBS.
9. Subsequent to washing, resuspend the cells in 50–100 mL complete RPMI 1640 medium.
10. Dispense 50 × 10^6 cells per 25 mL complete medium into 75-cm^3 cell culture flasks, and then incubate at 37°C for 1–2 h.
11. Carefully aspirate the medium and wash the adherent cells with 10 mL warmed RPMI 1640 complete medium.
12. Subsequent to washing, add 10–15 mL complete medium and carefully harvest the cells by scraping.

13. Transfer the cells to 50-mL falcon tubes, and then centrifuge at 800*g* for 5 min.
14. Remove the supernatant, resuspend the cells in complete RPMI 1640 medium, and then determine cell number.

3.2. DOTAP Transfection of Small RNAs

Mixing DOTAP with RNA results in spontaneously formed stable complexes that can be added directly into the tissue culture medium. We have recently demonstrated that RNA when delivered to the endosomes via DOTAP activates TLR7/8 leading to the production of proinflammatory cytokines and type I interferon (11).

1. Prior to transfection, plate purified monocytes or PBMC into 96-tissue culture plate at 1–2 × 10^5 cells/well/200 μL complete medium and incubate at 37°C for at least 1 h prior to transfection.
2. Mix in a 1.5-mL microcentrifuge tube, the desired concentration of RNA (1–8 μL) and transfection buffer (1–8 μL) to yield the desired RNA concentrations. For each concentration, the final volume should not exceed 16 μL.
3. In a separate microcentrifuge tube, mix 2 μL of DOTAP and 8 μL transfection buffer.
4. Mix both solutions and incubate at room temperature for 15–30 min.
5. Subsequently, add the mixed solutions to the cells and carefully rock the culture plate to ensure uniform distribution of the DOTAP/RNA complexes (*see* **Note 4**).
6. In the case of competition experiments, mix the immunosuppressive RNA oligonucleotide with 2′-modified RNA prior to mixing with DOTAP (*see* **Note 5**).

3.3. ELISA

1. On the day of transfection, coat the desired number of wells with 100 μL per well of capture antibody.
2. Seal the plate and incubate overnight at 4°C.
3. Aspirate the wells and wash two times with 200 μL/well with the washing buffer. After the last wash, invert plate and blot on absorbent paper to remove any residual washing buffer.
4. Add to each well 200 μL blocking buffer, and then incubate the plate at room temperature for at least 1 h.
5. Aspirate and wash the wells as in Step 3.
6. Pipette 100 μL of each sample and standard into appropriate wells and incubate the plate for 2 h at room temperature.

7. Aspirate and wash at least three times.
8. Add 100 μL of working detector to each well and incubate at room temperature for 1 h. Protect from light.
9. Aspirate and wash five times.
10. Add 100 μL of the substrate solution to each well, and incubate the plate for 10–30 min. Protect from light. If color change does not appear uniform, gently tap the plate to ensure thorough mixing.
11. Add 50 μL of stop solution to each well, and then read absorbance within 30 min using a microplate reader set to 450 nm (*see* **Note 6**).

It should be noted that the antagonist effect of 2′-modified RNAs was confirmed using microarray. As shown in **Fig. 25.2** treatment of peripheral blood mononuclear cells with the 21-mer unmodified RNA altered the expression of around 270 genes. Co-treatment of the cells with the 2′-*O*-methyl uridine-modified RNA inhibited the immunostimulatory effect of the unmodified RNA (**Fig. 25.2**). Also, co-treatment of immune cells with the poly-L-lysine agent inhibited the activation of innate immunity by both immunostimulatory RNAs and CpG DNA oligonucleotides (Mobergslien and Sioud, unpublished data).

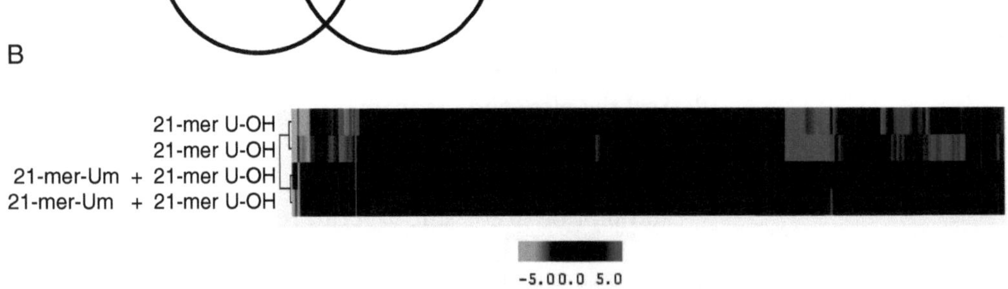

Fig. 25.2. Global gene expression in monocytes in response to unmodified and modified 21-mer oligonucleotides. The transcriptome changes after transfection with unmodified 21-mer (27H) alone or in combination with 21-mer (Um) (27 M) were measured relative to DOTAP-treated cells using oligo microarrays. In these experiments, cell pellets from two independent experiments were pooled before total RNA and probe preparations. Two arrays (dye-swapped) were performed for each group. All molecules were tested at 500 ng/mL. (**a**) Venn diagrams showing significantly ($p < 0.001$) upregulated and downregulated genes subsequent to the treatments. Each *circle* represents one treatment protocol as indicated. The intersections give a number of genes that are steadily differentially altered. (**b**) A cluster of 270 genes identified as differentially altered in response to 21-mer (27H).

4. Notes

1. Add 50 μL of stop solution to each well, and then read absorbance within 30 min using a microplate reader set to 450 nm (*see* **Note 7**).
2. Similar to monocytes, the same protocol can be used to transfect human primary cells such as dendritic cells.
3. The OpEIA™ kit for human TNF-α contains a detailed protocol and the materials necessary for developing ELISA for human TNF-α.
4. Avoid mixing of blood and Lymphoprep™.
5. The replacement of the culture medium prior to transfection is not required in general, but may increase transfection efficiency, depending on the particular application.
6. Usually we incubate the cells with the DOTAP overnight. However, incubation for only 6 h was found to be enough for cytokine production and secretion into culture supernatants.
7. If wavelength correction is available, set to 540 or 570 nm. If wavelength correction is not available, subtract reading at 540 or 570 nm from the reading at 450 nm. This subtraction will correct for optical imperfections in the plate. Reading made directly at 450 nm without correction may be higher and less accurate.

References

1. Medzhitov, R. and Janeway, C.A. Jr. (1997) Innate immunity: impact on the adaptive immune response. *Curr Opin Immunol*, **9**, 4–9.
2. Sioud, M. (2006) Innate sensing of self and non-self RNAs by Toll-like receptors. *Trends Mol Med*, **12**, 67–76.
3. Takeda, K. and Akira, S. (2005) Toll-like receptors in innate immunity. *Int Immunol*, **17**, 1–14.
4. Heil, F., Hemmi, H., Hochrein, H., Ampenberger, F., Kirschning, C., Akira, S., Lipford, G., Wagner, H., and Bauer, S. (2004) Species-specific recognition of single-stranded RNA via Toll-like receptor 7 and 8. *Science*, **303**, 1526–1529.
5. Beutler, B. (2009) Microbe sensing, positive feedback loops, and the pathogenesis of inflammatory diseases. *Immunol Rev*, **227**, 248–263.
6. Christensen, S.R., Shupe, J., Nickerson, K., Kashgarian, M., Flavell, R.A., and Shlomchik, M.J. (2006) Toll-like receptor 7 and TLR9 dictate autoantibody specificity and have opposing inflammatory and regulatory roles in a murine model of lupus. *Immunity*, **25**, 417–428.
7. Maynard, C.L. and Weaver, C.T. (2008) Diversity in the contribution of interleukin-10 to T-cell-mediated immune regulation. *Immunol Rev*, **226**, 219–233.
8. Sioud, M., Furset, G., and Cekaite, L. (2007) Suppression of immunostimulatory siRNA-driven innate immune activation by 2′-modified RNAs. *Biochem Biophys Res Commun*, **361**, 122–126.
9. Cekaite, L., Furset, G., Hovig, E., and Sioud, M. (2007) Gene expression analysis in blood cells in response to unmodified and 2′-modified siRNAs reveals

TLR-dependent and independent effects. *J Mol Biol*, Doi:10.1016/j.jb.2006.09.034.

10. Robbins, M., Judge, A., Liang, L., McClintock, K., Yaworski, E., and MacLachlan, I. (2007) 2′-O-methyl-modified RNAs act as TLR7 antagonists. *Mol Ther*, **15**, 1663–1669.

11. Sioud, M. (2005) Induction of inflammatory cytokines and interferon responses by double-stranded and single-stranded siRNAs is sequence-dependent and requires endosomal localization. *J Mol Biol*, **348**, 1079–1090.

Chapter 26

Modulation of Dendritic Cell Maturation and Function by siRNA-Bearing 5′-Triphosphate

Mouldy Sioud

Abstract

Small interfering RNAs (siRNAs) have emerged as a powerful technique for sequence-specific gene silencing in a wide variety of organisms. However, it has been shown that in vitro transcribed siRNA can activate innate immunity through RIG-1 helicase, a cytosolic sensor of viral RNA. We recently demonstrated that the activation of RIG-1 by siRNA-bearing 5′-triphosphate can be used to enhance dendritic cell maturation and function. Indeed, immature DCs that had been transfected with siRNA-bearing 5′-triphosphate activated T cells, indicating that, even in the absence of external stimuli such as TNF-α, those DCs were sufficiently mature to initiate T cell activation. Also, a recent study has shown that 5′-bearing siRNAs targeting Bcl-2 can induce apoptosis in tumour cells, further supporting the use of this strategy to combat cancer. This chapter describes the in vitro synthesis of siRNA targeting indoleamine 2,3-dioxygenase (IDO), an enzyme that plays a critical role in hindering immunity against tumours.

Key words: Indoleamine 2, 3-dioxygenase, siRNA, bifunctional siRNAs, dendritic cells, immunosuppression.

1. Introduction

The inactivation of gene function by reverse genetics is important for studying gene function and could also have impact on the treatment of diseases initiated and/or perpetuated by aberrant gene expression. Among the anti-gene strategies, antisenses oligonucleotides and ribozymes have been extensively applied (1). Recently, the discovery of RNA interference (RNAi) and related microRNA-mediated gene regulation (*see* **Chapters 4, 5, and 14**) has profoundly altered the understanding of gene regulation and offered an additional way to knock down the expression

M. Sioud (ed.), *RNA Therapeutics*, Methods in Molecular Biology 629,
DOI 10.1007/978-1-60761-657-3_26, © Springer Science+Business Media, LLC 2010

of virtually any gene in living cells (1, 2). RNAi is a powerful technique for sequence-specific gene silencing in a wide variety of organisms (2–5). It is initiated either by endogenous production of small RNAs (20–30 nt) or by experimentally introduced double-stranded RNAs, particularly small interfering RNAs (siRNAs, 21 nucleotides) (3). Although RNAi with chemically made siRNA provides a promising tool for studying gene functions and drug target validation (5, 6), the technique is limited by its high cost, since it requires the chemical synthesis of siRNAs without a guarantee that the designed molecules will be effective in gene silencing. Therefore, a successful gene-targeting project may require the design of many distinct siRNAs at a high cost. To overcome this problem in vitro transcription strategies have been used (7, 8). In contrast to chemically made siRNA, in vitro transcribed siRNA is sensed by the cytosolic antiviral helicase retinoic acid-induced protein I (RIG-1, encoded by *Ddx58* gene) (9, 10). The activation of innate immunity may represent an obstacle for the therapeutic applications of in vitro transcribed siRNAs.

One way to trigger innate immunity against cancer is to activate RIG-1 via the use of siRNA-bearing 5′-triphosphate. RIG-1 recognizes the 5′-triphosphate RNA produced by viral or phage polymerases such as SP6 and T7 RNA enzymes (9). Notably, triphosphate in the cytosol is recognized as a sign of infection. In a previous study, we have shown that cytoplasmic delivery of in vitro transcribed siRNA-bearing 5′-triphosphate inhibited IL-10 gene expression and simultaneously induced cytokine production through the activation of RIG-1 (11). This finding prompted us to test whether in vitro transcribed siRNAs targeting indoleamine 2,3-dioxygenase (IDO), an enzyme that is involved in hampering immunity against tumour cells (12), would silence its expression and enhance dendritic cell function by inducing cytokine production in dendritic cells (DC). Collectively, the data show that in vitro transcribed anti-IDO siRNA can effectively inhibit IDO expression and simultaneously induced TNF-α and IL-12 production (**Fig. 26.1a–f**) (13). In accordance with our early work, cytoplasmic delivery of chemically made siRNAs did not induce cytokine production because they lack the 5′-triphosphate signature (14).

Adjuvants play an essential role in the generation of immunity. They have long been known to enhance the generation of immune responses in part by stimulating DCs to fully mature and express co-stimulatory molecules. In the absence of such signals, DCs will present antigen in ways that lead to tolerance instead of immunity (15, 16). To mount an effective adaptive immune responses against infections, activation of Toll-like receptors and/or RIG-1-like receptors (RLR) is required. The expression of RIG-1 by DCs prompted us to test whether its activation by RNA-bearing 5′-triphosphate such as in vitro

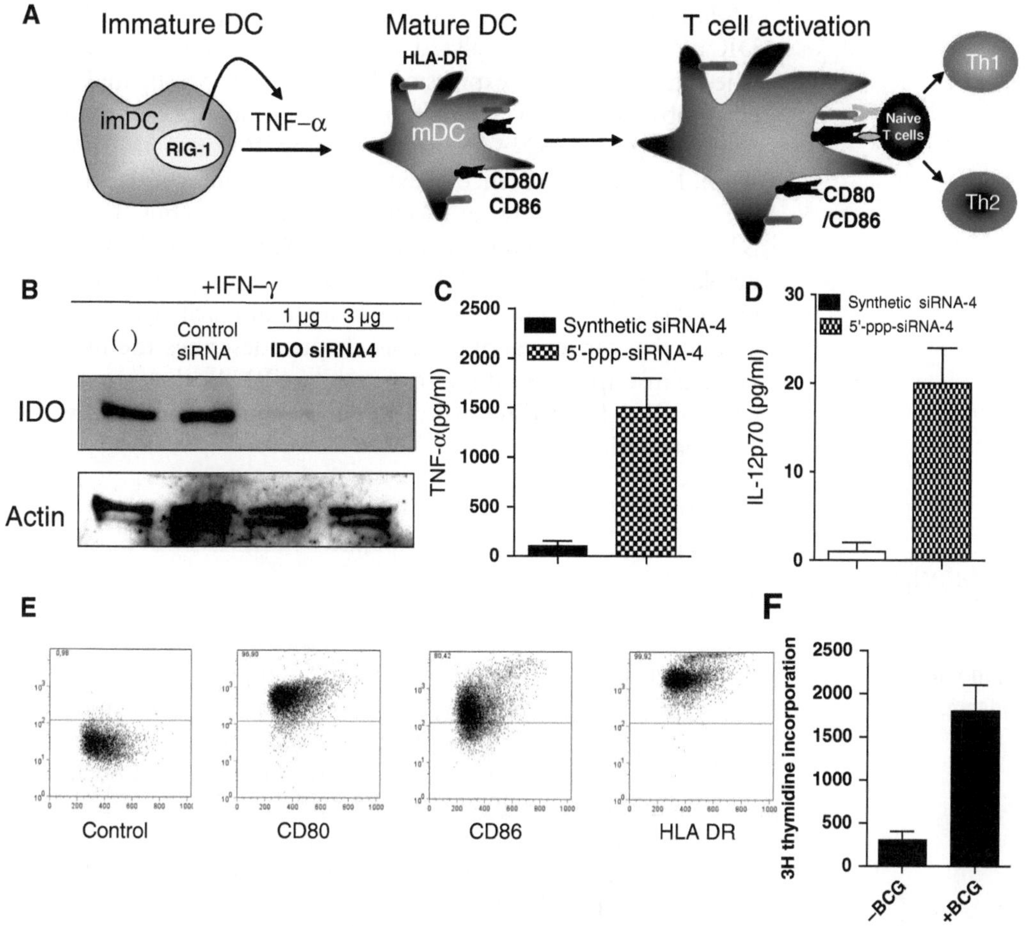

Fig. 26.1. Effects of a bifunctional siRNA on IDO expression, cytokine production, and DC maturation and function. (**a**) Schematic illustration of DC maturation protocol. (**b**) Gene silencing using in vitro transcribed anti-IDO siRNA. ImmoDC were transfected with in vitro transcribed siRNA (1 or 3 μg/75 or 225 pmol) using BTX electroporation method as described in **Section 3.5** and then they were stimulated with IFN-γ to induce IDO expression. Subsequent to 18 h transfection time, the expression of IDO and actin was analysed by Western blots. Notably, IDO expression is completely eliminated in siRNA-treated cells. (**c** and **d**) Effects of in vitro transcribed siRNA on TNF-α and IL-12 expressions. ImmoDC were transfected with in vitro transcribed siRNA (3 μg/225 pmol) as in (**b**). Subsequent to 18 h transfection time, TNF-α and IL-12 levels in culture supernatants were analysed by ELISA. Samples were assayed in triplicates and the results represent the mean ± SD of two independent experiments. (**e**) The expression of co-stimulatory molecules and HLA-DR in the transfected immature DCs was determined by flow cytometry after 48 h transfection time. The percentages of cells for this representative experiment are indicated. (**f**) The capacity of the 5′-triphosphate siRNA-transfected immature DCs to activate T cells was tested using BCG protein lysates. In these experiments, irradiated DCs (5×10^4) were incubated with autologous T cells (10^5) in the presence or absence of BCG protein extracts (5 μg/mL) for 5 days. Cell proliferation was measured by ^{3}H thymidine incorporation. Samples were analysed in triplicates and the results represent the mean ± SD of three independent experiments (12).

transcribed anti-IDO siRNA would promote DC maturation. In these experiments, immoDCs were electroporated with the 5′-triphosphate siRNA4 and then they were cultured for 48 h without the addition of external maturation cytokines such as

TNF-α. Subsequently, the expression of CD80, CD86, and HLA-DR molecules were analysed by flow cytometry (**Fig. 26.1e**). The data show high expression levels of the investigated markers. Subsequently, we have tested the stimulatory potential of 5′-triphosphate siRNA transfected immoDC on T cell proliferation. Even in the absence of external maturation cytokines, these DCs are able to activate T cells (**Fig. 26.1f**). Thus, the possibility of blocking IDO expression and simultaneously activating cytokine production and DC maturation using in vitro transcribed siRNAs is a novel strategy that should facilitate the design of effective cancer vaccines. This chapter describes the use of in vitro transcription strategy to generate siRNAs that are capable of knowing down IDO expression and simultaneously activating DC maturation.

2. Materials

2.1. In Vitro Transcription

1. Synthetic DNA oligonucleotide templates containing the minimal promoter sequence for efficient transcription by T7 phage polymerase. Oligonucleotides can be synthesized in-house or ordered from commercial oligonucleotide suppliers.
   - Oligonucleotides for the sense strand:
     5′-TAATACGACTCACTATA**GCCGTGAGTTTGT-CCTTTCAA**-3′
     5′-TTGAAAGGACAAACTCACGGCTATAGTGAGA-CGTATTA-3′
   - Oligonucleotides for the antisense strand:
     5′-TAATACGACTCACTATA**GAAAGGACAAACT-CACGGACT**-3′
     5′-AGACCGTGAGTTTGTCCTTTCTATAGTGAGA-CGTATTA-3
3. Optimized 5X transcription buffer: 200 mM Tris–HCl, pH 7.9, 30 mM $MgCl_2$.
4. Ribonucleotide triphosphates: adenosine 5′-triphosphate (ATP), guanosine 5′-triphosphate (GTP), cytidine 5′-triphosphate (CTP), and uridine 5′-triphosphate (UTP), 10 mM for each NTP. Store at –20°C in aliquots.
5. 100 mM DTT.
6. T7 RNA polymerase (17 U/μL).
7. Recombinant RNasin (40 U/μL).
8. RNase-free DNase I and ribonuclease T1.
9. 96 and 75% ethanol.

10. 3 M sodium acetate (Na-Ac), pH 7.
11. TE buffer: 10 mM Tris–HCl (pH 7.4), 1 mM EDTA.
12. Ethidium bromide (10 mg/mL).
13. Acrylamide/bis-acrylamide (39:1).
14. *N*,*N*,*N*′,*N*′-Tetramethylethylenediamine (TEMED).
15. 10% ammonium persulfate (APS) solution.
16. HROMA SPIN-10 column (Clontech).
17. 10X TBE stock solution (TBE): 108 g Tris base, 55 g boric acid, and 40 mL of 0.5 M EDTA, pH 8.0, dissolve in distilled water to make 1 L and autoclave.
18. Diethylpyrocarbonate (DEPC)-treated water (RNase-free water).
19. Phenol:chloroform:isoamyl alcohol (25:24:1), pH 6.6.
20. Polyacrylamide gel electrophoresis equipment.

2.2. Isolation of Blood Leukocytes and Generation of Dendritic Cells

1. Buffy coat from healthy adult donors.
2. RPMI 1640 medium.
3. RPMI 1640 medium supplemented with 10% fetal calf serum (FCS) and antibiotics (100 μg/mL penicillin and 100 μg/mL streptomycin).
4. Lymphoprep^{TM}.
5. Pasteur pipettes.
6. Z1 Coulter Particle Counter.
7. Cell scraper.
8. Granulocyte macrophage colony stimulating factor (GM-CSF).
9. Interleukin-4.
10. Tumour necrosis factor alpha (TNF-α).

3. Methods

3.1. Isolation of Peripheral Blood Mononuclear Cells

1. Transfer the buffy coat into a 75-cm^3 cell culture flask.
2. Add 50 mL RMPI and mix gently.
3. Carefully layer 20 mL of the diluted blood over 10 mL Lymphoprep^{TM} in a 50-mL falcon tube. Alternatively Lymphoprep^{TM} can be underlayered.
4. Centrifuge at 800*g* for 30 min in swing-out rotor.
5. Subsequent to centrifugation, the mononuclear cells form a distinct band at the sample/medium interface.

6. Remove the cells from the interface using a Pasteur pipette.
7. Dilute the harvested cells with medium and pellet the cells by centrifugation for 5 min at 800*g*.
8. Wash the cells twice with RPMI medium and resuspend in 50–100 mL RPMI.
9. Dispense 50 × 10^6 cells/25 mL complete medium into 75 cm^3 cell culture flask and then incubate at 37°C in a humidified atmosphere containing 5% CO_2 for 1–2 h.
10. Subsequently, carefully aspirate the medium and wash the adherent cells with 10 mL warmed RPMI complete medium.
11. Add 10–15 mL complete medium to the flask and carefully harvest the cells by scraping.
12. Transfer the cells to 50-mL falcon tubes and then centrifuge at 800*g* for 5 min.
13. Remove the supernatant, resuspend the cells in RPMI complete medium, and then determine the cell number.

3.2. Immature Dendritic Cells

To obtain immature dendritic cells, human monocytes were cultured in 25 cm^3 flasks at 5 × 10^5 per mL in complete RPMI medium supplemented with purified human recombinant cytokines GM-CFS (25 ng/mL) and IL-4 (50 ng/mL). Subsequent to 5 days in culture, the cells were washed, counted, and then electroporated with in vitro transcribed siRNAs.

3.3. T7 siRNAs: DNA Template for In Vitro Transcription

Because the first nucleotide incorporated into the RNA by T7 RNA polymerase is a guanosine (G), in vitro transcribed siRNA must start with a least one G. Therefore, the siRNA target sequence should start with G and end with C at position 19 (site length is 19 nt). The site requirement is **G**-N17-**C**. This sequence is frequently present in mRNAs. For each siRNA strand, combine equimolar amounts (1 μg each) of the DNA oligonucleotides (top and bottom strands) in 1X transfection buffer. Total volume 50 μL. Heat to 90°C, and then allow annealing at 37°C for 15–30 min.

3.4. Synthesis of Large Amounts of siRNA

For in vitro transcription, add to the annealed DNA template (50 μL) the following:

NTP mix (each 10 mM)	5 μL
5X transcription buffer	10 μL
RNasin	1 μL
T7 RNA polymerase	2 μL
DTT (100 mM)	10 μL

1. Add RNase-free water to 100 μL and incubate at 37°C for 2–3 h.
2. You can check at this point that the siRNA is properly transcribed by analysing 5 μL aliquot on a 15% polyacrylamide gel (*see* **Note 1**).
3. To eliminate DNA template, treat the reaction mixture with DNAse I (1 U) for 10 min at 37°C.
4. Subsequently, add an equal volume of phenol: chloroform:isoamylalcohol (25:24:1) to the in vitro transcription.
5. Mix (vortex) the contents of the tube until an emulsion forms.
6. Centrifuge the mixture at 12,000*g* for 2 min in a microcentrifuge at room temperature.
7. Transfer the upper (aqueous) phase containing RNA to a new eppendorf tube.
8. Add an equal volume of chloroform:isoamylalcohol (24:1).
9. Add one-tenth volume of 3 M sodium acetate, pH 5.2.
10. Add 2.5 volume of absolute (100–96%) EtOH and place at –20°C for 60 min. The RNA can be stored at this step at –20°C if desired.
11. Spin down for 15 min at 12,000*g* in a microcentrifuge at 4°C.
12. Wash the RNA pellet with 70% EtOH.
13. Air-dry the pellet.
14. Dissolve in 30–50 μL of RNase-free water.
15. Determine the concentration by measuring the optimum density (OD) (1 OD_{260} unit = 33 μg/mL). Always use gloves when handling RNA.

Notably, the entire procedure requires less than 5 h and multiple siRNAs can be synthesized simultaneously.

3.5. Transfection of Immature Dendritic Cells with In Vitro Transcribed siRNAs

To deliver siRNA to immature dendritic cells, we have used the BTX EMC electroporation device which uses square wave pulses. Prior to electroporation, wash the cells twice with RMPI and then resuspend in serum-free medium at 5–10 × 10^6 cells/ 0.5 mL.

1. Add to the cells (0.5 mL) anti-IDO siRNA (0.1–3 μg/transfection) and mix by carefully pipetting up and down. Include some controls (*see* **Note 2**).
2. Transfer the cells to a 4-mm gag cuvette, avoid air bubbles, and then incubate at 4°C for 5 min.

3. Set up the electroporation parameters:
   - Volt: 500
   - Pulse length: 2 ms
   - Pulse: 01
   - Interval: 01.0 s
   - Polarity: unipolar
4. After incubation at 4°C, place the metal walls in the cuvette against the metal walls in the electroporator and then pulse cells with the setting indicated above.
5. Transfer the cell immediately to 2.5-mL pre-warmed RPMI complete medium in a 6-well culture plate.
6. Incubate at 37°C in a humidified atmosphere containing 5% CO_2 for 18–72 h prior analysing siRNA effects on gene expression.
7. Cytoplasmic delivery of in vitro transcribed siRNAs is expected to activate RIG-1 and simultaneously inhibiting the expression of the targeted gene (e.g., IDO). RIG1 activation leads to the production of proinflammatory cytokines (e.g., TNF-α) and interferon β. Both factors can be measured by ELISA (*see* **Note 3**). The expression of co-stimulatory molecules such as CD80 and CD86 can be analysed by flow cytometry (*see* **Note 4**).
8. Normally IDO is not expressed by either immature or mature dendritic cells. To assess the anti-IDO siRNA silencing potency, add interferon γ (500 ng/mL) to cell culture in order to induce IDO expression (*see* **Note 5**).
9. Incubate the cells at 37°C in a humidified atmosphere containing 5% CO_2 for the desired time (15–20 h) prior to analysing siRNA effects.

4. Notes

1. To visualize the RNA, stain the gel after electrophoresis. Dilute a sample of the stock solution to 0.5 μg/mL with water and incubate the gel for 2 min. Destaining is usually not needed but can be carried out in water for 1 min if decreased background is necessary. The DNA bands can then be detected on a UV light box (254 nm wavelength). Ethidium bromide can also be incorporated into the gel and running buffer at 0.5 μg/mL and visualized immediately after electrophoresis.

2. One of the key experimental strategies to elucidate the function of a gene in vitro and in vivo is the specific ablation of its expression. To check for specificity, most siRNA suppliers recommend the use of negative siRNA controls with sequences that do not show sequence similarity to the human sequences such as the siRNA targeting β galactoside used in this chapter. Also, it is recommended to test multiple negative siRNA controls to identify ones that do not induce non-specific effects on gene expression. In addition, we think that it is central to use multiple siRNAs targeting different sites of the gene transcript being investigated. These siRNAs should first be analysed for effective reduction of target gene expression. Different siRNAs to the same target and with similar gene silencing efficiency should induce similar phenotypic effects. Any changes induced by one siRNA and not the other(s) could be attributed to bystander effects (e.g., off-target effects) and these siRNAs should not be used in therapy, particularly when bystander effects are not desired (12).
3. In vitro transcribed siRNAs were delivered using electroporation as described previously (7, 12). Cytokine contents were measured in culture supernatant 18 h after transfection using human TNF-α and IL-12 (p70) ELISA kit (BD Biosciences). In some cases, the expression of CD80, CD86, and HLA-DR molecules was analysed by flow cytometry.
4. The expression of the co-stimulatory molecules such as CD80 and CD86 can be analysed after 24, 48, or 72 h transfection time. In the experiments shown in **Fig. 26.1e**, immature DCs were transfected with in vitro transcribed anti-IDO siRNAs and then incubated at 37°C in 5% CO_2 for 2 days. Subsequently, they were stained with PE- or FITC-conjugated mAb against human CD80, CD86, and HLA-DR molecules. Briefly, the cells (10^5) were stained with respective PE-conjugated antibodies in PBS containing 1% FCS and 0.1% sodium azide (FACS buffer) for 30 min at 4°C. Negative controls were isotype-matched mAbs. After washing, the cells were resuspended in FACS buffer and surface expression was determined by flow cytometry (12).
5. In general IDO is not constitutively expressed by human monocytes. Previous studies have shown that IDO expression is induced by interferon-γ (IFN-γ)-dependent and -independent mechanisms. In human monocytes, we have shown that IDO can be induced by interferon and TLR7/8 ligands. However, in human DC cells, the expression of IDO is mainly under the IFN-γ, a factor that is secreted by activated T cells and NK cells.

References

1. Sioud, M. (2001) Nucleic acid enzymes as a novel generation of anti-gene agents. *Curr Mol Med*, **1**, 575–588.
2. Sharp, P.A. (2001) RNA interference. *Genes Dev*, **15**, 188–200.
3. Fire, A., Xu, S., Montgomery, M.K., Kostas, S.A., Driver, S.E., and Mello, C.C. (1998) Potent and specific genetic interference by double-stranded RNA in *Caenorhabditis elegans. Nature*, **391**, 806–811.
4. Sioud, M. (2004) Therapeutic siRNAs. *Trends Pharmacol Sci*, **25**, 22–28.
5. Meister, G. and Tuschl, T. (2004) Mechanisms of gene silencing by double-stranded RNA. *Nature*, **431**, 343–349.
6. Corey, D.R. (2007) Chemical modification: the key to clinical application of RNA interference? *J Clin Invest*, **117**, 3615–3622.
7. Leirdal, M. and Sioud, M. (2002) Gene silencing by preformed small RNA duplexes. *Biochem Biophys Res Commun*, **295**, 744–748.
8. Pichlmair, A., Schulz, O., Tan, C.P., Näslund, T.I., Liljeström, P., Weber, F., and Reisesousa, C. (2006) RIG-1-mediated antiviral responses to single-stranded RNA-bearing 5′-phosphates. *Science*, **314**, 997–1001.
9. Hornung, V., Ellegast, J., Kim, S., Brzózka, K., Jung, A., Kato, H., Poeck, H., Akira, S., Conzelmann, K.K., Schlee, M., Endres, S., and Hartmann, G. (2006) 5′-triphosphate RNA is the ligand for RIG-1. *Science*, **314**, 994–997.
10. Furset, G. and Sioud, M. (2007) Design of bifunctional siRNAs: combining immunostimulation and gene-silencing in one single siRNA molecule. *Biochem Biophys Res Commun*, **352**, 642–649.
11. Munn, D.H. and Mellor, A.L. (2004) IDO expression by dendritic cells: tolerance and tryptophan catabolism. *Nat Rev Immunol*, **4**, 762–774.
12. Furset Flatekval, G. and Sioud, M. (2009) Modulation of dendritic cell function and maturation with mono- and bifunctional small interfering RNAs targeting indoleamine 2,3-dioxygenase (p). *Immunology* DOI: 10.1111/j.1365–2567.2009.03093.x.
13. Poeck, H., Besch, R., Maihoefer, C., Renn, M., Tormo, D., Morskaya, S.S., Kirschnek, S., Gaffal, E., Landsberg, J., Hellmuth, J., Schmidt, A., Anz, D., Bscheider, M., Schwerd, T., Berking, C., Bourquin, C., Kalinke, U., Kremmer, E., Kato, H., Akira, S., Meyers, R., Häcker, G., Neuenhahn, M., Busch, D., Ruland, J., Rothenfusser, S., Prinz, M., Hornung, V., Endres, S., Tüting, T., and Hartmann, G. (2008) 5′-Triphosphate-siRNA: turning gene silencing and Rig-I activation against melanoma. *Nat Med*, **14**, 1256–1263.
14. Hunter, R.L. (2002) Overview of vaccine adjuvants: present and future. *Vaccine*, **20**, S7–S12.
15. Banchereau, J. and Steinman, R.M. (1998) Dendritic cells and the control of immunity. *Nature*, **392**, 245–252.
16. Kawai, T. and Akira, S. (2008) Toll-like receptor and RIG-1-like receptor signaling. *Ann NY Acad Sci*, **1143**, 1–20.

Chapter 27

Immunotherapy of Cancer with Dendritic Cells Loaded with Tumor Antigens and Activated Through mRNA Electroporation

An M.T. Van Nuffel, Jurgen Corthals, Bart Neyns, Carlo Heirman, Kris Thielemans, and Aude Bonehill

Abstract

Since decades, the main goal of tumor immunologists has been to increase the capacity of the immune system to mediate tumor regression. Considerable progress has been made in enhancing the efficacy of therapeutic anticancer vaccines. First, dendritic cells (DCs) have been identified as the key players in orchestrating primary immune responses. A better understanding of their biology and the development of procedures to generate vast amounts of DCs in vitro have accelerated the development of potent immunotherapeutic strategies for cancer. Second, tumor-associated antigens have been identified which are either selectively or preferentially expressed by tumor cells and can be recognized by the immune system. Finally, several studies have been performed on the genetic modification of DCs with tumor antigens. In this regard, loading the DCs with mRNA, which enables them to produce/process and present the tumor antigens themselves, has emerged as a promising strategy. Here, we will first overview the different aspects that must be taken into account when generating an mRNA-based DC vaccine and the published clinical studies exploiting mRNA-loaded DCs. Second, we will give a detailed description of a novel procedure to generate a vaccine consisting of tumor antigen-expressing dendritic cells with an in vitro superior capacity to induce anti-tumor immune responses. Here, immature DCs are electroporated with mRNAs encoding a tumor antigen, CD40 ligand (CD40L), CD70, and constitutively active (caTLR4) to generate mature antigen-presenting DCs.

Key words: Dendritic cells, mRNA, electroporation, cancer immunotherapy, vaccination, clinical trials.

1. Introduction

1.1. Background

Conventional treatment strategies for cancer have achieved major improvements. A single cell sensing mechanism that is able to detect emerging tumor cells is lacking so far. The immune system

M. Sioud (ed.), *RNA Therapeutics*, Methods in Molecular Biology 629,
DOI 10.1007/978-1-60761-657-3_27, © Springer Science+Business Media, LLC 2010

provides such a system and, as a result, the general belief is that therapeutic vaccination might provide a clinical benefit for cancer patients.

For active immunization against cancer, DCs are considered the most attractive candidates because of their ability to efficiently present and cross-present antigens, thus stimulating both $CD4^+$ T helper and $CD8^+$ cytotoxic T cells (1, 2). This gives them the capacity, in contrast to other antigen-presenting cells such as macrophages or B cells (3), to stimulate both naïve and memory T cells (4) and to generate immunologic memory (5). Moreover, it is feasible to isolate monocytes from the blood of cancer patients and to differentiate them into DCs in sufficient numbers to conduct a clinical trial (6–8) and these monocyte-derived DCs possess T-cell stimulatory functions in vitro (8, 9). Finally, vaccination with peptide-loaded DCs, but not with peptide in Freund's adjuvant or plasmid DNA, induced protective immunity against autologous melanoma in a mouse model (10).

In general, the DC's life cycle can be divided into two stages: the immature and the mature stage. Immature DCs reside in the periphery and are specialized in antigen recognition and uptake. Upon receiving a danger signal, DCs undergo several morphological, phenotypical, and functional changes and transform into mature cells. Thereby they decrease antigen uptake and increase antigen processing, presentation in both MHC class I and II molecules, and CCR7 expression. The latter mediates migration toward the lymph nodes, where the mature DCs can stimulate T cells to recognize and kill cells presenting epitopes from the antigens picked up in the periphery. To obtain an adaptive immune response, three main signals need to be delivered to the T cells. First, an HLA–peptide complex needs to be presented by the DCs and to be recognized by the T-cell receptor on T cells. Only fully matured DCs will be able to activate the specific T cells by concomitantly providing proper co-stimulatory signals (CD80, CD83, CD86, CD40, CD70, 4-1BB, OX-40 ligand, and ICOS ligand) to the T cells, which is the second signal. In response to T-cell receptor signaling, T helper cells (Th) will upregulate CD40L, leading to the licensing of the DC after interaction with its receptor CD40. This results in the third necessary signal: cytokine secretion to drive the adaptive immune response toward Th1 and induce cytotoxic T lymphocytes (CTL) (3, 11).

Considering the pivotal role of DCs in the generation of a robust immune response, the number of performed phase I and II clinical trials with tumor antigen-loaded DCs is gradually increasing (12). Unfortunately so far, DC immunotherapy with MHC class I- and II-restricted peptides did not demonstrate to be more effective than DTIC, a conventional chemotherapeutic agent in stage IV melanoma patients (13). In general limited clinical responses are achieved and because of many inter-study differences, the actual critical parameters or prediction markers for successful cancer immunotherapy with DCs are still undefined. The DC generation, the form and origin of antigens, the strategy to load the DCs with antigen, their maturation and activation status, and the dose, route, and scheme of DC administration are all variables that lead to results that are difficult to pool as a whole and so the body of data remains inconclusive (14, 15).

Therefore, we here first want to discuss the differences at each level of the vaccine preparation (*see* **Fig. 27.1**) between the reported cancer immunotherapy clinical trials using (m)RNA-loaded DCs.

1.2. DC Generation

Evidence is accumulating that the type of DC plays a critical role in determining the quality and quantity of the immune response (16). There are different precursors and differentiation methods to generate DCs in vitro (17).

In clinical trials with (m)RNA-loaded DCs only the non-proliferating blood monocyte-derived DCs have been used. Except for one study that used buffy coat (18), all groups used leukapheresis to obtain peripheral blood mononuclear cells (PBMCs) from patients (**Table 27.1**). On the one hand, this is because one single leukapheresis delivers enough PBMCs to generate sufficient numbers of DCs for vaccination purposes, even in children (6–8, 19, 20). On the other hand, leukapheresis provokes relatively minor "physical stress" to the patient. A first step in the handling of the leukapheresis product is the elimination of contaminating red blood cells and platelets by washing or by density gradient centrifugation. Next monocytes are enriched from the PBMCs, most often by plastic adherence (**Table 27.1**). When using cell factories, large numbers of DCs can be generated this way in a closed system according to "good manufacturing procedure" (GMP) conditions (9), a prerequisite for use in clinical trials. This, however, requires large amounts of medium, growth factors, and incubator space. Additionally, inefficient detachment of DCs from the plastic wall may lead to cell loss. Therefore, other techniques have been developed such as magnetic cell separation, and subsequent culture in suspension in gas permeable Teflon bags (19). Both positive selection of the $CD14^+$ monocytes (21) and depletion of $CD3^+$ T cells and $CD19^+$ B cells (20, 22) are being used in clinical trials. The latter has the advantage to leave the monocytes "untouched"; however, to our knowledge, no adverse events have been reported about using positively selected $CD14^+$ cells for DC differentiation. One trial depleted for CD3, CD16, CD56, and CD19 cells after plastic adherence (23). They reported a 50% decrease of DC yield and an increased production time, while no differences were seen between depleted and undepleted DCs in phenotype or in in vitro function.

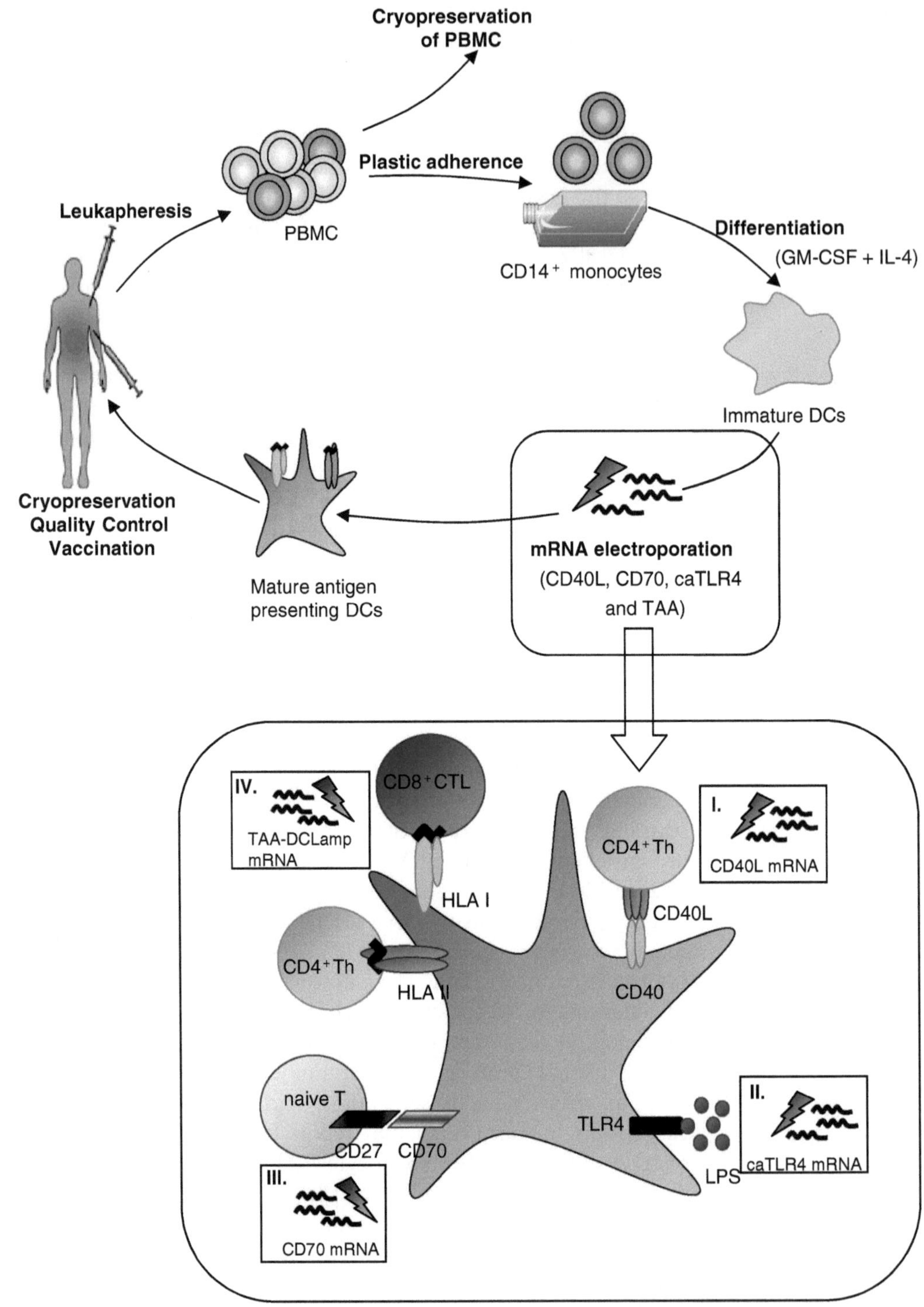

Fig. 27.1. Preparation of an mRNA-based DC vaccine. PBMCs are collected from the blood of cancer patients through leukapheresis. Part of the PBMCs is cryopreserved for immunomonitoring. DC precursors (CD14$^+$ monocytes) are isolated from the remaining PBMCs through a 2 h plastic adherence. During a 6 days culture in GM-CSF and IL-4, CD14$^+$ monocytes are differentiated into immature DCs. These immature DCs are then electroporated with mRNA-encoding

Therefore, the depletion step was omitted in the phase II study conducted by the same group (23).

An improved approach for monocyte purification is counterflow elutriation (8). This would lead to at least two times more and purer DCs than the gradient centrifugation/adherence-based monocyte enrichment, whereas morphology, maturation markers, viability, migratory capacity, possibility to electroporate, and priming capacity of MelanA-specific $CD8^+$ T cells remain identical (8). Moreover, no additional step like washing or gradient centrifugation is required when purifying monocytes by elutriation (8). Although no clinical trials using mRNA-loaded DCs generated by elutriation have been reported yet, several groups are currently using this attractive technique for their DC generation.

Although several cytokine combinations can be used to differentiate monocytes into immature DCs (17), in all completed RNA-DC clinical trials differentiation to DCs was performed before transfection, in the presence of IL-4 and GM-CSF (**Table 27.1**). The concentrations used, however, are not always mentioned as International Units, making it difficult to compare the different studies. Also, the medium used during each step for DC generation can affect the DC function (24). Basically any medium resulting in functional DCs can be used, supplemented with autologous plasma (protocol below). Often, serum-free medium like AIM-V, X-vivo 15, or CellGro DC medium is used in clinical grade protocols (6, 7, 20, 22, 23, 25–30). The latter is recommended to standardize the DC generation procedure and to avoid contamination with undefined plasma proteins (31).

1.3. Antigen Loading of DCs

1.3.1. Tumor Antigen Formulations

Antigens can be loaded onto DCs under many forms, among them are (HLA class I- or II-restricted) peptides, proteins, genetic material like DNA and RNA, or whole tumor cell content (lysate, apoptotic bodies, or DC–tumor cell fusions) (32, 16).

Up till now, defined peptides have been most widely used in clinical trials. However, their use has several disadvantages: they are dependent on prior identification and on HLA restriction and the HLA–peptide complexes have a short half-life (33).

Fig. 27.1. (continued) CD40L, caTLR4, CD70, and a TAA. Electroporation with CD40L (I) and caTLR4 (II) mRNA generates phenotypically mature, cytokine/chemokine-secreting DCs, while the electroporation of CD70 (III) provides a co-stimulatory signal to $CD27^+$ naive T cells by inhibiting activated T-cell apoptosis and by supporting T-cell proliferation. Concomitantly to electroporation with TriMix mRNA, DCs are loaded with TAA through mRNA electroporation (IV). Linking the TAA to the HLA class II targeting sequence of DC-LAMP ensures presentation of TAA-derived peptides in both HLA class I and class II molecules. After electroporation, DCs are cryopreserved after which an extensive quality control is performed. These DCs are then used for vaccination.

Table 27.1
Overview of clinical trials using (m)RNA-loaded DCs

	Generation	Antigen loading		Maturation	Administration	Vaccination scheme				
Reference number, author, year of publication, and cancer type	Origin of DC precursor, monocyte enrichment, and differentiation	Origin of mRNA + expressed antigens	Loading onto DC	Pre- or post mRNA-loading stimuli		Dose	Injections	Number of patients	Immune response	Clinical response
(18), Rains et al. (2001), colorectal cancer	Buffy coat plastic adherence RPMI 1640 + 5% fetal bovine serum + IL-4 (50 ng/mL) + GM-CSF (500 U/mL)	Autologous total tumor RNA (cryopreserved tumor) antigens unknown	Co-incubation 2 × 10E6 DC/mL + 25 µg RNA/mL + 50 µg/mL KLH 4 h	No[a]	i.v.	All DC from 1/2 buffy coat injection 2 and 4 ± half the dose of 1 and 3 as a result of cryopreservation	4 monthly	15	11/13: positive DTH skin reaction to KLH 2/13: negative DTH skin reaction to KLH 4/13: unaltered rate of CEA rise 2/13: reduced rate of CEA rise 7/13: CEA fall	3 patients died during vaccination 0/12: regression of disease 6/12: marked progression or development of new sites (short follow-up)
(25), Heiser et al. (2002), metastatic prostate cancer	Leukapheresis plastic adherence AIM-V (serum free) + IL-4 (500 U/mL) + GM-CSF (800 U/mL)	Defined mRNA PSA	Co-incubation 10 × 10E6 DC/mL + 15 µg mRNA/mL 45 min	No[a]	i.v. i.d.	$n = 3$: 10 × 10E6 DC $n = 4$: 30 × 10E6 DC $n = 6$: 50 × 10E6 DC + each time 10 × 10E6 DC	3 biweekly	13	8/8: increased PSA-specific IFNγ secretion frequency ranged from 1/10,000 to 1/2,000 PBMC increase was more pronounced in medium and high dose treated patients 9/9: increased lysis of CD8 T cells post-vaccination after 2 in vitro stimulations	Serum PSA levels measured as surrogate for clinical response 6/7: absolute PSA concentration continued to rise 1/7: decrease of absolute PSA level 1/7: unchanged log slope PSA 5/7: reduction of log slope PSA 1/7: decrease in log slope PSA hematagenous micrometastases: 3/3: transient elimination of tumor cells from peripheral blood

(27), Su et al. (2003), renal cell carcinoma	Leukapheresis plastic adherence AIM-V (serum free) + IL-4 (500 U/mL) + GM-CSF (800 U/mL)	Autologous total tumor RNA G250, hTERT, and OFA	Co-incubation 10 × 10E6 DC + 50 µg RNA/mL 1 h	No[a]	i.v. i.d.	$n = 8$: 10 × 10E6 DC $n = 2$: 30 × 10E6 DC + each time 10 × 10E6 DC	3 biweekly	15	6/7 expansion of tumor-specific T cells antigen-specific T-cell responses (G250, OFA, and hTERT) consistently inferior compared to overall anti-tumor response. Minor cross-reactivity to normal renal tissue antigens 1/1: increased cytotoxicity post- versus pre-vaccination against an autologous tumor cell line	5/15 withdrawal from study because rapid and symptomatic tumor progression 3/15 died after vaccination 7/15 stable or slowly progressing asymptomatic disease but 8/10 underwent other therapies after vaccination
(23), Morse et al. (2003) phase I: metastatic CEA expressing cancer	Leukapheresis plastic adherence AIM-V (serum free) + IL-4 (500 U/mL) + GM-CSF (800 U/mL) Phase I: + depletion with α-CD3, α-CD16, α-CD19, and α-CD56	Defined mRNA CEA	Co-incubation 5 × 10E6 DC/mL + 10 µg mRNA/mL 2–4 h	No[a]	i.v. i.d.	Phase I: $n = 7$: 10 × 10E6 DC (L) $n = 8$: 30 × 10E6 DC (IM) $n = 8$: 100 × 10E6 DC (H) + 1 × 10E6 DC for IM and H	Phase I: 4 weekly (L) or biweekly (IM and H)	Phase I: 29	Phase I: 16/18 retained recall antigen response 3/19: DTH reaction	Phase I: 1/23 complete response 2/23 minor response 3/23 stable disease 18/23 progressive disease
Phase II: colon cancer after resection of hepatic metastasis	Phase II: no extra depletion				i.v. i.d.	Phase II: 30 × 10E6 DC + 1 × 10E6 DC	Phase II: 4 biweekly	Phase II: 13	Phase II: 11/12 retained recall antigen response 6/13 DTH reaction 1/4 DTH reaction with unpulsed DC 5/5 DTH reaction on DC + HbsAg	Phase II: 3/13 died 9/13 relapsed 3/13 no evidence of disease

(continued)

Table 27.1 (continued)

	Generation	Antigen loading		Maturation	Administration	Vaccination scheme				
Reference number, author, year of publication, and cancer type	Origin of DC precursor, monocyte enrichment, and differentiation	Origin of mRNA + expressed antigens	Loading onto DC	Pre- or post mRNA-loading stimuli		Dose	Injections	Number of patients	Immune response	Clinical response
(6), Caruso et al. (2004), brain tumor	Leukapheresis plastic adherence AIM-V (serum free) + IL-4 (25 ng/mL) + GM-CSF (800 ng/mL)	Autologous total tumor RNA antigens unknown	Co-incubation 5 × 10E6 DC/mL + 25 μg RNA/mL 45 min	No[a]	i.v. i.d.	$n = 7$: 5 × 10E6 DC/m^2 $n = 0$: 15 × 10E6 DC/m^2 $n = 0$: 50 × 10E6 DC/m^2 + each 5 × 10E6 DC/m^2	Combined diphtheria and tetanus vaccine 1 week before DC vaccine: 3 biweekly If stable disease: 3 extra 3-monthly	9	0/4 tumor-specific IFNγ secretion 0/3 tumor-specific proliferation 2/5 modest tumor-specific humoral response	8 weeks after start DC vaccine: 4/7 stable disease 1/7 partial response 2/7 progressive disease 18 weeks after start DC vaccine: 2/5 stable disease 0/5 partial response 3/5 progressive disease
(26), Su et al. (2005), metastatic prostate cancer	Leukapheresis plastic adherence X-VIVO 15 (serum free) + IL-4 (500 U/mL) + GM-CSF (800 U/mL)	Defined mRNA hTERT or LAMP-1 hTERT	Electroporation 1 μg mRNA/1 × 10E6 DC exponential decay 300 V, 150 μF ViaSpan buffer 4-mm cuvette 5 min incubation before pulse	Post-IL-1β, TNFα IL-6, PGE2 for 20 h	i.d.	10 × 10E6 DC	6 LAMP patients: 3 6 TERT patients: 3 3 LAMP patients: 6 5 TERT patients: 6 Weekly	20	After 3 cycles of vaccination: Blood: 11/12: hTERT-specific CD8 T cells 9/12: hTERT-specific CD4 T cells higher frequency of specific CD4 T cells in LAMP-hTERT group DTH: 10/12 DTH reaction tends toward more pronounced DTH in LAMP-hTERT group After 6 cycles of vaccination: CD8 T-cell response peaked 2–4 weeks after the final (6th) dose	No objective clinical response LAMP-1: 4/4 reduced circulating tumor cells hTERT: 5/6 reduced circulating tumor cells in both only transiently

(7), Caruso et al. (2005), neuroblastoma	Leukapheresis plastic adherence AIM-V (serum free) + IL-4 (25 ng/mL) + GM-CSF (800 ng/mL)	Autologous total tumor RNA antigens unknown	Co-incubation 5 × 10E6 DC/mL + 25 μg RNA/mL 45 min	No[a]	i.v. i.d.	$n = 7$: 5 × 10E6 DC/m² $n = 0$: 15 × 10E6 DC/m² $n = 0$: 50 × 10E6 DC/m² + each 5 × 10E6 DC/m²	Chemotherapy, after 6 weeks rest combined diphtheria and tetanus vaccine after 1 week DC vaccine: 3 biweekly If stable disease: 3 extra 3-monthly	11	0/5 tumor-specific T-cell proliferation slightly increased proliferation to PHA 3/4 increased proliferation to SEB 2/3 tumor-specific humoral response 1/6 Ab titer for tetanus and diphtheria boosted	0/4 clinical response 10/11 death 1/11 stable disease
(22), Mu et al. (2005), androgen-resistant prostate cancer	Leukapheresis depletion of B and T cells with ISOLEX 300i magnetic cell selector CellGro DC medium (serum free) + IL-4 (1,000 U/mL) + GM-CSF (2,500 U/mL) in Teflon bags	Total mRNA of 3 allogeneic tumor cell lines, each line: 50 × 10E6 cells antigens unknown	Electroporation 0.2–1 μg/μL mRNA – 120 μL square wave	Post-IL-1β, TNFα IL-6, PGE2 for 48 h	i.n. i.d.	$n = 10$: 20 × 10E6 DC $n = 9$: 20 × 10E6 DC	4 weekly 5 patients: + boost vaccine	20	DTH: 3/7 i.n. DTH reaction 2/9 i.d. DTH reaction IFNγ ELISPOT and proliferation: 4/10 i.n. antivaccine-specific response 8/9 i.d. antivaccine-specific response	5/10 i.n. progressive disease 5/10 i.n. stable disease 7/10 i.n. decreased log slope PSA 3/9 i.d. progressive disease 6/9 i.d. stable disease 6/9 i.d. decreased log slope PSA
(28), Dannull et al. (2005), metastatic renal cell carcinoma (RCC) and ovarian carcinoma (OVA)	Leukapheresis plastic adherence X-VIVO 15 (serum free) + IL-4 (500 U/mL) + GM-CSF (800 U/mL)	Autologous total tumor RNA antigens unknown	Electroporation 5μg RNA/1 × 10E6 DC exponential decay 300 V, 150 μF ViaSpan buffer 4-mm cuvette 5 min incubation before pulse	Post-IL-1β, TNFα IL-6, PGE2 for 20 h	i.d.	10 × 10E6 DC	3 biweekly 7 received 18 μg/kg DAB_{389}IL-2 (ONTAK) (i.v.) 4 days in advance	10 RCC 1 OVA	10/11 specific anti-tumor CD4 and CD8 T cells Treg depletion results in higher numbers of tumor-specific CD8 T cells trend toward improved CD4 T-cell response CD8 T-cell response is prolonged and peaked 2 weeks after the final (third) dose stimulation Th1 CD4 response	DC vaccine alone: 2/4 alive with disease 2/4 death of disease Treg depletion + DC vaccine 6/7 alive with disease 1/7 death of disease

(continued)

Table 27.1 (continued)

Reference number, author, year of publication, and cancer type	Generation	Antigen loading		Maturation	Administration	Vaccination scheme				
	Origin of DC precursor, monocyte enrichment, and differentiation	Origin of mRNA + expressed antigens	Loading onto DC	Pre- or post mRNA-loading stimuli		Dose	Injections	Number of patients	Immune response	Clinical response
(20), Kyte et al. (2006), melanoma	Leukapheresis depletion of B and T cells CellGro DC medium (serum free) + IL-4 (1,000 U/mL) + GM-CSF (2,500 U/mL) in Teflon bags	Autologous tumor mRNA antigens unknown	Electroporation 2–324 µg mRNA/10E8 DC Single square wave pulse 800 V, 200 µs BTX 2-mm cuvette set cooled in tubes for 1 min after electroporation	Post-IL-1β, TNFα IL-6, PGE2 for 48 h	i.n.i.d.	$n = 9$: 20 × 10E6 DC/vial $n = 1$: 15 × 10E6 DC/vial $n = 1$: 8 × 10E6 DC/vial $n = 1$: 5 × 10E6 DC/vial $n = 8$: 20 × 10E6 DC/vial $n = 1$: 19 × 10E6 DC/vial $n = 1$: 12 × 10E6 DC/vial Actual amount administered not mentioned	4 weekly	22	Increased T-cell proliferation: 3/12 i.n. injected 7/10 i.d. injected DTH: 2/8 i.n. injected 6/10 i.d. injected	From the evaluable patients: stable disease: 1/12 i.n. injected 1/8 i.d. injected progressive disease: 11/12 i.n. injected 7/8 i.d. injected
(21), Markovic et al. (2006), stage IV melanoma	Leukapheresis CD14+ purification with CliniMACS X-VIVO 15 + 1% human AB serum + IL-4 (1,000 IU/mL) + GM-CSF (800 IU/mL) in polysterene flasks	Autologous tumor mRNA in vitro amplified 3 patients: all determined (gp100, Tyrosinase, and MART-1) 1 patient: all, except MART-1 1 patient: none the determined 1 patient: not determined	Electroporation 20–50 µg mRNA/mL 10×10^6 DC/mL two square wave pulses 50 µs each, 400 V, 500 ms pulse interval cytoporation formula R medium 4-mm cuvette	Post-TNFα, PGE2 for 48 h	s.c.	5 × 10E6 DC	4 3-weekly	6	Blood: No DC-induced proliferation or IFNγ release	0/5: objective clinical response 1/6: progressed prior to first injection 1/6: progressed at second injection 3/6: progressed at third injection 1/6: remained stable

(29), Nair et al. (2002), metastatic lung cancer adenocarcinoma	Leukapheresis plastic adherence AIM-V (serum free) + IL-4 (500 U/mL) + GM-CSF (800 U/mL)	Total tumor RNA antigens unknown (patient received earlier a CEA-mRNA-loaded DC vaccine)	5 × 10E6 DC/mL + 50 μg RNA	No	i.v. i.d.	30 × 10E6 DC + 1 × 10E6 DC	4 monthly (start 6 months after CEA vaccination)	1	Increased lytic activity of PBMC after total tumor RNA vaccination compared to CEA mRNA vaccination	No patient continued to have progressive disease
(30), Kyte et al. (2007)	Follow-up of 2 patients from the study by Kyte et al., (20)					2	First patient: 5 boosts 6, 15, 18, 21, and 23 months after vaccination Second patient: 1 boost after 3.5 months after 9 months: second round of DC vaccine (4 injections weekly)	2	Patient 1: after boost, sustained T-cell response for 3–9 months (proliferation + cytokines) boost resulted in more Th1-directed profile Patient 2: 36 clones antivaccine-specific from which 10/11 had different T-cell receptor mixed Th1/Th2 cytokines by monoclonal clones 11 clones recognized both mock and tumor antigen-expressing DC 47 clones were non-responsive (proliferation)	Patient 1: some pulmonary metastasis remained stable, moderate disease progression 23 months of DC vaccination: good general condition Thereafter: general disease progression death 31 months after study entry Patient 2: death 23 months after start vaccination due to loss of recognized antigens by tumor

(continued)

Table 27.1 (continued)

	Generation	Antigen loading		Maturation	Administration	Vaccination scheme				
Reference number, author, year of publication, and cancer type	Origin of DC precursor, monocyte enrichment, and differentiation	Origin of mRNA + expressed antigens	Loading onto DC	Pre- or post mRNA-loading stimuli		Dose	Injections	Number of patients	Immune response	Clinical response
Ongoing trial stage III and IV melanoma	Leukapheresis plastic adherence RPMI 1640 + 1% autologous plasma + IL-4 (500 U/mL) + GM-CSF (1,000 U/mL)	Defined mRNA sig-gp100-DC-LAMP sig-tyrosinase-DC-LAMP sig-mage-C2-DC-LAMP sig-mage-A3-DC-LAMP	Electroporation 50 × 10E6 DC 60 μg mRNA of each TAA 20 μg mRNA of each functional protein exponential decay 300 V, 450 μF, 99 Ω OptiMix buffer 4-mm cuvette	Concomitant CD40L, CD70, and caTLR4 mRNA	i.d.	50 × 10E6 DC + 0.5 × 10E6 DC	4 biweekly thereafter, every 8 weeks if absence of disease progression + for DTH together with the first injection and 1 week after the fourth vaccine	Ongoing		

The cancer type, generation, antigen loading, maturation, administration, and vaccination scheme of DCs as well as the immunological and clinical outcomes of the published trials are shown, representing the numerous differences between the completed studies. Studies are placed in chronological order of publication. From the last three studies mentioned, the first two are a summary of the follow-up of, respectively, one and two patients from the above-mentioned trials. The last study is our own approach, shown for comparison with the published trials.

DC: dendritic cell, KLH: keyhole limpet hemocyanin, CEA: carcinoembryonic antigen, HbsAg: hepatitis B surface antigen, L: low, IM: intermediate, H: high, OFA: oncofetal antigen, hTERT: human telomerase reverse transcriptase, LAMP-1: lysosome-associated membrane protein-1, DC-LAMP: dendritic cell lysosome-associated membrane protein, n: number of patients.

[a]It is shown that the uptake of RNA by simple co-incubation already results in some maturation of DCs, albeit suboptimal (25).

Therefore, the use of whole antigens sounds appealing. Indeed, DCs modified to express whole tumor antigens will present multiple, even unidentified epitopes in many HLA alleles, which will recruit a wide spectrum of T cells. DCs can be loaded with whole tumor antigen under the form of proteins, mRNA, and DNA.

Proteins being taken up by DCs enter the HLA class II presentation pathway for $CD4^+$ Th stimulation. Moreover, cross presentation leads to HLA class I presentation and activates $CD8^+$ CTL necessary for tumor eradication (1, 2).

On the contrary, when DNA or mRNA are used, antigens are produced endogenously by the DCs and will most efficiently be presented in HLA class I (34). To obtain Th stimulation by class II presentation, the genetic constructs can be modified, for instance, by flanking the antigen with HLA class II sorting signals, e.g., derived from the invariant chain, DC-LAMP, or LAMP-1 (35).

The use of mRNA is preferable to DNA. When producing mRNA, contamination with plasmid and/or genomic bacterial DNA is of no major concern as it is degraded during an incubation step with DNases after the mRNA transcription (36) and mRNA synthesis will only occur from the plasmid. Also, the use of mRNA bypasses the complex issues of transcriptional regulation, and the mRNA has easy access to the translation machinery upon entry into the cytoplasm, bypassing the need for nucleic factors (32, 37). Consequently, RNA leads to higher transfection efficiencies and lower cell mortality (37). Finally, the RNA will rapidly degrade after transfection and cannot integrate into the host genome (38) and is therefore allowed for application in the clinic. Consequently, mRNA is safer and easier to handle and is chosen by many groups for DC modification, including ours.

1.3.2. Origin of Tumor Antigen-Encoding (m)RNA

In general, two types of antigen-encoding mRNA are used to attempt triggering or boosting an anti-tumor immune response: defined mRNA, encoding a tumor antigen of choice, or total tumor-derived (m)RNA. The advantages and disadvantageous of using either synthetic mRNA or whole tumor (m)RNA are summarized in **Table 27.2**.

1.3.2.1. Defined or Synthetic mRNA

The discovery of tumor-associated antigens (39) formed a milestone that led to the use of defined mRNAs encoding a single antigen. By using the current knowledge of tumor-associated antigens a *tumor antigen-specific* immune response can be induced, possibly *reducing adverse events* associated with the use of total tumor RNA, such as autoimmunity (40), or negative effects caused by other proteins expressed by the tumor (21).

Several groups of antigens exist, among which to choose to load on DCs. A first subdivision that can be made is the

Table 27.2
Overview of the advantages and disadvantages of each of the two (m)RNA sources that can be used to load DCs. More detailed explanation of each topic is provided in the main text

Synthetic mRNA	Whole tumor (m)RNA
Advantages	*Disadvantages*
• No tumor biopsy required[a]	• Only patients with resectable tumor or metastasis (advanced stage)[b]
• Possibility to optimize mRNA (ARCA, UTR, A-tail length, and HLA-targetting sequences)	• Amounts of mRNA available is limited and highly variable in between patients[c]
	• mRNA from one metastasis might not bear all antigens of primary tumor
• High presentation of chosen antigens (subdominant antigens and antigens necessary for tumor cells)	• Lower expression/presentation of each antigen
• Fewer adverse effects due to autoimmunity expected	• Adverse events due to mRNA encoding suppressive factors or autoimmunity possible
• Possibility to generate large batches of mRNA at once	• Labor intensive (separate RNA preparation for each patient)[b]
Disadvantages	*Advantages*
• Presentation of limited number of TAA	• Presentation of all, known and unknown, tumor expressed antigens
• No patient specificity	• Presentation of the patient's TSA
• Not aware if TAA is expressed by the patient's tumor	• Applicable to any tumor type that is removable

ARCA: anti-reverse cap analog, UTR: untranslated region, HLA: human leukocyte antigen, TAA: tumor-associated antigen, TSA: tumor-specific antigen.

[a] In case identification of TSA is desired, a tumor biopsy will be necessary.

[b] Can be circumvented by using allogeneic tumor cell lines.

[c] Can be circumvented by in vitro amplification.

discrimination between tumor-associated antigens and tumor-specific antigens.

The *tumor-associated antigens* (TAAs) can be subdivided based on the potential tolerance for self-proteins. The three main groups are (1) cancer germline antigens, (2) differentiation antigens, and (3) overexpressed antigens (40).

The *cancer germline antigens*, also referred to as shared tumor specific, cancer testis (CT), or germinal antigens, are found in different human tumor types, in germ cells of healthy adult testis and fetal ovary, and sometimes in placental trophoblasts (41). Because their expression only takes places in healthy cells lacking HLA class I expression, they are "invisible" for CTL. An immune response toward this class of tumor antigens therefore solely targets tumor cells. The shared tumor-specific antigens include, among others, the members of the MAGE family (40). In the ongoing clinical trial for which the protocol is described below Mage-A3 and Mage-C2 are included.

(Tissue) differentiation antigens are not only expressed by the tumor but also in the normal tissue of origin of the malignancy. Antigens that belong to this group are, among others, MelanA/MART-1, gp100/Pmel17, Tyrosinase, CEA, and PSA. The latter four have been used in clinical trials with mRNA-loaded DCs (23, 25) and in the ongoing trial for which the protocol is described below.

Overexpressed antigens are present at a much higher level in cancer tissues compared to normal tissues (40–42). Examples are the Wilms' tumor 1 protein (WT1), proteinase 3 (Pr3), the tumor suppressor p53, survivin, and human telomerase reverse transcriptase (hTERT) (40–42). Only hTERT has already been used in published clinical trials with RNA-loaded DCs (**Table 27.1**) (26).

Thus, all three groups contain proteins expressed in normal tissues. This raises the concern that the repertoire of reactive T cells present in vivo could be limited due to self-tolerance mechanisms, especially for the high-avidity T cells (40, 42). The observation that all MAGE-A3 specific, HLA-A1-restricted CTL clones, derived from many patients after different immunotherapies, had a rather poor affinity strengthens this worry (40). However, for several other MAGE-derived epitopes high-affinity CTL clones could be derived (40). Moreover, in a recent study where peptide or ALVAC vaccination was compared with peptide-loaded mature DCs as immunotherapy adjuvants, it has been shown that high functional avidity was unlikely to be crucial for the capacity of antigen-specific CTL to trigger tumor regression in melanoma patients (43). Furthermore, antibodies and WT-1-specific $CD8^+$ CTL can be found in individuals after allogeneic hematopoietic stem cell transplantation, providing evidence that the overexpressed WT-1 antigen is still immunogenic (42) and no difference is observed in the induction of PSA-specific CTL between male and female healthy volunteers or cancer patients in vitro (44). These last observations, however, strengthen the worry for the occurrence of autoimmunity when using overexpressed antigens. This concern is less explicit when using cancer germline genes or differentiation antigens: the former completely circumvents possible autoimmunity, while the latter restricts rejection to the tissue of origin of the tumor (40). The severeness of the complication is then dependent on the type of healthy tissue that is rejected.

Besides the TAA, *tumor-specific antigens* (TSA) or unique antigens have been described, which are only expressed in cancer cells and characterize individual tumors. They can differ in tumor tissue fragments of the same tumor nodule and can be generated during tumor progression (45). TSA are novel proteins that arise as a result of somatic mutation, internal deletions, chromosome translocation, alterations in RNA splicing, or normal immunoglobulin (IG) or T-cell receptor (TCR) recombination events specific to leukemic B- or T-cell idiotypes (41)(45). Immune responses to TSA are found to be *associated with a clinical benefit* and bear potentially *stronger immunogenicity* (45). In melanoma patients the natural responses against TSA were predominant during the favorable course of the disease both for class I and class II HLA-restricted antigens, compared to shared antigens. Moreover, T-cell populations used for adoptive therapy leading to tumor regression were directed against TSA. Furthermore, patients who were free of disease after surgery or chemotherapy maintained a strong activity over time against TSA, while the activity against self-common antigens (TAA) faded away in the absence of tumor burden (45).

According to Parmiani et al. (45), several mechanisms may be responsible for more effective immunity against TSA than against shared antigens: (1) the exquisite tumor specificity of TSA, (2) lack of any possible form of tolerance as compared with TAA, (3) multiple expression by a single tumor, and (4) resistance to host immunoselection in case the unique antigen is essential to the maintenance of the neoplastic conditions (45). Because of their tumor specificity, using TSA also eliminates autoimmunity issues.

To date, no RNA-DC clinical trial has made use of TSA under the form of defined antigens (**Table 27.1**), most probably because of the difficulty to identify and molecularly characterize unique antigens at the single tumor/patient level (45). The use of autologous whole tumor derived material, as described below, is therefore a more clinically applicable approach to make use of the advantages of TSA.

For successful immunization, the antigen must be carefully chosen, keeping several considerations in mind. First, it is evident that immunotherapy with DCs loaded with mRNA encoding only one tumor antigen leads to a high risk on tumor escape due to immunoselection. This process involves mutation or loss of the recognized antigen by the tumor, as a consequence of immunologic pressure, allowing tumor variants to grow (46, 47). Therefore, it is advisory to *target multiple antigens* (31). Targeting multiple antigens also leads to a greater chance that the targeted tumor actually expresses the chosen antigens in case determination of expressed tumor antigens is not possible in advance (for example, in case the tumor mass is inaccessible for biopsy, in case of minimal residual disease after radiaton or chemotherapy, or in case of early-stage cancer treatment). Moreover, individuals may vary dramatically in their available T-cell repertoire. Providing a broad source of TAA immunogens therefore might allow them to mount responses against the tumor (41). Using synthetic mRNA to target multiple antigens has the advantage, compared to using whole tumor mRNA, that it is possible to make a mix of DCs presenting different antigens in which each DC only presents the epitopes derived from a particular antigen (cf. protocol below). This allows the presentation of *subdominant antigens* to the T cells by avoiding the concept of "priority of the first" (41, 48).

Second, the characteristics of the chosen antigens are of importance for successful immunization against cancer. *Proteins necessary for tumor cell maintenance or metabolism* will remain conserved during the evolution of the disease avoiding tumor escape by immunoselection (31, 45). Therefore, an immune response directed toward proteins crucial for tumor cell survival may have a better chance to lead to clinical benefit (45). In a follow-up study, a broad polyclonal immune response present after vaccination with total tumor mRNA-loaded DCs could not prevent tumor escape by loss of the recognized antigens, which indicates the importance of the characteristics of the presented tumor antigens (30). Proteins with a role in the oncogenic process are often found in the category of the overexpressed antigens and TSA (42, 45). Examples are survivin and telomerase.

When using defined tumor antigen-encoding mRNA, the mRNA is usually synthesized in vitro. To overcome the intrinsic instability of this synthetic RNA, a lot of *optimizations* have been introduced in order to obtain a more efficient translation or higher stability of the mRNA (*see* also **Sections 3** and **4**) and consequently more and longer protein expression, which is believed to be beneficial. This includes adjusting the poly-A tail length, the absence of restriction site-derived nucleotides at the very end of the poly-A tail, the origin, location, and amount of untranslated regions (49), optimizations of the codon sequence for translation by human cells (50), and the use of anti-reverse cap analog (ARCA) (51, 52).

When antigen loading of the DCs is performed with defined mRNAs, mRNA encoding a fusion protein of a tumor antigen with a HLA class II targeting signal (e.g. from invariant chain, DC-LAMP and LAMP-1) can be used (35). This results in antigen presentation in HLA class II, necessary for the induction of an effective CD4$^+$ T-cell response, as well as in an enhanced antigen presentation in the context of HLA class I (35, 53). hTERT-LAMP-1 mRNA-electroporated mature DCs led in vivo to significantly higher frequencies of hTERT-specific CD4$^+$ T cells and an increased cytotoxicity of the CD8$^+$ T cells (26). Furthermore, this clinical trial shows that the LAMP-1 sequence does not carry any intrinsic immunogenicity, so the improved TAA-expressing mRNA does not induce cross-reactivities against other LAMP-expressing cells (26).

1.3.2.2. Total Tumor-Derived (m)RNA

Because of the genetically unstable nature of cancer cells (54), single-antigen vaccines may lead to selection of genetic variants of tumor cells that can escape the too "narrow" immune response (55). Loading a broad spectrum of antigens rather than a single antigen onto DCs potentially may give rise to more effective anticancer immune responses (34, 56).

In this regard *total tumor (m)RNA* has the major advantage to allow antigen presentation of *all, known and unknown, tumor antigens*, including the *patient-specific tumor antigens* and even extends the scope of vaccination to *rare cancer forms* (34, 48, 56).

Delivering total tumor RNA to DCs can be performed in the form of extracted total tumor RNA, enriched poly-adenylated mRNA, or tumor-DC hybrids (34, 48, 57). Since in the latter approach antigen presentation is not solely due to the mRNA present, we will not discuss this in further detail. Although loading mature DCs with total tumor RNA has been shown to result in antigen presentation (48) and to induce immunological and clinical responses in vivo (6, 20, 27), the technique has some drawbacks.

First, it is unavoidable that many self-proteins also are presented by the DCs and therefore this technique relies on the capacity of the immune system to differentiate between self-proteins and tumor-specific antigens. So far, no *autoimmune reactions* have been observed in clinical trials (6, 7, 18, 21–23, 27, 28) although cellular and clinical responses have been found (**Table 27.1**). It must, however, be mentioned that at least in three studies, patients with a history of autoimmune disease were excluded (23, 27, 28). Absence of autoimmunity is promising for the use of total tumor mRNA; however, when vaccine formulations improve, the chance on autoimmune responses might increase and should be closely followed up (40, 58). Especially because it has been reported that IFN-γ secretion by activated T cells after vaccination with DCs loaded with renal tumor RNA, increased consistently above background levels when challenged with benign renal epithelium mRNA-loaded DCs. This suggests minor cross-reactivity of the vaccine-induced T cells (27). Also after vaccination of melanoma patients with autologous total tumor mRNA-loaded DCs, mock-transfected DCs responsive T-cell clones could be generated (30). Furthermore, in in vitro experiments with prostate cancer patient material, it has been shown that total tumor RNA-transfected DCs can induce CTLs, which kill normal prostatic tissue antigen-expressing target cells, through the recognition of prostate-specific antigen (PSA) (56).

Second, loading of the DCs needs to be performed with a *minimal amount of tumor mRNA* (56, 59, 60). Dendritic cells electroporated with mRNA derived from melanoma cell lines with high telomerase activity did not result in any significantly measurable telomerase activity by the DCs. Of note, electroporation of synthetic hTERT-mRNA did result in a significant telomerase activity in DCs. This points to an important concern for the use of total tumor RNA in cancer vaccines; each antigen will be present at minute concentrations (61). In addition, the CTL stimulatory capacity of DCs loaded by co-incubation with different ratios of tumor-derived RNA versus PBMCs-derived RNA demonstrated in vitro that a certain threshold of tumor RNA is required to stimulate tumor-specific T cells. This emphasizes the need for pure tumor-derived RNA and tumor enrichment of samples with limited tumor content (56).

The amount of RNA isolated from each gram of tumor was usually quite *variable* between the different patients vaccinated in one trial (6, 20). Nevertheless, one study reported no statistically significant correlation between the vaccine responders and nonresponders and the mRNA concentrations when all available mRNA was used for each patient (20). About a decade ago, in vitro *amplification* of total tumor-derived RNA was reported to be a feasible technique in order to enlarge the RNA amount and even to enrich the mRNA fraction (62). In a clinically applicable fashion, nonlimiting amounts of tumor mRNA can even be generated starting from cryosections of microscopic tumor samples with laser capture microdissection. This technique also allows selective procurement of tumor cells. It is followed by RNA extraction, RT-PCR-based amplification, and in vitro transcription to obtain full-length mRNA suitable for DC transfection (56, 62). Thus, in case too low mRNA yields are achieved, in vitro amplification of tumor-derived mRNA can be a possibility (56, 63). This will lead to enough tumor mRNA of all patients where biopsies are available from, allowing to use equal quantities to load the DCs of each patient. Amplified tumor RNA has been used in a study with advanced melanoma patients (21). Here, however, the viability of immature electroporated DCs with in vitro amplified mRNA is lower after freezing–thawing. Although the authors did not look into further detail about what caused this observation, they suggest that some in vitro amplified RNA species might interfere with DC recovery (21). Indeed, it is not unthinkable that the functional proteins expressed by the tumor cells, and after loading with total tumor mRNA also by the DCs, have a deleterious effect on the latter.

Finally, the *requirement of a tumor biopsy* for total tumor (m)RNA extraction makes it only feasible for patients with resectable disease (32). This is in contrast with the theory on immunosurveillance, where it is postulated that DC-based immunotherapy might have its highest potential in early stages of disease or in patients with minimal residual disease (32, 58). In cases autologous tumor material was not available, mRNA derived from three *allogeneic tumor cell lines* revealed to be a good alternative for immunotherapy in androgen-resistant prostate cancer patients (22). In vitro, CTLs generated with DCs loaded with RNA from one tumor cell line were able to lyse other tumor cell lines (64).

Of course, the advantage of patient-specific tumor antigen presentation is lost by this approach but on the other hand, using allogeneic tumor cell lines might allow the production of an "off-the-shelf" product suitable for all patients with a particular tumor type (32).

Recently, a major effort has been made to examine the best choice of tumor antigen origin. In this report all clinical DC immunotherapy studies with clinically evaluable patients were included independent of the antigen formulation. Compiling the results of 3,444 patients in 173 published trials showed that a higher percentage of patients mounted an objective clinical response (8.1%) when immunotherapy was applied with whole tumor or tumor extracts as antigens, compared to the use of molecularly defined antigens (3.6% of patients) (p <0.0001, chi-square test). No distinction was made between the antigen constructs used in the different studies (41).

To exclude intrinsic reasons of tumor type causing the beneficial outcome of total tumor-derived antigens, the same analysis was performed only on advanced stage melanoma patients, leading to the same conclusion. Finally, the response rate to autologous versus allogeneic tumor material was investigated, ending up being not significantly different ($p = 0.15$), suggesting no role for direct stimulation by TSA (41).

The same conclusion is drawn in another study where 626 patients with malignant melanoma from 38 immunotherapy trials were subjected to analysis. Additionally, in this study it was stated that the use of defined antigens in the form of peptides is associated with a high incidence of stable disease, whereas vaccination with autologous tumor antigen has a slightly higher, but not statistically significant, tendency to result in a complete or a partial response (14).

1.3.3. Loading DCs with mRNA

There are different ways to load DCs with mRNA for vaccination. A first distinction that can be made is in vivo versus ex vivo loading of DCs. In this chapter, we will mainly focus on the latter approach.

Direct in vivo administration of cell-free mRNA as vaccine formulation is attractive because it is simpler, less laborious, and cheaper than ex vivo loading of DCs. Four mRNA formulations for in vivo delivery to immature DCs have been described (reviewed in (36)): naked mRNA directly injected into the tissues, mRNA entrapped in liposomes, and mRNA loaded on gold particles administered by gene gun (36). However, efficient delivery to the DCs in vivo and concomitant delivery of a maturation stimulus are still at question (65).

Therefore, the most common approach to load DCs with RNA is ex vivo loading. Here, the DCs can be subjected to maturation-inducing agents in vitro, mimicking an inflammatory environment. Ex vivo RNA loading can be performed in different ways: simple co-incubation, receptor-mediated endocytosis, lipofection, electroporation, and particle-mediated transfer (37, 63). So far, only simple co-incubation and electroporation have been used in clinical trials (**Table 27.1**).

Simple *co-incubation* (passive pulsing) of immature DCs with RNA is the easiest way for antigen delivery, but usually the uptake of mRNA is suboptimal (66, 67). Between 10 and 50 μg/mL RNA is added to the DCs for 45 min – 4 h (**Table 27.1**). Generally, this results in undetectable or very low protein expression (66, 67). Thus, it is no surprise that contradictory results have been obtained with respect to the T-cell stimulatory potential of these DCs. Monocyte-derived DCs passively pulsed with Mart-1/MelanA mRNA did not lead to any measurable IFN-γ secretion by an HLA-matched MelanA-specific CTL clone, as measured by ELISA (66). Using a different T-cell clone and the more sensitive ELISPOT assay, some, however, very low antigen presentation by the DCs was observed, resulting in the priming of very low numbers of autologous naïve T cells (67, 68). In contrast, stimulation of PBMCs with DCs passively pulsed with defined PSA mRNA or total tumor RNA is reported to induce antigen-specific CTL in vitro from both healthy individuals as well as from patients with up to 40% of target cell lysis (44, 56). Moreover, several studies report immune responses after vaccination using passively pulsed DCs (**Table 27.1**).

On the other hand, *electroporation* has been validated by different groups as an efficient technique for loading DCs with antigen-encoding mRNA leading to a higher transfection efficiency in comparison to simple co-incubation or lipofection (66–69). Up to now, five clinical studies using mRNA-electroporated DCs have been reported (20–22, 26, 28) and several others are being conducted at this moment (12). mRNA electroporation is a rapid and relatively simple technique that modifies DCs without altering their immunophenotype, their maturation potential, their migration capacity, and their T-cell stimulatory potential (21, 33, 67, 68, 70). Furthermore, electroporated mature DCs can be cryopreserved without loss of these characteristics (68, 71). A possible drawback of electroporation is the reduction of IL-12p70 secretion by DCs (72). Recently, it has been shown that this can be counterbalanced by electroporating IL-12p70-encoding mRNA. Such DCs induce a strong antigen-specific $CD8^+$ T-cell response with a high avidity and also enhance the effector function of natural killer cells (52, 73). Furthermore, they switch the Th2 cytokine profile (IL-5, IL-13) toward a Th1 cytokine profile (IFN-γ, TNF-α) in blood-derived T cells and TIL from melanoma patients (74).

Besides cytokine-encoding mRNA, other exogenous mRNA can be included to ameliorate DC function. RNAs used in vitro include mRNA-encoding E/L selectin to improve DC migration (75) and mRNA-encoding functional proteins, such as GITRL (76), CD83 (77), CD40L, CD70, and a constitutive active form of TLR4 (78), acting on the T-cell stimulatory capacity of DCs. Not only mRNA-encoding tumor antigens, cytokines, or other functional proteins are being electroporated into DCs. Other RNAs such as the maturation-inducing double-stranded RNA analog Ampligen (79) or siRNA inhibiting the expression of proteins like A20 or of cytokines like IL-10 (80) can be electroporated into DCs as well.

Currently, a trial is running using siRNA for inducible immunoproteasome β subunits. Here, monocytes, being DC precursors are modified (81). Monocytes can also be electroporated with mRNA and subsequently differentiated with GM-CSF for 24 h and maturated with polyI:C for another 24 h. These DCs, generated in 48 h, can stimulate autologous influenza-specific T cells, but have a rather low electroporation efficiency (82). Another group has shown that DCs electroporated in the monocyte stage led to a higher yield of mature DCs and to more IL-12p70 and less IL-10 secretion (83). The electroporation efficiency with enhanced green fluorescent protein (eGFP) mRNA was still 51% at day 8, the time at which DCs were fully matured by this protocol (83). It is, however, not unthinkable that such an extended time between transfection and the ready-to-administer vaccine may lead to lack of sufficient expression for those proteins with a shorter half-life than eGFP. So far, no completed clinical trials with DCs electroporated in the monocyte stage have been published.

The electroporation procedure itself is constituted of many variables: (1) the number of cells transfected per electroporation, (2) the final concentration of mRNA used, (3) the buffer in which the electroporation takes places, (4) pre-incubation (or not) of the cells with the mRNA, (5) the distance between the electrodes (1-, 2-, and 4-mm gap cuvettes) which influences the optimal pulse conditions, (6) the type of electroporation pulse given (square wave versus exponential decay pulse), and (7) the DC density during the electroporation pulse. All these factors vary between the different clinical studies reported so far. They are listed in **Table 27.1**.

It has been shown that a difference in distance of the electrodes (4 versus 1 mm) does not make a major difference in electroporation efficiency and viability of the DCs for the square wave pulse (21). For the amount of (m)RNA used for electroporation mass units are used. However, the same mass of mRNA results in a different number of absolute mRNA molecules depending on the length of the encoded tumor antigens, and it has been shown that the amount of (m)RNA plays a role in the amount of protein expressed per DC (68). As different studies often use different tumor antigens, expressing the amount of (m)RNA in molarity would ease comparison of trials and allow setting up similar study designs. Finally, different groups reported different optimal conditions concerning the electroporation pulse (19, 21). Recently, it has been shown that DCs subjected to prolonged electrical square wave pulses of 4 ms upregulated the level of transgene expression equipping these DCs to be more potent in inducing influenza matrix protein M1-specific T cells. This longer pulse was, however, at the expense of low survival rates (84).

1.4. DC Maturation

Mature DCs induce a stronger functional in vivo immune response compared to immature DCs (85, 86) and immature DCs have been shown to induce T-cell anergy and to stimulate peripheral regulatory T cells (87–90). Therefore, maturation of DCs is a critical factor in DC immunotherapy. In stage IV melanoma patients, vaccination with peptide- and keyhole limpet hemocyanin (KLH)-pulsed immature DCs did not lead to the induction of proliferative T cells, of humoral responses or of delayed-type hypersensitivity (DTH) reactions. This is in contrast to their mature counterparts. Mature DCs vaccinated patients also showed a longer stable disease (86). In another study with stage IV melanoma patients, immature and mature DCs were injected in different lymph nodes of the same patients. There, proliferation in response to recall antigens as well as tumor antigen-specific IFN-γ secretion and cytotoxicity was seen with both immature and mature DCs, but was more abundant with mature DCs (85).

In the 11 completed clinical trials with RNA-loaded DCs only six added maturation agents to mature the DCs before administration to the patient (**Table 27.1**). However, none of the other studies used completely immature DCs, because the cells were loaded with mRNA by co-incubation and this has been shown to lead to a low level of maturation of the DCs (25).

Many in vitro experiments are being performed in an attempt to identify the most potent maturation stimulus and time course of administration. In 1997, Jonuleit et al. proposed an inflammatory cytokine cocktail (91), which is the most widely used maturation method for RNA-based DC vaccines and consists of TNF-α, IL-6, IL-1β, and PGE2 (19, 20, 22, 26, 28). One study used a slightly different cocktail, leaving IL-6 and IL-1β behind (21). Since then, different studies have been published questioning the use of this maturation cocktail. First, although PGE2 improves DC migration (92, 93) it also decreases the IL-12p70 secretion by DCs (92, 94, 95). Second, cytokine cocktail-matured DCs have been shown to expand immunosuppressive regulatory T cells (96). Therefore, the use of poor immunogenic cytokine cocktail-matured DCs might possibly explain why no dramatic differences have been seen in the clinical and immunological outcomes between studies using immature versus mature DCs (14, 15) (**Table 27.1**).

More recently, the use of Toll-like receptor ligands has been proposed as an efficient way to induce in vitro DC maturation prior to vaccination (88, 97). Unfortunately, DCs matured with TLR ligands like LPS and polyI:C are not clinically applicable or produce toxic side effects (98). Moreover, it has been shown that DCs matured with polyI:C fail to express the transgene efficiently after mRNA electroporation (99). Therefore alternatives such as caTLR4 and Ampligen (poly(I:C_{12}U)) have been developed (98, 100). Furthermore, TLR ligands are being combined (concomitantly or sequentially) with inflammatory cytokines, PGE2, or CD40L in vitro (93, 97, 101).

Maturation can be performed before as well as after mRNA transfection of the DCs. Based on in vitro experiments, it has been reported that electroporation of DCs with tumor antigen after maturation with the cytokine cocktail leads to a higher T-cell stimulatory capacity of the DCs and might therefore be preferable (70, 84, 102). However, in all the clinical studies with mRNA-loaded DCs so far, the electroporation was performed before maturation (**Table 27.1**).

Another important factor that must be taken into consideration is the time span of maturation in vitro. Maturation-inducing culture varied from 20 h (26, 28) to 48 h (20–22) in the reported studies. However, it has been shown that long in vitro maturation might possibly exhaust the DCs, resulting in insufficient cytokine secretion at the time of reaching the lymph nodes and consequently in poor T-cell stimulation in vivo (103, 104). This problem can be resolved by maturing DCs through electroporation with mRNA encoding functional proteins (78). The concept of "TriMix DCs," for which a clinical grade "immune-therapy trial"-scale generation protocol is described below, combines the DC maturation–activation stimuli with the loading of tumor antigen to avoid injection of exhausted DCs. Other approaches could be to mature mRNA-loaded DCs in vivo, for example, with agonistic antibodies for CD40 (105, 106) or by local treatment of the injection site with TNF-α or imiquimod, a TLR7/8 ligand (65, 107).

The data obtained from the clinical trials conducted so far with (m)RNA-loaded DCs prove that a lot of work still needs to be performed in order to determine which maturation strategy is preferable. Obviously, this step in the vaccine generation process still holds a lot of room for improvement and eventually new clinical trials will need to be performed in order to determine whether optimally matured DCs will be the true key to a better DC vaccine.

1.5. Quality Control (QC)

To allow an adequate interpretation of the results of a study, vigorous quality controls need to be performed (31). In clinical trials performed with (m)RNA-loaded DCs, two main vaccine production components require extensive control: the DCs and the (m)RNA.

1.5.1. QC of the DCs

As a first step, thorough preclinical in vitro work with focus on the DC phenotype, their cytokine secretion profile (e.g., IL-6, IL-10, IL-12p70, and IFN-α), their ability to present antigens, to induce antigen-specific T-cell responses, and their migration capacity toward the CCR7 ligand CCL19 needs to be performed. Second, the relevant criteria for quality control need to be predefined.

Because each vaccine is patient specific, QC of the DCs needs to be performed at the end of each vaccine preparation cycle, and under the same form as the cells will be injected (31, 32). Based on the current ratio, "minimum release criteria" for DC vaccines have been proposed by Figdor and colleagues (31) and the strategies to achieve these criteria have been described (32). We here want to place the published clinical trials with mRNA-loaded DCs against these criteria.

Since the *T-cell stimulatory capacity* of the DCs is a key factor, it needs to be extensively controlled. Nevertheless, full functional characterization of each patients' DC preparation is very labor intensive. Therefore, most groups perform a phenotypical analysis of co-stimulatory molecules and MHC surface expression as an alternative to investigate the maturation state of their DC preparation. Unfortunately, so far, no strong correlation has been revealed between vaccination with phenotypically mature DCs and the objective clinical responses after vaccination of melanoma patients (14). So it seems that a mature phenotype and functionality do not always go hand in hand, and at least cytokine secretion, in particular IL-12p70, needs to be included in the quality control (90).

Phenotypical characterization by flow cytometry is performed in all studies, but none of these studies looks consistently at all proposed markers, being $CD83^+$, $CD80^+$, $CD86^+$, MHC class I^+ and II^+, and $CCR7^+$ for mature monocyte-derived DCs and $CD14^{-/lo}$, $CD83^-$, $CD80^{-/lo}$, $CD86^{lo}$, MHC class I^+ and II^+, and $CCR5^+$ for immature monocyte-derived DCs (31). Only two studies mention the threshold values that need to be achieved to release the product for vaccine use (6, 7), and four others give the percentages of the controlled markers (20–23). The remaining studies either only mention the markers controlled or give an indication of their presence by "+" and "–" or "high" and "low," without further explanation of what is considered high or low.

This of course gives an indication of the DC product, but again makes it difficult to compare the different studies.

The *purity* of the DC population is determined in 11 out of 13 studies. This is done by staining for the DC lineage marker CD11c (6, 7, 18) together or not with one or more of the following non-DC lineage markers: CD3 (T cells), CD14 (monocytes), CD19 (B cells), and CD16/CD56 (NK cells) (18, 20–22, 25–27, 30). Only in five studies percentages are indicated (18, 20, 22, 23, 30). Of these studies, one does not mention how the purity is defined (23). Possibly, it might be determined by flow cytometric light scatter properties. Except one group, which considered that only a DC product with less than 15% of monocytes (based on CD14 staining) were allowed for administration, none of the trials set up a "required level of purity" (6, 7). More than 80% of pure DCs is proposed as a minimum (31). In general this still remains to be established and will largely be dictated by the overall immunopotency of the final product (32).

To determine the *viability* of the DCs right before injection, trypan blue exclusion is mainly used in the mRNA-loaded DC immunotherapy trials (18, 20, 30). If a two laser extended flow cytometer is available, propidium iodide (PI), 7-aminoactinomycin (7-AAD), or viability dye (ViD) staining can be good alternatives (32, 108). The first two stain death cells by intercalating their DNA, whereas the latter reacts with the amine groups in the cytoplasm (108). For more precise information, this can be combined with Annexin V, which will also indicate the amount of early apoptotic cells that did not lose membrane integrity yet (108).

Evidently, the *sterility* of the end product needs to be guarded (32). However, it may be assumed that this was performed in every study, only 5 out of 13 studies mention it (18, 20–22, 27). Of these, only two studies define absence of bacterial, fungal, mycoplasma, and/or endotoxin contamination (18, 27).

Besides these "minimum release criteria," additional "not too labor intensive" validation of the vaccine consists of confirming the antigen-loaded state of the DCs and the stability of the DC phenotype (31).

The first reflects the efficiency of (m)RNA loading. In case defined antigens or total tumor mRNA-containing known antigens are used, DC loading efficiency of tumor antigens can be controlled by intracellular staining and flow cytometric analysis or by immunohistochemistry for each separate tumor antigen. Up to date, no RNA-loaded DC trials performed this control by direct antigen staining. In case co-electroporation occurs with functional proteins, one of those can be used to determine the electroporation efficiency. The clinical trial based on the protocol below quantifies CD70 expression as electroporation efficiency control. It must be mentioned, however, that it has been shown that antigen presentation can take place without detectable presence of the whole tumor antigen by the DCs (109).

Therefore, a more direct control of the tumor antigen loading and of the consequent tumor antigen presentation of DCs consists of stimulating antigen-specific T cells from T-cell lines or from reporter cell lines transfected with T-cell receptor and reporter constructs (31).

DCs with a fully mature phenotype maintain their morphological and phenotypical characteristics after cytokine withdrawal, which can be examined using the washout test. Here, DCs are cultured for 2 days in medium without cytokines (31).

1.5.2. QC of the (m)RNA

Whether DCs are modified with an mRNA encoding a defined antigen, total tumor antigens, or functional proteins, the quality of the mRNA is responsible for the actual presence of protein and therefore for antigen presentation or functionality of the mRNA-loaded DCs. Again, inter-study and intra-study comparisons are difficult, and therefore standardization is a must (14, 31). For these reasons, both the *intactness* and the *concentration* of the mRNA for DC loading are important. Notwithstanding this, in some studies the applied mRNA quality control is not mentioned (7, 23, 28, 29). Here we discuss the advantages and disadvantages of the methods used in the other clinical studies with mRNA-loaded DCs.

A classical method to measure the concentration of RNA, independent of its source is spectrophotometry. For defined mRNAs, capillary electrophoresis (RNA nano chip) or gel electrophoresis, stained with either ethidium bromide or the more sensitive GelStar, as used in some RNA-DC-loaded clinical trials, followed by densitometry scanning can be applied. The first, spectrophotometry, is the least labor intensive and allows an estimate of protein contamination by the ratio $A_{260\ \mathrm{nm}}/A_{280\ \mathrm{nm}}$. However, any contamination with UV absorbing components such as DNA or phenol will give a wrong measurement of the RNA concentration (110). Furthermore, the presence and amount of DNA contamination as well as RNA degradation cannot be measured. In contrast, electrophoresis cannot give an estimate of the protein contamination but visualizes the presence of contaminating DNA and of degraded mRNA. The combination of RNA separation and concentration measurement has the advantage that the desired mRNA can be quantified correctly independently of the DNA contamination and mRNA degradation. To dissociate the presence of contaminating DNA from degraded mRNA, PCR is necessary for adequate purity control. Therefore, combining spectrophotometry, gel electrophoresis, and PCR is recommended for proper mRNA quality control.

In studies where total tumor mRNA is used, control of the intactness of the mRNA is impossible as both agarose gel and capillary electrophoresis will result in a smear of mRNAs. When total RNA is isolated, the rRNA/mRNA fraction can be determined. Furthermore, RT-PCR and PCR of β-actine can be performed to evaluate RNA quality and DNA contamination, respectively.

1.6. Route of Administration

DCs can be administered subcutaneously (s.c.), intradermally (i.d.), intravenously (i.v.), intralymphatically (i.l.), intranodally (i.n.), and/or intratumorally (i.t.) (15, 65). Except for intratumoral and intralymphatic injection, all these injection routes have been applied in (m)RNA-loaded DC studies and were well tolerated (6, 7, 18, 20–23, 25–28).

Evaluation of the best administration route of DCs can be considered from two perspectives: direct tracking of the migration ability of the DCs to the lymph nodes or analyzing the immunological/clinical response they provoke.

A commonly used method to track DC migration in vivo is by In-111 labeling and subsequent imaging of these radioactive DCs. Indium-111-labeled mRNA-loaded DCs injected i.v. accumulate primarily in the lungs and redistribute subsequently to the liver, spleen, and bone marrow, but not to any lymph node or tumor mass. Subcutaneously injected DCs did not appear in any regional lymph node either. Only i.d. injected DCs could migrate to draining lymph nodes, albeit in minor quantities (0.1–0.4% of activity compared to the activity at the injection site) (111). A possible explanation for this lack of in vivo migration after s.c. injection might be found in the fact that the DCs used in this study were not fully mature. Indeed, maturation of DCs has been shown to induce upregulation of CCR7, one of the necessary components for migration toward the lymph nodes (65, 92). One study where mature tumor lysate-loaded DCs were injected, showed migration after s.c. injection, albeit threefold lower than after i.d. injection (112). Even after maturation of the DCs, a significant amount of i.d. injected DCs remained at the injection site, although a higher percentage of migration was seen compared to immature DC (mean ± SD, 1.8 ± 1.1% mature DC versus 0.3 ± 0.1% immature DC). Additionally, mature DCs distributed over more lymph nodes, whereas immature never migrated to more than one lymph node (113).

Different attempts have been made to improve DC migration toward lymph nodes. The i.d. injection site can be pretreated prior to injection with TNF-α or TLR ligands to ameliorate maturation (65). This will enhance CCR7 expression and thus also migration. For i.v. injection, co-electroporation of the DCs with E/L selectin-encoding mRNA might be a promising strategy (75). On the other hand, applying i.n. or i.l. injections circumvent the need for migration from the skin or blood stream. It is believed that i.n. injection might partially destroy the lymph node architecture, but on average more than 10% of intranodally administered peptide-loaded DCs migrate to vicinal lymph nodes, which is a much higher percentage than achieved after i.d. injection. This occurs independently of the maturation of DCs, but mature DCs were found deep in the T-cell areas of the lymph nodes, whereas immature DCs remained in the lymph node periphery. So also for i.n. injection, maturation of DCs is important (113).

A major drawback for both i.l. and i.n. injections is the difficulty to administer the DCs correctly. It has been shown through magnetic resonance imaging (MRI) that despite ultrasound guidance, an experienced radiologist correctly injected the DCs in the lymph nodes in about 50% of the cases (114). Consequently, the incorrectly administered DCs did not show any migration (114).

Taking the data together from the direct tracking of DC migration in vivo in humans, properly performed i.n. injection results in a quick distribution of the highest amount of DCs toward different lymph nodes and i.d. injection leads to much less DCs arriving at the lymph node. However, it is not known how many DCs are necessary to induce an immune response. If low amounts are sufficient, i.d. injection might have the advantage that the required migration is an extra selection step between full mature and immature DCs and might therefore lead to a better immune response (115).

To fully determine the most proper administration route, evaluation of the induced immune response and clinical outcome is required. Two clinical studies conducted with mature mRNA-electroporated DCs had the intention to compare the i.n. versus i.d. injection on the induced immune response and clinical outcome (20, 22). In the first study, a response was demonstrated in 3 out of 12 melanoma patients vaccinated in an inguinal lymph node and in 7 out of 10 patients vaccinated intradermally into the region draining to these lymph nodes (20). From the 19 prostate cancer patients that completed four vaccinations in the second study, 4 out of 10 of the i.n. injected patients showed an immune response and 8 out of 9 of the i.d. injected patients. Six out of 9 intradermal injected patients showed stable disease, where only 5 out of 10 i.n. injected patients had stable disease. All other patients had progressive disease (22). Based on the higher migration after i.n. injection seen by direct imaging of DC migration in vivo in humans, the outcome of both studies is unexpected (113). Unfortunately, the successfulness of the i.n. injections was measured as increased width of the cortex/paracortex region by ultrasound which does not give enough detailed anatomic information to verify correct performance, as mentioned above (114, 116). Moreover, the DCs used for i.n. injection expressed less CD83 in both studies which could affect the immunological response rates (61). Therefore, definite conclusions cannot be drawn about the superior route for injection between i.n. and i.d. administration from these two studies. This shows again the importance of quality control at each step of the vaccination.

The location of the tumor that is affected by the immune response also differs depending on the administration route. In mice, i.v. and s.c. vaccinations are preferential for viscerally and nonviscerally located metastases, respectively (117). Also the nature of the immune response is affected, a Th1 immunity with greater frequency was induced when peptide-loaded DCs were administered i.d. or i.l. compared to i.v. The latter injection route was associated with a significantly higher titer of Ag-specific Abs (118). Therefore, when a Th1 response is desired, i.d. injection might be advantageous because of the simplicity of i.d. injection compared to i.l. injection. On the other hand, such observations also led to the idea that combination of different routes of administration might be beneficial (31, 65).

Several mRNA-loaded DC studies combine different injection sites (6, 7, 23, 25, 27, 29). In all these studies the i.d. injected dose is lower than the i.v., but the minimal quantity of DCs necessary to stimulate T cells is not yet defined.

Based on the difficulties of i.l. and i.n. injections and the poor migration and immune responses obtained after s.c. injection, the patients are vaccinated in the clinical study for which the DC protocol is described below by i.d. injection. In the future, this might be combined with i.v. injections.

1.7. Vaccination Scheme

A vaccination scheme comprises an amount of DC injected (dose) at different time points (inter-injection period). Both variables will be discussed hereafter.

1.7.1. Injected Dose of DCs

The fact that the highest dose of DCs is not necessarily the most effective one (as is the case with conventional cytostatic therapy) together with the negligible toxicity of DC vaccines makes that conventional dose-escalating phase I trials with toxicity as the primary end-point do not result in the selection of the optimal dose to be used in further clinical settings (15). Therefore, for immunotherapy, dose-escalating studies should have immune responses as the primary end-point rather than toxicity.

The doses used until now vary from 5×10^6 to 100×10^6 (**Table 27.1**), (21, 23). The absolute minimal adequate dose is still undefined, despite the effort of comparing different doses by five RNA-loaded DC studies (6, 7, 23, 25, 27). Lack of sufficient amounts of DCs, the low number of patients in each subgroup for comparison as well as the general low immune and clinical responses might be responsible for this failure.

Nevertheless, in one study a higher frequency of IFN-γ secreting PBMCs post-vaccination compared to pre-vaccination seemed to be present in the patients receiving an intermediate and high dose compared to those injected with low doses of DCs (**Table 27.1**), (25).

So far, doses up to 100 × 10^6 RNA-transfected DCs/injection are applied and well tolerated (23). Doses as low as 10 × 10^6 DCs/injection in adults induced immune responses detectable in blood and DTH reaction and decreased transiently the amount of circulating tumor cells (**Table 27.1**), (26, 28). One study used only 5 × 10^6 DCs/injection and did not observe any immune or clinical response (21). Since the necessary dose is most probably dependent on all the above discussed factors (origin of DCs, maturation, antigen presentation, and administration route), it is subject to discussion whether in this study (21) the absence of responses can be attributed to the relative low dose only.

1.7.2. Number of Injections and Inter-injection Period

Three to six injections are applied for RNA-loaded DC vaccines, with intervals varying from once a week to once a month (**Table 27.1**). Immune monitoring at different time points during vaccination revealed a general increase in $CD8^+$ T-cell expansion, continuing for about 2 weeks after the final vaccine (26, 28). Responses thereafter often decline.

The number of antigen-specific T cells in the blood increased for up to 20 injections as observed in one patient vaccinated with peptide-loaded mature DCs in a time span of 22 months. During the six following vaccines the antigen-specific T-cell precursor frequency remained stable, extending over an additional year (119).

The long-term follow-up of one patient receiving the more conventional four DC injections at a weekly interval, showed almost undetectable tumor antigen-specific proliferation after 6 months. After a single boost vaccine, the patient's response remained prominent for 9 months (30). This, together with the observation by others that the presence of a measurable hTERT-specific $CD8^+$ T cells response in the blood showed in general an inverse pattern with the presence of circulating tumor cells, favors continued vaccination or further boosting (26). In addition, one or two boost vaccines also led to a negative PSA log slope in patients whose rising PSA slope only slightly decreased or stabilized after four weekly DC injections (22).

Based on these observations, inclusion of boost vaccine administrations in a vaccine schedule is advisory to maintain or to increase the immune response over time. Effective cancer therapy will not merely require a strong initial immune response, but will also depend on prolonged T-cell activity and on memory T cells able to respond to a tumor relapse. The minimal dose of DCs, repetition of injections, and the optimal time between the injections necessary to maintain the immune response are most probably a consequence of the capacity of DCs to induce effector and memory T cells. This in turn will be dependent on the type of DC, their maturation status, antigen-presenting capacity, administration route, and the immunosuppressive activity of the patient's tumor. As none of these parameters are fixed in DC immunotherapy, the dose and administration required will probably be vaccine and patient specific. This fact creates yet another challenge on the long road to effective RNA-loaded DC-based tumor immunotherapy.

1.8. Our Approach

We have recently developed a novel approach to generate a DC-based vaccine with a superior capacity to induce anti-tumor immune responses in vitro (78, 120). This method consists of giving the DCs a maturation/activation signal and loading them with TAA, both through mRNA electroporation.

The maturation/activation signal is given to the DCs through electroporation with a so-called TriMix of mRNA encoding the functional CD40L, CD70, and constitutively active TLR4 (caTLR4) proteins. The combination of CD40L and caTLR4 expression mimics CD40 ligation (121) and TLR4 signaling of the DCs (100), respectively, and generates phenotypically mature, cytokine/chemokine-secreting DCs, as has been shown for CD40 and TLR4 ligation through addition of soluble CD40L and LPS (122). The introduction of CD70 into the DCs provides a co-stimulatory signal to $CD27^+$ naive T cells by inhibiting activated T-cell apoptosis and by supporting T-cell proliferation (123). Providing the DCs with a maturation signal through mRNA electroporation offers the advantage that the DCs do not need to be pre-incubated for up to 48 h with soluble maturation signals like pro-inflammatory cytokines or TLR ligands to achieve DC activation, which can render the cells "exhausted" and inferior for vaccination purposes (103). As a result, TriMix-electroporated DCs, which can be injected into the patient within a few hours after electroporation, will mature and secrete most of their immunostimulatory cytokines and chemokines *in situ*. It has also been postulated that maturation of DCs *in situ* resembles more closely the physiological process involved in response to pathogen infection and may therefore lead to enhanced T-cell immunity (107).

Another major advantage of activating DCs through mRNA electroporation is the fact that it can be combined with TAA loading in the same electroporation procedure. This results in less hands-on time and manipulation steps and therefore in reduced cell loss and contamination risk. Furthermore, by using full-length TAA-encoding mRNA all possible antigenic epitopes of the TAA will be presented by the DCs. As a result, this strategy could induce a broad TAA-specific T-cell response and is not dependent on the knowledge of each patient's HLA haplotype or on the prior identification of TAA-derived epitopes (124). Moreover, the TAA-encoding plasmid used for in vitro transcription of capped mRNA can be genetically modified by adding an

HLA class II targeting sequence. This not only routes the TAA to the HLA class II compartments for processing and presentation of HLA class II-restricted TAA-derived peptides but also enhances processing and presentation in the context of HLA class I molecules (35, 53).

We have shown that such TriMix DCs co-electroporated with whole TAA mRNA can stimulate antigen-specific T cells both in vitro and through vaccination and thus form a promising new approach for anti-tumor immunotherapy. A schematic overview of the different steps taken in the preparation of such a DC vaccine with an emphasis on the different mRNAs used for electroporation is given in **Fig. 27.1**. These different steps will be described in more detail in the following sections.

2. Materials

2.1. Production of Capped mRNA

1. pST1-CD40L, pST1-CD70, pST1-caTLR4, and pST1-sig-TAA-DCLAMP plasmids (**Fig. 27.2**).
2. GenElute HP Endotoxin-Free Plasmid Maxiprep Kit.
3. *Sap*I restriction enzyme.
4. Proteinase K (10 mg/mL).
5. 10% sodium dodecyl sulfate.
6. Phenol–chloroform–isoamylalcohol.
7. Chloroform–isoamylalcohol.
8. 3 M sodium acetate, pH 5.2 (nuclease free).
9. 100 and 70% ethanol (EtOH).
10. Sterile, nuclease- and endotoxin-free water.
11. mMESSAGE mMACHINE T7 Ultra Kit.

2.2. Generation of Immature DCs

1. Peripheral blood mononuclear cells (PBMCs) collected through leukapheresis (*see* **Note 1**).
2. 4-Tray cell factories (*see* **Note 2**).
3. X-VIVO-15 medium (Lonza, Verviers, Belgium).
4. Phosphate-buffered saline (PBS).
5. RPMI 1640 medium.
6. GM-CSF and IL-4 (*see* **Note 3**).
7. Heat-inactivated autologous plasma (AP) (*see* **Note 4**).

2.3. Electroporation of DC for Vaccination

1. Serum-free RPMI 1640 medium.
2. Optimix washing solution A.

3. 2X Optimix solution B.
4. Sterile, nuclease- and endotoxin-free water.
5. 4-mm electroporation cuvettes.
6. EasyjecT Plus Multipurpose Electroporation System.
7. RPMI 1640 medium supplemented with 1% AP, 1,000 U/mL GM-CSF, and 500 U/mL IL-4.
8. 50-mL tubes.
9. Ultra-low attachment T-75 flask.
10. Cryopreservation medium consisting of AP, 10% DMSO, and 2% glucose.
11. 1-mL cryopreservation tubes.
12. Cryofreezing container (Cryo 1°C freezing container, rate of cooling –1°C/min).
13. PBS.
14. Human serum albumin (HSA).
15. 1-mL syringes.

3. Methods

3.1. Production of Capped mRNA

For the production of capped mRNA suitable for electroporation, the following procedures must be performed: (1) cloning of the gene(s) of interest into a suitable vector and production of plas-

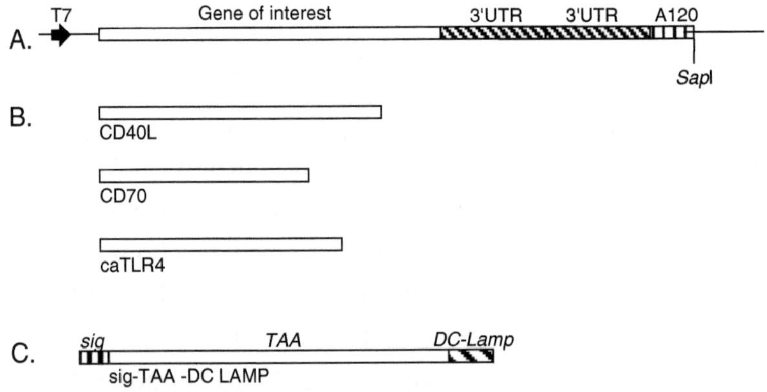

Fig. 27.2. Schematic representation of the pST1 plasmids. (**a**) The T7 promoter, human β-globin 3′ UTRs, A120-stretch and the unique *Sap*I site are shown. (**b**) Schematic representation of the pST1 plasmids used for preparation of TriMix mRNA. (**c**) Schematic representation of the pST1 plasmid used for preparation of tumor antigen-encoding mRNA. The signal sequence and the HLA class II targeting signal of DC-LAMP are shown.

mid DNA, (2) linearization of the plasmid DNA, and (3) in vitro transcription of capped mRNA.

3.1.1. DNA Plasmids

For preparation of TriMix mRNA, the DNA sequences encoding the CD40L, CD70, and caTLR4 proteins were cloned into the pST1 vector (49) (provided by Dr. U. Sahin, Johannes Gutenberg University, Mainz, Germany) by using standard molecular cloning techniques. The caTLR4 DNA consists of the intracellular and transmembrane fragments of TLR4 (100). For preparation of TAA mRNA, the DNA sequence encoding full-length TAA (*see* **Note 5**), flanked by the signal sequence and the HLA class II targeting sequence of DC-LAMP (35), was cloned into the pST1 vector by using standard molecular cloning techniques. The signal sequence will translocate newly synthesized protein to the ER, while the DC-LAMP targeting sequence will provide transport to the HLA class II compartments (*see* **Note 6**).

The pST1 vector (*see* **Note 7**) contains a bacteriophage T7 promotor, which controls in vitro transcription. It also contains a stretch of 120 adenines downstream of the insert to add a poly(A) tail to the in vitro transcribed mRNA. Between the coding region and the poly(A) tail, two sequential human β-globin 3′ untranslated regions (UTR) cloned head to tail are present. At the 3′ end of the poly(A) tail, a unique *Sap*I restriction site is present to allow linearization of the plasmid DNA before in vitro transcription. After linearization with *Sap*I, the poly(A) tail will be unmasked with a free 3′ end.

After cloning, sufficient amounts of plasmid DNA were prepared with the GenElute HP Endotoxin-Free Plasmid Maxiprep Kit according to the manufacturer's instructions.

3.1.2. Linearization of Plasmid DNA and Subsequent Purification Through Phenol/Chloroform Extraction and Ethanol Precipitation

1. Digest 100 μg of the plasmid DNA with 100 U of *Sap*I restriction enzyme in a volume of 500 μL.
2. Incubate overnight at 37°C (*see* **Note 8**).
3. Add 5 μL of proteinase K and 5 μL of 10% SDS and incubate for 30 min at 50°C.
4. Add an equal volume of phenol/chloroform (pH 8.0) and mix well.
5. Centrifuge for 10 min at 16,000*g* in an Eppendorf microcentrifuge.
6. Transfer the upper layer to a clean, new tube.
7. Add an equal volume of chloroform, vortex and centrifuge again.
8. Transfer the upper layer to a clean, new tube
9. Add a 1:10 volume of 3 M sodium acetate (pH 5.2) and two volumes of 100% EtOH and mix well.

10. Incubate overnight at –20°C.
11. Centrifuge for 15 min at 16,000*g*.
12. Remove supernatant, add 1 mL of 70% EtOH and centrifuge for 5 min at 16,000*g*.
13. Resuspend in 100 μL sterile, nuclease- and endotoxin-free water.

3.1.3. Production of In Vitro Transcribed Capped mRNA

In vitro transcription of capped mRNA is performed with T7 RNA polymerase by using the mMESSAGE mMACHINE T7 Ultra Kit according to the manufacturer's instructions. This kit uses an anti-reverse cap analog (ARCA), which allows T7 RNA polymerase to synthesize RNAs capped exclusively in the correct orientation. Substitution of traditional cap analog with ARCA allows for synthesis of capped RNAs that are 100% functional, in contrast to transcription reactions using traditional cap analog where only half of the cap analog is incorporated in the correct orientation. As a result, translation of the mRNA is more effective and much higher protein expression levels can be achieved with the same amount of mRNA (51). After transcription, remaining plasmid DNA is removed by DNase degradation to reduce the risk of introducing foreign DNA into the cells. Then size and integrity of the mRNA are checked by gel electrophoresis and quantity and purity are determined by spectrophotometry. Good quality mRNA is then stored at –20°C in small aliquots (*see* **Note 9**).

3.2. Generation of Immature DCs

1. Adjust the concentration of the washed PBMCs to 5×10^6 cells/mL X-VIVO-15 medium supplemented with 2% AP.
2. Bring 400 mL of the cell suspension into one 4-tray Cell Factory.
3. Allow plastic adherence of the DC precursors ($CD14^+$ monocytes) for 1.5 h at 37°C (*see* **Note 10**).
4. Remove the non-adherent cells for cryopreservation and wash the Cell Factory once with 250 mL PBS.
5. Bring 640 mL of RPMI 1640 medium supplemented with 1% AP, 1,000 U/mL GM-CSF, and 500 U/mL IL-4 into the Cell Factory.
6. Incubate the cells at 37°C and 5% CO_2 in a humidified incubator.
7. On days 2 and 4 of DC culture, add the same amount of GM-CSF and IL-4 to the cells as on day 0 in 40 mL RPMI 1640 medium supplemented with 1% AP.
8. On day 6 of DC culture, the cells are harvested for subsequent vaccine preparation.

3.3. Electroporation of DC for Vaccination

1. Prepare a 50-mL Falcon tube with 25 mL RPMI 1640 medium supplemented with 1% AP, 1,000 U/mL GM-CSF, and 500 U/mL IL-4 (at 37°C).
2. Adjust the physical parameters of the Easyject Plus electroporator as follows: voltage, 300 V; capacitance, 450 μF; resistance, 99 Ω (*see* **Note 11**).
3. Wash 50×10^6 DCs once with 10 mL serum-free RPMI 1640 medium and once with 5 mL Optimix washing solution A.
4. While performing the last washing step, prepare the following mRNA mix for electroporation in a final volume of 300 μL sterile, nuclease- and endotoxin-free water:
   - 20 μg CD40L mRNA
   - 20 μg CD70 mRNA
   - 20 μg caTLR4 mRNA
   - 60 μg sig-TAA-DC-LAMP mRNA
5. Add 300 μL 2X Optimix solution B.
6. Resuspend the washed DCs in the mRNA-Optimix solution B mix and transfer into a 4-mm electroporation cuvette.
7. Insert the cuvette into the electroporation chamber and trigger the pulse.
8. Immediately after the electroporation, transfer the DCs to the 50-mL Falcon tube with 25 mL RPMI 1640 medium supplemented with 1% AP, 1,000 U/mL GM-CSF, and 500 U/mL IL-4 and rinse the electroporation cuvette with 600 mL DC culture medium.
9. If you have performed different electroporations (with different TAA), pool the different DCs into one T-75 flask and incubate the electroporated DCs for 1 h at room temperature (*see* **Note 12**).
10. Harvest the electroporated DCs and freeze them at $\pm$ 12.5×10^6 per cryotube (*see* **Note 13**).
11. Perform quality control (*see* **Note 14**).
12. Thaw four cryotubes of electroporated DCs (*see* **Note 13**) and let them rest for 2 h.
13. For vaccination, resuspend $\pm$ 50×10^6 electroporated DCs in 250 μL PBS supplemented with 1% HAS.
14. Transfer the DCs to a sterile 1-mL syringe.
15. The DCs can now be used for vaccination.

3.4. Patients, Vaccination Schedule, Clinical Evaluation, and Immunomonitoring

We vaccinate patients with recurrent stage III or stage IV melanoma. These patients are incorporated in the study after written informed consent and with approval of the study protocol by the institutional ethical committee and national competent authorities.

DC vaccines are administered by four biweekly intradermal injections and every 8 weeks thereafter in the absence of disease progression. Injections are performed at two different sites (axillar and/or inguinal region).

Patients undergo two leukaphereses, one before vaccination and one after the fourth vaccine. From both leukaphereses, vaccines are prepared and PBMCs are frozen for immunomonitoring.

Also prior to vaccination and after the fourth vaccine, the clinical status of the patients is evaluated. At similar time points, DTH injections are given for immunomonitoring.

Immune monitoring is performed on both DTH infiltrating lymphocytes and on blood lymphocytes. A DTH reaction is induced by injecting 5×10^5 TriMix DC per antigen 1 week after the fourth vaccine. A skin biopsy is performed 48–72 h later. After 2 weeks of in vitro culture with low amounts of IL-2 (100 IU/mL), the DTH infiltrating lymphocytes are analyzed for specificity. Their antigen-specific activation (CD137 upregu-

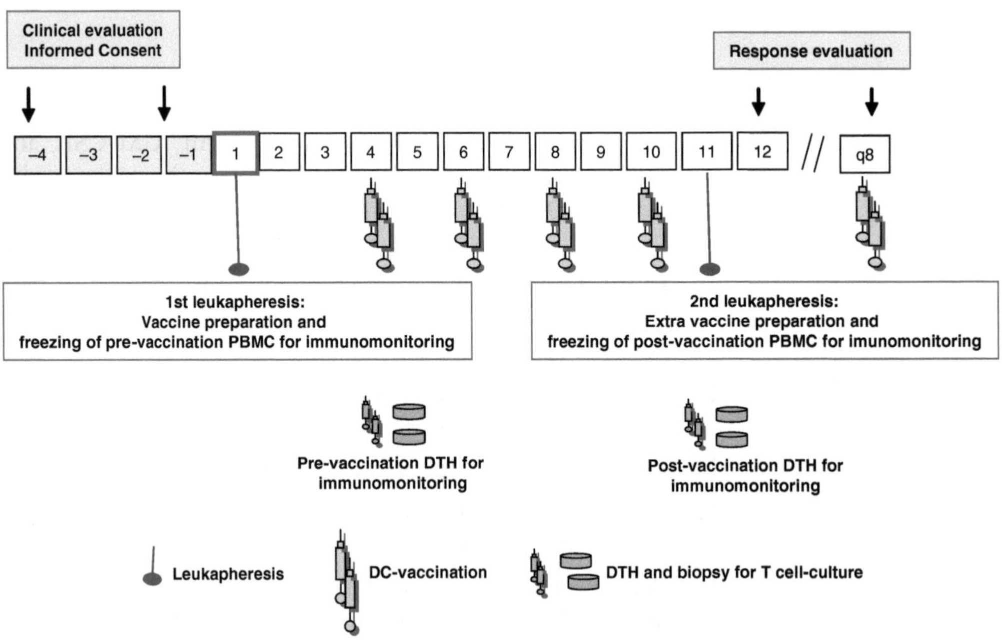

Fig. 27.3. Schedule of vaccine administrations, leukaphereses, clinical evaluation, and immunomonitoring assays. Patients undergo two leukaphereses, one before vaccination and one after the fourth vaccine. From both leukaphereses, vaccines are prepared and PBMCs are frozen for immunomonitoring. DC vaccines are administered by four biweekly intradermal injections and every 8 weeks thereafter in the absence of disease progression. Prior to vaccination and after the fourth vaccine, the clinical status of the patients is evaluated. At similar time points, DTH injections are given for immunomonitoring. Immune monitoring is performed on both DTH infiltrating lymphocytes and on blood lymphocytes.

lation (125)), cytolytic capacity (CD107a upregulation (126)), and cytokine release (IFN-γ and TNF-α) in response to autologous EBV-B cells expressing the antigens present in the TriMix-DC vaccines is assessed. On the other hand, immune responses are also analyzed in the blood of the vaccinated patients. Therefore, PBMCs are stimulated for 1 week with pools of 10 overlapping 15-mer peptides covering the whole sequence of the vaccinal TAA. After 1 week, the PBMCs are restimulated overnight with the same peptide pools and IFN-γ ELISPOT screening is performed.

A schedule of the vaccine administrations, leukaphereses, clinical evaluation, and immunomonitoring assays is given in **Fig. 27.3**.

4. Notes

1. Leukapheresis is performed with a COBE Spectra (Cobe, Denver, CO) and approximately 10–15 L of blood is processed. The leukapheresed PBMCs are then washed with a COBE Cell-Processor 2991 to remove contaminating platelets. Samples from the washed cell suspension are tested for hematocrit, total white blood cells, and platelet count.
2. DCs for vaccination must be produced in a clean room following the current guidelines of good manufacturing practice (GMP). We have therefore developed a reproducible method to produce large amounts of DCs suitable for clinical use, by culturing them in a closed system using Cell Factories (9): we designed tubing sets with sterile connections and septa for injections to transfer the PBMCs to the culture vessel, to perform the necessary washing steps, the addition of cytokines and AP, and the harvesting. Typically, 4–5 Cell Factories can be filled with the PBMCs of one single leukapheresis.
3. GM-CSF and IL-4 are prepared in-house but are also commercially available. The cytokines prepared in-house are animal protein-free and endotoxin-free. Their biological activity was titrated against standards obtained from the National Institute for Biological Standards and Controls (NIBSC, South Lills, UK).
4. AP is collected from each patient and decomplemented at 56°C for 50 min. Then plasma is centrifuged at 23,000*g* for 20 min at 4°C. As an alternative to AP, human AB serum can be used.

5. As TAA, we use the shared tumor-specific antigens Mage-A3 and Mage-C2, as well as the differentiation antigens Tyrosinase and gp100. But basically any TAA-encoding sequence can be used for cloning. We have successfully cloned open reading frames ranging from 400 to 3,000 bp. Alternatively, virus-specific antigens such as HIV gag, tat, rev, or nef could be used for the immunotherapy of infectious diseases.
6. We have successfully used the HLA class II targeting sequences of the DC-LAMP as well as the LAMP-1 protein. In contrast, linking the full-length TAA to the first 80 amino acids of the invariant chain protein as a targeting signal was far less beneficial in our experience, both for HLA class I and class II presentation of TAA-derived epitopes (35).
7. Other vectors can be used for the in vitro transcription of mRNA such as the pGEM-vector. This vector contains a 64 adenine poly(A) tail, *Not*I or *Spe*I restriction sites for linearization (which will confer a poly(A) tail extended with unrelated nucleotides) and in some embodiments the 5′ and 3′ untranslated regions of the *Xaenopus laevis* β-globin gene. It has been shown, however, that a poly(A) tail measuring 120 nucleotides compared with a shorter one, an unmasked poly(A) tail with a free 3′ end rather than one extended with unrelated nucleotides, and two sequential β-globin 3′ untranslated regions cloned head to tail between the coding region and the poly(A) tail each independently enhanced RNA stability and translational efficiency (49). We have confirmed that pST1-derived mRNA is superior to pGEM-derived mRNA.
8. Check whether plasmid linearization was complete by gel electrophoresis. Linearization is very important prior to in vitro transcription, because linearized templates give transcripts of defined length and higher yields. Circular plasmid DNA templates will generate extremely long, heterogenous RNA transcripts.
9. Gel electrophoresis of the transcribed mRNA should confer one single, sharp band. If not, mRNA might be degraded. mRNA quantity is measured at 260 nm. The ratio of the readings at 260 and 280 nm provides an estimate of the purity of the mRNA and should be between 1.9 and 2.1. If not, RNA might be contaminated with protein. After aliquoting and freezing, avoid freeze-thawing of the mRNA.

10. Plastic adherence of the PBMCs in Cell Factories is not feasible in RPMI 1640 medium. We therefore use X-VIVO-15 medium.
11. The Easyject Plus apparatus electroporated the DCs with an exponential decay pulse. Other machines can electroporate with a square wave pulse, but at our end, the exponential decay pulse works best.
12. We typically perform eight electroporations. The $2 \times 50 \times 10^6$ DCs are electroporated with TriMix mRNA + TAA (either gp100, Mage-A3, Mage-C2, or Tyrosinase). This yields enough electroporated DCs for one vaccination cycle.
13. DCs are frozen in cryotubes at $\pm$ 12.5 $\times 10^6$ DCs per tube in 1 mL AP with 10% DMSO and 2% glucose. The DCs are slowly frozen to –80°C using a cryofreezing container and subsequently stored in liquid nitrogen until use. For thawing, DCs are quickly thawed in a 37°C water bath, transferred to culture medium, and pelleted at room temperature. The thawed DCs are then resuspended in 5 mL pre-warmed culture medium. Cell number and viability are determined with trypan blue. Typically, cells are >80% viable and >90% of the frozen cells are recuperated after thawing.
14. For quality control, the DCs are characterized for phenotype, function, purity, and sterility. The endpoints in this analysis are the number of DC, purity and viability, electroporation efficiency (measured by the % of cells expressing CD70), the immunophenotype (including CD40, CD80, CD83, CCR7, and CD14 expression), and functional characterization by IL-12p70 secretion at 0–24 and 24–48 h. Vaccine samples are analyzed for sterility by long-term microbiological culture.

References

1. Villadangos, J.A. and Schnorrer, P. (2007) Intrinsic and cooperative antigen-presenting functions of dendritic-cell subsets in vivo. *Nat Rev Immunol*, **7**, 543–555.
2. Melief, C.J. (2008) Cancer immunotherapy by dendritic cells. *Immunity*, **29**, 372–383.
3. Reis e Sousa, C. (2006) Dendritic cells in a mature age. *Nat Rev Immunol*, **6**, 476–483.
4. Prlic, M., Williams, M.A., and Bevan, M.J. (2007) Requirements for CD8 T-cell priming, memory generation and maintenance. *Curr Opin Immunol*, **19**, 315–319.
5. Smith, C.M., Wilson, N.S., Waithman, J., Villadangos, J.A., Carbone, F.R., Heath, W.R., and Belz, G.T. (2004) Cognate CD4(+) T cell licensing of dendritic cells in CD8(+) T cell immunity. *Nat Immunol*, **5**, 1143–1148.
6. Caruso, D.A., Orme, L.M., Neale, A.M., Radcliff, F.J., Amor, G.M., Maixner, W., Downie, P., Hassall, T.E., Tang, M.L., and Ashley, D.M. (2004) Results of a phase 1 study utilizing monocyte-derived dendritic cells pulsed with tumor RNA in children and young adults with brain cancer. *Neuro Oncol*, **6**, 236–246.
7. Caruso, D.A., Orme, L.M., Amor, G.M., Neale, A.M., Radcliff, F.J., Downie, P., Tang,

M.L., and Ashley, D.M. (2005) Results of a Phase I study utilizing monocyte-derived dendritic cells pulsed with tumor RNA in children with Stage 4 neuroblastoma. *Cancer*, **103**, 1280–1291.

8. Erdmann, M., Dorrie, J., Schaft, N., Strasser, E., Hendelmeier, M., Kampgen, E., Schuler, G., and Schuler-Thurner, B. (2007) Effective clinical-scale production of dendritic cell vaccines by monocyte elutriation directly in medium, subsequent culture in bags and final antigen loading using peptides or RNA transfection. *J Immunother*, **30**, 663–674.
9. Tuyaerts, S., Noppe, S.M., Corthals, J., Breckpot, K., Heirman, C., De Greef, C., Van Riet, I., and Thielemans, K. (2002) Generation of large numbers of dendritic cells in a closed system using Cell Factories. *J Immunol Methods*, **264**, 135–151.
10. Schreurs, M.W., Eggert, A.A., de Boer, A.J., Vissers, J.L., van Hall, T., Offringa, R., Figdor, C.G., and Adema, G.J. (2000) Dendritic cells break tolerance and induce protective immunity against a melanocyte differentiation antigen in an autologous melanoma model. *Cancer Res*, **60**, 6995–7001.
11. Gilboa, E. (2007) DC-based cancer vaccines. *J Clin Invest*, **117**, 1195–1203.
12. http://www.clinicaltrials.gov/ct2/search
13. Schadendorf, D., Ugurel, S., Schuler-Thurner, B., Nestle, F.O., Enk, A., Brocker, E.B., Grabbe, S., Rittgen, W., Edler, L., Sucker, A., Zimpfer-Rechner, C., Berger, T., Kamarashev, J., Burg, G., Jonuleit, H., Tuttenberg, A., Becker, J.C., Keikavoussi, P., Kampgen, E., and Schuler, G. (2006) Dacarbazine (DTIC) versus vaccination with autologous peptide-pulsed dendritic cells (DC) in first-line treatment of patients with metastatic melanoma: a randomized phase III trial of the DC study group of the DeCOG. *Ann Oncol*, **17**, 563–570.
14. Engell-Noerregaard, L., Hansen, T.H., Andersen, M.H., Thor Straten, P., and Svane, I.M. (2009) Review of clinical studies on dendritic cell-based vaccination of patients with malignant melanoma: assessment of correlation between clinical response and vaccine parameters. *Cancer Immunol Immunother*, **58**, 1–14.
15. Lesterhuis, W.J., Aarntzen, E.H., De Vries, I.J., Schuurhuis, D.H., Figdor, C.G., Adema, G.J., and Punt, C.J. (2008) Dendritic cell vaccines in melanoma: from promise to proof? *Crit Rev Oncol Hematol*, **66**, 118–134.
16. Tyagi, R.K., Mangal, S., Garg, N., and Sharma, P.K. (2009) RNA-based immunotherapy of cancer: role and therapeutic implications of dendritic cells. *Expert Rev Anticancer Ther*, **9**, 97–114.
17. Tuyaerts, S., Aerts, J.L., Corthals, J., Neyns, B., Heirman, C., Breckpot, K., Thielemans, K., and Bonehill, A. (2007) Current approaches in dendritic cell generation and future implications for cancer immunotherapy. *Cancer Immunol Immunother*, **56**, 1513–1537.
18. Rains, N., Cannan, R.J., Chen, W., and Stubbs, R.S. (2001) Development of a dendritic cell (DC)-based vaccine for patients with advanced colorectal cancer. *Hepatogastroenterology*, **48**, 347–351.
19. Mu, L.J., Gaudernack, G., Saeboe-Larssen, S., Hammerstad, H., Tierens, A., and Kvalheim, G. (2003) A protocol for generation of clinical grade mRNA-transfected monocyte-derived dendritic cells for cancer vaccines. *Scand J Immunol*, **58**, 578–586.
20. Kyte, J.A., Mu, L., Aamdal, S., Kvalheim, G., Dueland, S., Hauser, M., Gullestad, H.P., Ryder, T., Lislerud, K., Hammerstad, H., and Gaudernack, G. (2006) Phase I/II trial of melanoma therapy with dendritic cells transfected with autologous tumor-mRNA. *Cancer Gene Ther*, **13**, 905–918.
21. Markovic, S.N., Dietz, A.B., Greiner, C.W., Maas, M.L., Butler, G.W., Padley, D.J., Bulur, P.A., Allred, J.B., Creagan, E.T., Ingle, J.N., Gastineau, D.A., and Vuk-Pavlovic, S. (2006) Preparing clinical-grade myeloid dendritic cells by electroporation-mediated transfection of in vitro amplified tumor-derived mRNA and safety testing in stage IV malignant melanoma. *J Transl Med*, **4**, 35.
22. Mu, L.J., Kyte, J.A., Kvalheim, G., Aamdal, S., Dueland, S., Hauser, M., Hammerstad, H., Waehre, H., Raabe, N., and Gaudernack, G. (2005) Immunotherapy with allotumour mRNA-transfected dendritic cells in androgen-resistant prostate cancer patients. *Br J Cancer*, **93**, 749–756.
23. Morse, M.A., Nair, S.K., Mosca, P.J., Hobeika, A.C., Clay, T.M., Deng, Y., Boczkowski, D., Proia, A., Neidzwiecki, D., Clavien, P.A., Hurwitz, H.I., Schlom, J., Gilboa, E., and Lyerly, H.K. (2003) Immunotherapy with autologous, human dendritic cells transfected with carcinoembryonic antigen mRNA. *Cancer Invest*, **21**, 341–349.
24. Peng, J.C., Thomas, R., and Nielsen, L.K. (2005) Generation and maturation of dendritic cells for clinical application under serum-free conditions. *J Immunother*, **28**, 599–609.

25. Heiser, A., Coleman, D., Dannull, J., Yancey, D., Maurice, M.A., Lallas, C.D., Dahm, P., Niedzwiecki, D., Gilboa, E., and Vieweg, J. (2002) Autologous dendritic cells transfected with prostate-specific antigen RNA stimulate CTL responses against metastatic prostate tumors. *J Clin Invest*, **109**, 409–417.
26. Su, Z., Dannull, J., Yang, B.K., Dahm, P., Coleman, D., Yancey, D., Sichi, S., Niedzwiecki, D., Boczkowski, D., Gilboa, E., and Vieweg, J. (2005) Telomerase mRNA-transfected dendritic cells stimulate antigen-specific CD8+ and CD4+ T cell responses in patients with metastatic prostate cancer. *J Immunol*, **174**, 3798–3807.
27. Su, Z., Dannull, J., Heiser, A., Yancey, D., Pruitt, S., Madden, J., Coleman, D., Niedzwiecki, D., Gilboa, E., and Vieweg, J. (2003) Immunological and clinical responses in metastatic renal cancer patients vaccinated with tumor RNA-transfected dendritic cells. *Cancer Res*, **63**, 2127–2133.
28. Dannull, J., Su, Z., Rizzieri, D., Yang, B.K., Coleman, D., Yancey, D., Zhang, A., Dahm, P., Chao, N., Gilboa, E., and Vieweg, J. (2005) Enhancement of vaccine-mediated antitumor immunity in cancer patients after depletion of regulatory T cells. *J Clin Invest*, **115**, 3623–3633.
29. Nair, S.K., Morse, M., Boczkowski, D., Cumming, R.I., Vasovic, L., Gilboa, E., and Lyerly, H.K. (2002) Induction of tumor-specific cytotoxic T lymphocytes in cancer patients by autologous tumor RNA-transfected dendritic cells. *Ann Surg*, **235**, 540–549.
30. Kyte, J.A., Kvalheim, G., Lislerud, K., thor Straten, P., Dueland, S., Aamdal, S., and Gaudernack, G. (2007) T cell responses in melanoma patients after vaccination with tumor-mRNA transfected dendritic cells. *Cancer Immunol Immunother*, **56**, 659–675.
31. Figdor, C.G., de Vries, I.J., Lesterhuis, W.J., and Melief, C.J. (2004) Dendritic cell immunotherapy: mapping the way. *Nat Med*, **10**, 475–480.
32. Nicolette, C.A., Healey, D., Tcherepanova, I., Whelton, P., Monesmith, T., Coombs, L., Finke, L.H., Whiteside, T., and Miesowicz, F. (2007) Dendritic cells for active immunotherapy: optimizing design and manufacture in order to develop commercially and clinically viable products. *Vaccine*, **25**(Suppl 2), B47–B60.
33. Bonehill, A., Heirman, C., Tuyaerts, S., Michiels, A., Zhang, Y., van der Bruggen, P., and Thielemans, K. (2003) Efficient presentation of known HLA class II-restricted MAGE-A3 epitopes by dendritic cells electroporated with messenger RNA encoding an invariant chain with genetic exchange of class II-associated invariant chain peptide. *Cancer Res*, **63**, 5587–5594.
34. Heiser, A., Maurice, M.A., Yancey, D.R., Coleman, D.M., Dahm, P., and Vieweg, J. (2001) Human dendritic cells transfected with renal tumor RNA stimulate polyclonal T-cell responses against antigens expressed by primary and metastatic tumors. *Cancer Res*, **61**, 3388–3393.
35. Bonehill, A., Heirman, C., Tuyaerts, S., Michiels, A., Breckpot, K., Brasseur, F., Zhang, Y., Van Der Bruggen, P., and Thielemans, K. (2004) Messenger RNA-electroporated dendritic cells presenting MAGE-A3 simultaneously in HLA class I and class II molecules. *J Immunol*, **172**, 6649–6657.
36. Pascolo, S. (2004) Messenger RNA-based vaccines. *Expert Opin Biol Ther*, **4**, 1285–1294.
37. Ponsaerts, P., Van Tendeloo, V.F., and Berneman, Z.N. (2003) Cancer immunotherapy using RNA-loaded dendritic cells. *Clin Exp Immunol*, **134**, 378–384.
38. Breckpot, K., Heirman, C., Neyns, B., and Thielemans, K. (2004) Exploiting dendritic cells for cancer immunotherapy: genetic modification of dendritic cells. *J Gene Med*, **6**, 1175–1188.
39. van der Bruggen, P., Traversari, C., Chomez, P., Lurquin, C., De Plaen, E., Van den Eynde, B., Knuth, A., and Boon, T. (1991) A gene encoding an antigen recognized by cytolytic T lymphocytes on a human melanoma. *Science*, **254**, 1643–1647.
40. Lucas, S. and Coulie, P.G. (2008) About human tumor antigens to be used in immunotherapy. *Semin Immunol*, **20**, 301–307.
41. Neller, M.A., Lopez, J.A., and Schmidt, C.W. (2008) Antigens for cancer immunotherapy. *Semin Immunol*, **20**, 286–295.
42. Van Driessche, A., Gao, L., Stauss, H.J., Ponsaerts, P., Van Bockstaele, D.R., Berneman, Z.N., and Van Tendeloo, V.F. (2005) Antigen-specific cellular immunotherapy of leukemia. *Leukemia*, **19**, 1863–1871.
43. Connerotte, T., Van Pel, A., Godelaine, D., Tartour, E., Schuler-Thurner, B., Lucas, S., Thielemans, K., Schuler, G., and Coulie, P.G. (2008) Functions of Anti-MAGE T-cells induced in melanoma patients under different vaccination modalities. *Cancer Res*, **68**, 3931–3940.
44. Heiser, A., Dahm, P., Yancey, D.R., Maurice, M.A., Boczkowski, D., Nair, S.K., Gilboa,

E., and Vieweg, J. (2000) Human dendritic cells transfected with RNA encoding prostate-specific antigen stimulate prostate-specific CTL responses in vitro. *J Immunol*, **164**, 5508–5514.
45. Parmiani, G., De Filippo, A., Novellino, L., and Castelli, C. (2007) Unique human tumor antigens: immunobiology and use in clinical trials. *J Immunol*, **178**, 1975–1979.
46. Zitvogel, L., Tesniere, A., and Kroemer, G. (2006) Cancer despite immunosurveillance: immunoselection and immunosubversion. *Nat Rev Immunol*, **6**, 715–727.
47. Campoli, M., Chang, C.C., and Ferrone, S. (2002) HLA class I antigen loss, tumor immune escape and immune selection. *Vaccine*, **20**(Suppl 4), A40–A45.
48. Muller, M.R., Grunebach, F., Nencioni, A., and Brossart, P. (2003) Transfection of dendritic cells with RNA induces CD4- and CD8-mediated T cell immunity against breast carcinomas and reveals the immunodominance of presented T cell epitopes. *J Immunol*, **170**, 5892–5896.
49. Holtkamp, S., Kreiter, S., Selmi, A., Simon, P., Koslowski, M., Huber, C., Tureci, O., and Sahin, U. (2006) Modification of antigen-encoding RNA increases stability, translational efficacy, and T-cell stimulatory capacity of dendritic cells. *Blood*, **108**, 4009–4017.
50. Hart, D.P., Xue, S.A., Thomas, S., Cesco-Gaspere, M., Tranter, A., Willcox, B., Lee, S.P., Steven, N., Morris, E.C., and Stauss, H.J. (2008) Retroviral transfer of a dominant TCR prevents surface expression of a large proportion of the endogenous TCR repertoire in human T cells. *Gene Ther*, **15**, 625–631.
51. Mockey, M., Goncalves, C., Dupuy, F.P., Lemoine, F.M., Pichon, C., and Midoux, P. (2006) mRNA transfection of dendritic cells: synergistic effect of ARCA mRNA capping with Poly(A) chains in cis and in trans for a high protein expression level. *Biochem Biophys Res Commun*, **340**, 1062–1068.
52. Bontkes, H.J., Kramer, D., Ruizendaal, J.J., Kueter, E.W., van Tendeloo, V.F., Meijer, C.J., and Hooijberg, E. (2007) Dendritic cells transfected with interleukin-12 and tumor-associated antigen messenger RNA induce high avidity cytotoxic T cells. *Gene Ther*, **14**, 366–375.
53. Kreiter, S., Selmi, A., Diken, M., Sebastian, M., Osterloh, P., Schild, H., Huber, C., Tureci, O., and Sahin, U. (2008) Increased antigen presentation efficiency by coupling antigens to MHC class I trafficking signals. *J Immunol*, **180**, 309–318.
54. Sjoblom, T., Jones, S., Wood, L.D., Parsons, D.W., Lin, J., Barber, T.D., Mandelker, D., Leary, R.J., Ptak, J., Silliman, N., Szabo, S., Buckhaults, P., Farrell, C., Meeh, P., Markowitz, S.D., Willis, J., Dawson, D., Willson, J.K., Gazdar, A.F., Hartigan, J., Wu, L., Liu, C., Parmigiani, G., Park, B.H., Bachman, K.E., Papadopoulos, N., Vogelstein, B., Kinzler, K.W., and Velculescu, V.E. (2006) The consensus coding sequences of human breast and colorectal cancers. *Science*, **314**, 268–274.
55. Dunn, G.P., Old, L.J., and Schreiber, R.D. (2004) The immunobiology of cancer immunosurveillance and immunoediting. *Immunity*, **21**, 137–148.
56. Heiser, A., Maurice, M.A., Yancey, D.R., Wu, N.Z., Dahm, P., Pruitt, S.K., Boczkowski, D., Nair, S.K., Ballo, M.S., Gilboa, E., and Vieweg, J. (2001) Induction of polyclonal prostate cancer-specific CTL using dendritic cells transfected with amplified tumor RNA. *J Immunol*, **166**, 2953–2960.
57. Parkhurst, M.R., DePan, C., Riley, J.P., Rosenberg, S.A., and Shu, S. (2003) Hybrids of dendritic cells and tumor cells generated by electrofusion simultaneously present immunodominant epitopes from multiple human tumor-associated antigens in the context of MHC class I and class II molecules. *J Immunol*, **170**, 5317–5325.
58. Nestle, F.O., Banchereau, J., and Hart, D. (2001) Dendritic cells: On the move from bench to bedside. *Nat Med*, **7**, 761–765.
59. Ludewig, B., McCoy, K., Pericin, M., Ochsenbein, A.F., Dumrese, T., Odermatt, B., Toes, R.E., Melief, C.J., Hengartner, H., and Zinkernagel, R.M. (2001) Rapid peptide turnover and inefficient presentation of exogenous antigen critically limit the activation of self-reactive CTL by dendritic cells. *J Immunol*, **166**, 3678–3687.
60. Iezzi, G., Scotet, E., Scheidegger, D., and Lanzavecchia, A. (1999) The interplay between the duration of TCR and cytokine signaling determines T cell polarization. *Eur J Immunol*, **29**, 4092–4101.
61. Kyte, J.A. and Gaudernack, G. (2006) Immuno-gene therapy of cancer with tumour-mRNA transfected dendritic cells. *Cancer Immunol Immunother*, **55**, 1432–1442.
62. Boczkowski, D., Nair, S.K., Nam, J.H., Lyerly, H.K., and Gilboa, E. (2000) Induction of tumor immunity and cytotoxic T lymphocyte responses using dendritic cells transfected with messenger RNA amplified from tumor cells. *Cancer Res*, **60**, 1028–1034.

63. Grunebach, F., Muller, M.R., Nencioni, A., and Brossart, P. (2003) Delivery of tumor-derived RNA for the induction of cytotoxic T-lymphocytes. *Gene Ther*, **10**, 367–374.
64. Nencioni, A., Muller, M.R., Grunebach, F., Garuti, A., Mingari, M.C., Patrone, F., Ballestrero, A., and Brossart, P. (2003) Dendritic cells transfected with tumor RNA for the induction of antitumor CTL in colorectal cancer. *Cancer Gene Ther*, **10**, 209–214.
65. Adema, G.J., de Vries, I.J., Punt, C.J., and Figdor, C.G. (2005) Migration of dendritic cell based cancer vaccines: in vivo veritas? *Curr Opin Immunol*, **17**, 170–174.
66. Van Tendeloo, V.F., Ponsaerts, P., Lardon, F., Nijs, G., Lenjou, M., Van Broeckhoven, C., Van Bockstaele, D.R., and Berneman, Z.N. (2001) Highly efficient gene delivery by mRNA electroporation in human hematopoietic cells: superiority to lipofection and passive pulsing of mRNA and to electroporation of plasmid cDNA for tumor antigen loading of dendritic cells. *Blood*, **98**, 49–56.
67. Kalady, M.F., Onaitis, M.W., Padilla, K.M., Emani, S., Tyler, D.S., and Pruitt, S.K. (2002) Enhanced dendritic cell antigen presentation in RNA-based immunotherapy. *J Surg Res*, **105**, 17–24.
68. Tuyaerts, S., Michiels, A., Corthals, J., Bonehill, A., Heirman, C., de Greef, C., Noppe, S.M., and Thielemans, K. (2003) Induction of Influenza Matrix Protein 1 and MelanA-specific T lymphocytes in vitro using mRNA-electroporated dendritic cells. *Cancer Gene Ther*, **10**, 696–706.
69. Saeboe-Larssen, S., Fossberg, E., and Gaudernack, G. (2002) mRNA-based electrotransfection of human dendritic cells and induction of cytotoxic T lymphocyte responses against the telomerase catalytic subunit (hTERT). *J Immunol Methods*, **259**, 191–203.
70. Schaft, N., Dorrie, J., Thumann, P., Beck, V.E., Muller, I., Schultz, E.S., Kampgen, E., Dieckmann, D., and Schuler, G. (2005) Generation of an optimized polyvalent monocyte-derived dendritic cell vaccine by transfecting defined RNAs after rather than before maturation. *J Immunol*, **174**, 3087–3097.
71. Ponsaerts, P., Van Tendeloo, V.F., Cools, N., Van Driessche, A., Lardon, F., Nijs, G., Lenjou, M., Mertens, G., Van Broeckhoven, C., Van Bockstaele, D.R., and Berneman, Z.N. (2002) mRNA-electroporated mature dendritic cells retain transgene expression, phenotypical properties and stimulatory capacity after cryopreservation. *Leukemia*, **16**, 1324–1330.
72. Dullaers, M., Breckpot, K., Van Meirvenne, S., Bonehill, A., Tuyaerts, S., Michiels, A., Straetman, L., Heirman, C., De Greef, C., Van Der Bruggen, P., and Thielemans, K. (2004) Side-by-side comparison of lentivirally transduced and mRNA-electroporated dendritic cells: implications for cancer immunotherapy protocols. *Mol Ther*, **10**, 768–779.
73. Bontkes, H.J., Kramer, D., Ruizendaal, J.J., Meijer, C.J., and Hooijberg, E. (2008) Tumor associated antigen and interleukin-12 mRNA transfected dendritic cells enhance effector function of natural killer cells and antigen specific T-cells. *Clin Immunol*, **127**, 375–384.
74. Minkis, K., Kavanagh, D.G., Alter, G., Bogunovic, D., O'Neill, D., Adams, S., Pavlick, A., Walker, B.D., Brockman, M.A., Gandhi, R.T., and Bhardwaj, N. (2008) Type 2 Bias of T cells expanded from the blood of melanoma patients switched to type 1 by IL-12p70 mRNA-transfected dendritic cells. *Cancer Res*, **68**, 9441–9450.
75. Dorrie, J., Schaft, N., Muller, I., Wellner, V., Schunder, T., Hanig, J., Oostingh, G.J., Schon, M.P., Robert, C., Kampgen, E., and Schuler, G. (2008) Introduction of functional chimeric E/L-selectin by RNA electroporation to target dendritic cells from blood to lymph nodes. *Cancer Immunol Immunother*, **57**, 467–477.
76. Tuyaerts, S., Van Meirvenne, S., Bonehill, A., Heirman, C., Corthals, J., Waldmann, H., Breckpot, K., Thielemans, K., and Aerts, J.L. (2007) Expression of human GITRL on myeloid dendritic cells enhances their immunostimulatory function but does not abrogate the suppressive effect of CD4+CD25+ regulatory T cells. *J Leukoc Biol*, **82**, 93–105.
77. Aerts-Toegaert, C., Heirman, C., Tuyaerts, S., Corthals, J., Aerts, J.L., Bonehill, A., Thielemans, K., and Breckpot, K. (2007) CD83 expression on dendritic cells and T cells: correlation with effective immune responses. *Eur J Immunol*, **37**, 686–695.
78. Bonehill, A., Tuyaerts, S., Van Nuffel, A.M., Heirman, C., Bos, T.J., Fostier, K., Neyns, B., and Thielemans, K. (2008) Enhancing the T-cell stimulatory capacity of human dendritic cells by co-electroporation with CD40L, CD70 and constitutively active TLR4 encoding mRNA. *Mol Ther*, **16**, 1170–1180.
79. Michiels, A., Breckpot, K., Corthals, J., Tuyaerts, S., Bonehill, A., Heirman, C., Thielemans, K., and Aerts, J.L. (2006) Induction of antigen-specific CD8+ cytotoxic

T cells by dendritic cells co-electroporated with a dsRNA analogue and tumor antigen mRNA. *Gene Ther*, **13**, 1027–1036.
80. Breckpot, K., Aerts-Toegaert, C., Heirman, C., Peeters, U., Beyaert, R., Aerts, J.L., and Thielemans, K. (2009) Attenuated expression of A20 markedly increases the efficacy of double-stranded RNA-activated dendritic cells as an anti-cancer vaccine. *J Immunol*, **182**, 860–870.
81. http://www.clinicaltrials.gov/ct2/show/NCT00672542?term=melanoma+AND+dendritic+cells+AND+siRNA&rank=1
82. Ponsaerts, P., Van den Bosch, G., Cools, N., Van Driessche, A., Nijs, G., Lenjou, M., Lardon, F., Van Broeckhoven, C., Van Bockstaele, D.R., Berneman, Z.N., and Van Tendeloo, V.F. (2002) Messenger RNA electroporation of human monocytes, followed by rapid in vitro differentiation, leads to highly stimulatory antigen-loaded mature dendritic cells. *J Immunol*, **169**, 1669–1675.
83. Milano, F., van Baal, J.W., Rygiel, A.M., Bergman, J.J., Van Deventer, S.J., Kapsenberg, M.L., Peppelenbosch, M.P., and Krishnadath, K.K. (2007) An improved protocol for generation of immuno-potent dendritic cells through direct electroporation of CD14+ monocytes. *J Immunol Methods*, **321**, 94–106.
84. Met, O., Eriksen, J., and Svane, I.M. (2008) Studies on mRNA electroporation of immature and mature dendritic cells: effects on their immunogenic potential. *Mol Biotechnol*, **40**, 151–160.
85. Jonuleit, H., Giesecke-Tuettenberg, A., Tuting, T., Thurner-Schuler, B., Stuge, T.B., Paragnik, L., Kandemir, A., Lee, P.P., Schuler, G., Knop, J., and Enk, A.H. (2001) A comparison of two types of dendritic cell as adjuvants for the induction of melanoma-specific T-cell responses in humans following intranodal injection. *Int J Cancer*, **93**, 243–251.
86. de Vries, I.J., Lesterhuis, W.J., Scharenborg, N.M., Engelen, L.P., Ruiter, D.J., Gerritsen, M.J., Croockewit, S., Britten, C.M., Torensma, R., Adema, G.J., Figdor, C.G., and Punt, C.J. (2003) Maturation of dendritic cells is a prerequisite for inducing immune responses in advanced melanoma patients. *Clin Cancer Res*, **9**, 5091–5100.
87. Steinman, R.M. and Nussenzweig, M.C. (2002) Avoiding horror autotoxicus: the importance of dendritic cells in peripheral T cell tolerance. *Proc Natl Acad Sci U S A*, **99**, 351–358.
88. Lutz, M.B. and Schuler, G. (2002) Immature, semi-mature and fully mature dendritic cells: which signals induce tolerance or immunity? *Trends Immunol*, **23**, 445–449.
89. Cools, N., Van Tendeloo, V.F., Smits, E.L., Lenjou, M., Nijs, G., Van Bockstaele, D.R., Berneman, Z.N., and Ponsaerts, P. (2008) Immunosuppression induced by immature dendritic cells is mediated by TGF-beta/IL-10 double-positive CD4+ regulatory T cells. *J Cell Mol Med*, **12**, 690–700.
90. Enk, A.H. (2005) Dendritic cells in tolerance induction. *Immunol Lett*, **99**, 8–11.
91. Jonuleit, H., Kuhn, U., Muller, G., Steinbrink, K., Paragnik, L., Schmitt, E., Knop, J., and Enk, A.H. (1997) Pro-inflammatory cytokines and prostaglandins induce maturation of potent immunostimulatory dendritic cells under fetal calf serum-free conditions. *Eur J Immunol*, **27**, 3135–3142.
92. Scandella, E., Men, Y., Gillessen, S., Forster, R., and Groettrup, M. (2002) Prostaglandin E2 is a key factor for CCR7 surface expression and migration of monocyte-derived dendritic cells. *Blood*, **100**, 1354–1361.
93. Boullart, A.C., Aarntzen, E.H., Verdijk, P., Jacobs, J.F., Schuurhuis, D.H., Benitez-Ribas, D., Schreibelt, G., van de Rakt, M.W., Scharenborg, N.M., de Boer, A., Kramer, M., Figdor, C.G., Punt, C.J., Adema, G.J., and de Vries, I.J. (2008) Maturation of monocyte-derived dendritic cells with Toll-like receptor 3 and 7/8 ligands combined with prostaglandin E2 results in high interleukin-12 production and cell migration. *Cancer Immunol Immunother*, **57**, 1589–1597.
94. van der Pouw Kraan, T.C., Boeije, L.C., Smeenk, R.J., Wijdenes, J., and Aarden, L.A. (1995) Prostaglandin-E2 is a potent inhibitor of human interleukin 12 production. *J Exp Med*, **181**, 775–779.
95. Kalinski, P., Vieira, P.L., Schuitemaker, J.H., de Jong, E.C., and Kapsenberg, M.L. (2001) Prostaglandin E(2) is a selective inducer of interleukin-12 p40 (IL-12p40) production and an inhibitor of bioactive IL-12p70 heterodimer. *Blood*, **97**, 3466–3469.
96. Banerjee, D.K., Dhodapkar, M.V., Matayeva, E., Steinman, R.M., and Dhodapkar, K.M. (2006) Expansion of FOXP3high regulatory T cells by human dendritic cells (DCs) in vitro and after injection of cytokine-matured DCs in myeloma patients. *Blood*, **108**, 2655–2661.
97. Mailliard, R.B., Wankowicz-Kalinska, A., Cai, Q., Wesa, A., Hilkens, C.M., Kapsenberg, M.L., Kirkwood, J.M., Storkus, W.J., and Kalinski, P. (2004) alpha-type-1 polarized dendritic cells: a novel immunization

tool with optimized CTL-inducing activity. *Cancer Res*, **64**, 5934–5937.

98. Jasani, B., Navabi, H., and Adams, M. (2009) Ampligen: a potential toll-like 3 receptor adjuvant for immunotherapy of cancer. *Vaccine*, **27**(25–26), 3401–3404.
99. Schuurhuis, D.H., Lesterhuis, W.J., Kramer, M., Looman, M.G., van Hout-Kuijer, M., Schreibelt, G., Boullart, A.C., Aarntzen, E.H., Benitez-Ribas, D., Figdor, C.G., Punt, C.J., de Vries, I.J., and Adema, G.J. (2008) Polyinosinic polycytidylic acid prevents efficient antigen expression after mRNA electroporation of clinical grade dendritic cells. *Cancer Immunol Immunother*, **58**(7), 1109–1115.
100. Cisco, R.M., Abdel-Wahab, Z., Dannull, J., Nair, S., Tyler, D.S., Gilboa, E., Vieweg, J., Daaka, Y., and Pruitt, S.K. (2004) Induction of human dendritic cell maturation using transfection with RNA encoding a dominant positive toll-like receptor 4. *J Immunol*, **172**, 7162–7168.
101. Zobywalski, A., Javorovic, M., Frankenberger, B., Pohla, H., Kremmer, E., Bigalke, I., and Schendel, D.J. (2007) Generation of clinical grade dendritic cells with capacity to produce biologically active IL-12p70. *J Transl Med*, **5**, 18.
102. Michiels, A., Tuyaerts, S., Bonehill, A., Corthals, J., Breckpot, K., Heirman, C., Van Meirvenne, S., Dullaers, M., Allard, S., Brasseur, F., van der Bruggen, P., and Thielemans, K. (2005) Electroporation of immature and mature dendritic cells: implications for dendritic cell-based vaccines. *Gene Ther*, **12**, 772–782.
103. Langenkamp, A., Messi, M., Lanzavecchia, A., and Sallusto, F. (2000) Kinetics of dendritic cell activation: impact on priming of TH1, TH2 and nonpolarized T cells. *Nat Immunol*, **1**, 311–316.
104. Camporeale, A., Boni, A., Iezzi, G., Degl'Innocenti, E., Grioni, M., Mondino, A.,, and Bellone, M. (2003) Critical impact of the kinetics of dendritic cells activation on the in vivo induction of tumor-specific T lymphocytes. *Cancer Res*, **63**, 3688–3694.
105. Vonderheide, R.H., Flaherty, K.T., Khalil, M., Stumacher, M.S., Bajor, D.L., Hutnick, N.A., Sullivan, P., Mahany, J.J., Gallagher, M., Kramer, A., Green, S.J., O'Dwyer, P.J., Running, K.L., Huhn, R.D., and Antonia, S.J. (2007) Clinical activity and immune modulation in cancer patients treated with CP-870,893, a novel CD40 agonist monoclonal antibody. *J Clin Oncol*, **25**, 876–883.
106. Turner, J.G., Rakhmilevich, A.L., Burdelya, L., Neal, Z., Imboden, M., Sondel, P.M., and Yu, H. (2001) Anti-CD40 antibody induces antitumor and antimetastatic effects: the role of NK cells. *J Immunol*, **166**, 89–94.
107. Nair, S., McLaughlin, C., Weizer, A., Su, Z., Boczkowski, D., Dannull, J., Vieweg, J., and Gilboa, E. (2003) Injection of immature dendritic cells into adjuvant-treated skin obviates the need for *ex vivo* maturation. *J Immunol*, **171**, 6275–6282.
108. Perfetto, S.P., Chattopadhyay, P.K., Lamoreaux, L., Nguyen, R., Ambrozak, D., Koup, R.A., and Roederer, M. (2006) Amine reactive dyes: an effective tool to discriminate live and dead cells in polychromatic flow cytometry. *J Immunol Methods*, **313**, 199–208.
109. Yewdell, J.W. and Nicchitta, C.V. (2006) The DRiP hypothesis decennial: support, controversy, refinement and extension. *Trends Immunol*, **27**, 368–373.
110. Ricicova, M. and Palkova, Z. (2003) Comparative analyses of *Saccharomyces cerevisiae* RNAs using Agilent RNA 6000 Nano Assay and agarose gel electrophoresis. *FEMS Yeast Res*, **4**, 119–122.
111. Morse, M.A., Coleman, R.E., Akabani, G., Niehaus, N., Coleman, D., and Lyerly, H.K. (1999) Migration of human dendritic cells after injection in patients with metastatic malignancies. *Cancer Res*, **59**, 56–58.
112. Ridolfi, R., Riccobon, A., Galassi, R., Giorgetti, G., Petrini, M., Fiammenghi, L., Stefanelli, M., Ridolfi, L., Moretti, A., Migliori, G., and Fiorentini, G. (2004) Evaluation of in vivo labelled dendritic cell migration in cancer patients. *J Transl Med*, **2**, 27.
113. De Vries, I.J., Krooshoop, D.J., Scharenborg, N.M., Lesterhuis, W.J., Diepstra, J.H., Van Muijen, G.N., Strijk, S.P., Ruers, T.J., Boerman, O.C., Oyen, W.J., Adema, G.J., Punt, C.J., and Figdor, C.G. (2003) Effective migration of antigen-pulsed dendritic cells to lymph nodes in melanoma patients is determined by their maturation state. *Cancer Res*, **63**, 12–17.
114. de Vries, I.J., Lesterhuis, W.J., Barentsz, J.O., Verdijk, P., van Krieken, J.H., Boerman, O.C., Oyen, W.J., Bonenkamp, J.J., Boezeman, J.B., Adema, G.J., Bulte, J.W., Scheenen, T.W., Punt, C.J., Heerschap, A., and Figdor, C.G. (2005) Magnetic resonance tracking of dendritic cells in melanoma patients for monitoring of cellular therapy. *Nat Biotechnol*, **23**, 1407–1413.
115. Schaft, N., Birkholz, K., Hofmann, C., Schmid, M., Theiner, G., and Dorrie, J. (2008) Dendritic cell vaccination and other strategies to tip the balance of the immune system: DC2007 5th International Meeting,

July 1618, Bamberg, Germany. *Cancer Immunol Immunother*, **57**, 913–928.

116. Verdijk, P., Scheenen, T.W., Lesterhuis, W.J., Gambarota, G., Veltien, A.A., Walczak, P., Scharenborg, N.M., Bulte, J.W., Punt, C.J., Heerschap, A., Figdor, C.G., and de Vries, I.J. (2007) Sensitivity of magnetic resonance imaging of dendritic cells for in vivo tracking of cellular cancer vaccines. *Int J Cancer*, **120**, 978–984.
117. Mullins, D.W., Sheasley, S.L., Ream, R.M., Bullock, T.N., Fu, Y.X., and Engelhard, V.H. (2003) Route of immunization with peptide-pulsed dendritic cells controls the distribution of memory and effector T cells in lymphoid tissues and determines the pattern of regional tumor control. *J Exp Med*, **198**, 1023–1034.
118. Fong, L., Brockstedt, D., Benike, C., Wu, L., and Engleman, E.G. (2001) Dendritic cells injected via different routes induce immunity in cancer patients. *J Immunol*, **166**, 4254–4259.
119. Carrasco, J., Van Pel, A., Neyns, B., Lethe, B., Brasseur, F., Renkvist, N., van der Bruggen, P., van Baren, N., Paulus, R., Thielemans, K., Boon, T., and Godelaine, D. (2008) Vaccination of a melanoma patient with mature dendritic cells pulsed with MAGE-3 peptides triggers the activity of nonvaccine anti-tumor cells. *J Immunol*, **180**, 3585–3593.
120. Bonehill, A., Van Nuffel, A.M.T., Corthals, J., Tuyaerts, S., Heirman, C., Françios, V., Colau, D., Van Der Bruggen, P., Neyns, B., and Thielemans, K. (2009) Single step antigen loading and activation of dendritic cells by mRNA electroporation for the purpose of therapeutic vaccination in melanoma patients. *Clin Cancer Res*, **15**, 3366–3375.
121. Kikuchi, T., Moore, M.A., and Crystal, R.G. (2000) Dendritic cells modified to express CD40 ligand elicit therapeutic immunity against preexisting murine tumors. *Blood*, **96**, 91–99.
122. Lapointe, R., Toso, J.F., Butts, C., Young, H.A., and Hwu, P. (2000) Human dendritic cells require multiple activation signals for the efficient generation of tumor antigen-specific T lymphocytes. *Eur J Immunol*, **30**, 3291–3298.
123. Borst, J., Hendriks, J., and Xiao, Y. (2005) CD27 and CD70 in T cell and B cell activation. *Curr Opin Immunol*, **17**, 275–281.
124. Van Driessche, A., Ponsaerts, P., Van Bockstaele, D.R., Van Tendeloo, V.F., and Berneman, Z.N. (2005) Messenger RNA electroporation: an efficient tool in immunotherapy and stem cell research. *Folia Histochem Cytobiol*, **43**, 213–216.
125. Wolfl, M., Kuball, J., Ho, W.Y., Nguyen, H., Manley, T.J., Bleakley, M., and Greenberg, P.D. (2007) Activation-induced expression of CD137 permits detection, isolation, and expansion of the full repertoire of CD8+ T cells responding to antigen without requiring knowledge of epitope specificities. *Blood*, **110**, 201–210.
126. Betts, M.R., Brenchley, J.M., Price, D.A., De Rosa, S.C., Douek, D.C., Roederer, M., and Koup, R.A. (2003) Sensitive and viable identification of antigen-specific CD8+ T cells by a flow cytometric assay for degranulation. *J Immunol Methods*, **281**, 65–78.

Chapter 28

Non-MHC-Dependent Redirected T Cells Against Tumor Cells

Hilde Almåsbak, Marianne Lundby, and Anne-Marie Rasmussen

Abstract

Adoptive transfer of T cells with restricted tumor specificity provides a promising approach to immunotherapy of cancers. However, the isolation of autologous cytotoxic T cells that recognize tumor-associated antigens is time consuming and fails in many instances. Alternatively, gene modification with tumor antigen-specific T-cell receptors (TCR) or chimeric antigen receptors (CARs) can be used to redirect the specificity of large numbers of immune cells toward the malignant cells. Chimeric antigen receptors are composed of the single-chain variable fragment (scFv) of a tumor-recognizing antibody cloned in frame with human T-cell signaling domains (e.g., CD3ζ, CD28, OX40, 4-1BB), thus combining the specificity of antibodies with the effector functions of cytotoxic T cells. Upon antigen binding, the intracellular signaling domains of the CAR initiate cellular activation mechanisms including cytokine secretion and cytolysis of the antigen-positive target cell.

In this chapter, we provide detailed protocols for large-scale ex vivo expansion of T cells and manufacturing of medium-scale batches of CAR-expressing T cells for translational research by mRNA electroporation. An anti-CD19 chimeric receptor for the targeting of leukemias and lymphomas was used as a model system. We are currently scaling up the protocols to adapt them to cGMP production of a large number of redirected T cells for clinical applications

Key words: mRNA electroporation, chimeric antigen receptor, gene therapy, T lymphocytes.

1. Introduction

1.1. Adoptive Cell Therapy

Advances in the understanding of immune surveillance and mechanisms underlying tumor escape have led to the development of novel strategies to enhance antitumor immunity in cancer patients (1). Among these, T-cell-based therapies have shown promising results. Clinical trials have established that immune effector T cells are capable of eradicating tumors.

M. Sioud (ed.), *RNA Therapeutics*, Methods in Molecular Biology 629,
DOI 10.1007/978-1-60761-657-3_28, © Springer Science+Business Media, LLC 2010

Indeed, treatment of patients with leukemia after allogeneic stem cell transplantation by donor lymphocyte infusion (DLI) induced complete remissions through graft-versus-leukemia (GVL) reactivity (*see* **Chapter 30**). Also, injection of tumor-infiltrating lymphocytes (TIL) or antigen-specific $CD8^+$ T-cell clones to patients with cancer resulted in cancer regression. (2–7). While DLI can only be used after allogeneic transplantation, the use of TIL has been hampered by the difficulty to obtain a significant number of TIL from tumor specimens for re-infusion. Furthermore, the generation of large numbers of cloned T cells requires multiple activation cycles and prolonged ex vivo expansion periods resulting in generation of terminally T-differentiated cells. These cells have short telomers and exhibited poor in vivo persistence and tumor clearance (8–10). In addition to the limitations mentioned above, most tumor antigens are self-proteins that are not expected to induce efficient immune activation due to peripheral tolerance mechanisms and thymic deletion of autoreactive T cells (11, *see* **Chapter 29**). Studies with adoptive T-cell (ATC) transfer have clearly indicated that there is a need for optimization (12).

1.2. Redirecting T-Cell Specificity

The advent of clinically acceptable and scalable methods for isolation, genetic engineering, and ex vivo expansion of T cells in the past decades has paved the way for novel experimental therapies. These involve the uniform redirection of the specificity of large numbers of T cells to tumors by transferring high-affinity αβ T-cell receptor (TCR) genes cloned from tumor-reactive cytotoxic T lymphocytes (CTL) (13, *see* **Chapter 29**) or tumor-specific antibody fusion genes (14–17). By redirecting T cells through genetic manipulation it is possible to circumvent some of the major obstacles of T-cell therapy, such as the isolation of tumor-specific T cells and their in vitro expansion (18). However, one disadvantage endowing T cells with antitumor specificity by TCR gene transfer is the major histocompatability complex (MHC) restriction mechanism. Therefore, the gene construct can only be used in patients expressing a certain MHC haplotype. Also, the TCR transfer is not relevant for targeting malignant cells that do not display MHC–peptide complexes on the cell surface (19). The gene transfer of αβ-TCRs was comprehensively reviewed by Uckert et al. (20, 21).

Chimeric antibody receptors (CARs) offer significant advantages over TCR transfer strategy. Such immune receptors act independently of MHC, implying that a single vector can be used universally to confer recognition regardless of recipient MHC class, only restricted to patients having malignant cells that express the particular tumor-associated antigen (TAA) or self-antigen specified by the antibody. Since antigen recognition by a CAR is not limited by MHC presentation, redirected T cells might also be used to target nonpeptide antigens, such as carbohydrates and phospholipids (22–25). Introduction of CARs into $CD4^+$ and $CD8^+$ T cells generates both helper (Th) and cytotoxic (Tc) tumor-specific cells, which might be important for optimal immune function and sustained antitumor response (26). Furthermore, CAR-expressing $CD4^+CD25^+Foxp3^+$ regulatory T cells have shown promising results in mouse models for the treatment of autoimmune diseases (27, 28) and finally; CARs have also been used to re-direct NK and γδ-T cells toward tumor targets (29–32). Thus, the broad versatility and flexibility of chimeric receptors might render them valuable as tools for the immunotherapy of a wide spectrum of diseases in the future.

1.3. The Modular Composition of Chimeric Antigen Receptors

The chimeric antigen receptor, introduced by Zelig Eshhar as the T body, is designed as a single-polypeptide chain composed of an extracellular binding module derived from an antibody single-chain variable fragment (scFv) coupled via a transmembrane anchor to an intracellular signaling domain, usually the CD3ς chain of the TCR/CD3 complex (33) (**Figs. 28.1 and 28.2**). Receptor binding and cross-linking lead to phosphorylation of endogenous SRC/SYK family kinases, ultimately activating the CAR expressing cell. Activation results in cytokine secretion and exertion of cytolysis by granzyme–perforin-mediated mechanisms

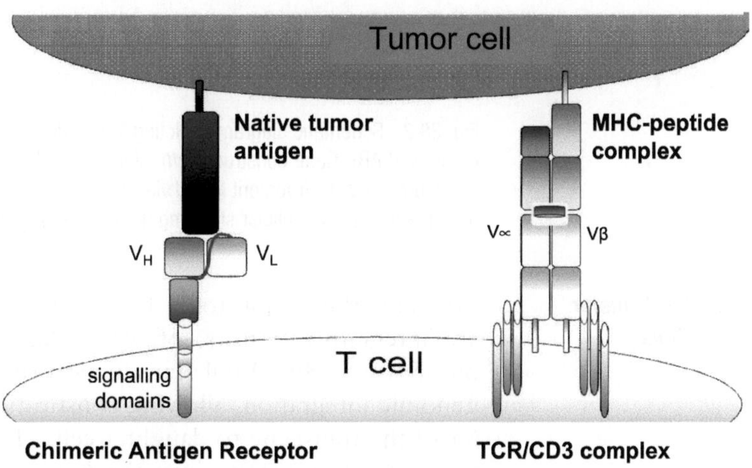

Fig. 28.1. Schematic diagram depicting the organization of the chimeric antigen receptor (CAR) and the endogenous TCR/CD3 complex. The scFv antibody fragment of the CAR binds to an exposed native tumor-associated antigen and binding is propagated through a "combined signal pathway" composed of linked human lymphocyte stimulatory and costimulatory molecules. T-cell activation by CARs is MHC independent, which is not the case with the TCR, recognizing a peptide in complex with the MHC molecule. Upon TCR binding, the activation is mediated through CD3ς cytoplasmic domains, and further enhanced by binding to molecules associated with different costimulatory mechanisms.

directed toward the antigen-positive malignant cell. For optimal T-cell activation, additional costimulatory molecules are required (34, 35). Therefore, several studies have incorporated elements encoding signaling domains from various lymphocyte costimulatory molecules, such as CD28, CD137 (4-1BB), and CD134 (OX40) (36–38). Importantly, the combinations of costimulatory domains together with the CD3ς have shown to be superior in sustaining persistence of the transferred T cells in vivo (16, 23, 36).

On the molecular level, the chimeric receptor construct consists of linked domains; the immunoglobulin heavy-chain leader peptide, the specified scFv fragment (alternatively), a spacer derived from a human immunoglobulin hinge region, connecting the ectodomain to the transmembrane and intracytoplasmic domains encoding the different stimulatory and costimulatory lymphocyte signaling parts into a single open reading frame, as shown schematically in **Fig. 28.2**.

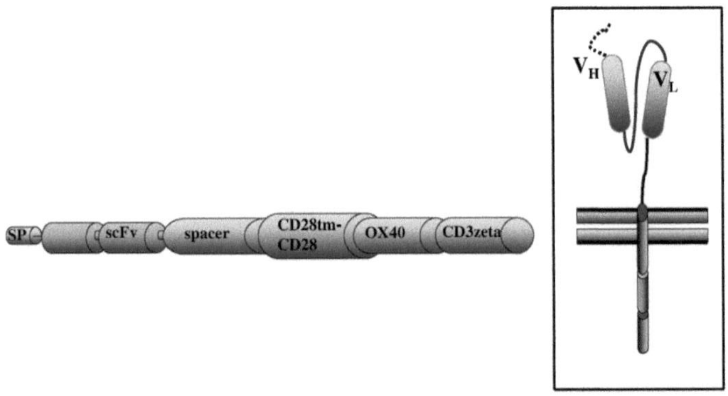

Fig. 28.2. Schematic diagram depicting the modular composition of a chimeric antigen receptor (CAR). Gene construct (*left*): signal peptide (SP) for transport to endoplasmic reticulum, the scFv fragment linked via a spacer molecule (human Ig-domains) to transmembrane and intracellular signaling domains (e.g., CD28tm-CD28-Ox40-CD3ς).

1.4. Gene Transfer into T Cells

To date, the major tools for genetic modification of T cells are integrating vectors, of which retroviruses are most commonly used (39). Viral transduction of T cells results in stable genomic integration, allowing constitutive expression and transfer of the transgene to daughter cells. The ability of retroviruses to randomly integrate into the host cell chromosome poses a potential risk of insertional mutagenesis (40, 41). Nevertheless, work with T-cell gene therapy has not established evidence for oncogenesis (42–44). Furthermore, the lack of cellular transformation has also been confirmed in long-term animal studies (45, 46).

Malignant transformation appears only as a hypothetical issue and as a consequence does not represent a major concern in gene therapy with peripheral T cells. Probably, a more relevant aspect of permanent modification is the risk of potential damage to normal tissues through unexpected autoaggressive behavior of the modified T cells. This has been exemplified in a recent clinical trial in renal carcinoma using autologous modified T cells that express a CAR specific for carboxy anhydrase IX (47). Although initially well tolerated, patients who received the T-cell therapy developed liver enzyme abnormalities explained by autoreactivity with low levels of the antigen expressed on the bile duct epithelium. One way to address these potential safety issues is to cotransduce the T cells with a suicide agent, such as herpes simplex virus thymidine kinase gene (48), to offer the possibility to eliminate clonally expanded oncogenic or autoreactive cells upon administration of the suicide-activating agent. Although proven effective, incorporation of suicide agents has some limitations and increases the complexity of the gene therapy protocol. To avoid these problems, we are using an RNA transfection strategy based on the electroporation of T cells with in vitro transcribed mRNA. A large number of clinical studies using mRNA-transfected DC for cellular vaccination have demonstrated the feasibility and safety of this strategy (49–51). Unlike viral and DNA vectors, mRNA does not integrate into recipient chromosome and due to the intrinsic instability of the mRNA molecule, receptor expression is only transient. Transfection efficiency is very high and expression is evident in nearly the entire cell population (**Fig. 28.3**). The onset of receptor expression is quick (surface expression of receptor is detected as early as 1–2 h postelectroporation) and the duration of expression of transiently expressed CARs and TCR is at least 8 days postelectroporation (52–55). Notably, receptor expression is in general dependent on the half-life of the mRNA, the stability of the translated protein, and cell division.

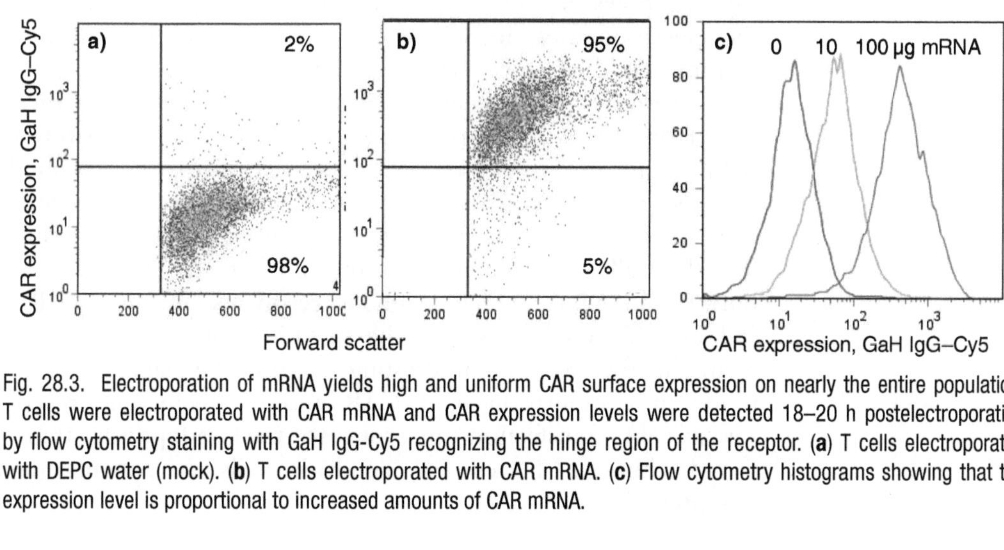

Fig. 28.3. Electroporation of mRNA yields high and uniform CAR surface expression on nearly the entire population. T cells were electroporated with CAR mRNA and CAR expression levels were detected 18–20 h postelectroporation by flow cytometry staining with GaH IgG-Cy5 recognizing the hinge region of the receptor. (**a**) T cells electroporated with DEPC water (mock). (**b**) T cells electroporated with CAR mRNA. (**c**) Flow cytometry histograms showing that the expression level is proportional to increased amounts of CAR mRNA.

The transient nature of receptor expression might represent a disadvantage of this technology as studies with adoptive transfer of T cells for cancer therapy have clearly demonstrated that there is a correlation between persistence of the transferred gene-modified T cells and tumor regression (56). Thus, to compensate for the lack of longevity of the retargeting program with the mRNA approach, it might be necessary to repeat T-cell infusions over time to achieve the desired antitumor effect. Ultimately, the time and cost required for viral vector production as well as the complex intellectual property (IP) landscape and considerable regulatory requirements accompanying the use of this technology seems justified to address the clinical efficiency of mRNA electroporation in cases where stable transgene expression is not required or undesired. Moreover, the lower cost and workload associated with this technique, as well as the reproducible generation of large amounts of synthetic mRNA, indicate that our strategy would be the preferred method of choice for short-term in vitro studies addressing conceptual mechanisms. mRNA electroporation might emerge as an attractive alternative to viral transduction for the rapid screening of sets of receptors and coreceptors prior to choosing the final candidate for stable transduction for use in clinical trials.

It should be noted that the electroporation technology offers a straightforward way of introducing novel functionalities to T cells simultaneously along with the CAR. Electroporation of mRNA encoding chemokine receptors, e.g., CCR7, CXCR4, CXCR2 (57, 58), can readily be used to transiently direct T cells to the selected tumor microenvironment (*see* **Chapter 28**). Furthermore, the function of T cells or dendritic cells can be altered by small interfering RNAs (*see* **Chapters 5** and **26**). Previous studies have shown that the electroporation technique can deliver active siRNAs into mammalian cells (59, 60). Therefore, the blocking of apoptotic pathways in T cells by Bax-specific siRNA or targeting downstream mediators in the TGF-β signaling pathway (61) might promote prolonged survival of the transplanted T cells. Finally, electroporation of ferumoxides has recently been reported as a method to label cells for adoptive transfer, thus allowing for in vivo tracking of the T-cell distribution and migration by magnetic resonance imaging (MRI) (62–64). To further explore the versatility of electroporation, it might therefore be interesting to use this labeling technique in combination with CAR and/or chemokine receptor mRNA transfection in order to track adoptively transferred chimeric T cells or dendritic cells in patients (*see* **Chapter 27**).

1.5. Manufacturing Transiently Redirected T Cells

1.5.1. Preparing Large Numbers of T Cells for Redirection

Compared to the lengthy process of ex vivo isolation, characterization, and expansion of T cells with a native specificity for a tumor-associated antigen (TAA), genetic redirection of polyclonal T cells provides large numbers of tumor-specific T cells within a few weeks. Although a standard leukapheresis typically yields more than 2×10^9 T cells, it is questioned whether this quantity is sufficient to generate "therapeutic" relevant cell numbers of gene-modified cells. Moreover, it is anticipated that a therapeutic strategy based on transiently redirected T cells may require repeated rounds of adoptive transfer, which further augments the need for development of adequate ex vivo expansion protocols to generate large numbers of T cells. There are no clinical trials to date investigating transiently redirected T cells and stably gene-modified T cells have only recently entered the clinic in early phase I studies, implying that specifications regarding cell dosage, as well as timing and numbers of reinfusions, are still not established. Such specifications will ultimately rely on future clinical safety and efficacy studies based on dose-escalation designs. Additionally, some of the early clinical trials with stably modified T cells were presumably conducted with functionally impaired T cells (8), which limits the utility of extrapolation to future studies using more "fit" T cells. Nevertheless, current protocols for clinical studies with permanently retargeted T cells specify dosage-escalation designs with infused T-cell numbers ranging from 2×10^7 to 2×10^8 cells/m^2 corresponding to approximately 3.6×10^7–3.6×10^8 cells (for 1.8 m^2) (65) to 10^9–10^{11} cells (66).

The first clinical trials with transfer ATC for malignant melanoma patients using TIL were reported in 1988 (67). TIL were generated by culturing cells in high-dose IL-2 without antigen stimulation until more than 10^{11} T cells were obtained. This protocol has its limitations since most tumor-specific T cells need regular restimulation with antigen by DCs or nonspecific activation through allogeneic PBMC plus anti-CD3 antibody or mitogens (e.g., PHA). The limitation of generating large amounts of autologous DCs as stimulators for the T cells initiated the production of cGMP-compliant artificial antigen-presenting cells (artificial APC), based on paramagnetic beads co-coupled with monoclonal antibodies to CD3 and CD28 (68–71). Anti-CD3 and anti-CD28-coated Dynabeads (CD3/CD28 beads) mimic the interaction between T cells and APCs in vivo and they have been demonstrated as a feasible and effective method for clinical grade production of T cells (72–77). The CD3/CD28 beads support long-term proliferation of T cells in culture, while maintaining desirable biological characteristics (70, 71, 78). Importantly, during the ex vivo bead-driven activation, T cells are induced to produce key immunomodulatory molecules, including IL-2, IFN-γ, and TNF-α (75), and in settings where T cells are anergic or nonresponsive, activation and expansion have shown to reverse anergy and restore responsiveness to mitogenic or antigen stimulation (79–81). Several clinical trials have been conducted using bead-based T-cell expansion protocols and several hundred patients have been treated with CD3/CD28 bead-expanded T cells (79, 82–85).

1.5.2. T-Cell Phenotype and Function

Naive T cells express high levels of CCR7, CD62L, CD27, and CD28 and are equipped with long telomers. During the linear differentiation that follows antigen stimulation, expression of these molecules decline. Terminally differentiated T cells have short telomers and have lost the expression of CCR7, CD62L, CD27, and CD28 and have gained the expression of CD57, which is a molecule associated with senescence and clonal exhaustion and a lack of protective function. It has been demonstrated in several clinical trials that the fraction of transferred T cells that persisted were phenotypically less-differentiated clonotypes ($CD27^+CD28^+CD57^-CD45RO^+$). Furthermore, recent results have established a link between T-cell persistence and tumor regression following adoptive transfer (56, 86). In vitro expansion protocols for generation of TIL for ATC usually take 20–40 days and induce differentiation toward a late effector state resulting in phenotypic and functional changes that make T cells less fit to mediate antitumor responses in vivo. The adoptive transfer of ex vivo expanded $CD8^+$ T cells in the absence of CD4 help has been demonstrated to result in low persistence of the transferred T cells (87). It remains, however, to be determined whether the CD4 help needs to be homospecific or heterospecific to rescue $CD8^+$ T cells from AICD (88). The adoptive transfer of a mixture of CAR-transfected $CD4^+$ and $CD8^+$ T cells has the benefit of contributing the useful properties of CAR-expressing $CD8^+$ T cells targeting tumor cell killing and CD4 helper cells expressing the same CAR, which will produce Th1 cytokines that are important for $CD8^+$ T-cell survival. Following 10 days expansion with CD3/CD28 beads, T cells demonstrate a Th1/Tc1 cytokine profile, which has been shown to be favorable for achieving an antitumor response upon adoptive transfer. The CD4/CD8 ratio is decreased following 10 days expansion with Dynabeads, indicating a growth advantage for $CD8^+$ T cells (89). Since no animal or clinical data are published using CAR mRNA-transfected T cells, we cannot conclude whether the transfer of both $CD4^+$ and $CD8^+$ T cells expressing CAR will lead to better tumor cell killing than by transferring CAR $CD8^+$ T cells alone. Furthermore, it is well documented that following tumor cell lysis mediated by, e.g., TAA-specific vaccination, tumor antigens are taken up by APC and presented to T cells under inflammatory signals, which may lead to induction of new tumor-specific T cells with the potential to kill residual tumor cells by epitope-spreading mechanisms (90). All CAR-transfected T cells coexpress endogenous TCRs, of which some might be directed toward tumor antigen. These tumor-specific T cells can still retain functionality following CAR downregulation provided that they express the central memory or effector memory phenotype.

Messenger RNA transfection leads to transient expression of CARs implying that the transferred T cells must have the capability to home to the tumor site and perform CAR-mediated killing within a few days following transfer. Following 10 days expansion with CD3/CD28-beads, the CAR-transfected T cells contain a mixture of central memory, effector memory, and effector cells (**Fig. 28.4**). L-selectin, CD62L, which is a prerequisite for homing to lymph nodes (LN) is highly expressed on expanded T cells and the number of cells expressing the terminally differentiation marker CD57 is significantly reduced. However, only few T cells express the CCR7 chemokine receptor, which is important for guiding cells to and within lymphoid organs. Here, mRNA transfection technology opens the prospects of introducing different receptors into the same T cells in order to achieve several features, such as killing through CARs and homing to LN mediated by CCR7.

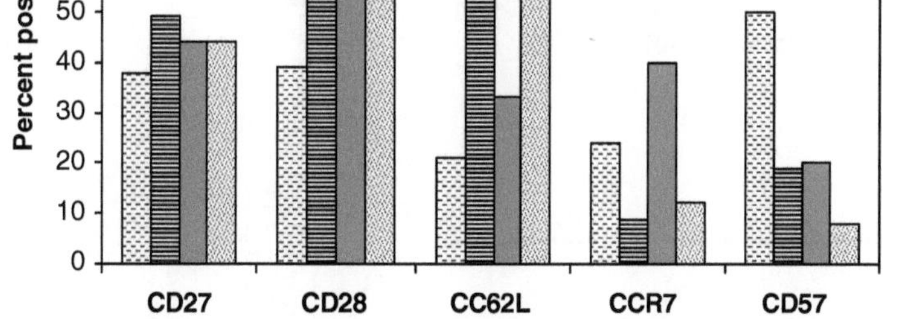

Fig. 28.4. Following 10 days expansion with Dynabeads *ClinExVivo* CD3/CD28, the T-cell population contains a mixture of central memory (CM: $CD28^{+}/CCR7^{+}/CD62L^{+}$), effector memory (EM: $CD28^{+/-}/CCR7^{-}$), and a few effector cells (EC: $CD28^{-}/CCR7^{-}$). The phenotype of $CD4^{+}$ and $CD8^{+}$ T cells has been analyzed by flow cytometry to assess the differentiation status of the final expanded T-cell product and compare it to the starting cell sample. The data are from one donor and representative of three independent experiments.

1.5.3. Depletion of Regulatory T Cells

$CD4^+CD25^+$ regulatory T cells (T_{regs}) represent 2–5% of peripheral blood lymphocytes and are the main contributors to the maintenance of immune homeostasis, preventing autoimmunity and mediating peripheral tolerance. Increased number of T_{regs} have been detected in peripheral blood and in the tumor microenvironment of late-stage cancer patients and this increase correlated with reduced overall survival (91–94). T_{regs} recognize self-peptides presented by MHC class II molecules on APC and given that many TAA are of self-origin, T_{regs} may suppress T-cell activity against TAA and thereby facilitate tumor escape. Introduction of CAR into T_{regs} may facilitate T_{reg}-mediated suppression of CAR-transfected effector T cells, supporting inclusion of a T_{reg}-depletion step prior to T-cell expansion.

Several lines of evidence suggest that nonmyeloablative therapy can enhance the efficacy of adoptive T-cell therapy by depleting T_{regs} and cells competing for activating cytokines, and

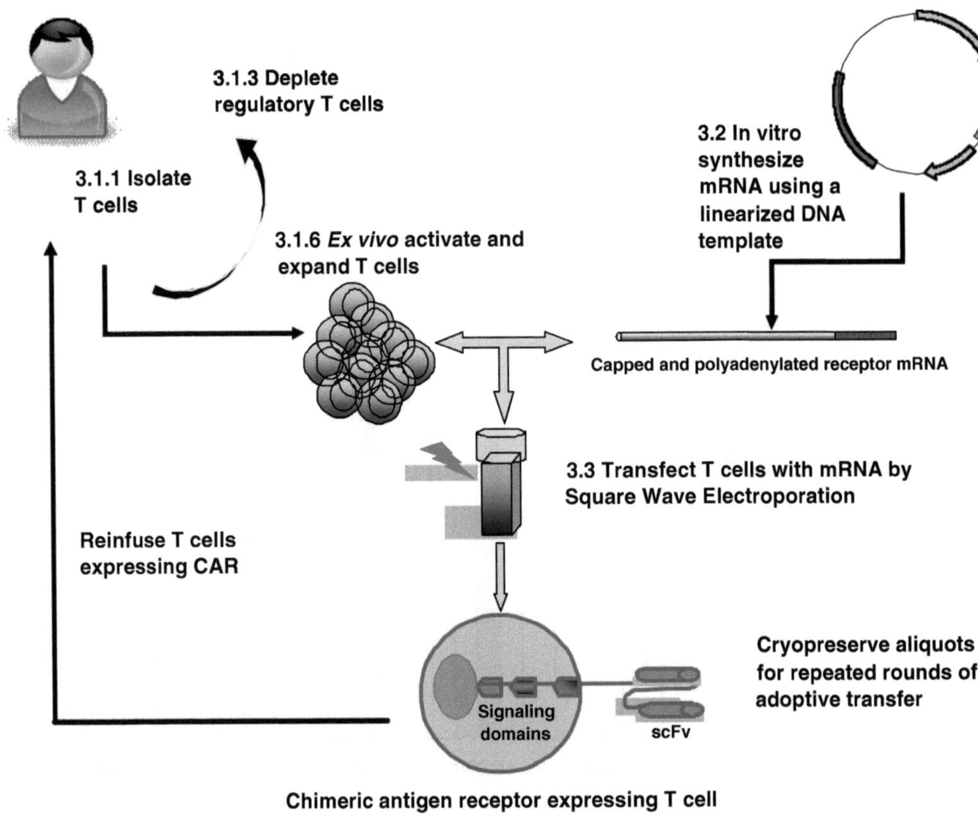

Fig. 28.5. Manufacturing redirected T cells for adoptive immunotherapy by electroporation of mRNA encoding a chimeric antigen receptor. Overview; peripheral blood lymphocytes are isolated from a leukapheresis product by elutriation and subsequently depleted for regulatory T cells before *ex vivo* expansion for 10 days using Dynabeads *ClinExVivo* CD3/CD28 beads in a Wave Bioreactor. The expanded cells are further gene modified to express a chimeric antigen receptor by electroporation with synthetic mRNA. Following appropriate quality controls, CAR expressing T cells are reinfused to the patient.

increase function and availability of APC (12, 95). In vivo depletion of T_{regs} using chemotherapy or ONTAK (diphtheria toxin-conjugated CD25 monoclonal antibody) combined with cancer vaccines has demonstrated enhanced antitumor immunity in a number of patients (96–98). A recent study by Cheadle et al. (99) demonstrated that adoptive transfer of CD19-targeted T cells following cyclophosphamide treatment could eradicate malignant B cells in a SCID/beige mouse model. Taken together, these data suggest that depletion of T_{reg} in vitro and in vivo may facilitate improved clinical responses in patients receiving adoptive transfer of redirected T cells.

1.6. Manufacturing Process Overview

The flowchart presented in **Fig. 28.5** gives an overview of the process that we developed for the generation of redirected T cells for clinical studies. Starting with 5×10^8 T_{reg}-depleted T cells from peripheral blood, the final T-cell product is expanded approximately 200-fold in 10 days. The expanded T cells contain a mixed population of central and effector memory T-cell phenotype. We further developed a medium-scale, nonviral gene transfer method based on electroporation of chimeric antigen receptor mRNA up to 1×10^8 T cells in one cuvette. We are currently evaluating alternative methods allowing for scaling up the electroporation.

For clinical studies, all steps have to be performed under current good manufacturing procedure (cGMP) conditions, implying that requirements for reagents, facilities, personnel, and procedures are set up in accordance with the appropriate guidelines.

2. Materials

2.1. Preparing T Cells for Gene Modification

2.1.1. Isolation of T Cells from Peripheral Blood

1. Leukapheresis system (Cobe Spectra auto PBSC, Caridian BCT) with disposables as required.
2. Elutra Cell Separation System with disposables as required (Caridian BCT).
3. Buffer 1: Dulbecco's phosphate-buffered saline D-PBS without Ca^{2+}/Mg^{2+} and with 1% human serum albumin (HSA).

2.1.2. Determination of Viability and Cell Concentration

1. Hemacytometer (VWR), microscope.
2. Trypan blue, 0.4% solution (Invitrogen).

2.1.3. Depletion of Regulatory T Cells

1. Transfer bag 600 mL (Baxter).
2. Variable Speed Rocker (Heidolph Polymax 1040).
3. Flow cytometer, BD LSR II, or equivalent.

4. Dynabeads *ClinExVivo* CD25 (4×10^8 beads/mL).
5. Dynal *ClinExVivo* MPC.
6. Dynal MPC-15.
7. Buffer 2: 500 mL D-PBS, 2% HSA (w/v), 0.4% sodium citrate (w/v).
8. Staining buffer (SB): D-PBS, 0.1% HSA, 0.1% NaN_3.
9. CD8-FITC, CD4-PE, and CD25-APC.

2.1.4. Cryopreservation of T Cells

1. Cryocyte freezing container, 50 mL.
2. Controlled rate freezer, Planer Kryo K10/16.
3. Freezing medium (100 mL): 15 mL DMSO, 12 mL Plasma-Lyte A, 33 mL Hespan, and 40 mL HSA, 200 mg/mL.

2.1.5. Thawing of Cryopreserved Cells

1 Transfer bag 150 mL (Baxter).

2 Thawing buffer: D-PBS, 2% HSA, 0.2% sodium citrate (Baxter).

3 Cell culture medium 1: CellGro DC medium, 5% inactivated human serum, 1% Glutamax, 0.8% Mucomyst, 100 U/mL IL2.

2.1.6. Large-Scale Manufacturing of T Cells

1. Vuelifebag FEP Teflon bag (Cell Genix)
2. Transfer bags, 600 and 2,000 mL
3. 10 L perfusion bags, cutoff 1.2 μm (Sartorius, Wave Biotech AG)
4. 20 L waste bag (Wave Biotech AG)
5. Wave Bioreactor 2/10 EH with CO_2 mix 20 and CO_2 (PhEur) (GE Healthcare)
6. Cobe Cell Processor 2991 (Cardian BCT) or equivalent
7. Cell mixer (Heldolph Polymax 1040)
8. Dynabeads *ClinExVivo* CD3/CD28 (4×10^8 beads/mL)
9. Dynal *ClinExVivo* MPC
10. Dynal MPC-15
11. Buffer 3: D-PBS, 0.1% HSA
12. Buffer 2 (*See* **Section 2.1.3.7**)
13. Cell culture medium 1 (*See* **Section 2.1.5**)
14. Cell culture medium 2: Cell culture medium 1 supplemented with 0.2% Pluronic F-68
15. Trypan blue
16. Easy Lyse (Dako)
17. CD8-FITC, CD4-PE, CD25-APC, and CD3-FITC or PE

2.2. Preparing mRNA for Transfection

2.2.1. Preparation of the DNA Template

1. DNA vector containing the gene of interest
2. *Mfe*I restriction enzyme
3. NEB buffer 4
4. Wizard SV Gel and PCR Clean-Up System or similar purification kit
5. 1% agarose gel
6. Electrophoresis equipment

2.2.2. Synthesis and Purification of mRNA

1. Linearized DNA template
2. Ribomax large-scale RNA production system-T7
3. RNA CAP structure analogue
4. RQ1 RNase-free DNase
5. MegaClear Ion Exchange Spin Column
6. NanoDrop

2.3. Transfection of mRNA

2.3.1. Electroporation

1. ECM 830 Square Wave Electroporator (BTX Harvard Apparatus)
2. BTX Electroporation Cuvettes PLUS, 4 mm Gap
3. Culture flask or six-well culture plate
4. CellGro DC medium (Cell Genix)
5. Purified mRNA
6. Cell culture medium 1 (*See* **Section 2.1.5**)
7. Optional (freezing) 5,100 Cryo 1°C Freezing Container, "Mr. Frosty" (Nalgene)

2.3.2. Verification of Receptor Expression by Flow Cytometry

1. Flow Cytometer LSR II
2. Staining buffer (*see* **Section 2.1.3**)
3. Cy5-GaH IgG (Fc-γ)(Jackson ImmunoResearch)
4. Propidium iodide (100 μg/mL). Keep the solution tightly closed at 4°C protected from light

2.3.3. Verification of T Functionality: Cell Killing

1. 96-Well round-bottom plates
2. LumaPlate-96 (Packard)
3. Microplate beta counter
4. CAR-expressing T cells
5. $CD19^{+}$ target cells, e.g., SupB15 (DSMZ ACC 389) or NALM6 (DSMZ ACC 128)
6. Cell culture medium 1 (*see* **Section 2.1.5**)
7. Trypan blue
8. Chromium-51 Radionuclide; sodium chromate solution

9. Fetal calf serum
10. D-PBS buffer
11. 5% Triton-X 100

2.3.4. Verification of T Functionality: Degranulation and Cytokine Production

1. Flow cytometer, LSR II
2. 48-Well polystyrene plates
3. GolgiStop (Monensin) (BD Biosciences)
4. Brefeldin A
5. Saponin: 0.25 g saponin, 1 mL FCS, dilute to 50 mL with D-PBS
6. 4% paraformaldehyde (PFA): Dissolve 0.4 g paraformaldehyde in 100 mL D-PBS in a conical flask and seal with film. On a heater (56–60°C) inside a fume hood, mix until the solution turns colorless. Allow to cool before handling
7. 0.5 M EDTA pH 8.0: Dissolve 186.1 g EDTA (disodium salt) in 700 mL dH_2O. While stirring, gradually add 20 g NaOH, adjust with water to 950 mL, adjust the pH to 8.0 with 1.0 M NaOH, and then complete to 1 L with distilled water. Autoclave
8. 20 mM EDTA: Make a 1:25 dilution of 0.5 M EDTA stock solution in D-PBS
9. Intracellular staining buffer (ISB): D-PBS supplemented with 2% FCS
10. CD4-PE, CD8-PC7, and CD107a-PC5
11. IFN-γ FITC
12. PC5- or FITC-conjugated irrelevant isotype-matched control antibodies
13. Aggregated γ-globulin

3. Methods

We assume that the reader has access to an appropriate cell culture facility with the necessary equipment such as microscopes, centrifuges, culture hoods, and incubators. We also expect that operators are well skilled in standard molecular biology as well as cell culturing and flow cytometry analysis.

3.1. Preparing T Cells for Gene Modification

3.1.1. Isolation of T Cells from Peripheral Blood

Peripheral blood is routinely the source of T cells for adoptive transfer. Blood is assessed by vein puncture (small scale) or leukapheresis (large scale) and subsequently enriched in a mononuclear fraction by Ficoll gradient separation (small scale) or by using closed, semi-automated devices, such as the Elutra Cell Separation System (Cardian BCT) (large scale). Various ex vivo cell fractionation devices can further be used downstream to enrich for the cell population for genetic redirection. The separation principle utilized by such devices is based on recognition of cell-specific surface molecules by monoclonal antibodies that are immobilized on paramagnetic beads and subsequently captured on a magnet (100). Clinical devices used routinely in cell separations are Dynabeads and Dynal Magnetic Particle Concentrators (MPCs) (Invitrogen), Isolex 300i (Baxter), and CliniMACS (Miltenyi Biotech). These devices are mostly operated in facilities providing therapeutic apheresis and cell therapy procedures, which require that the operators are extensively trained in the specified procedures. Further details are thus not described here.

3.1.2. Determination of Cell Viability and Concentrations

This protocol describes a simple method for the determination of cell viability and numbers based on discrimination between viable versus nonviable cells using the Trypan blue exclusion test (*see* **Note 1**).

1. Resuspend the cells thoroughly to achieve a uniform suspension and take out 20 μL of the suspension.
2. Add 20 μL of Trypan blue (0.4%). Make replicates as required. Do not leave cells with the dye for more than 15 min, as viable cells will begin to stain.
3. Transfer 10 μL of homogenous suspension to the hemacytometer.
4. View the cells under a microscope and count both blue (dead) and transparent (live) cells. Count 4 × 1 mm^2 areas of the counting chamber. For accurate determination, count at least 100–200 cells.
5. The cell concentration is calculated as follows:

$$\text{Cell concentration per mL} = \text{total cell count in four squares} \times 2{,}500 \times 2(\text{dilution})$$

6. Cell viability (%) = (total viable/total viable and nonviable) × 100%.

3.1.3. Depletion of Regulatory T Cells

We included a magnetic depletion step of T_{regs} prior to expansion to minimize potential inhibition of immune responses. The magnetic beads were custom manufactured for us by Invitrogen through the coupling of Zenapax (Daclizumab, CD25 humanized antibody, Roche Pharmaceuticals) to clinical grade Dynabeads under cGMP conditions.

1. Following elutriation, centrifuge approximately 5 × 10^9 cells from the lymphocyte fraction and discard the

supernatant. Resuspend the cells in 70–90 mL of Buffer 2 in a 600 mL transfer bag.

2. Take out a sample (approximately 1 mL) for cell counting and flow cytometry analysis.
3. Count viable cells as described in **Section 3.1.2**. Subsequently, stain 1×10^6 cells with CD8-FITC, CD4-PE, and CD25-APC and analyze the cells using flow cytometry. Determine the percentage of $CD4^+CD25^{high}$ and total T cells ($CD4^+CD8^+$).
4. Calculate the total number of $CD4^+CD25^{high}$ cells (T_{reg}).
5. Calculate the amount of CD25-Dynabeads required for T_{reg} depletion. Perform the depletion step of T_{reg} using five CD25-Dynabeads per one $CD4^+CD25^{high}$ cell using the subsequent protocol.
6. Dilute the cells in Buffer 2 to 0.5×10^8/mL. Carefully remove the foam and add 50 mL of air to the bag.
7. Take out the required volume of CD25-Dynabeads from the vial and transfer to a 15 mL tube. Wash the beads once in 10 mL of Buffer 2 by placing the tube in a magnet (MPC-15) for 1 min and then discard the supernatant.
8. Resuspend the CD25-Dynabeads in 1–2 mL Buffer 2 and transfer the washed beads to the cell bag with a syringe.
9. Incubate the bag on a variable speed rocker for 30 min at 18–22°C (2–3 rpm). Turn the bag every 5 min to prevent sedimentation of the CD25-Dynabeads (*see* **Note 2**).
10. After incubation, connect a transfer bag (buffer bag) containing 200 mL of Buffer 2 and a transfer bag (collection bag) to the bag containing the cells and beads (capture bag).
11. Transfer approximately 100 mL buffer from the buffer bag to the capture bag. Clamp the outlet tube and mix carefully. Place the capture bag on the large-scale magnet (Step 5 of **Section 2.1.3**) with an angle of 0° for 1 min. Wrap the outlet tube firmly around the capture magnet four times. Change the angle of the magnet to 30° for 1 min.
12. Open the outlet tube from the capture bag and drain the supernatant with the T_{reg} depleted cells to the collection bag. Clamp the tubing.
13. Add the remaining buffer from the buffer bag to the capture bag to wash the bead–cell complexes by careful agitation (*see* **Note 3**).
14. Place the bag on the magnet as described above to capture bead–cell complexes. Drain the noncaptured cells into the collection bag from Step 12.

15. Seal the collection bag with the T_{reg} depleted cells. Take out a sample for analysis with a syringe. Count the cells and stain approximately 1×10^6 cells with CD8-FITC, CD4-PE, and CD25-APC to monitor the efficiency of the T_{reg} depletion.
16. Weigh the bag to calculate the approximate volume of the cell suspension (1 g $\approx$ 1 mL). Calculate the total number of cells in the collection bag.
17. The cells can now be cryopreserved as described below. Alternatively, they can be processed directly according to the protocol for large-scale T-cell expansion (*see* **Section 3.1.6**).

3.1.4. Cryopreservation of T Cells

1. Cool the freezing medium to 2–8°C (*see* **Note 4**).
2. Centrifuge the cells at 400×*g* for 10 min.
3. Discard the supernatant and resuspend the cells in a small volume of cold PlasmaLyte A.
4. Count viable cells as described in **Section 3.1.2**.
5. Dilute cells to a final concentration of 4×10^8 cells/mL in PlasmaLyte A. Transfer the cells to a freezing tube container.
6. Cool the cells to 2–8°C.
7. Add the same volume of freezing medium using a syringe. Freeze the cells in a controlled rate freezer, and then transfer the cells to a N_2 tank for storage.

3.1.5. Thawing and Washing of Cryopreserved T Cells

We recommend the use of cell-processing devices as the Cobe Spectra (Cardian) or the Cytomate (Baxter) for buffer exchange and concentration.

1. Thaw the bag containing the cells in a 37°C water bath until complete disappearance of ice crystals. Alternatively, you can use a thawing device like the Barkey plasmatherm.
2. Connect a 600 mL transfer bag with thawing buffer to the bag containing the frozen cells. The volume of thawing buffer should be 10 times the volume of the cells.
3. Carefully add 10 times the volume of the cells of thawing buffer.
4. Centrifuge the bag at 600×*g* for 10 min. Discard the supernatant.
5. Resuspend the cells to approximately 2×10^7 cells/mL in cell culture medium 1 for leukapheresis product containing monocytes or product containing low percentage of T cells (*see* **Note 5**).
6. Take out approximately 1 mL cells and determine the number of viable cells.

3.1.6. Large-Scale Manufacturing of T Cells

In this protocol, the T_{reg}-depleted T cells are expanded approxi mately 100-200-fold in 10 days using Dynabeads CD3/CD28 and a Wave Bioreactor. The cells serve as an excellent source for the generation of gene-modified therapeutic T cells (*see* **Note 6**). The protocol described uses elutra fractionated lymphocytes as the starting material. If the cell product contains monocytes or low numbers of T cells (e.g., CLL patients), *see* **Note 5**.

3.1.6.1. Activation and Expansion of T Cells with Artificial APC

1. The lymphocyte fraction after elutriation usually contains 70–85% T cells, and monocytes are retained in a separate fraction. Remove 1 mL of the thawed elutra sample from the bag and perform flow analysis to determine the number of $CD3^+$ cells.
2. Transfer 5×10^8 $CD3^+$ cells to a Teflon bag (see **Section 2.1.6**) in 500 mL culture medium 1.
3. Wash 1.25 mL Dynabeads *ClinExVivo* CD3/CD28 corresponding to 5×10^8 beads according to the manufacturer's instructions (Dynal, Invitrogen). The number of Dynabeads corresponds to 1 bead per $CD3^+$ cell.
4. Transfer the washed beads to the bag using a syringe.
5. Carefully mix the cells and beads.
6. Remove all air in the bag.
7. Place the bag in an incubator at 37°C/5% CO_2. Leave the cells undisturbed for 3 days.
8. After 3 days static culture, the cells are ready for transfer to a Wave Bioreactor. Gently mix the bag to help dissociate the cell:bead complexes.
9. Remove 1 mL of cell suspension from the bag, resuspend well and dilute an aliquot of the sample 1:1 with trypan blue, and count the cells in a hemacytometer without removing the beads.
10. Remove the beads from the sample with a magnet, and then stain the cells with anti-CD3-FITC to determine the percentage of $CD3^+$ T cells. Analyze by flow cytometry.
11. Measure the volume in the bag by weighing (use 1 g $\approx$ 1 mL) and calculate the total number of T cells in the bag using the cell concentration and percentage of T cells.
12. Install the perfusion bag on the Wave Bioreactor as described in the manufacturer's instructions (Wave/GE Healthcare). Fill the bag with CO_2/air accordingly (*see* **Note 7**).
13. Set maximum volume to 5,000 mL and temperature to 37°C.

14. Set perfusion volume to 1,800 mL with shot volume 50 mL.
15. Add 200 mL of cell culture medium 2 to the 10 L perfusion bag prior to transferring the cell-bead complexes from the Teflon bag.
16. Add culture medium 2 (*see* **Note 8**) to the bag to obtain a final concentration of 0.5×10^6 $CD3^+$ T cells/mL.
17. Connect a 20 L waste bag to the outlet port in the perfusion bag and place the waste bag at a lower level than the Wave Bioreactor. Waste will be drained into the waste bag during exchange of media.
18. Start rocking at 8–10 rpm. Leave the cells on the Wave Bioreactor overnight.
19. Before taking out a sample for counting, rocking is increased to 25 rpm for 2 min to resuspend the cells. Take out a sample (use syringe and the sample port, see **Fig. 28.6**) and then determine the cell number and viability. Do not remove the beads.
20. Subsequently, add cell culture medium 2 to the cells in order to adjust the cell concentration to 0.5×10^6/mL.
21. Count cells every day. Add fresh medium to maintain a cell concentration at 0.5×10^6 cells/mL.

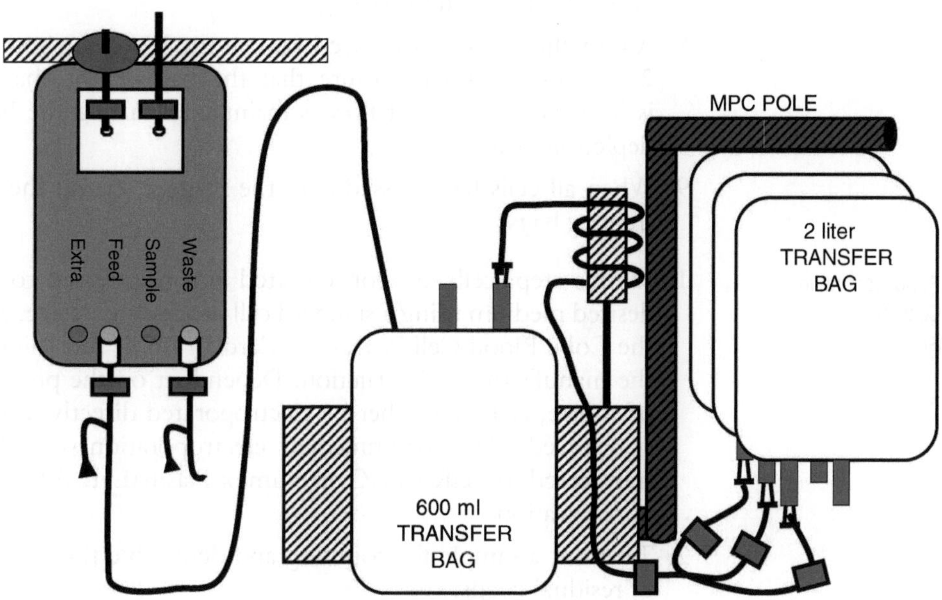

Fig. 28.6. Debeading the final T-cell product following the *ex vivo* expansion using the Dynal Invitrogen magnetic particle concentrator (MPC). Schematic overview of the equipment configuration for the debeading step for removal of CD3/CD28-beads using the Dynal *ClinExVivo* MPC.

22. When the volume reaches 5,000 mL (usually on day 5), start perfusion. The medium is now exchanged automatically during expansion until day 10. Perfusion starts at 1,800 mL/day and is then increased to 3,600 mL/day from day 7 to day 10 (*see* **Note 9**). Increase the concentration of IL-2 to 200 U/mL on day 5 and to 500 U/mL on day 8.

3.1.6.2. De-beading and Harvesting T Cells After Expansion

This step is undertaken to remove Dynabeads from the cells. The depletion step usually takes approximately 1 h. **Figure 28.6** shows the schematic setup for de-beading process using a large-scale magnet, the Dynal *ClinExVivo* MPC.

1. Connect a 600 mL transfer bag to the Wave perfusion bag.
2. Serially connect 3 × 2 L transfer bags to the 600 mL transfer bag and hang on a pole. Ensure that the 2 L transfer bags are closed during the next step.
3. Place the 600 mL bag on the magnet (angle = 0°).
4. Hang the Wave bag on a pole.
5. Open the clamp from the Wave bag and fill the bag on the magnet (lid closed).
6. Drain air from the bag back into the Wave bag by tilting the magnet.
7. Rotate the magnet to 30°. Open clamps to the serial connected 2 L transfer bags and start draining the bead-depleted cells into the 2 L transfer bags.
8. Adjust the height difference between the Wave bag and the 2 L transfer bags to ensure that the bag on the magnet is filled and a constant flow is maintained during the bead depletion process.
9. When all cells have passed over the magnet, clamp the 2 L transfer bags.

3.1.6.3. Concentrating the T Cells After Expansion

1. In this step, cells are concentrated and transferred to the desired medium using a standard cell processing device, e.g., the Cobe Blood Cell Processor (Cardian BCT) according to the manufacturer's instruction. Dependent on the protocol of choice, cells can either be electroporated directly or cryopreserved. Cells for immediate electroporation should be transferred to CellGro DC medium or PlasmaLyte A for cryopreservation.
2. Take out a sample for counting and determine the number of residual beads.

3.1.6.4. Counting of Residual Beads in the Final T-Cell Product

Count residual beads ($n = 3$) in the concentrated T-cell product.

1. Take out approximately 1×10^6 cells in each of the three sample tubes

2. Add 1 mL Buffer 2.
3. Centrifuge at 800×*g* for 4 min. Place the tube in a magnet to prevent loss of beads.
4. Pipette off and discard the supernatant.
5. Resuspend in the residual buffer (typically 10–20 μL).
6. Count residual beads in the whole buffer volume using a hemacytometer as described for cells. Do not add trypan blue.

3.1.6.5. Analyzing and Releasing Expanded T Cells

The product release tests for the expanded T cells must follow appropriate regulatory requirements. The tests specified in our protocol and their accompanying release criteria are listed below.

- Sterility: No growth of bacteria
- Mycoplasma: Negative
- Endotoxin: ≤1 EU/mL
- T-cell purity: ≥80% $CD3^+$ T cells after 10 days expansion, determined by flow cytometry (typically >90–95%)
- Cell viability: 80%
- Residual beads: ≤100 beads per 3×10^6 cells (74, 75, 101)

3.2. Preparing mRNA for Transfection

3.2.1. Preparation of the DNA Template

The expression cassette for the chimeric receptor was isolated from a retroviral SFG vector (kindly provided by Dr. Martin Pulè, UCL) by digestion with *Eco*RI and *Sal*I and subsequently recloned into the $pClpA_{102}$ mRNA expression vector (102) (*see* **Note 10**). This vector inserts a T7 promoter upstream and a 102 poly-A/T stretch downstream of the gene. The poly-A tail directly precedes a unique *Mfe*I restriction site to allow the linearization of the DNA template vector (*see* **Note 11**).

1. Digest 20 μg CAR-$pClpA_{102}$ plasmid with 25 U (2.5 μL) *Mfe*I in 10 μL NEB Buffer 4 in a total volume of 100 μL dH_2O for 1.5–2 h at 37°C. Check the completeness of the linearization by agarose gel electrophoresis (one single band of appropriate size should be visible).
2. Purify the linearized DNA with Wizard® SV Gel and PCR Clean-Up System according to manufacturers' instructions. The system is based on the ability of DNA to bind to silica membranes in the presence of chaotropic salts and is designed to purify DNA fragments of 100 bp–10 kb. The membrane can bind up to 40 μg of DNA and allows >90% recovery of isolated DNA fragments or PCR products in as little as 15 min.
3. Add an equal volume of the membrane-binding solution to the digested DNA and then transfer to the assembled mini-column. Continue as described by the promega protocol. Elute the DNA in 50 μL nuclease-free water.

4. Determine DNA concentration and purity by NanoDrop (*see* **Note 12**). One digest yields approximately 17–18 μg linearized DNA. This amount of template is sufficient for the synthesis of approximately 0.5–1 mg mRNA in 200–300 μL reaction volume using the large-scale Ribomax kit.
5. Store the linearized vector at –20°C.

3.2.2. Synthesis and Purification of mRNA

The in vitro transcription of the linearized DNA template proceeds by using an adequate RNA polymerase in the presence of the four NTPs and a CAP analogue (*see* **Note 13**). Ensure proper working techniques to prevent RNase contamination (*see* **Note 14**). The procedure described here is slightly modified from that described in the Promega Ribomax protocol (*see* **Note 15**). A single 100 μL reaction typically produces 0.2–0.3 μg CAR mRNA in 4–6 h (one kit contains sufficient reagents for a 1 mL reaction). The reaction mixture can be scaled up or down.

1. Dissolve the CAP analogue to 60 mM in DEPC water and then set up the reaction by mixing the following (*see* **Note 16**):

20.0 μL	T7 transcription buffer (5×) (completely dissolved at 37°C)
7.5 μL	rATP (100 mM)
7.5 μL	rCTP (100 mM)
7.5 μL	rUTP (100 mM)
3.0 μL	rGTP (100 mM)
20.0 μL	G-cap analogue (60 mM)
10 μg	X μL template DNA

   Add RNase-free water up to 90 μL and 10 μL T7 enzyme mix (RNA polymerase, RNase inhibitor, inorganic pyrophosphatase). The final volume is 100 μL.
2. Incubate 4–6 h at 37°C and proceed directly to the DNase treatment step described below.
3. Subsequent to transcription, remove the DNA template by enzymatic digestion. Add 1 U DNase for each microgram of DNA used in the mRNA synthesis and incubate at 37°C for 15 min.
4. Purify the synthesized mRNA by ion exchange (*see* **Note 17**). MegaClear columns are designed specifically for the purification of large-scale transcription reactions and efficiently separate RNA from the unincorporated NTP and CAP molecules, enzymes, and buffer components. One column has a capacity adequate for purification of 100 μL synthesis reaction. Mix the transcription reaction with an equal volume of binding solution, pass over the RNA-binding filter, and wash according to manufacturers' protocols.

5. Elute the purified mRNA in 50 μL DEPC water. Repeat elution once to increase yield (*see* **Note 18**).
6. Quality control: Determine concentration and purity of the synthetic mRNA by spectrophotometry using NanoDrop. Run an agarose gel electrophoresis to verify the length and the integrity of the synthesized mRNA. Alternatively, check mRNA integrity by the Agilent Bioanalyzer (*see* **Note 19**). For clinical purposes, analyze for sterility using the appropriate pharmacopeia method.
7. Aliquot the mRNA into the appropriate volumes for transfection of T cells (*see* **Note 20**).
8. Store mRNA at −80°C and avoid repeated freeze and thaw cycles.

3.3. Transfection of mRNA

Here we describe a protocol for transfecting 7–10 million cells (10–15 million cells/mL) in a single 4 mm gap disposable cuvette by square wave electroporation with 70 μg CAR mRNA (*see* **Notes 21–23**). Controls include DEPC-water (mock) electroporated cells and cells treated as the mock control but not pulsed in order to evaluate the electroporation procedure on cell viability.

3.3.1. Electroporation

1. Count and determine viability of the concentrated T cells. Alternatively, you may use a thawed T-cell aliquot. Take out the desired cell volume.
2. Spin the cells at 400×*g* for 5 min and wash twice in CellGro medium without additives. Use the largest volume possible for the washing.
3. Resuspend the cells in the calculated volume with CellGro medium. The cell concentration should be approximately 10–15 million cells/mL.
4. Thaw the mRNA preparation and use 70 μg mRNA/-transfection. Place the required mRNA amount in a sterile Eppendorf tube and incubate in ice (*see* **Note 24**). The volume should not exceed 70 μL.
5. Add 0.7 mL of the homogenous cell suspension of the mRNA and carefully mix by pipetting up and down (6–8 times).
6. Immediately transfer the cells to a 4 mm gap cuvette, avoid air bubbles.
7. Pulse cells in the electroporator with settings 500 V for 2 ms (*see* **Note 24**).
8. Transfer the cells immediately to 5.6 mL prewarmed (20–37°C) culture medium (*see* **Note 25**) in a six-well sterile culture plate or a small culture flask. The final concentration of T cells should be around 1–2 million cells/mL.

9. Incubate overnight in a humidified incubator at 37°C to allow maximal expression of the chimeric receptor (18–20 h) (*see* **Note 26**).
10. Perform QC as described in **Sections 3.3.2, 3.3.3 and 3.3.4.**
11. Cryopreserve CAR-expressing cells for later experiments or infusions (*see* **Note 27**).

3.3.2. Verification of Receptor Expression by Flow Cytometry

1. Count the cells, transfer approximately 300,000 cells (in triplicates) to a flow tube, and then wash with 1 mL SB buffer.
2. Add 50 μL Cy-5-conjugated GaH IgG antibody (0.75 mg/mL) diluted 1:100 in SB buffer (*see* **Note 28**).
3. Incubate the cells with the antibody at 4°C for 30 min.
4. Wash the cells once with cold SB and resuspend in 300 μL SB buffer.
5. Analyze by flow cytometry. Prior to analysis add 3 μL PI in order to gate for living cells.

3.3.3. Verification of Functionality by Analyzing the Killing of Target Cells

Cytotoxicity is the key function of effector T cells in antitumor immune responses and thus assessment of the killing potential of the transfected cells is of special interest. The chromium-51 release assay is a widely used assay to test cytolytic capacity of T cells and it is easily performed, very sensitive, and suitable for high throughput. For this assay, target $CD19^+$ leukemia or lymphoma cells are loaded with the radiochrome and mixed with effector cells for 4–6 h. Cytotoxicity is measured by counting the radioactivity released in the supernatant (*see* **Note 29**). To control for specificity, CD19-negative cells (e.g., KG1A) should be used. The KG1A cells must be labeled and treated in the same way as the $CD19^+$ target cells. It should be noted than the present CAR is targeted to CD19-expressing cells.

1. Target $CD19^+$ SupB15 cells are maintained in exponential growth phase by adjusting the cell concentration with fresh medium every 2–3 days. One day before the assay, the cell culture should be adjusted to approximately 0.6–0.8 × 10^6 cells/mL.
2. On the day of the assay, perform a viable cell count and transfer 2 × 10^6 cells to a tube with lid. Spin down cells for 5 min at 400×*g*. Discard the supernatant and resuspend the cells in 460 μL FCS.
3. Add 7.5 MBq ^{51}Cr (40 μL) to the cell suspension. Pipette tips and disposables containing chromium should be discarded as required for radioactive waste.

4. Incubate the cells for 1 h at 37°C in a 5% CO_2 humidified incubator. Rotate cells carefully every 15 min. In the meantime, count the effector T cells and resuspend to 0.5×10^6 cells/mL.
5. Prepare serial dilutions of the effector cells in medium to yield different effector-to-target cell ratios (E:T) of 25:1 and 12.5:1 and 6.25:1. Make dilutions of effector cells of 0.5, 0.25, and 0.125×10^6 cells/mL.
6. Dispense 100 μL of effectors in triplicates in a 96-well round-bottom microtiter plate. Keep cells at 37°C in a 5% CO_2 humidified incubator until target cells are ready.
7. Subsequent to incubation, wash the target cells three times with 10 mL of ice-cold PBS, spin 300×g for 5 min. Resuspend the cells in 1 mL PBS and count viable cells.
8. Resuspend the cells in cell culture medium 1–20,000 cells/mL.
9. Carefully mix the labeled SupB15 target cells and add 100 μL to each well containing effector cells. Dispense two additional triplicate series of labeled target cells (not containing effector cells).
10. To one triplicate series of labeled target cells add 25 μL of 5% Triton-X 100 and 75 μL medium (maximum release wells). To a second series of triplicate-labeled target cells add 100 μL of medium (spontaneous release wells).
11. Incubate the plate for 4–6 h at 37°C in the 5% CO_2 humidified incubator.
12. After incubation, harvest 50 μL of supernatant from the wells without disturbing the cell pellet. Transfer to a Luma plate.
13. Dry in heating cabinet at 45°C for 1.5 h, or overnight at room temperature in a hood.
14. Add top seal and assay ^{51}Cr release on a gamma counter using the appropriate program.
15. Calculate specific lysis as described below:

$$\text{Specific lysis}(\%) = \frac{(\text{cpm experimental release} - \text{mean cpm spontaneous release})}{(\text{mean cpm maximum release} - \text{mean cpm spontaneous release})} \times 100$$

3.3.4. Verification of Functionality by Analyzing CD107a and IFN-γ Expression

When effector T cells recognize their target cell, they secrete the contents of lytic granules by degranulation into the immunological synapsis with the target cell. As a result CD107a (lysosome-associated membrane protein-1, LAMP) is exposed on the surface of the effector cell. CD107a can be detected by flow cytometry (103, 104). The flow assay also allows for simultaneous analysis of relevant phenotype markers (CD4, CD8, CD56, etc.), as well as detection of intracellular IFN-γ production.

The experimental setup should include appropriate controls: (1) mock-transfected effector cells incubated with target cells, (2) CAR-transfected effector cells incubated with CD19-negative control cells (e.g., KG1A cells), and (3) cell staining with irrelevant isotype control antibodies for proper gating.

1. Harvest the target cells (SupB15) and adjust the cell density to 2×10^6 cells/mL in cell culture medium 1 without IL-2.
2. Harvest the effector cells and adjust the concentration to 4×10^6 cells/mL in the medium described above.
3. Add 250 μL target cells per well in a 48-well plate.
4. Prepare a stock solution of GolgiStop and Brefeldin A (*see* **Note 30**) by mixing GolgiStop (0.33 μL/sample) and Brefeldin A (10 μL/sample of a 1:10 dilution).
5. Add 10.3 μL of the stock solution to each sample.
6. Add 20 μL of CD107a, as well as an irrelevant PE-Cy5 antibody to the wells.
7. Add 250 μL responder cell suspension per well.
8. Incubate for 5 h at 37°C in the 5% CO_2 humidified incubator.
9. Add 50 μL 20 mM EDTA (Step 8 of **Section 2.3.4**) to each well.
10. Incubate for 5 min at 37°C.
11. Transfer to Eppendorf tubes.
12. Spin down at 21,000×*g* for 17 s.
13. Remove the supernatant and resuspend cells in 500 μL ISB.
14. Spin down at 21,000×*g* for 17 s and remove the supernatant. Resuspend in residual volume.
15. Add 10 μL of 10 mg/mL aggregated γ-globulin to each sample.
16. Add 20 μL of CD8-PECy7 and 20 μL of CD4-PE. Following this step, keep samples in the dark.
17. Incubate at 4°C for 30 min.
18. Add 0.8 mL ISB, spin as above (12), and discard the supernatant.
19. Resuspend cells in 150 μL ISB and fix cells with 50 μL 4% PFA (Step 6 of **Section 2.3.4**).
20. Incubate cells for 10 min or optional, store samples overnight at 4°C.

21. For intracellular staining, all incubations are performed at room temperature.
22. Spin down at 21,000×*g* for 23 s. Fixed cells are more buoyant than live cells, spin therefore for 23 s instead of 17 s. Discard the supernatant.
23. Resuspend cells in 500 μL 0.5% saponin
24. Incubate for 10 min at RT.
25. Spin down, remove the supernatant carefully, and resuspend in residual volume (approximately 50 μL) + 2 μL IFN-γ-FITC (or irrelevant FITC as control) and incubate for 20 min at RT.
26. Add 500 μL of 0.5% saponin in PBS + 2% FCS.
27. Centrifuge maximum in Eppendorf for 23 s.
28. Discard the supernatant carefully.
29. Wash 1× with 0.5 mL PBS + 1% HSA.
30. Discard the supernatant carefully.
31. Resuspend cells in 200 μL PBS + 1% HSA for flow analysis on a BD LSR II flow cytometer.
32. Gate on CD4$^+$ and CD8$^+$ cells, respectively, and then evaluate surface expression of CD107a and intracellular IFN-γ. An example of functional activity is shown in **Fig 28.7**.

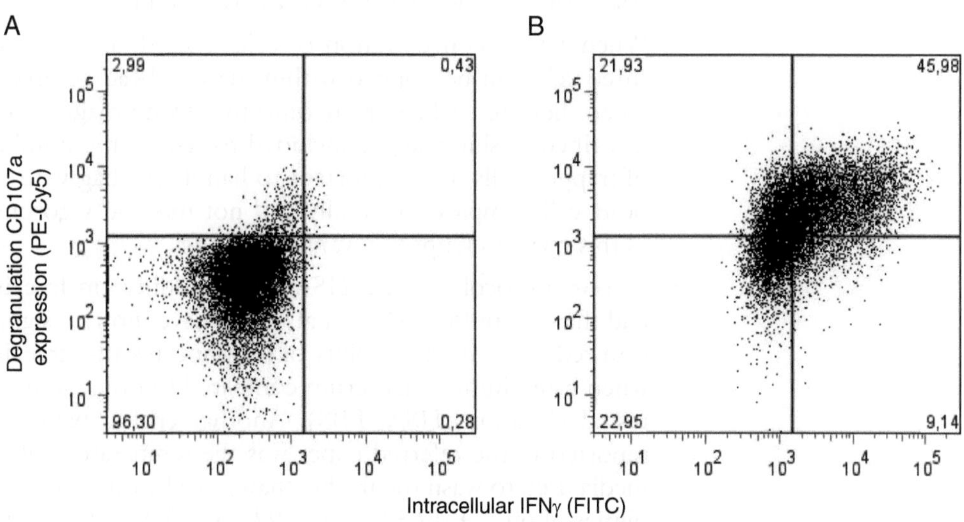

Fig. 28.7. CAR-expressing T cells kill CD19$^+$ lymphoma cells (SupB15). Functional activity was assessed by flow cytometry detection of degranulation (CD107a expression) and production of the inflammatory cytokine IFN-γ (intracellular staining). The CD8$^+$-gated cytograms show the activity of (**a**) T cells electroporated with DEPC-water in co-culture with lymphoma cells. (**b**) T cells electroporated with CAR mRNA.

4. Notes

1. Microscopy is a rapid and simple method for determination of cell concentration and viability but the technique has major shortcomings due to subjective determination of cell count, as well as manual, time-consuming steps, and several alternative automated methods are available. Counting devices based on the Coulter principle, such as the Beckman Coulter Counters, measure changes in electrical resistance of a micro-channel containing an electrolyte solution when a cell flows through. The resistance change correlates to cell size, mobility, and concentration. Coulters do not distinguish viable from nonviable cells. Viability must be assessed separately either with trypan blue staining or microscopy, alternatively with propidium iodine staining and flow cytometry to record the percentage of dead cells. Recently, several automated bench-top cell counting devices have entered the market, such as the Vi-CELL (Beckman Coulter) and the Countess (Invitrogen) systems. These devices counts live and dead cells and measures average cell size and the detection principle is based on trypan blue staining combined with a sophisticated image analysis algorithm to produce accurate cell and viability counts.
2. Too high rocking speed may prevent efficient capture of cells and a low rocking speed may lead to sedimentation of the beads. To optimize the capturing process, we turn the bag with the bead–cell complexes every 5 min.
3. When a high concentration of cells is used, also noncaptured cells can be trapped within the cell–bead complexes when they are packed firmly onto the strong magnet. The described washing step is included to reduce the numbers of trapped cells. It is important to handle the bag with the bead–cell complexes carefully and not mix too vigorously as this may interrupt the wanted binding.
4. In our protocol, we use HSA as the additive in freezing and thawing buffers. Disis et al. (105) have shown a significant reduction in the viability of lymphocytes after freezing when using human AB serum compared to other additives tested (Dextran, HSA, FBS). Another critical parameter reported in the referred paper was the temperature of the media used to wash the freshly thawed cells, with mean viabilities of 69.7 ± 12.5% (4°C), 92.55 ± 3.1% (25°C), and 95.11 ± 2.5% (37°C). This trend has been verified in our hands, and we prewarm thawing buffers to 37°C in our small-scale thawing procedures.

5. If the starting material is a leukapheresis product containing monocytes that digest beads when incubated at 37°C or the product contains low numbers of T cells (e.g., CLL patients), $CD3^+$ T cells should be enriched by a capture step with Dynabeads ClinExVivo CD3/CD28 prior to expansion (101). The details for performing this alternative protocol are not described in this chapter but are delineated in the package insert for Dynabeads ClinExVivo CD3/CD28.
6. A relevant aspect of the T-cell activation process prior to transfection is associated with a general observation that transfection of nonstimulated T cells by mRNA electroporation is highly inefficient (57, 106). While experiments conducted in our laboratory are in accordance with the published observations, others have reported high transfection efficiencies also in nonstimulated T cells (55, 107). This discrepancy might be attributed to differences in the T-cell preparation, as the second group includes a positive magnetic selection step prior to electroporation and they indicate that the interaction with magnetic beads might be sufficient to stimulate the cells.
7. To prevent cells from escaping through the filter in the bag during rocking and exchange of media, it is important that the cutoff for the filter in the bag is below 2 μm. We recommend using a 10 L perfusion bag (Sartorius) from Wave Biotech AG equipped with filters with cutoff of 1.2 μm.
8. Mucomust (*N*-acetylcysteine, NAC) is a radical scavenger of reactive oxygen intermediates and addition of 10 mM to the T-cell culture medium results in significant increased expansion of $CD3^+$ cells (76). Pluroronic F-68 is added to buffers for the Wave expansion to prevent foaming during rocking.
9. When perfusion is performed as described in this protocol, the maximum concentration of cells that can be obtained is 20×10^6/mL. The maximum culture volume is 5 L in a 10 L perfusion bag.
10. The DNA template encoding the gene to be transcribed must have a promoter that is recognized by a bacteriophage polymerase, which binds to the promoter sequence and initiates and guides the transcription. SP6, T3, and T7 phage RNA polymerases have all high specificity for their respective 23 base promoters and the development of cloning vectors containing promoters for these polymerases has made in vitro synthesis of RNA molecules a routine laboratory procedure. Cloning vectors, e.g., pSP, pGEM, and pTZ plasmids, are commonly and commercially available.

The vector for mRNA transcription should preferably also contain a poly-(A/T) stretch of defined length (>100 nt) downstream of the cloning site to increase the in vivo stability and translation efficiency of the transfected mRNA. The poly-(A) tail might alternatively be attached enzymatically by terminal polyadenylation instead of being encoded in the template; however, a template encoded poly-(A) tail results in a more well-defined and homogenous RNA species as the length of the enzymatically synthesized poly-(A) tail is dependent on many reaction parameters that cannot easily be controlled (108). Rabinovich et al. (53) describe an alternative procedure allowing for mRNA synthesis by a vector-free method from polymerase chain reaction (PCR) generated DNA template. In this method, the primers are encoding the T7 promoter and a poly-(T) stretch. Here templates for RNA synthesis can be generated from a variety of substrates without prior cloning. The disadvantage of this method is the risk of PCR-based introduction of errors, as well as scale of template production as mRNA synthesis typically requires micrograms of DNA.

11. The plasmid to be transcribed must first be linearized at a convenient restriction site so that the polymerase runs off at the end of the template, yielding a discrete transcript. The yield of the full-length transcripts decreases if template DNA is incompletely linearized due to a read-through reaction and accumulation of longer transcripts of variable length.

12. NanoDrop is a full-spectrum spectrophotometer that measures 1 μL samples with high accuracy and reproducibility without any need for cuvettes. Concentration of either DNA or RNA is calculated directly and the instrument gives you the ratios of the absorbance values for 280 vs 260 and 260 vs 230 nm, allowing for evaluation of purity of the nucleic acids.

13. Capping of RNA promotes efficient initiation of protein synthesis as well as increases stability and processing of mRNA in vivo. Uncapped RNA degrades rapidly by cellular RNases after the transfection. While a natural CAP (N7-methylguanosine) is present on the 5′ end of most eukaryotic mRNAs, m7G(5′)ppp(5′)G is used as a standard CAP analogue in synthetic transcripts. This CAP can be incorporated into the RNA in both forward and reverse orientation leading to the synthesis of two populations of RNA (109), where RNA capped with reverse 5′ CAPs cannot be translated, resulting in up to 50% reduction in the expected yield of protein. To prevent incorporation of the CAP analogue in the reverse orientation, an Anti-Reverse

Cap Analog (ARCA) was synthesized (110) and here, one of the 3′ OH groups in CAP is replaced with -OCH_3. This modification leads to generation of uniform mRNAs with CAP analogues incorporated only in the functional orientation and thus synthesis of mRNAs that are completely translatable. The use of ARCA ensures uniform and reproducible capped mRNA production and now ARCA is incorporated as the standard CAP analogue in several kits, as in mMESSAGE mMACHINE Ultra. It must be noted, however, that synthesis reactions using ARCA produce lower amounts of mRNA than with CAP when using equimolar amounts of the two reagents and may require further optimization of the concentrations. The reaction mixture contains four times more CAP analogue than of the GTP and this ratio results in a final product that is approximately 80% capped. The CAP analogue adds significant costs to the transcription reaction and it is possible to substitute the CAP by incorporation of an Internal Ribosome Entry Site (IRES) in the vector (111). We have compared protein expression kinetics in activated T cells electroporated with IRES-EGFP and CAP-EGFP and although translation initiation was similar for the constructs, maximum protein expression was twofold higher with the capped mRNA, likely reflecting the increased stability of capped mRNA compared to noncapped IRES mRNA. In applications where very high protein expression and prolonged duration are not mandatory, substituting CAP with IRES might be a viable approach. The use of an IRES also ensures that all mRNA molecules are homogenous (all contain IRES) and are translateable.

14. Ribonucleases are stable enzymes that hydrolyze RNA and it is important to take precautions to prevent RNase contamination; wear gloves, avoid using equipment and areas that have been exposed to RNases. Use autoclaved DEPC-treated water and sterile tubes and micro pipette tips designated for RNA work. Linearized DNA templates for transcription should be RNase free, if presence of RNase is suspected, it is strongly advised to treat the DNA with proteinase K to remove RNases prior to transcription. The subsequent transcription reaction can also preferably be set up using a ribonuclease inhibitor (included in the buffers in some of the commercial kits).
15. We have used two different kits; Promega Ribomax and Ambion mMESSAGE mMACHINE, both specifically designed for high yield synthesis of RNA. We typically obtain higher yield with an optimized Ribomax method compared to the standard protocol using mMESSAGE

mMACHINE. However, the homogeneity of the mRNA seems to be better with the Ambion kit, as we sometimes have experienced the appearance of a weak smear with Ribomax kit, probably due to accumulation of longer transcripts caused by read-through transcription of small amounts of nonlinearized plasmid. The mRNA generated with the Ambion kit based on the same linearized plasmid appears as a single band of the expected size. Higher yields of longer capped transcripts may be achieved by increasing concentrations of rGTP and CAP accordingly (keep ratio constant). The reaction mixture in the protocol presented is optimized to incorporate increased concentrations of GTP and CAP compared to the protocol given by Promega (102). For smaller capped transcripts, increase incubation time as well as amount of the polymerase and the DNA template to optimize yield.

16. The reaction mixture should be prepared at room temperature because DNA may precipitate in the presence of spermidine (a component of the Transcription 5× buffer) at 4°C. It is important to add reagents in the order given.
17. Anion exchange purification of mRNA using Megaclear columns/Ambion (or RNeasy Ion Exchange/Qiagen) yields very pure mRNA, which also repeatedly has shown to be more functional in our hands compared to gel filtration (NucAway) and standard chloroform/trizol purification and ethanol precipitation. If chloroform/trizol purification is preferred, we recommend the use of Eppendorf Gel Lock Heavy tubes to facilitate the separation of the organic and the aqueous phase. Any organic solvent contamination will compromise the quality of the mRNA, as well as negatively affect the accuracy of the concentration determination.
18. The second elution is important for high recovery of pure mRNA and increases the yield with approximately 20%. However, it results in a lower overall concentration of the mRNA which is not always desirable. Optimization of purification (incubation time/volumes) may be necessary.
19. Due to the presence of RNases and the instability of mRNA, proper quality control including verification of mRNA integrity prior to clinical applications is essential. In addition to (or replacing) standard electrophoresis, transcripts should be assayed with the Agilent Bioanalyzer. This device is a versatile tool for the analysis of nucleic acids as well as a range of other macromolecules, it requires only small volumes (1 μL) and utilizes a microfluids-based electrophoresis separation principle resulting in high sensitivity.

20. The optimal mRNA concentration, likewise, the dose of T cells for transfection and subsequent infusion must be determined separately for each mRNA product and according to the specific application.
21. The BTX EMC electroporation device uses square-wave pulses, keeping a preset voltage constant in a defined time. This square-wave pulse shaping leads to high viability and has been shown to be more efficient in intracellular delivery of oligonucleotides compared to conventional exponential decay electroporation methods (112).
22. A disadvantage with the BTX device in our application is the relatively low cell volumes that can be transfected in a single pulse; the larger cuvette can only accommodate approximately 0.8 mL cell suspension. BTX recommends optimal transfections using cell concentrations in the range of 5–10 million cells/mL. Our application may potentially require gene modification in the order of billions of T cells, which obviously is not a feasible approach using the BTX device. We therefore investigated the effect of increasing the cell concentration during pulsing and we were able to electroporate up to 200 million cells/mL without observing any change in viability. Transfection efficiency correlated negatively with cell concentration and to maintain protein expression at a comparable high level as the initial, we increased the amount of mRNA as well as optimized electroporation conditions (increase pulse length and/or voltage).
23. The companies, MaxCyte (Gaithersburg, MD, USA) and CytoPulse Sciences Inc (Glen Burnie, MD, USA) are developing and documenting automated, dynamic large-scale transfection devices for clinical uses. In such systems, cells and transfectants are processed in electroporation chambers in static increments. The systems are sterile and closed, and cell volumes do not limit the application.
24. The amount of mRNA and electroporation conditions must be optimized for each cell type, mRNA species and application, and protein expression levels as well as the effect of the desired functionality and cell viability should be evaluated under the optimization process. Our protocol has been optimized with regard to maximize expression as well as functionality of 10 days expanded T cells modified with the specific anti-CD19 CAR construct (in vector with a 102 adenosine tail) as described (113). Dilute the mRNA in DEPC-water to 1 μg/μL. The mRNA solution should not exceed 10% (vol/vol) of the medium of cells to be electroporated.

25. Even though electroporation-associated toxicity generally is low (viability typically >90%), activated T cells are more fragile than unstimulated cells and the addition of antioxidants such as NAC in the medium increases survival. Antioxidants function by decreasing electro-induced oxidative stress by scavenging free radicals formed by the electroporation.
26. The kinetics of protein expression following electroporation of synthetic mRNA must be established for each mRNA construct and cell type. We detect expression of our anti-CD19 CAR shortly after electroporation (1–2 h), expression peaks after 18–20 h and decays rapidly until it is below detection limit of the flow assay by day 7. The T cells still retain some lymphoma cell killing potential at day 8 after receptor expression is undetectable by flow cytometry.
27. CAR-transfected T cells can be cryopreserved without compromising receptor expression, cytotoxic potential, or secretion of IFN-γ (113). Freeze the cells in aliquots of appropriate volumes suited for single infusions or experiments and use media as described (**Section 3.1.4**) and cryocontainers (tubes or bags) with proper size. Alternatively, for small-scale freezing using cryotubes, we recommend "Mr. Frosty" freezing device, which provides the critical, repeatable –1°C/min cooling rate required for successful cell cryopreservation and recovery.
28. The detection method is dependent on the construct to be translated. The Cy5-conjugate specified here detects the spacer region (human IgG Fc) of the receptor. Any human serum immunoglobulins present in the sample will potentially block the detection and we use either FCS or HSA as the additive in staining buffers. Alternatively, our receptor can be detected by using antimouse F(ab)-specific antibodies conjugated to a fluorochrome of choice.
29. An obvious disadvantage of the radiochrome release assay is that it requires handling radioactive material and all the instrumentation for measuring the isotopes. CD107a/b expression is an alternative to the radiochrome assays. Some advantages of the Cr-release method are that researchers have used this method for years and are familiar with the use and it makes it easy to compare data with others, it offers high throughput. The Cr- release assay is most widely used as it is readily performed and sensitive. As an alternative to this assay the release of any molecule from the lysed cells can be used. Most often the release of enzymes is analyzed such as lactate dehydrogenase whose enzymatic activ-

ity can easily be measured in colorimetric enzyme assays. However, these enzyme-based assays lack the linearity of the radiochrome-release assay. Cytolysis assays are not only employed to monitor the activity of the T cells but also as a read-out for analyzing peptides for their capacity to trigger T cells and, thereby, is instrumental in the search for new T-cell epitopes.

30. GolgiStop (Monensin) inhibits distal Golgi function by preventing acidification of endocytic vessels. This hinders degradation of internalized proteins (CD107a). Brefeldin A prevents the exocytosis of cytokine-containing vesicles allowing for the visualization of cytokine production following stimulation.

Acknowledgments

This work was supported in part by CHILDHOPE (EU contract #037381) and the authors would like to thank Dr. Martin Pulè at UCL Cancer Institute, University College London, UK, for providing the vector for the chimeric antigen receptor. The authors would also thank Dr. Stein Sæbøe-Larssen at our laboratory for providing the pCIpA$_{102}$ expression vector and Dr. Rainer Loew at EUFETS, Idar-Oberstein, Germany, for the IRES-EGFP construct.

References

1. June, C.H. (2007) Adoptive T cell therapy for cancer in the clinic. *J Clin Invest*, **117**(6), 1466–1476.
2. Dudley, M.E., Wunderlich, J.R., Yang, J.C., Sherry, R.M., Topalian, S.L., Restifo, N.P., Royal, R.E., Kammula, U., White, D.E., Mavroukakis, S.A. et al. (2005) Adoptive cell transfer therapy following non-myeloablative but lymphodepleting chemotherapy for the treatment of patients with refractory metastatic melanoma. *J Clin Oncol*, **23**(10), 2346–2357.
3. Leen, A.M., Rooney, C.M., and Foster, A.E. (2007) Improving T cell therapy for cancer. *Annu Rev Immunol*, **25**, 243–265.
4. Porter, D.L. and Antin, J.H. (2006) Donor leukocyte infusions in myeloid malignancies: new strategies. *Best Pract Res Clin Haematol*, **19**(4), 737–755.
5. Rosenberg, S.A. and Dudley, M.E. (2004) Cancer regression in patients with metastatic melanoma after the transfer of autologous antitumor lymphocytes. *Proc Natl Acad Sci USA*, **101**(Suppl 2), 14639–14645.
6. Rosenberg, S.A., Restifo, N.P., Yang, J.C., Morgan, R.A., and Dudley, M.E. (2008) Adoptive cell transfer: a clinical path to effective cancer immunotherapy. *Nat Rev Cancer*, **8**(4), 299–308.
7. Yee, C., Thompson, J.A., Byrd, D., Riddell, S.R., Roche, P., Celis, E., and Greenberg, P.D. (2002) Adoptive T cell therapy using antigen-specific CD8+ T cell clones for the treatment of patients with metastatic melanoma: in vivo persistence, migration, and antitumor effect of transferred T cells. *Proc Natl Acad Sci USA*, **99**(25), 16168–16173.

8. June, C.H. (2007) Principles of adoptive T cell cancer therapy. *J Clin Invest*, **117**(5), 1204–1212.
9. Berger, C., Jensen, M.C., Lansdorp, P.M., Gough, M., Elliott, C., and Riddell, S.R. (2007) Adoptive transfer of effector CD8 T cells derived from central memory cells establishes persistent T cell memory in primates. *J Clin Invest*, **118**(1), 294–305.
10. Perret, R. and Ronchese, F. (2008) Memory T cells in cancer immunotherapy: which CD8(+) T-cell population provides the best protection against tumours? *Tissue Antigens*, **72**(3), 187–194.
11. de Visser, K.E., Schumacher, T.N., and Kruisbeek, A.M. (2003) CD8+ T cell tolerance and cancer immunotherapy. *J Immunother (1997)*, **26**(1), 1–11.
12. Gattinoni, L., Powell, D.J., Jr., Rosenberg, S.A., and Restifo, N.P. (2006) Adoptive immunotherapy for cancer: building on success. *Nat Rev Immunol*, **6**(5), 383–393.
13. Morgan, R.A., Dudley, M.E., Wunderlich, J.R., Hughes, M.S., Yang, J.C., Sherry, R.M., Royal, R.E., Topalian, S.L., Kammula, U.S., Restifo, N.P. et al. (2006) Cancer regression in patients after transfer of genetically engineered lymphocytes. *Science*, **314**(5796), 126–129.
14. Cooper, L.J., Al-Kadhimi, Z., DiGiusto, D., Kalos, M., Colcher, D., Raubitschek, A., Forman, S.J., and Jensen, M.C. (2004) Development and application of CD19-specific T cells for adoptive immunotherapy of B cell malignancies. *Blood Cells Mol Dis*, **33**(1), 83–89.
15. Eshhar, Z. (2008) The T-body approach: redirecting T cells with antibody specificity. *Handb Exp Pharmacol*, **181**, 329–342.
16. Emtage, P.C.R., Lo, A.S.Y., Gomes, E.M., Liu, D.L., Gonzalo-Daganzo, R.M., and Junghans, R.P. (2008) Second-generation anti-carcinoembryonic antigen designer T cells resist activation-induced cell death, proliferate on tumor contact, secrete cytokines, and exhibit superior antitumor activity in vivo: a preclinical evaluation. *Clin Cancer Res*, **14**, 8112–8122.
17. Yu, K., Hu, Y., Tan, Y., Shen, Z., Jiang, S., Qian, H., Liang, B., and Shan, D. (2008) Immunotherapy of lymphomas with T cells modified by anti-CD20 scFv/CD28/CD3zeta recombinant gene. *Leuk Lymphoma*, **49**(7), 1368–1373.
18. Sadelain, M., Riviere, I., and Brentjens, R. (2003) Targeting tumours with genetically enhanced T lymphocytes. *Nat Rev Cancer*, **3**(1), 35–45.
19. Aptsiauri, N., Cabrera, T., Garcia-Lora, A., Lopez-Nevot, M.A., Ruiz-Cabello, F., and Garrido, F. (2007) MHC Class I antigens and immune surveillance in transformed cells. In: Kwang, W.J. (Ed.) *International Review of Cytology a Survey of Cell Biology*. Academic Press, New York, pp. 139–189.
20. Engels, B. and Uckert, W. (2007) Redirecting T lymphocyte specificity by T cell receptor gene transfer – a new era for immunotherapy. *Mol Aspects Med*, **28**(1), 115–142.
21. Uckert, W. and Schumacher, T.N. (2009) TCR transgenes and transgene cassettes for TCR gene therapy: status in 2008. *Cancer Immunol Immunother*, **58**(5), 809–822.
22. Mezzanzanica, D., Canevari, S., Mazzoni, A., Figini, M., Colnaghi, M.I., Waks, T., Schindler, D.G., and Eshhar, Z. (1998) Transfer of chimeric receptor gene made of variable regions of tumor-specific antibody confers anticarbohydrate specificity on T cells. *Cancer Gene Ther*, **5**(6), 401–407.
23. Krause, A., Guo, H.F., Latouche, J.B., Tan, C., Cheung, N.K., and Sadelain, M. (1998) Antigen-dependent CD28 signaling selectively enhances survival and proliferation in genetically modified activated human primary T lymphocytes. *J Exp Med*, **188**(4), 619–626.
24. Yun, C.O., Nolan, K.F., Beecham, E.J., Reisfeld, R.A., and Junghans, R.P. (2000) Targeting of T lymphocytes to melanoma cells through chimeric anti-GD3 immunoglobulin T-cell receptors. *Neoplasia*, **2**(5), 449–459.
25. Westwood, J.A., Murray, W.K., Trivett, M., Haynes, N.M., Solomon, B., Mileshkin, L., Ball, D., Michael, M., Burman, A., Mayura-Guru, P. et al. (2009) The Lewis-Y carbohydrate antigen is expressed by many human tumors and can serve as a target for genetically redirected T cells despite the presence of soluble antigen in serum. *J Immunother*, **32**(3), 292–301.
26. Kennedy, R. and Celis, E. (2008) Multiple roles for CD4+ T cells in anti-tumor immune responses. *Immunol Rev*, **222**, 129–144.
27. Elinav, E., Waks, T., and Eshhar, Z. (2008) Redirection of regulatory T cells with predetermined specificity for the treatment of experimental colitis in mice. *Gastroenterology*, **134**(7), 2014–2024.
28. Eran, E., Nitzan, A., Tova, W., and Eshhar, Z. (2009) Amelioration of colitis by genetically engineered murine regulatory T cells redirected by antigen-specific chimeric receptor. *Gastroenterology*, **136**(5), 1721–1731.
29. Boissel, L., Betancur, M., Wels, W.S., Tuncer, H., and Klingemann, H. (2009) Transfection with mRNA for CD19 specific chimeric

antigen receptor restores NK cell mediated killing of CLL cells. *Leukemia Res*, **33**(9), 1255–1259.

30. Uherek, C., Tonn, T., Uherek, B., Becker, S., Schnierle, B., Klingemann, H.G., and Wels, W. (2002) Retargeting of natural killer-cell cytolytic activity to ErbB2-expressing cancer cells results in efficient and selective tumor cell destruction. *Blood*, **100**(4), 1265–1273.
31. Muller, T., Uherek, C., Maki, G., Chow, K.U., Schimpf, A., Klingemann, H.G., Tonn, T., and Wels, W.S. (2008) Expression of a CD20-specific chimeric antigen receptor enhances cytotoxic activity of NK cells and overcomes NK-resistance of lymphoma and leukemia cells. *Cancer Immunol Immunother*, **57**(3), 411–423.
32. Rischer, M., Pscherer, S., Duwe, S., Vormoor, J., Jurgens, H., and Rossig, C. (2004) Human gammadelta T cells as mediators of chimaeric-receptor redirected anti-tumour immunity. *Br J Haematol*, **126**(4), 583–592.
33. Eshhar, Z., Waks, T., Gross, G., and Schindler, D.G. (1993) Specific activation and targeting of cytotoxic lymphocytes through chimeric single chains consisting of antibody-binding domains and the gamma or zeta subunits of the immunoglobulin and T-cell receptors. *Proc Natl Acad Sci USA*, **90**(2), 720–724.
34. Hombach, A. and Abken, H. (2007) Costimulation tunes tumor-specific activation of redirected T cells in adoptive immunotherapy. *Cancer Immunol Immunother*, **56**(5), 731–737.
35. Finney, H.M., Lawson, A.D., Bebbington, C.R., and Weir, A.N. (1998) Chimeric receptors providing both primary and costimulatory signaling in T cells from a single gene product. *J Immunol*, **161**(6), 2791–2797.
36. Pule, M.A., Straathof, K.C., Dotti, G., Heslop, H.E., Rooney, C.M., and Brenner, M.K. (2005) A chimeric T cell antigen receptor that augments cytokine release and supports clonal expansion of primary human T cells. *Mol Ther*, **12**(5), 933–941.
37. Carpenito, C., Milone, M.C., Hassan, R., Simonet, J.C., Lakhal, M., Suhoski, M.M., Varela-Rohena, A., Haines, K.M., Heitjan, D.F., Albelda, S.M. et al. (2009) Control of large, established tumor xenografts with genetically retargeted human T cells containing CD28 and CD137 domains. *Proc Natl Acad Sci USA*, **106**(9), 3360–3365.
38. Finney, H.M., Akbar, A.N., and Lawson, A.D. (2004) Activation of resting human primary T cells with chimeric receptors: costimulation from CD28, inducible costimulator, CD134, and CD137 in series with signals from the TCR zeta chain. *J Immunol*, **172**(1), 104–113.
39. Cotrim, A.P. and Baum, B.J. (2008) Gene therapy: some history, applications, problems, and prospects. *Toxicol Pathol*, **36**, 97–103.
40. Hacein-Bey-Abina, S., Garrigue, A., Wang, G.P., Soulier, J., Lim, A., Morillon, E., Clappier, E., Caccavelli, L., Delabesse, E., Beldjord, K. et al. (2008) Insertional oncogenesis in 4 patients after retrovirus-mediated gene therapy of SCID-X1. *J Clin Invest*, **118**(9), 3132–3142.
41. Ott, M.G., Schmidt, M., Schwarzwaelder, K., Stein, S., Siler, U., Koehl, U., Glimm, H., Kuhlcke, K., Schilz, A., Kunkel, H. et al. (2006) Correction of X-linked chronic granulomatous disease by gene therapy, augmented by insertional activation of MDS1-EVI1, PRDM16 or SETBP1. *Nat Med*, **12**, 401–409.
42. Hege, K.M. and Roberts, M.R. (1996) T-cell gene therapy. *Curr Opin Biotechnol*, 7(6), 629–634.
43. Muul, L.M., Tuschong, L.M., Soenen, S.L., Jagadeesh, G.J., Ramsey, W.J., Long, Z., Carter, C.S., Garabedian, E.K., Alleyne, M., Brown, M. et al. (2003) Persistence and expression of the adenosine deaminase gene for 12 years and immune reaction to gene transfer components: long-term results of the first clinical gene therapy trial. *Blood*, **101**, 2563–2569.
44. Muul, L.M. and Candotti, F. (2007) Immune responses to gene-modified T cells. *Curr Gene Ther*, 7(5), 361–368.
45. Newrzela, S., Cornils, K., Li, Z., Baum, C., Brugman, M.H., Hartmann, M., Meyer, J., Hartmann, S., Hansmann, M.L., Fehse, B. et al. (2008) Resistance of mature T cells to oncogene transformation. *Blood*, **112**, 2278–2286.
46. Westwood, J.A., Murray, W.K., Trivett, M., Shin, A., Neeson, P., Macgregor, D.P., Haynes, N.M., Trapani, J.A., Mayura-Guru, P., Fox, S. et al. (2008) Absence of retroviral vector-mediated transformation of gene-modified T cells after long-term engraftment in mice. *Gene Ther*, **15**, 1056–1066.
47. Lamers, C.H., Sleijfer, S., Vulto, A.G., Kruit, W.H., Kliffen, M., Debets, R., Gratama, J.W., Stoter, G., and Oosterwijk, E. (2006) Treatment of metastatic renal cell carcinoma with autologous T-lymphocytes genetically retargeted against carbonic anhydrase IX: first clinical experience. *J Clin Oncol*, **24**(13), e20–e22.

48. Bonini, C., Bondanza, A., Perna, S.K., Kaneko, S., Traversari, C., Ciceri, F., and Bordignon, C. (2007) The suicide gene therapy challenge: how to improve a successful gene therapy approach. *Mol Ther*, **15**(7), 1248–1252.
49. Sullenger, B.A. and Gilboa, E. (2002) Emerging clinical applications of RNA. *Nature*, **418**, 252–258.
50. Gilboa, E. and Vieweg, J. (2004) Cancer immunotherapy with mRNA-transfected dendritic cells. *Immunol Rev*, **199**, 251–263.
51. Kyte, J.A. and Gaudernack, G. (2006) Immuno-gene therapy of cancer with tumour-mRNA transfected dendritic cells. *Cancer Immunol Immunother*, **55**(11), 1432–1442.
52. Zhao, Y., Zheng, Z., Cohen, C.J., Gattinoni, L., Palmer, D.C., Restifo, N.P., Rosenberg, S.A., and Morgan, R.A. (2006) High-efficiency transfection of primary human and mouse T lymphocytes using RNA electroporation. *Mol Ther*, **13**(1), 151–159.
53. Rabinovich, P.M., Komarovskaya, M.E., Ye, Z.J., Imai, C., Campana, D., Bahceci, E., and Weissman, S.M. (2006) Synthetic messenger RNA as a tool for gene therapy. *Hum Gene Ther*, **17**(10), 1027–1035.
54. Yoon, S.H., Lee, J.M., Cho, H.I., Kim, E.K., Kim, H.S., Park, M.Y., and Kim, T.G. (2008) Adoptive immunotherapy using human peripheral blood lymphocytes transferred with RNA encoding Her-2//neu-specific chimeric immune receptor in ovarian cancer xenograft model. *Cancer Gene Ther*, **16**, 489–497.
55. Birkholz, K., Hombach, A., Krug, C., Reuter, S., Kershaw, M., Kampgen, E., Schuler, G., Abken, H., Schaft, N., and Dorrie, J. (2009) Transfer of mRNA encoding recombinant immunoreceptors reprograms CD4+ and CD8+ T cells for use in the adoptive immunotherapy of cancer. *Gene Ther*, **16**, 596–604.
56. Robbins, P.F., Dudley, M.E., Wunderlich, J., El-Gamil, M., Li, Y.F., Zhou, J., Huang, J., Powell, D.J., Jr., and Rosenberg, S.A. (2004) Cutting edge: persistence of transferred lymphocyte clonotypes correlates with cancer regression in patients receiving cell transfer therapy. *J Immunol*, **173**(12), 7125–7130.
57. Mitchell, D.A., Karikari, I., Cui, X., Xie, W., Schmittling, R., and Sampson, J.H. (2008) Selective modification of antigen-specific T cells by RNA electroporation. *Hum Gene Ther*, **19**(5), 511–521.
58. Kershaw, M.H., Wang, G., Westwood, J.A., Pachynski, R.K., Tiffany, H.L., Marincola, F.M., Wang, E., Young, H.A., Murphy, P.M., and Hwu, P. (2002) Redirecting migration of T cells to chemokine secreted from tumors by genetic modification with CXCR2. *Hum Gene Ther*, **13**(16), 1971–1980.
59. Yin, J., Ma, Z., Selliah, N., Shivers, D.K., Cron, R.Q., and Finkel, T.H. (2006) Effective gene suppression using small interfering RNA in hard-to-transfect human T cells. *J Immunol Methods*, **312**(1–2), 1–11.
60. Mantei, A., Rutz, S., Janke, M., Kirchhoff, D., Jung, U., Patzel, V., Vogel, U., Rudel, T., Andreou, I., Weber, M. et al. (2008) siRNA stabilization prolongs gene knockdown in primary T lymphocytes. *Eur J Immunol*, **38**(9), 2616–2625.
61. Thomas, D.A. and Massagu, J. (2005) TGF-[beta] directly targets cytotoxic T cell functions during tumor evasion of immune surveillance. *Cancer Cell*, **8**, 369–380.
62. Walczak, P., Ruiz-Cabello, J., Kedziorek, D.A., Gilad, A.A., Lin, S., Barnett, B., Qin, L., Levitsky, H., and Bulte, J.W.M. (2006) Magnetoelectroporation: improved labeling of neural stem cells and leukocytes for cellular magnetic resonance imaging using a single FDA-approved agent. *Nanomed Nanotechnol Biol Med*, **2**, 89–94.
63. Long, C.M. and Bulte, J.W. (2009) In vivo tracking of cellular therapeutics using magnetic resonance imaging. *Expert Opin Biol Ther*, **9**(3), 293–306.
64. drup-Link, H.E., Meier, R., Rudelius, M., Piontek, G., Piert, M., Metz, S., Settles, M., Uherek, C., Wels, W., Schlegel, J. et al. (2005) In vivo tracking of genetically engineered, anti-HER2/neu directed natural killer cells to HER2/neu positive mammary tumors with magnetic resonance imaging. *Eur Radiol*, **15**(1), 4–13.
65. CD19 Chimeric Receptor Expressing T Lymphocytes in B-Cell Non Hodgkin's Lymphoma and Chronic Lymphocytic Leukemia. (NCT00586391). 2008. Baylor College of Medicine. Ref Type: Case
66. Roger Williams Medical Center. Trial of Anti-PSMA Designer T Cells in Advanced Prostate Cancer after Non-Myeloablative Conditioning. ClinicalTrials.gov. 2008. Ref Type: Electronic Citation
67. Rosenberg, S.A., Packard, B.S., Aebersold, P.M., Solomon, D., Topalian, S.L., Toy, S.T., Simon, P., Lotze, M.T., Yang, J.C., Seipp, C.A. et al. (1988) Use of tumor-infiltrating lymphocytes and interleukin-2 in the immunotherapy of patients with metastatic melanoma. A preliminary report. *N Engl J Med*, **319**(25), 1676–1680.

68. Levine, B.L., Cotte, J., Small, C.C., Carroll, R.G., Riley, J.L., Bernstein, W.B., Van Epps, D.E., Hardwick, R.A., and June, C.H. (1998) Large-scale production of CD4+ T cells from HIV-1-infected donors after CD3/CD28 costimulation. *J Hematother*, 7(5), 437–448.
69. Levine, B.L., Mosca, J.D., Riley, J.L., Carroll, R.G., Vahey, M.T., Jagodzinski, L.L., Wagner, K.F., Mayers, D.L., Burke, D.S., Weislow, O.S. *et al.* (1996) Antiviral effect and *ex vivo* CD4+ T cell proliferation in HIV-positive patients as a result of CD28 costimulation. *Science*, **272**(5270), 1939–1943.
70. Levine, B.L., Bernstein, W.B., Connors, M., Craighead, N., Lindsten, T., Thompson, C.B., and June, C.H. (1997) Effects of CD28 costimulation on long-term proliferation of CD4+ T cells in the absence of exogenous feeder cells. *J Immunol*, **159**, 5921–5930.
71. Bonyhadi, M., Frohlich, M., Rasmussen, A., Ferrand, C., Grosmaire, L., Robinet, E., Leis, J., Maziarz, R.T., Tiberghien, P., and Berenson, R.J. (2005) In vitro engagement of CD3 and CD28 corrects T cell defects in chronic lymphocytic leukemia. *J Immunol*, **174**(4), 2366–2375.
72. Garlie, N.K., LeFever, A.V., Siebenlist, R.E., Levine, B.L., June, C.H., and Lum, L.G. (1999) T cells coactivated with immobilized anti-CD3 and anti-CD28 as potential immunotherapy for cancer. *J Immunother*, **22**(4), 336–345.
73. Porter, D.L., Levine, B.L., Bunin, N., Stadtmauer, E.A., Luger, S.M., Goldstein, S., Loren, A., Phillips, J., Nasta, S., Perl, A. et al. (2006) A phase 1 trial of donor lymphocyte infusions expanded and activated *ex vivo* via CD3/CD28 costimulation. *Blood*, **107**(4), 1325–1331.
74. Thompson, J.A., Figlin, R.A., Sifri-Steele, C., Berenson, R.J., and Frohlich, M.W. (2003) A phase I trial of CD3/CD28-activated T cells (Xcellerated T cells) and interleukin-2 in patients with metastatic renal cell carcinoma. *Clin Cancer Res*, **9**(10 Pt 1), 3562–3570.
75. Hami, L.S., Green, C., Leshinsky, N., Markham, E., Miller, K., and Craig, S. (2004) GMP production and testing of Xcellerated T cells for the treatment of patients with CLL. *Cytotherapy*, **6**(6), 554–562.
76. Kalamasz, D., Long, S.A., Taniguchi, R., Buckner, J.H., Berenson, R.J., and Bonyhadi, M. (2004) Optimization of human T-cell expansion *ex vivo* using magnetic beads conjugated with anti-CD3 and anti-CD28 antibodies. *J Immunother (1997)*, **27**(5), 405–418.
77. Hami, L.S., Harjinder, C., Yan, V., and Craig, S. (2003) Comparison of a static process and the bioreactor-based process for the GMP manufacture of autologous Xcellerated T cells for clinical trials. *BioProcessing J*, **2**, 1–10.
78. Berger, C., Blau, C.A., Clackson, T., Riddell, S.R., and Heimfeld, S. (2003) CD28 costimulation and immunoaffinity-based selection efficiently generate primary gene-modified T cells for adoptive immunotherapy. *Blood*, **101**, 476–484.
79. Levine, B.L., Bernstein, W.B., Aronson, N.E., Schlienger, K., Cotte, J., Perfetto, S., Humphries, M.J., Ratto-Kim, S., Birx, D.L., Steffens, C. et al. (2002) Adoptive transfer of costimulated CD4+ T cells induces expansion of peripheral T cells and decreased CCR5 expression in HIV infection. *Nat Med*, **8**(1), 47–53.
80. Laport, G.G., Levine, B.L., Stadtmauer, E.A., Schuster, S.J., Luger, S.M., Grupp, S., Bunin, N., Strobl, F.J., Cotte, J., Zheng, Z. et al. (2003) Adoptive transfer of costimulated T cells induces lymphocytosis in patients with relapsed/refractory non-Hodgkin lymphoma following CD34+-selected hematopoietic cell transplantation. *Blood*, **102**(6), 2004–2013.
81. Rapoport, A.P., Levine, B.L., Badros, A., Meisenberg, B., Ruehle, K., Nandi, A., Rollins, S., Natt, S., Ratterree, B., Westphal, S. et al. (2004) Molecular remission of CML after autotransplantation followed by adoptive transfer of costimulated autologous T cells. *Bone Marrow Transplant*, **33**(1), 53–60.
82. Paulos, C.M., Suhoski, M.M., Plesa, G., Jiang, T., Basu, S., Golovina, T.N., Jiang, S., Aqui, N.A., Powell, D.J., Jr., Levine, B.L. et al. (2008) Adoptive immunotherapy: good habits instilled at youth have long-term benefits. *Immunol Res*, **42**(1–3), 182–196.
83. Mitsuyasu, R.T., Anton, P.A., Deeks, S.G., Scadden, D.T., Connick, E., Downs, M.T., Bakker, A., Roberts, M.R., June, C.H., Jalali, S. et al. (2000) Prolonged survival and tissue trafficking following adoptive transfer of CD4zeta gene-modified autologous CD4(+) and CD8(+) T cells in human immunodeficiency virus-infected subjects. *Blood*, **96**(3), 785–793.
84. Thompson, J.A., Figlin, R.A., Sifri-Steele, C., Berenson, R.J., and Frohlich, M.W. (2003) A phase I trial of CD3/CD28-activated T cells (Xcellerated T cells) and interleukin-2 in

patients with metastatic renal cell carcinoma. *Clin Cancer Res*, **9**(10 Pt 1), 3562–3570.
85. Rapoport, A.P., Stadtmauer, E.A., Aqui, N., Badros, A., Cotte, J., Chrisley, L., Veloso, E., Zheng, Z., Westphal, S., Mair, R. et al. (2005) Restoration of immunity in lymphopenic individuals with cancer by vaccination and adoptive T-cell transfer. *Nat Med*, **11**(11), 1230–1237.
86. Pule, M.A., Savoldo, B., Myers, G.D., Rossig, C., Russell, H.V., Dotti, G., Huls, M.H., Liu, E., Gee, A.P., Mei, Z. et al. (2008) Virus-specific T cells engineered to coexpress tumor-specific receptors: persistence and antitumor activity in individuals with neuroblastoma. *Nat Med*, **14**, 1264–1270.
87. Rosenberg, S.A., Yang, J.C., Robbins, P.F., Wunderlich, J.R., Hwu, P., Sherry, R.M., Schwartzentruber, D.J., Topalian, S.L., Restifo, N.P., Filie, A. et al. (2003) Cell transfer therapy for cancer: lessons from sequential treatments of a patient with metastatic melanoma. *J Immunother (1997)*, **26**(5), 385–393.
88. de Goer de Herve, M.-G., Cariou, A., Simonetta, F., and Taoufik, Y. (2008) Heterospecific CD4 help to rescue CD8 T cell killers. *J Immunol*, **181**, 5974–5980.
89. Rasmussen, A.M. et al. Ex vivo expansion protocol for human tumor specific T cells for adoptive T cell therapy. (submitted).
90. Butterfield, L.H., Ribas, A., Dissette, V.B., Amarnani, S.N., Vu, H.T., Oseguera, D., Wang, H.J., Elashoff, R.M., McBride, W.H., Mukherji, B. et al. (2003) Determinant spreading associated with clinical response in dendritic cell-based immunotherapy for malignant melanoma. *Clin Cancer Res*, **9**(3), 998–1008.
91. Wada, J., Yamasaki, A., Nagai, S., Yanai, K., Fuchino, K., Kameda, C., Tanaka, H., Koga, K., Nakashima, H., Nakamura, M. et al. (2008) Regulatory T-cells are possible effect prediction markers of immunotherapy for cancer patients. *Anticancer Res*, **28**(4C), 2401–2408.
92. Curiel, T.J. (2008) Regulatory T cells and treatment of cancer. *Curr Opin Immunol*, **20**(2), 241–246.
93. Griffiths, R.W., Elkord, E., Gilham, D.E., Ramani, V., Clarke, N., Stern, P.L., and Hawkins, R.E. (2007) Frequency of regulatory T cells in renal cell carcinoma patients and investigation of correlation with survival. *Cancer Immunol Immunother*, **56**(11), 1743–1753.
94. Wolf, A.M., Wolf, D., Steurer, M., Gastl, G., Gunsilius, E., and Grubeck-Loebenstein, B. (2003) Increase of regulatory T cells in the peripheral blood of cancer patients. *Clin Cancer Res*, **9**(2), 606–612.
95. Dudley, M.E., Wunderlich, J.R., Yang, J.C., Hwu, P., Schwartzentruber, D.J., Topalian, S.L., Sherry, R.M., Marincola, F.M., Leitman, S.F., Seipp, C.A. et al. (2002) A phase I study of nonmyeloablative chemotherapy and adoptive transfer of autologous tumor antigen-specific T lymphocytes in patients with metastatic melanoma. *J Immunother (1997)*, **25**(3), 243–251.
96. Dannull, J., Su, Z., Rizzieri, D., Yang, B.K., Coleman, D., Yancey, D., Zhang, A., Dahm, P., Chao, N., Gilboa, E. et al. (2005) Enhancement of vaccine-mediated antitumor immunity in cancer patients after depletion of regulatory T cells. *J Clin Invest*, **115**(12), 3623–3633.
97. Thistlethwaite, F.C., Elkord, E., Griffiths, R.W., Burt, D.J., Shablak, A.M., Campbell, J.D., Gilham, D.E., Austin, E.B., Stern, P.L., and Hawkins, R.E. (2008) Adoptive transfer of T(reg) depleted autologous T cells in advanced renal cell carcinoma. *Cancer Immunol Immunother*, **57**(5), 623–634.
98. Mahnke, K., Schonfeld, K., Fondel, S., Ring, S., Karakhanova, S., Wiedemeyer, K., Bedke, T., Johnson, T.S., Storn, V., Schallenberg, S. et al. (2007) Depletion of CD4+CD25+ human regulatory T cells in vivo: kinetics of Treg depletion and alterations in immune functions in vivo and in vitro. *Int J Cancer*, **120**(12), 2723–2733.
99. Cheadle, E.J., Hawkins, R.E., Batha, H., Rothwell, D.G., Ashton, G., and Gilham, D.E. (2009) Eradication of established B-cell lymphoma by CD19-specific murine T cells is dependent on host lymphopenic environment and can be mediated by CD4+ and CD8+ T cells. *J Immunother*, **32**(3), 207–218.
100. Neurauter, A.A., Bonyhadi, M., Lien, E., Nokleby, L., Ruud, E., Camacho, S., and Aarvak, T. (2008) Cell isolation and expansion using dynabeads (r). *Adv Biochem Eng Biotechnol*, **106**, 41–73.
101. Hollyman, D., Stefanski, J., Przybylowski, M., Bartido, S., Borquez-Ojeda, O., Taylor, C., Yeh, R., Capacio, V., Olszewska, M., Hosey, J. et al. (2009) Manufacturing validation of biologically functional T cells targeted to CD19 antigen for autologous adoptive cell therapy. *J Immunother*, **32**(2), 169–180.
102. Saeboe-Larssen, S., Fossberg, E., and Gaudernack, G. (2002) mRNA-based electrotransfection of human dendritic cells and induction of cytotoxic T lymphocyte responses against the telomerase catalytic subunit

(hTERT). *J Immunol Methods*, **259**(1–2), 191–203.
103. Betts, M.R., Brenchley, J.M., Price, D.A., De Rosa, S.C., Douek, D.C., Roederer, M., and Koup, R.A. (2003) Sensitive and viable identification of antigen-specific CD8+ T cells by a flow cytometric assay for degranulation. *J Immunol Methods*, **281**(1–2), 65–78.
104. Mittendorf, E.A., Storrer, C.E., Shriver, C.D., Ponniah, S., and Peoples, G.E. (2005) Evaluation of the CD107 cytotoxicity assay for the detection of cytolytic CD8+ cells recognizing HER2/neu vaccine peptides. *Breast Cancer Res Treat*, **92**(1), 85–93.
105. Disis, M.L., Dela, R.C., Goodell, V., Kuan, L.Y., Chang, J.C., Kuus-Reichel, K., Clay, T.M., Kim, L.H., Bhatia, S., Ghanekar, S.A. et al. (2006) Maximizing the retention of antigen specific lymphocyte function after cryopreservation. *J Immunol Methods*, **308**(1–2), 13–18.
106. Smits, E., Ponsaerts, P., Lenjou, M., Nijs, G., Van Bockstaele, D.R., Berneman, Z.N., and Van, T.V. (2004) RNA-based gene transfer for adult stem cells and T cells. *Leukemia*, **18**(11), 1898–1902.
107. Schaft, N., Dorrie, J., Muller, I., Beck, V., Baumann, S., Schunder, T., Kampgen, E., and Schuler, G. (2006) A new way to generate cytolytic tumor-specific T cells: electroporation of RNA coding for a T cell receptor into T lymphocytes. *Cancer Immunol Immunother*, **55**(9), 1132–1141.
108. Holtkamp, S., Kreiter, S., Selmi, A., Simon, P., Koslowski, M., Huber, C., Tureci, O., and Sahin, U. (2006) Modification of antigen-encoding RNA increases stability, translational efficacy, and T-cell stimulatory capacity of dendritic cells. *Blood*, **108**, 4009–4017.
109. Pasquinelli, A.E., Dahlberg, J.E., and Lund, E. (1995) Reverse 5′ caps in RNAs made in vitro by phage RNA polymerases. *RNA*, **1**(9), 957–967.
110. Stepinski, J., Waddell, C., Stolarski, R., Darzynkiewicz, E., and Rhoads, R.E. (2001) Synthesis and properties of mRNAs containing the novel "anti-reverse" cap analogs 7-methyl(3′-O-methyl)GpppG and 7-methyl (3′-deoxy)GpppG. *RNA*, 7(10), 1486–1495.
111. Tan, X. and Wan, Y. (2008) Enhanced protein expression by internal ribosomal entry site-driven mRNA translation as a novel approach for in vitro loading of dendritic cells with antigens. *Hum Immunol*, **69**(1), 32–40.
112. Liu, Y. and Bergan, R. (2001) Improved intracellular delivery of oligonucleotides by square wave electroporation. *Antisense Nucleic Acid Drug Dev*, **11**, 7–14.
113. Almåsbak, H. et al. Transiently redirected T cells for adoptive transfer: coelectroporation of chimeric antigen and chemokine receptor mRNAs. (in preparation).

Chapter 29

Overcoming Self-Tolerance to Tumour Cells

Mouldy Sioud

Abstract

Over the past decade, immunotherapy has emerged as a promising alternative form of cancer treatment with the potential to eradicate tumour metastasis. However, its curative potential is in general limited by peripheral tolerance mechanisms and the elimination of self-reactive T cells via thymic negative selection. Unlike infectious challenges, tumour cells arise endogenously, and therefore the majority of tumour antigens are recognized as self. Under appropriate conditions, however, tumour reacting T cells can be activated through a mechanism of molecular mimicry, which involves the recognition of cross reactive foreign antigens mimicking tumour antigens. Moreover, dendritic cells can be reprogrammed by RNA interference to present self-antigens and activate anti-tumour T cells. This review highlights some of the strategies used to break self-tolerance against solid and blood tumour cells. Also, the possibility of reprogramming DC and/or lymphocyte functions using small interfering RNAi (siRNA) is discussed.

Key words: Tolerance, cancer vaccines, tumour antigens, high-affinity T cells, RNA interference.

1. Introduction

The therapeutic use of the immune system to specifically attach tumours has been a longstanding vision among tumour immunologists. However, tumour regressions after anticancer vaccination in clinical trials have been rare (1). Therefore, the identification of potent tumour rejection antigens and/or new vaccination strategies is essential for the development of effective cancer vaccines. Traditionally, vaccines have been effective in the induction of protective immunity to bacteria and virus based on the recognition of foreign (non-self), antigens expressed by these pathogens. In contrast to pathogens, cancer cells arise from one's

M. Sioud (ed.), *RNA Therapeutics*, Methods in Molecular Biology 629,
DOI 10.1007/978-1-60761-657-3_29, © Springer Science+Business Media, LLC 2010

own tissue (self) and this poses a challenge in the development of effective active immunotherapies for cancer (2). Therapeutic vaccines present a substantial challenge, given that they must bypass immune-regulatory mechanisms that have already led to tumour

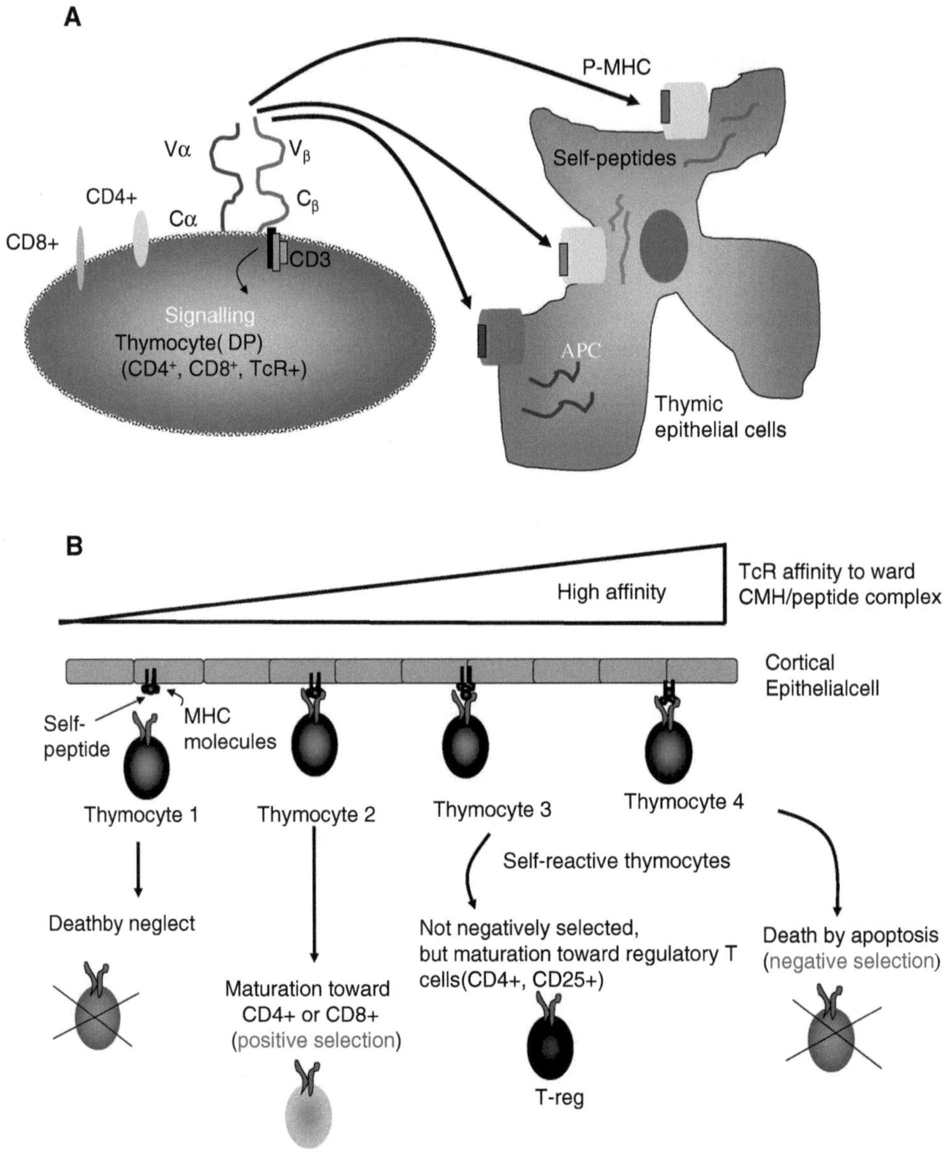

Fig. 29.1. MHC restriction and selection of T-cell repertoire; (**a**) subsequent to TcR gene rearrangement and functional assembly, immature double positive (CD4+, CD8+) progenitors look for an appropriate self-peptide-MHC complex on which they can be positively selected. (**b**) The affinity of T-cell receptor for self-peptide-MHC ligands is the crucial parameter that drives developmental outcome in the thymus. Thymocytes that have no affinity die by neglect. This is thought to be the fate of most thymocytes. The thymocytes carrying TcR with low/medium affinity for self-peptide-MHC survive and differentiate a process that is known as positive selection, whereas high-affinity self-reactive cells are deleted by a process known as negative selection. A pool of autoreactive thymocytes with high affinity is allowed to differentiate into regulatory T cells that play an important role in peripheral tolerance.

tolerance. The enormity of this challenge is underscored by the lack of effective vaccination strategies to treat chronic infectious diseases, despite the large number of foreign antigens present in infectious organisms.

Under normal conditions, the immune system is able to react against foreign antigens while remaining unresponsive to self-antigens. This self-tolerance is acquired and maintained by combination of central and peripheral tolerance (3, 4). Central tolerance refers to those events in the early life of a lymphocyte that focus the adaptive immune system on pathogens and turn it away from healthy tissue. In the case of T cells, central tolerance occurs during thymic development with self-reactive T cells eliminated during thymic negative selection. Less self-reactive cell-bearing TcRs with low affinity to self-antigen are positively selected on self-MHC molecules (4). However, it is becoming increasingly clear that thymic selection is not a perfect process, and a significant number of self-reactive T cells are exported into the periphery (**Fig. 29.1a, b**). The immune repertoire therefore contains autoreactive immune cells that may reject tumours, when activated properly (see below).

To control the activation of peripheral autoreactive T cells, a number of suppressive mechanisms operate in the periphery. These include the induction of anergy, cytokines, and immunosuppressor cells such as innate CD25+ regulatory T cells (Tregs), which develop in the thymus and express the transcription factor forkhead box P3 (Foxp3) (5, 6). The commitment to Foxp3-expressing cells requires the expression of high-affinity self-reactive receptors and therefore represents an alternative fate to negative selection. Despite the presence of these control mechanisms, in autoimmune diseases self-reactive T and B cells become aggressive and cause tissue injury, in some cases leading to severe injuries. Understanding the mechanisms that activate autoreactive T cells in autoimmunity may facilitate the generation of tumour-specific cytotoxic T cells.

2. Autoimmunity and T-Cell Activation

As mentioned above, a key feature of the adaptive immune system is its ability to respond to foreign antigens while failing to react to antigens derived from self-tissues. An important question in autoimmunity is how autoimmunization against, for example, synovial membrane, nucleosomes, and myelin basic protein, occurs in rheumatoid arthritis, systemic lupus erythematosus, and multiple sclerosis patients, respectively (7). The fact that these autoantigens are ubiquitous and present in normal individuals raises the question of why autoimmunity does not happen more often. Therefore, an essential defect in the mechanisms that maintain immunologic tolerance to self-constituents must play an important role. The degeneracy of antigen recognition by the T-cell receptor (TcR), such as a T cell can be activated by different peptides bound to one or even several MHC molecules, implies that responses to microbial antigens could result in the activation of T cells that are cross-reactive with self-antigens (8) (**Fig. 29.2**). This flexibility of TcR recognition is thought to be central to many immunological processes, including thymic selection and the ability of TcRs to recognize nearly all pathogen-derived peptides.

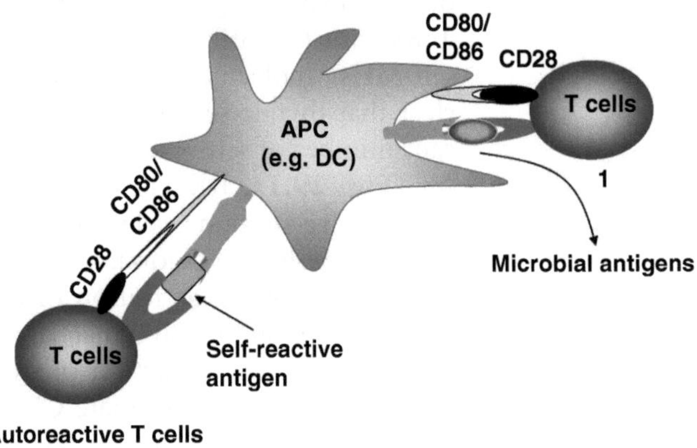

Fig. 29.2. Epitope mimicry and tumour immunity: the activation of high-avidity T cells with xenogeneic antigens may cross-react with host tissue antigens and eventually break down self-tolerance to tumour antigens. The activation of low-affinity T cells by tumour antigens may not be sufficient to mount a protective immunity (for more details see text).

Although recent studies indicated that the predisposition of individuals to several common autoimmune diseases, such as rheumatoid arthritis, systemic lupus erythematosus, and multiple sclerosis, is genetically linked to certain human MHC class II molecules (9), the involvement of infections has been implicated in the onset and/or perpetuation of autoimmunity (10–12). The assumption is that an infectious agent (e.g., parasite, bacteria, and virus) displays epitopes immunologically resembling host determinant and due to the mirror antigen differences between the two, the pathogen's epitope subsequently induces an immune response that eventually breaks tolerance to the host epitope. Once the immune system becomes primed to attack the invader, it might eventually destroy normal tissues. Notably, the capacity of an epitope mimic to induce an autoimmune disease depends on its appropriate presentation by the host antigen presenting cells, thereby supporting the association between major histocompatibility complex (MHC) products and autoimmune diseases indicated above (9). Indeed, the binding association between a given

MHC molecule and its peptide, whether it be a self-peptide or foreign peptide involved in either a class I or II interaction, is a genetically controlled event. Because of unique MHC polymorphisms within a population, peptides exhibiting binding capacity in a given host may be completely non-reactive in other individuals of a particular species who lack those MHC alleles (9). With regard to rheumatoid arthritis, epitopes derived from different pathogens such as *Mycobacterium tuberculosis* activated synovial T cell (13). In addition, most of the described mimic epitopes (mimotopes) induce autoimmune disease in animal models (14).

3. Epitope Spreading

Studies in animal models have made it clear that in autoimmunity T-cell response can spread to involve T cells specific for other self-antigens in a process known as epitope spreading (15, 16). Epitope spreading may occur either within a single antigen (intramolecular spreading), different antigens (intermolecular spreading) from the same tissue structure, or within different antigens that are not physically linked (17–19). Activating a broader set of T cells through epitope spreading should be beneficial during an anti-pathogen or anti-tumour immune response, because the pathogen or tumour cannot easily escape immune control with a single mutation in an immunogenic epitope. The phenomenon of epitope spreading (both intermolecular and intramolecular) was noted in few cases of clinical responders to peptide vaccines in trials involving patients with melanoma (20).

4. Breaking Self-Tolerance Against Self- and/or Tumour-Specific Antigens

Many, if not all, tumours overexpress normal proteins with relatively low endogenous expression, and these overexpressed proteins can be targets for single-antigen vaccines (21–25). This overexpression is expected to enhance the range of self-peptides presented by MHC molecules, thus allowing the activation of T cell against tumours. However, as indicated above high-avidity T cells to self-antigens undergo negative selection, and peripheral tolerance mechanisms diminish the number or eliminate autoreactive self-peptide-specific CTLs (4). This is a major limitation to activate high-affinity T cells against tumours. To treat cancers, therefore it is essential to break self-tolerance and prime T lymphocytes with high affinity against self- or tumour antigens (26).

Table 29.1
Potential mechanisms of breaking immune tolerance to tumours

Antigen overexpression
Difference in HLA molecules needed to present shared antigens
Cross-reactive antigens (molecular mimicry)
Inhibition of immunosuppressive mechanisms
Control of endosome maturation and acidification

To activate high-affinity T cells against tumour cells, in previous studies we have explored the therapeutic potential of self-antigen/foreign antigen cross-reactivity (*see* **Table 29.1**) in an immunocompetent rat glioma model (27). Specifically, the strategy induced high-avidity cross-reactive cytotoxic T cells that killed glioma cells and enhanced rat survival. Our goal now is to use the identified cross-reactive recombinant proteins and synthetic peptides in order to activate high-avidity T cells against tumour cells. In most studies, vaccination with xenogeneic or altered antigens showed higher immune reactivity when compared to syngeneic immunization. We expect even more anti-tumour effect by combining the most active proteins and/or peptides in a cocktail vaccine.

In addition to solid tumours, we have been interested in developing new therapies against blood cancers such as leukaemia (28, 29). Notably, the recognition that the graft-versus-leukaemia (GVL) effect is important for eradication of the malignant cells shows that immunomodulation is possible, particularly in acute myeloid leukaemia (AML) (30). To break self-tolerance against leukaemia cells, several groups have generated allo-restricted CTLs against self-antigens such as CD19, CD20, CD33, and CD45 using a strategy described nearly 10 years ago by Stauss and colleagues (31). This strategy takes the advantage of the disparity in HLA molecules needed to present shared antigens rather than relying on a difference in antigenic expression between host and donor (*see* **Chapter 30**). As discussed by Stauss (31), in patients who undergo HLA-mismatched haematopoietic stem cell transplantation, the strategy could be used by stimulating donor T cells with APCs carrying the host HLA restriction, pulsed with a peptide from a haematopoietic protein, to generate donor T cells that recognize self-antigens in the context of host HLA molecules. The generated CTLs are expected to kill patient leukaemic cells, which express the haematopoietic antigen and the appropriate HLA molecules, but they would not kill host non-haematopoietic tissue or donor haematopoietic cells because they do not express or present the antigen, respectively (31–33). Most of the studies have focused on HLA-A0201 because it is expressed in around 50% of the white population. Using APC from HLA-A0201+ patients or T2 cell expressing HLA0201 molecules, several studies have reported on the isolation and characterization of allo-restricted T cells with high-avidity T-cell receptor towards self-peptides from CD molecules (34, 35). Also, the possibility to redirect the specificity of recipient T lymphocytes by transfer of an allo-antigen-specific TcR to T cells has been evaluated (36). Despite the in vitro activity of the selected CTLs and the feasibility of T-cell receptor gene transfer, this strategy is limited because of several problems and risks amongst which are rapid tolerance of allo-reactive T cells, TcR revision, defect in antigen presentation by leukaemic cells, and the formation of mixed TcR dimers. Furthermore, the graft-versus-leukaemia effect seen in HSCT patients is more likely mediated by more than one T-cell clone. Therefore, there is a need for developing new strategies and/or combining several T-cell specificies. Also, it is important to generate T-cell and antibody responses in patients.

5. The Endosomes and Self-Tolerance

As discussed in **Chapter 3**, the sensing of pathogen nucleic acids occurs in the endosomal compartments via Toll-like receptors (TLRs). Upon sensing bacteria or viruses, TLRs trigger inflammatory and antiviral responses by inducing the activation of transcription factors, such as nuclear factor NF-κB, or interferon regulatory factors, which ultimately leads to the eradication of the invading pathogen. These findings indicate that the endosomes are important in innate immunity. The endosomes are also essential for switching on the adaptive immune response towards pathogens (37). Extracellular antigens are taken up by antigen presenting cells (APC) and processed into peptides within the so-called MHC class II compartment. Processed peptides are loaded onto pre-assembled MHC class II complexes to replace the invariant chain (li), which is a type II transmembrane protein that is responsible for the proper folding of nascent MHC class II molecules in the endoplasmic reticulum and sorting through the trans-golgi network to the low pH endosomal compartments in the APC (**Fig. 29.3**). As MHC class II molecules move towards vesicular compartments enriched for exogenous antigens, li is gradually cleaved away by proteases, leaving behind a short peptide named class II-associated invariant chain peptide (CLIP), which remains bound to the peptide-binding groove (36). It should be noted that cysteine cathepsins play an important role in proteolytic processing of the invariant chain and foreign antigens

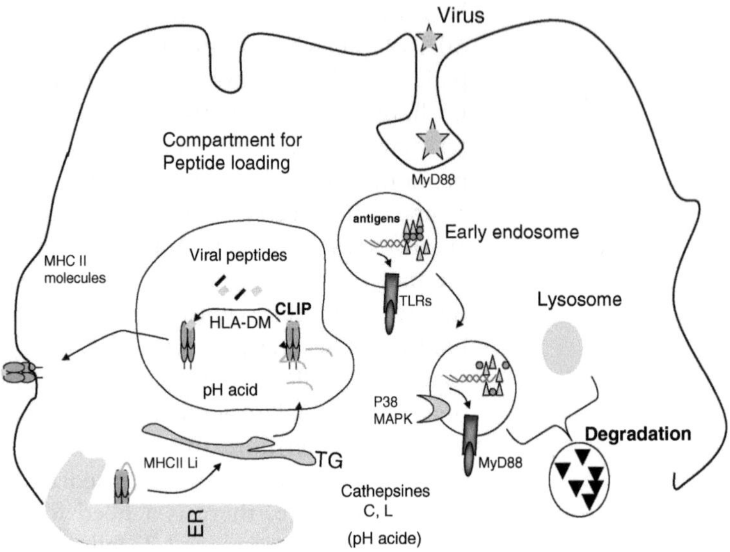

Fig. 29.3. Schematic representationof T-cell activation by viral peptides; class II pathway presents mainly peptides derived from the proteins which enter the endocytic pathway of the cell, in particular endocytosed extracellular proteins. Following viral infection, for example, viral proteins are processed and presented to CD4 T cells. Viral RNAs activate endosomal TLRs leading to endosome maturation and induction inflammatory cytokines. The acidification of endosomes is essential for the function of the endosomes and antigen processing.

in a number of antigen presenting cells (38). In the endosomal compartments specialized for the capture of exogenous antigenic peptides, HLA-DM plays a crucial role in exchanging CLIP for an exogenous antigen-derived peptide (37, 39). Both cysteine cathepsins and HLA-DM require pH acid for full activity. This might explain why the inhibition of endosome maturation/acidification can hinder antigen presentation and subsequent activation of adaptive immunity. Thus, under physiological conditions DCs will not properly process and present self-antigens to T cells. Apparently, the activation of endosomal toll-like receptors is required for both non-self- and self-antigen processing and presentation by DCs (40). In the absence of danger signals, endosomal TLRs are not activated and consequently self-antigens are not effectively presented. Gene profiling in response to immunostimulatory RNAs identified several genes some of which are involved in endosome maturation(41).

Although TLR agonists enhance immune responses, they also induce expression of DC-derived immunosuppressive factors such as interleukin-10 and SOCS-1 proteins. These inhibitory feedback mechanisms may hamper effective immunity against tumour cells. Therefore, the development of agents that stimulate DC maturation through TLRs and simultaneously inhibit the expression of immunosuppressive factors and/or regulatory T-cell function may break self-tolerance and activate high-affinity autoreactive T cells that have escaped central tolerance. To test this hypothesis, first we have refined RNAi technology by designing dual (bifunctional) siRNA molecules that cleave the mRNA of the target genes and increase the production of cytokines and type I interferon by binding and stimulating endosomal TLRs (42). Second, we applied a novel DC-based vaccine (i.e. DC loaded with leukaemia antigens that have been transfected with an IL-10 siRNA capable of coordinately activating DC via TLR7/8) in a rat model of AML (**Fig. 29.4a–c**). Interestingly, leukaemic rats treated with this new vaccine had less leukaemic cell mass in their bone marrows and less extramedullar dissemination of the leukaemic disease post mortem compared with rats given the control vaccine (43). Collectively, the data demonstrate the possible usefulness of dual siRNAs as an immunomodulatory drug with anti-leukaemic properties. Using the same strategy of vaccination, we also found that it is possible to break self-tolerance against self-antigens, including vimentin, Hsp70, and CD33, by targeting the

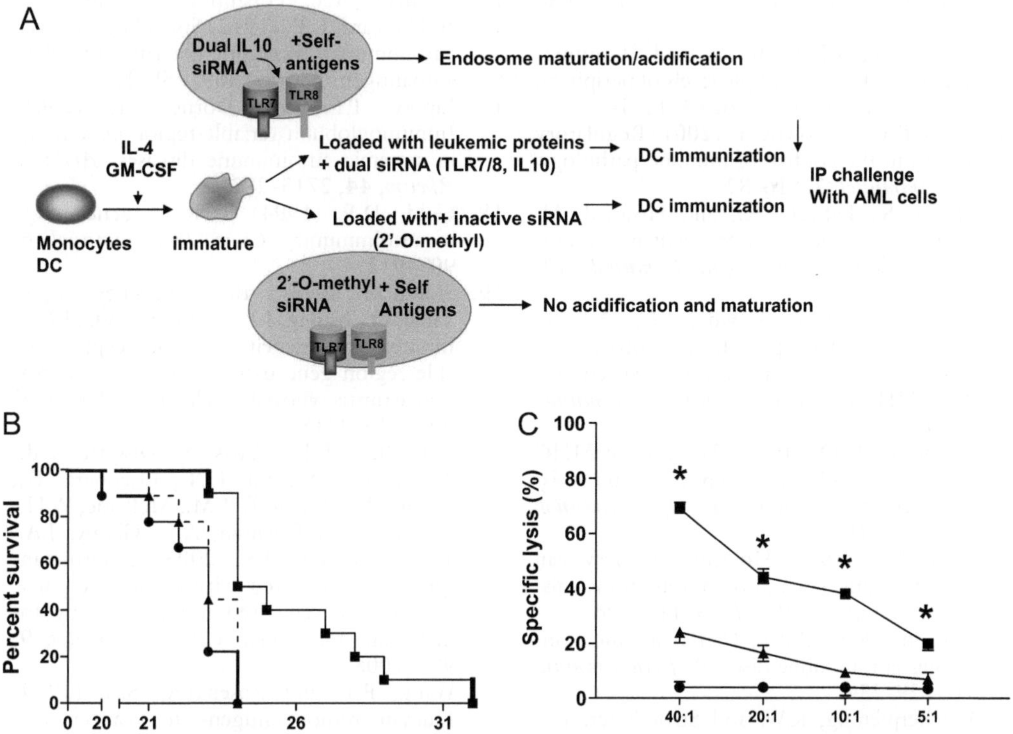

Fig. 29.4. Targeting siRNA and tumour antigens to the same endosome increases tumour immunity. (**a**) Schematic representation of DC vaccine. (**b**) Increased survival upon treatment with active siRNA. Leukaemic rats treated with DCs transfected with active siRNA (*thick filled line*) lived longer compared with either those given inactive siRNA (*thin solid line*) or untreated rats (*dashed line*). (**c**) T cells from vaccinated rats efficiently killed leukaemic cells in vitro (*square*) when compared to controls.

antigens and immunostimulatory siRNAs to the same endosomes (unpublished data). This new concept of vaccination is applicable to any self-specific and/or tumour-specific antigens. Notably, the currently used DC vaccination protocols are clearly not yet optimized, and generating more effective vaccines continues to be an area of active research.

References

1. Blattman, J.N. and Greenberg, P.D. (2004) Cancer immunotherapy: a treatment for the masses. *Science*, **305**, 200–205.
2. Marincola, F.M., Jaffee, E.M., Hicklin, D.J., and Ferrone, S. (2000) Escape of human solid tumors from T-cell recognition: molecular mechanisms and functional significance. *Adv Immunol*, **74**, 181–273.
3. Parkin, J. and Cohen, B. (2001) An overview of the immune system. *Lancet*, **357**, 1777–1789.
4. Hogquist, K.A., Baldwin, T.A., and Jameson, S.C. (2005) Central tolerance: learning self-control in the thymus. *Nat Rev Immunol*, **5**, 772–782.
5. Arnold, B., Schonrich, G., and Harmmerling, G.J. (1993) Multiple levels of peripheral tolerance. *Immunol Today*, **14**, 12–14.
6. Malo, K.J. and Powrie, F. (2001) Regulatory T cells in the control of immune pathology. *Nat Immunol*, **2**, 816–822.
7. Mocci, S., Lafferty, K., and Howard, M. (2000) The role of autoantigen in autoimmune disease. *Curr Opin Immunol*, **12**, 725–730.
8. Marrack, P., Scott-Browne, J.P., Dai, S., Gapin, L., and Kappler, J.W. (2008) Evolutionarily conserved amino acids that control TCR-MHC interaction. *Annu Rev Immunol*, **26**, 171–203.
9. McDevitt, H.O. (1998) The role of MHC class II molecules in susceptibility and resistance to autoimmunity. *Curr Opin Immunol*, **10**, 677–681.
10. Davis, J.M. (1997) Molecular mimicry: can epitope mimicry induce autoimmune disease?. *Immunol Cell Biol*, **75**, 113–126.
11. Fujinami, R.S. (2001) Can virus infections trigger autoimmune disease?. *J Autoimmun*, **16**, 229–234.
12. Wucherpfennig, K.W. and Strominger, J.L. (1995) Molecular mimicry in T cell-mediated autoimmunity: viral peptides activate human T cell clones specific for myelin basic protein. *Cell*, **80**, 695–705.
13. Quale, A.J., Wilson, K.B., Li, S.G., Kjeldsen-Kragh, J., Oftung, F., Shinnick, T., Sioud, M., Førre, Ø., Capra, J.D., and Natvig, J.B. (1992) Peptide recognition, T cell receptor usage and HLA restriction elements of human heat-shock protein (hsp) 60 and mycobacterial 65-kDa hsp-reactive T cell clones from rheumatoid synovial fluid. *Eur J Immunol*, **22**, 1315–1322.
14. Myers, L.K., Rosloniec, E.F., Cremer, M.A., and Kang, A.H. (1997) Collagen-induced arthritis, an animal model for autoimmunity. *Life Sci*, **61**, 1861–1878.
15. Gordon, T.P. (1998) Determinant spreading: lessons from animal models and human disease. *Immunol Rev*, **164**, 209–229.
16. Lehmann, P.V., Forsthuber, T., Miller, A., and Sercarz, E.E. (1992) Spreading of T-cell autoimmunity to cryptic determinants of an autoantigen. *Nature*, **359**, 155–157.
17. Lipsky, P.E. and Dorner, T. (2001) Immunoglobulin variable-region gene usage in systemic autoimmune diseases. *Arthritis Rheum*, **44**, 2715–2727.
18. Gold, D.P. (1994) TCR V gene usage in autoimmunity. *Curr Opin Immunol*, **6**, 907–912.
19. Sioud, M., Kjeldsen-Kragh, J., Suleyman, S., Vinje, O., Natvig, J.B., and Førre, Ø. (1992) Limited heterogeneity of T cell receptor variable region gene usage in juvenile rheumatoid arthritis synovial T cells. *Eur J Immunol*, **22**, 2413–2418.
20. Butterfield, L.H., Ribas, A., Dissette, V.B., Amarnani, S.N., Vu, H.T., Oseguera, D., Wang, H.J., Elashoff, R.M., McBride, W.H., Mukherji, B., Cochran, A.J., Glaspy, J.A., and Economou, J.S. (2003) Determinant spreading associated with clinical response in dendritic cell-based immunotherapy for malignant melanoma. *Clin Cancer Res*, **9**, 998–1008.
21. Wang, R.F. and Rosenberg, S.A. (1999) Human tumor antigens for cancer vaccine development. *Immunol Rev*, **170**, 85–100.
22. Houghton, A.N. (1994) Cancer antigens: immune recognition of self and altered self. *J Exp Med*, **180**, 1–4.
23. Hansen, M.H., Østenstad, B., and Sioud, M. (2001) Antigen-specific IgG antibodies

in stage IV long-term survival breast cancer patients. *Mol Med*, 7, 230–239.
24. Hansen, M.H., Østenstad, B., and Sioud, M. (2001) Identification of immunogenic antigens using a phage-displayed cDNA library from an invasive ductal breast carcinoma tumour. *Int J Oncol*, **19**, 1303–1309.
25. Sioud, M. and Hansen, M.H. (2001) Profiling the immune response in patients with breast cancer by phage-displayed cDNA libraries. *Eur J Immunol*, **31**, 716–725.
26. Sioud, M. (2002) How does autoimmunity cause tumor regression? A potential mechanism involving cross-reaction through epitope mimicry. *Mol Med*, **8**, 115–119.
27. Sioud, M. and Sørensen, D. (2003) Generation of an effective anti-tumor immunity after immunization with xenogeneic antigens. *Eur J Immunol*, **33**, 38–45.
28. Iversen, P.O., Emanuel, P.D., and Sioud, M. (2002) Targeting Raf-1 gene expression by a DNA enzyme inhibits juvenile myelomonocytic leukemia cell growth. *Blood*, **99**, 4147–4153.
29. Iversen, P.O. and Sioud, M. (1998) Modulation of granulocyte-macrophage colony-stimulating factor gene expression by a tumor necrosis factor specific ribozyme in juvenile myelomonocytic leukemic cells. *Blood*, **92**, 4263–4268.
30. Kolb, H.J., Schattenberg, A., Goldman, J.M., Hertenstein, B., Jacobsen, N., Arcese, W., Ljungman, P., Ferrant, A., Verdonck, L., Niederwieser, D., van Rhee, F., Mittermueller, J., de Witte, T., Holler, E., and Ansari, H.; European Group for Blood and Marrow Transplantation Working Party Chronic Leukemia. (1995) Graft-versus-leukemia effect of donor lymphocyte transfusions in marrow grafted patients. *Blood*, **86**, 2041–2050.
31. Stauss, H.J. (1999) Immunotherapy with CTLs restricted by nonself MHC. *Immunol Today*, **20**, 180–183.
32. Amrolia, P.J., Reid, S.D., Gao, L., Schultheis, B., Dotti, G., Brenner, M.K., Melo, J.V., Goldman, J.M., and Stauss, H.J. (2003) Allorestricted cytotoxic T cells specific for human CD45 show potent antileukemic activity. *Blood*, **101**, 1007–1014.
33. Gao, L., Yang, T.H., Tourdot, S., Sadovnikova, E., Hasserjian, R., and Stauss, H.J. (1999) Allo-major histocompatibility complex-restricted cytotoxic T lymphocytes engraft in bone marrow transplant recipients without causing graft-versus-host disease. *Blood*, **94**, 2999–3006.
34. Bae, J., Martinson, J.A., and Klingemann, H.G. (2005) Identification of CD19 and CD20 peptides for induction of antigen-specific CTLs against B-cell malignancies. *Clin Cancer Res*, **11**, 1629–1638.
35. Heemskerk, M.H., Hoogeboom, M., de Paus, R.A., Kester, M.G., van der Hoorn, M.A., Goulmy, E., Willemze, R., and Falkenburg, J.H. (2003) Redirection of antileukemic reactivity of peripheral T lymphocytes using gene transfer of minor histocompatibility antigen HA-2-specific T-cell receptor complexes expressing a conserved alpha joining region. *Blood*, **102**, 3530–3540.
36. Marijt, W.A., Heemskerk, M.H., Kloosterboer, F.M., Goulmy, E., Kester, M.G., van der Hoorn, M.A., van Luxemburg-Heys, S.A., Hoogeboom, M., Mutis, T., Drijfhout, J.W., van Rood, J.J., Willemze, R., and Falkenburg, J.H. (2003) Hematopoiesis-restricted minor histocompatibility antigens HA-1- or HA-2-specific T cells can induce complete remissions of relapsed leukemia. *Proc Natl Acad Sci USA*, **100**, 2742–2747.
37. Trombetta, E.S. and Mellman, I. (2005) Cell biology of antigen processing *in vitro* and *in vivo*. *Annu Rev Immunol*, **23**, 975–1028.
38. Shi, G.P., Bryant, R.A., Riese, R., Verhelst, S., Driessen, C., Li, Z., Bromme, D., Ploegh, H.L., and Chapman, H.A. (2000) Role for cathepsin F in invariant chain processing and major histocompatibility complex class II peptide loading by macrophages. *J Exp Med*, **191**, 1177–1186.
39. Sadegh-Nasseri, S., Chen, M., Narayan, K., and Bouvier, M. (2008) The convergent roles of tapasin and HLA-DM in antigen presentation. *Trends Immunol*, **29**, 141–147.
40. Blander, J.M. and Medzhitov, R. (2004) Regulation of phagosome maturation by signals from toll-like receptors. *Science*, **304**, 1014–1018.
41. Cekaite, L., Furset, G., Hovig, E., and Sioud, M. (2007) Gene expression analysis in blood cells in response to unmodified and 2′-modified siRNAs reveals TLR-dependent and independent effects. *J Mol Biol*, **365**, 90–108.
42. Furset, G. and Sioud, M. (2007) Design of bifunctional siRNAs: combining immunostimulation and gene-silencing in one single siRNA molecule. *Biochem Biophys Res Commun*, **352**, 642–649.
43. Iversen, P.O., Semaeva, E., Sørensen, D.R., Wiig, H.,, and Sioud, M. (2009) Dendritic cells loaded with tumour antigens and a dual immunostimulatory and anti-interleukin-10-specific small interference RNA prime T lymphocyte against leukemic cells. *Transl Oncol*, **4**, 242–246.

Chapter 30

Recent Advances in Hematopoietic Stem Cell Transplantation and Perspectives of RNAi Applications

Yngvar Fløisand and Mouldy Sioud

Abstract

In adults, the bone marrow compartment contains hematopoietic stem cells (HSCs) which can differentiate into progenitors with more restricted lineage potential and generate all cellular elements of the blood. HSCs for stem cell transplantation can be obtained by bone marrow collection, mobilization into peripheral blood followed by apheresis, or use of stem cells from cord blood. Currently, hematopoietic stem cell transplantation (SCT) is used to treat patients with various hematological diseases. Although substantial progress has been made, a number of challenges can limit the efficacy of HSC transplantation, including the occurrence of graft-versus-host disease (GvHD) in allogeneic stem cell transplantation (ASCT), the susceptibility of patients to opportunistic infections and relapse of malignancies after SCT. Recent studies indicate that small interfering RNAs (siRNAs) can specifically and efficiently interfere with the expression of oncogenic genes. Therefore, the possibility of interfering with the expression of these proteins in hematopoietic cells may offer a new option to correct cell differentiation and function. In addition to the generation of T cells restricted by nonself MHC as reviewed by Stauss and colleagues in 1999, the modulation of NK cell receptor expression and T-cell activation is a new strategy that could limit GvHD. This chapter reviews the recent advances in ASCT and discusses the potential application of RNAi in hematopoietic cells.

Key words: Hematopoietic stem cells, graft-versus-host disease, NK cells, RNA interference, oncogenes.

1. Introduction

The bone marrow compartment contains the cellular elements of hematopoiesis in humans. Mature blood cells originate from a self-renewing population of multipotential hematopoietic stem cells that become committed to differentiate along the erythroid,

M. Sioud (ed.), *RNA Therapeutics*, Methods in Molecular Biology 629,
DOI 10.1007/978-1-60761-657-3_30,

megakaryocytic, granulocytic, monocytic, and lymphocytic lineages. In the process of acquiring lineage identity, the hematopoietic progenitors lose their self-renewal potential.

The exact phenotype of the stem cell in humans is not known, but there are several markers where expression is gained or lost during the differentiation of primitive human hematopoietic stem and progenitor cells (HSC). Among these markers are the molecules CD34, CD28, CD117, and CD133. No marker is expressed exclusively by HSC, but the CD34 marker is often used as a defining parameter of early progenitor cells, capable of long-term repopulation in transplanted patients. $CD34^+$ cells comprise between 0.5 and 2% of the mononuclear cells in the normal bone marrow. CD34 is a highly glycosylated type I transmembrane glycoprotein believed to function as an adhesion molecule (1, 2).

Several studies suggest that leukemia is a stem cell disease, in which the stem cell self-renewal mechanisms are preserved, but the tight control is lost due to malignant transformation. Chemotherapy is targeted to the bulk of malignant cells and does not necessarily eliminate the leukemic stem cell, which in turn may cause disease relapse. Eradication of malignant cells, including leukemic stem cells through immunologically active donor lymphocytes is achieved through high-dose chemotherapy with or without radiotherapy and subsequent allogeneic stem cell transplantation.

Allogeneic stem cell transplantation (ASCT) is a well-documented treatment for a variety of malignant and nonmalignant disorders of the immunohematopoietic system. In addition to being a treatment for acute myeloid and lymphoid leukemia as well as chronic myeloid leukemias, this modality of treatment is also used for nonmalignant hematological disorders such as, e.g., aplastic anemia, paroxysmal nocturnal hematuria (PNH), congenital immunodeficiencies, and recently for the treatment of relapsed lymphomas and chronic lymphocytic leukemia with reduced intensity conditioning (RIC). After the advent of potent tyrosine kinase inhibitors (TKIs) for the treatment of chronic myeloid leukemia (CML), ASCT is now regarded as indicated only in CML patients where TKI therapy has failed.

2. Donors and Stem Cell Sources

Approximately one-third of patients will have an available matched sibling donor. In cases where no sibling donor is available, HLA-matched unrelated donors (MUD) can be used. Through donor registries, more than 11 million volunteer donors are available worldwide and EBMT data show that in 2007 the proportion of MUD was slightly higher than the number of HLA identical sibling donors (3). The outcome with regard to disease-free survival (DFS) after transplantation with MUD compared to sibling donor is similar both after myeloablative and reduced intensity conditioning and success rates have improved as HLA matching has become more sophisticated (4, 5).

Traditionally, hematopoietic stem cells (HSC) were harvested from the iliac crests under general anesthesia. After the availability of growth factors available for use, G-CSF or GM-CSF has been used for mobilization of stem cells into the peripheral blood for harvest through apheresis. Since the 1990s, cord blood cells have been used for HLA-matched and HLA-mismatched allogeneic stem cell transplant in children and more recently, in adults.

Each of these sources of HSC has its significant advantages and limitations. With bone marrow, collection must be done under general anesthesia and the graft is a mix of different cells contained within the bone marrow. G-CSF-mobilized peripheral blood stem cells (PBSC) can easily be harvested through a vein without the need for general anesthesia, but with a potential for unwanted side effects of G-CSF with muscle and bone pain. There is generally no problem collecting a sufficient number of stem cells, but the graft will also contain a higher proportion of T cells. PBSCs are being increasingly used in patients with hematolo gical malignancies and are now used in over 70% of ASCT (3), especially when an unrelated donor is used. PBSCs provide faster engraftment compared to bone marrow and is probably associated with fewer relapses, but is associated with an increased risk of chronic GvHD (6).

Umbilical cord blood cells (UCB) are becoming an important source of stem cells for patients without matched related or unrelated donors. UCB are easy to collect and there is a potential immediate availability of cryopreserved units. Due to the nature of stem cells contained in the graft, a partial mismatch is accepted. Transplantation with single UCB units in adults has been associated with high transplant-related mortality (TRM). This high TRM has been due to slow engraftment and immunoincompetence with infectious complications. Another limiting factor is the potential number of cells contained in each unit, which will generally be insufficient for adult recipients. This may be overcome by using two units of cord blood (7–9). Furthermore, treatment with donor lymphocyte infusions is not possible after transplantation with UCB. Tomizawa et al. described a possible solution to this by expanding $CD4^+$ T cells ex vivo from the cord blood residues in an infused bag, thus providing a means of obtaining donor lymphocytes (10).

3. Conditioning Regimens

Conditioning therapy before stem cell transplantation has several goals, namely eradicating remaining malignant cells and inducing immunosuppression to permit engraftment of allogeneic cells. The traditional method of conditioning has been by myeloablation through chemotherapy with or without radiation therapy. The two most commonly used myeloablative regimens are a combination of Busulfan and Cyclophosphamide (BuCy) and total body irradiation (TBI) followed by Cyclophosphamide (TBI-Cy).

The concept of "reduced intensity conditioning" (RIC) is based on the idea of avoiding the high morbidity and mortality associated with standard myeloablative conditioning in patients with advanced disease and/or comorbidities. This involves a shift in the focus of conditioning from primary myeloablation to immunoablation. The goal of RIC HSCT is not tumor eradication or destruction of host hematopoiesis and malignancy by chemotherapy, but rather via immune-mediated mechanisms. The reduced immediate toxicity of RIC transplants enables patients otherwise ineligible for myeloablative conditioning the option of ASCT. The safety and effectiveness of these RIC regimens have allowed for their increasing use over the past decade, mainly for the treatment of low-grade non-Hodgkin lymphoma and chronic lymphocytic leukemia (11–16). The use of RIC transplants for other malignancies, such as acute myelogenous leukemia is currently under investigation. The lower mortality rate offered by RIC transplant is offset by a higher incidence of relapse, especially in advanced and aggressive disease (15, 17). Furthermore, there are a variety of RIC regimens in use and it is currently not known which of these provide the optimal conditioning and disease control.

4. Complications Following ASCT

ASCT is associated with a high treatment-related morbidity and mortality, the most important risk factors being marrow aplasia after high-dose chemotherapy, mucositis, infections with bacterial, viral and fungal agents, and acute and chronic graft-versus-host disease.

4.1. Mucositis

Mucositis following chemotherapy-induced damage to the mucosal lining of the gastrointestinal tract is a common conditioning-related toxicity, especially following myeloablative conditioning. This is usually not life-threatening, but is a risk factor for the need of analgesics, total parenteral nutrition and is strongly associated with infectious complications due to Gram-negative bacilli such as *Escherichia coli*, *Pseudomonas aeruginosa* and fungal infections with *Candida* species. Furthermore, mucositis probably contributes to the development and maintenance of acute graft-versus-host disease through production of inflammatory cytokines such as TNF-α and IL-1 that enhance activation of host antigen-presenting cells. Also, translocation of lipopolysaccharide (LPS) across the intestinal mucosa activates the innate immune system and promotes the production of inflammatory cytokines (18, 19). Considerable research has been done in search of therapeutic agents for mucositis, the most promising candidate being Palifermin, a recombinant human keratinocyte growth factor (KGF). Palifermin exerts effects on keratinocytes, fibroblasts, and endothelial cells as well as attenuating the effect of TNF-α and the expression of adhesion molecules (20, 21).

4.2. Graft-Versus-Host Disease

Graft-versus-host disease (GvHD) is a major cause of morbidity and mortality following ASCT. Acute GvHD, by definition, appears during the first 100 days posttransplant and is caused by donor T cells activated by recipient antigen-presenting cells (APC). However, the distinction of acute and chronic GvHD by day of onset is no longer always applicable, as symptoms of acute GvHD may present beyond day 100 in the RIC ASCT setting. Thus, new definitions of GvHD based on specificity of signs and symptoms rather than time of onset have been proposed (22). Risk factors for the development of GvHD include HLA disparities between recipient and donor, use of unrelated donors, female donor to male recipient and older age of the recipient. The occurrence and severity of acute GvHD has been linked to a number of other factors, such as gene polymorphisms affecting IL-1, IL-6, IL-10, and TNF-α (23, 24). Furthermore, some mutations in nucleotide oligomerization domain 2/caspase recruitment domain 15 (NOD2/CARD15) are associated with severe acute GvHD and increased treatment-related mortality (25, 26). In one study done with patients suffering from acute leukemia, NOD2/CARD15 SNPs were associated with a reduction in overall survival due to an increase in disease relapse (27). Mutations both on donor and recipient side appear to play a role (28). Based on this observation, it has been proposed that genotyping potential donors might be a useful strategy to reduce the risk of GvHD and contribute to an antileukemic effect. More recently, we have found that hematopoietic progenitor cells express NOD2/CARD15 receptor and response to its ligands by secreting cytokines and differentiating into myeloid cells (29).

GvHD can occur despite immunosuppressive prophylaxis and even in well-matched settings with HLA-identical sibling donors, demonstrating that other factors, such as minor histocompatibi lity antigens, are also of importance (30). The incidence of GvHD can be reduced through T-cell depletion of the graft before transplant. This beneficial effect may be offset by an increased risk of graft rejection, high rates of viral infections, and an increased risk of leukemia relapse. Delayed donor lymphocyte infusions (DLI) may be able to amend some of these problems inducing remissions in relapsed leukemias after ASCT and augmenting antiviral immunity, but is also associated with potentially life-threatening GvHD.

In the last 5 years, there has been increasing focus on the role of regulatory T cells (Tregs) in all aspects of immuno logy, including ASCT. Tregs are linked to autoinflammatory disease and play an important role in immune responses to self- and foreign antigens. In animal models, Tregs can prevent graft rejection and autoimmune diseases (31–33). A recent study by Rieger and colleagues showed that acute and chronic GvHD correlated with reduction or absence of Tregs in GvHD target organs, whereas increased numbers were found in non-GvHD-associated inflammation, suggesting that the suppressive effect of Tregs is impaired during GvHD due to insufficient numbers (34). The current pharmacological strategy to avoid GvHD is based on calcineurin inhibitors, usually in combination with low-dose methotrexate given shortly after transplant. Newer pharmacolo gical agents include Mycofenolate Mofetil (MMF) and sirolimus. MMF has been administered as part of a prospective randomized trial and showed significantly less mucositis and more rapid neutrophil engraftment compared to methotrexate while the frequency and severity remained similar. The study was closed early because of superiority of MMF with respect to mucositis and neutrophil engraftment and MMF is currently widely used in RIC transplants and UCB transplants (35, 36). Sirolimus has been shown to be effective in combination with Tacrolimus, but has been associated with unacceptable rates of veno-occlusive liver disease (VOD) when myeloablative regimens containing Busulfan are used (37).

Treatment of severe acute GvHD remains a challenging task and is mainly based on high-dose corticosteroids and calcineurin inhibitors. Treatment options of steroid refractory GvHD are unsatisfactory as no single agent has proven to be effective, although some studies report positive effects from TNF-α inhibitors, extracorporeal photopheresis and Rituximab (38–43). Mesenchymal stem cells (MSC) pose an exciting new alternative for the treatment of GvHD. MSCs are immunomodulatory and have been promising in a phase II trial in patients with steroid refractory GvHD, where infusion of ex vivo expanded MSCs from the original stem cell donor or from a third party produced 55% complete responses (44).

4.3. Infections

As a consequence of high-dose chemotherapy with bone marrow aplasia and damage of mucosal barriers, pharmacological immunosuppression and complications such as GvHD with immunomodulatory effects, patients undergoing ASCT have an increased risk of serious infections with bacterial, viral, and fungal agents. RIC ASCT is associated with a lower frequency of early infections compared to myeloablative regimens, but there does not seem to be a reduced risk of late infections. During the neutropenic phase following conditioning and before engraftment, patients predictably become septic with bacterial infections and require broad-spectrum antibiotic therapy. Invasive fungal infections with yeast or molds pose even greater challenges as they are difficult to diagnose and even more difficult to treat if they become clinically relevant. Although mortality remains high, there has been an improvement in clinical outcome after the use of routine antifungal prophylaxis, earlier diagnostics with radiological imaging and new, more potent antifungal agents (45, 46).

In immunosuppressed patients, and especially in patients receiving high-dose corticosteroids for the treatment of GvHD, reactivation of viral infections can pose a significant problem. This is true particularly for cytomegalovirus (CMV), which may cause potentially life-threatening systemic disease, polyomavirus causing hemorrhagic cystitis, and EBV-related posttransplant lymphoproliferative disease. PCR monitoring of CMV viral load and preemptive treatment of CMV infections with antivirals have reduced the incidence and mortality rate of CMV infection in the ASCT setting (47). Still, viral infections continue to cause morbidity and mortality in ASCT patients and antiviral therapy has potential for unwanted side effects like bone marrow suppression, and better therapies are needed. Adoptive T-cell therapy with specific T cells has been tried with good results in CMV infection and EBV-related lymphoma (48–51).

5. Graft-Versus-Malignancy Effect

Originally, it was believed that the major antileukemic effect of ASCT was due to high-dose chemotherapy and radiation. It has become increasingly clear that there is also a significant graft-versus-malignancy effect (52). The first reports of lower incidences of relapse in patients with GvHD came in 1979 (53) and later patients with chronic GvHD were shown to have a survival benefit (54, 55). The antileukemic effect of graft was shown to be T-cell dependent when increased relapse rates were demonstrated in patients receiving T-cell-depleted grafts (56).

This effect is dependent on disease, e.g., it is more pronounced in chronic myeloid leukemia than in acute lymphoblastic leukemia. Further evidence of T-cell activity against the underlying malignancy was provided in showing that relapse of leukemia can be cured by reduction of immunosuppression and donor lymphocyte infusions (DLI) (57). Again, the best results are seen after DLI in patients with relapsed CML. Up to 80% of patients treated with DLI for relapse CML will experience a complete cytogenetic response, i.e., disappearance of the Philadelphia chromosome in cytogenetic analysis, and for a majority of patients, these remissions are durable. The results are better in less advanced stages of disease (58). The main side effect of DLI is GvHD and myelosuppression. Although the graft-versus-leukemia (GVL) effect is associated with GvHD, it is preferable to transfuse donor lymphocytes in escalating doses to avoid triggering serious and potentially lethal GvHD (59). Other strategies for avoiding deleterious GvHD after DLI has been attempted by T-cell subset depletion of $CD8^{+}$ T cells and transfer of suicide genes into T cells, making it possible to pharmacologically ablate selective T cells (60–62). Immunotherapeutic options such as DLI can prevent or treat relapsed disease, but at the cost of potentially deleterious GvHD. Despite the potency of DLI, relapse remains a major problem after ASCT. Further refinement of treatment is needed, especially minimizing side effects such as GvHD, while maximizing the antileukemic effect. In this respect, Stauss and colleagues described the possibility of generating allorestricted cytotoxic T lymphocytes (CTLs) that can mediate a graft-versus-leukemia reaction without causing graft-versus-host disease (63). The existence of an allorestricted repertoire raises the possibility of generating CTLs reactive against synthetic self-peptides bound to nonself-MHC molecules, because tolerance to self-antigens is self-MHC restricted (*see* **Chapter 29**). It could be feasible to generate CTLs against self-antigens for immunotherapy and/or TcR gene transfer as reviewed by Stauss and colleagues in 1999. Indeed, it has been shown by several investigators that alloreactive CTLs against tumor antigens and CD molecules such as CD19, CD20, CD33, CD45 can be established (64–69). The GvL effect is most likely generated by several different T-cell clones and tumor cells have been shown to inactivate high-affinity T cells. Thus, it seems likely that adoptive transfer of single peptide-specific T cells will not provide an efficient response.

5.1. Alloreactive NK Cells

ASCT using HLA-haploidentical donors is an increasingly used and studied alternative for patients without an available HLA-matched donor. Haploidentical ASCT takes advantage of donor-versus-recipient NK-cell allo- and antileukemic activity. This donor-versus-recipient NK cell alloreactivity derives from a mismatch between inhibitory killer immunoglobulin-like (KIR) receptors for self-MHC class I molecules on donor NK cells and the MHC class I ligands on recipient cells. Donor NK cells sense the missing expression of self-MHC class I molecules and mediate allo-reactions (missing self-recognition) (70, 71), which may result in potent antileukemic responses with low rates of graft failure, relapse, and GvHD (72–74). Nonrelapse mortality, mainly due to infection, remains a problem in haploidentical transplants (75).

6. Potential Modulation of GvHD by Indoleamine 2,3 Dioxygenase

2,3 Indoleamine dioxygenase (IDO) is a key immunomodulatory enzyme that catalyzes the first and rate-limiting step of tryptophan catabolism. Decreased tryptophan and/or increased metabolite concentrations induce a stress response in nearby responding T cells, leading to anergy or apoptosis (76, 77). IDO expression has been shown to be critical for allogeneic fetal tolerance, tumor tolerance, and downregulation of autoimmunity (78–81). These immunosuppressive functions make IDO an interesting target for transplantation. The lung and intestine, which are GvHD target organs, constitutively express IDO and can upregulate expression during inflammation. Immune cells such as monocytes, macrophages, and dendritic cells can be induced to express IDO by certain Toll-like receptor ligands and ligation of costimulatory molecules. In addition, IDO activity has been linked to the generation of T-regulatory cells. Strategies to upregulate IDO in GvHD target organs could be useful tools for gut protection (82). In this respect, a study showed that patients with severe acute GvHD were less able to upregulate IDO on exposure to IFN-γ than healthy donors or those with milder GvHD (83). Locally induced IDO could block the activation of autoreactive T cells without interfering with GvL. This possibility is currently being studied in animal models.

7. Autologous Stem Cell Transplantation and Gene Therapy

Primary immunodeficiencies, such as severe combined immuno deficiency (SCID) are invariably fatal without treatment. ASCT is usually successful if a matched related or unrelated donor is available. This treatment is not available for all patients, and efforts have been done to provide effective gene therapy through transduction of $CD34^+$ hematopoietic progenitor cells with a retroviral vector containing a functional version of the gene. Gene therapy has been shown to be effective in some patients, but its use has been limited by the development of malignancy due to insertional mutagenesis caused by the retroviral vector. Recent reports however, describe the safety and feasibility of gene therapy in SCID patients with deficiency in adenosine deaminase (ADA). In one study, 10 patients with SCID due to ADA deficiency were infused with autologous $CD34^+$ cells transduced with a retroviral vector containing the ADA gene after nonmyeloablative conditioning. After a median of 4 years follow-up, all patients are alive and eight do no longer require enzyme-replacement therapy (84–88).

8. RNA Interference and HSC

RNAi is a fundamental cellular mechanism for silencing gene expression that can be harnessed for the development of new drugs (*see* **Chapters 4, 5, 6, 7, 8, 9, 10,** and **11**). The reduced expression of pathological proteins through RNAi is applicable to all classes of molecular targets, including those that are difficult to modulate selectively with traditional drugs such as noncoding RNAs (*see* **Chapters 14** and **16**). With respect to cancer, an ideal target for therapeutic approach would be a gene that plays an essential role in tumor biology and which is exclusively expressed in the tumors, but not in normal cells. Based on this, leukemic fusion genes could constitute good targets for RNAi.

Chronic myeloid leukemia (CML) is a relatively well-differentiated myeloproliferative disorder originating from transformed hematopoietic stem cells. The disease arises as a consequence of a rare mutational event resulting in a reciprocal translocation between the long arms of chromosomes 9 and 22 (89). This translocation creates the chimeric oncogene *bcr-abl* with the protein product BCR-ABL, a constitutively active tyrosine kinase. The expression of *bcr-abl* has been shown to be necessary and sufficient for the transformed phenotype of CML cells. Also, the chromosomal translocation t(8;21) is associated with 10–15% of all cases of acute myeloid leukemias (AML). The resultant fusion protein AML1-ETO interferes with hematopoietic gene expression and is an important regulator of leukemogenesis (90). Such fusion oncogenes would fulfill the requirement of an ideal target. Since siRNAs bind to the transcript, they can be targeted against almost any gene. Previous studies have shown that siRNA targeting *bcr-abl* or *aml-eto* can bring about leukemia cell proliferation in vitro and in vivo (91, 92). This may become a useful strategy for eliminating leukemic cells, alone or in combination with other therapies.

In addition to systemic or local delivery of chemically synthesized siRNAs, introduction of antifusion proteins shRNA into autologous CD34+ hematopoietic stem cells for transplantation into humans could be an attractive approach for the elimination of fusion proteins that are required for leukemic cell proliferation and survival. In the case of bcr-abl oncogene, such a strategy should lead to the constitutive expression of shRNAs in immature and mature lymphoid and myeloid cells, rendering the kinase inactive in most cells. Rossi and colleagues employed lentivirus-based gene transfer vectors to deliver anti-HIV-1 shRNAs to CD34+ progenitor cells (93).

As indicated in **Section 5.1**, NK cells can be beneficial in bone marrow transplantation, depending on their genotype and activation status. They do not express T-cell receptors or immunoglobulin genes. In humans, NK cells are usually defined as CD3- $CD56^+$ lymphocytes and comprise about 5–20% of peripheral blood lymphocytes. Previous studies have shown that NK cell recognition and activation is due to a variety of receptors that bind to specific ligands on tumor cells and normal cells (94). A common feature of several inhibitory NK receptors is the capability to bind MHC class I molecules as predicted by the effector inhibition model within the missing self-hypothesis of recognition by NK cells (70). Important inhibitory NK receptors are members of the KIR family in humans and of the C-type lectin-like Ly49 receptors in mice, both sensing the expression of various allelic variants of classical MHC class I molecules. Although HLA and KIR genes are inherited independently, most individuals posses a full complement of KIR genes for the three major class ligands (groups 1 and 2 HLA-C alleles and HLA-Bw4 alleles). Therapeutic strategies for cancer therapy are being developed based on preventing NK cell inhibition or using NK cell receptors to activate these cells. Notably, there are some clinical data from studies of bone marrow transplantation that support the idea that preventing NK cell inhibition by HLA class I molecules can be a means to promote graft-versus-leukemia effects and limit GvHD in AML and CML patients (95). One way to prevent NK cell inhibition by HLA class I molecules would be the inhibition of KIR receptor expression using RNA interference. Silencing the expression of KIR receptors may lead to the generation of a large number of alloreactive NK cells for adaptive cancer immunotherapy. In addition to KIR receptors, RNA interference can be used to downregulate the expression of HLA expression. Allele-specific HLA inhibition could help to overcome limitations imposed by extensive HLA polymorphism that limits the availability of suitable donors.

References

1. Civin, C.I., Strauss, L.C., Brovall, C., Fackler, M.J., Schwartz, J.F., and Shaper, J.H. (1984) Antigenic analysis of hematopoiesis. III. A hematopoietic progenitor cell surface antigen defined by a monoclonal antibody raised against KG-1a cells. *J Immunol*, **133**, 157–165.
2. Egeland, T., Tjonnfjord, G., Steen, R., Gaudernack, G., and Thorsby, E. (1993) Positive selection of bone marrow-derived CD34 positive cells for possible stem cell transplantation. *Transplant Proc*, **25**, 1261–1263.
3. Gratwohl, A., Baldomero, H., Schwendener, A., Rocha, V., Apperley, J., Frauendorfer, K., and Niederwieser, D. (2009) The EBMT activity survey 2007 with focus on allogeneic HSCT for AML and novel cellular therapies. *Bone Marrow Transplant*, **43**, 275–291.
4. Uzunel, M., Remberger, M., Sairafi, D., Hassan, Z., Mattsson, J., Omazic, B., Barkholt, L., and Ringden, O. (2006) Unrelated versus related allogeneic stem cell transplantation after reduced intensity conditioning. *Transplantation*, **82**, 913–919.
5. Remberger, M., Mattsson, J., Hausenberger, D., Schaffer, M., Svahn, B.M., and Ringden, O. (2008) Genomic tissue typing and optimal antithymocyte globuline dose using unrelated donors results in similar survival and relapse as HLA-identical siblings in haematopoietic stem-cell transplantation for leukaemia. *Eur J Haematol*, **80**, 419–428.
6. Heldal, D., Brinch, L., Tjonnfjord, G., Solheim, B.G., Egeland, T., Albrechtsen, D., Aamodt, G., and Evensen, S.A. (2003) Fewer relapses and increased chronic GVHD in patients transplanted with blood stem cells: a 5-year follow-up in a single centre study. *Bone Marrow Transplant*, **32**, 257–264.
7. Ballen, K.K., Spitzer, T.R., Yeap, B.Y., McAfee, S., Dey, B.R., Attar, E., Haspel, R., Kao, G., Liney, D., Alyea, E., Lee, S., Cutler, C., Ho, V., Soiffer, R., and Antin, J.H. (2007) Double unrelated reduced-intensity umbilical cord blood transplantation in adults. *Biol Blood Marrow Transplant*, **13**, 82–89.
8. Ballen, K.K. (2005) New trends in umbilical cord blood transplantation. *Blood*, **105**, 3786–3792.
9. Barker, J.N., Weisdorf, D.J., DeFor, T.E., Blazar, B.R., McGlave, P.B., Miller, J.S., Verfaillie, C.M., and Wagner, J.E. (2005) Transplantation of 2 partially HLA-matched umbilical cord blood units to enhance engraftment in adults with hematologic malignancy. *Blood*, **105**, 1343–1347.
10. Tomizawa, D., Aoki, Y., Nagasawa, M., Morio, T., Kajiwara, M., Sekine, T., Shimizu, N., Kato, M., Yachie, A., and Mizutani, S. (2005) Novel adopted immunotherapy for mixed chimerism after unrelated cord blood transplantation in Omenn syndrome. *Eur J Haematol*, **75**, 441–444.
11. Dreger, P., Corradini, P., Kimby, E., Michallet, M., Milligan, D., Schetelig, J., Wiktor-Jedrzejczak, W., Niederwieser, D., Hallek, M., and Montserrat, E. (2007) Indications for allogeneic stem cell transplantation in chronic lymphocytic leukemia: the EBMT transplant consensus. *Leukemia*, **21**, 12–17.
12. Sorror, M.L., Storer, B.E., Sandmaier, B.M., Maris, M., Shizuru, J., Maziarz, R., Agura, E., Chauncey, T.R., Pulsipher, M.A., McSweeney, P.A., Wade, J.C., Bruno, B., Langston, A., Radich, J., Niederwieser, D., Blume, K.G., Storb, R., and Maloney, D.G. (2008) Five-year follow-up of patients with advanced chronic lymphocytic leukemia treated with allogeneic hematopoietic cell transplantation after nonmyeloablative conditioning. *J Clin Oncol*, **26**, 4912–4920.
13. Dreger, P., Brand, R., Hansz, J., Milligan, D., Corradini, P., Finke, J., Deliliers, G.L., Martino, R., Russell, N., Van Biezen, A., Michallet, M., and Niederwieser, D. (2003) Treatment-related mortality and graft-versus-leukemia activity after allogeneic stem cell transplantation for chronic lymphocytic leukemia using intensity-reduced conditioning. *Leukemia*, **17**, 841–848.
14. Khouri, I.F., Saliba, R.M., Giralt, S.A., Lee, M.S., Okoroji, G.J., Hagemeister, F.B., Korbling, M., Younes, A., Ippoliti, C., Gajewski, J.L., McLaughlin, P., Anderlini, P., Donato, M.L., Cabanillas, F.F., and Champlin, R.E. (2001) Nonablative allogeneic hematopoietic transplantation as adoptive immunotherapy for indolent lymphoma: low incidence of toxicity, acute graft-versus-host disease, and treatment-related mortality. *Blood*, **98**, 3595–3599.
15. Vigouroux, S., Michallet, M., Porcher, R., Attal, M., Ades, L., Bernard, M., Blaise, D., Tabrizi, R., Garban, F., Cassuto, J.P., Chevalier, P., Facon, T., Ifrah, N., Renaud, M., Tilly, H., Vernant, J.P., Kuentz, M., Bourhis, J.H., Bordigoni, P., Deconinck, E., Lioure, B., Socie, G., and Milpied, N. (2007) Long-term outcomes after reduced-intensity conditioning allogeneic stem cell transplantation for low-grade lymphoma: a survey by the French Society of Bone Marrow Graft

Transplantation and Cellular Therapy (SFGM-TC). *Haematologica*, **92**, 627–634.
16. Rezvani, A.R., Storer, B., Maris, M., Sorror, M.L., Agura, E., Maziarz, R.T., Wade, J.C., Chauncey, T., Forman, S.J., Lange, T., Shizuru, J., Langston, A., Pulsipher, M.A., Sandmaier, B.M., Storb, R., and Maloney, D.G. (2008) Nonmyeloablative allogeneic hematopoietic cell transplantation in relapsed, refractory, and transformed indolent non-Hodgkin's lymphoma. *J Clin Oncol*, **26**, 211–217.
17. Robinson, S.P., Goldstone, A.H., Mackinnon, S., Carella, A., Russell, N., de Elvira, C.R., Taghipour, G., and Schmitz, N. (2002) Chemoresistant or aggressive lymphoma predicts for a poor outcome following reduced-intensity allogeneic progenitor cell transplantation: an analysis from the Lymphoma Working Party of the European Group for Blood and Bone Marrow Transplantation. *Blood*, **100**, 4310–4316.
18. Niscola, P., Romani, C., Cupelli, L., Scaramucci, L., Tendas, A., Dentamaro, T., Amadori, S., and de Fabritiis, P. (2007) Mucositis in patients with hematologic malignancies: an overview. *Haematologica*, **92**, 222–231.
19. Goldberg, J., Jacobsohn, D.A., Zahurak, M.L., and Vogelsang, G.B. (2005) Gastrointestinal toxicity from the preparative regimen is associated with an increased risk of graft-versus-host disease. *Biol Blood Marrow Transplant*, **11**, 101–107.
20. Stiff, P.J., Emmanouilides, C., Bensinger, W.I., Gentile, T., Blazar, B., Shea, T.C., Lu, J., Isitt, J., Cesano, A., and Spielberger, R. (2006) Palifermin reduces patient-reported mouth and throat soreness and improves patient functioning in the hematopoietic stem-cell transplantation setting. *J Clin Oncol*, **24**, 5186–5193.
21. Spielberger, R., Stiff, P., Bensinger, W., Gentile, T., Weisdorf, D., Kewalramani, T., Shea, T., Yanovich, S., Hansen, K., Noga, S., McCarty, J., LeMaistre, C.F., Sung, E.C., Blazar, B.R., Elhardt, D., Chen, M.G., and Emmanouilides, C. (2004) Palifermin for oral mucositis after intensive therapy for hematologic cancers. *N Engl J Med*, **351**, 2590–2598.
22. Filipovich, A.H., Weisdorf, D., Pavletic, S., Socie, G., Wingard, J.R., Lee, S.J., Martin, P., Chien, J., Przepiorka, D., Couriel, D., Cowen, E.W., Dinndorf, P., Farrell, A., Hartzman, R., Henslee-Downey, J., Jacobsohn, D., McDonald, G., Mittleman, B., Rizzo, J.D., Robinson, M., Schubert, M., Schultz, K., Shulman, H., Turner, M., Vogelsang, G., and Flowers, M.E. (2005) National Institutes of Health consensus development project on criteria for clinical trials in chronic graft-versus-host disease: I. Diagnosis and staging working group report. *Biol Blood Marrow Transplant*, **11**, 945–956.
23. Markey, K.A., MacDonald, K.P., and Hill, G.R. (2008) Impact of cytokine gene polymorphisms on graft-vs-host disease. *Tissue Antigens*, **72**, 507–516.
24. Rocha, V., Franco, R.F., Porcher, R., Bittencourt, H., Silva, W.A. Jr., Latouche, A., Devergie, A., Esperou, H., Ribaud, P., Socie, G., Zago, M.A., and Gluckman, E. (2002) Host defense and inflammatory gene polymorphisms are associated with outcomes after HLA-identical sibling bone marrow transplantation. *Blood*, **100**, 3908–3918.
25. Holler, E., Rogler, G., Brenmoehl, J., Hahn, J., Greinix, H., Dickinson, A.M., Socie, G., Wolff, D., Finke, J., Fischer, G., Jackson, G., Rocha, V., Hilgendorf, I., Eissner, G., Marienhagen, J., and Andreesen, R. (2008) The role of genetic variants of NOD2/CARD15, a receptor of the innate immune system, in GvHD and complications following related and unrelated donor haematopoietic stem cell transplantation. *Int J Immunogenet*, **35**, 381–384.
26. Dickinson, A.M., Middleton, P.G., Rocha, V., Gluckman, E., and Holler, E. (2004) Genetic polymorphisms predicting the outcome of bone marrow transplants. *Br J Haematol*, **127**, 479–490.
27. Mayor, N.P., Shaw, B.E., Hughes, D.A., Maldonado-Torres, H., Madrigal, J.A., Keshav, S., and Marsh, S.G. (2007) Single nucleotide polymorphisms in the NOD2/CARD15 gene are associated with an increased risk of relapse and death for patients with acute leukemia after hematopoietic stem-cell transplantation with unrelated donors. *J Clin Oncol*, **25**, 4262–4269.
28. Elmaagacli, A.H., Koldehoff, M., Hindahl, H., Steckel, N.K., Trenschel, R., Peceny, R., Ottinger, H., Rath, P.M., Ross, R.S., Roggendorf, M., Grosse-Wilde, H., and Beelen, D.W. (2006) Mutations in innate immune system NOD2/CARD 15 and TLR-4 (Thr399Ile) genes influence the risk for severe acute graft-versus-host disease in patients who underwent an allogeneic transplantation. *Transplantation*, **81**, 247–254.
29. Sioud, M. and Floisand, Y. (2009) NOD2/CARD15 on bone marrow CD34+ hematopoietic cells mediates induction of cytokines and cell differentiation. *J Leukoc Biol*, **85**, 939–946.

30. Ferrara, J.L., Levine, J.E., Reddy, P., and Holler, E. (2009) Graft-versus-host disease. *Lancet*, **373**, 1550–1561.
31. Graca, L., Thompson, S., Lin, C.Y., Adams, E., Cobbold, S.P., and Waldmann, H. (2002) Both CD4(+)CD25(+) and CD4(+)CD25 (−) regulatory cells mediate dominant transplantation tolerance. *J Immunol*, **168**, 5558–5565.
32. Hara, M., Kingsley, C.I., Niimi, M., Read, S., Turvey, S.E., Bushell, A.R., Morris, P.J., Powrie, F., and Wood, K.J. (2001) IL-10 is required for regulatory T cells to mediate tolerance to alloantigens in vivo. *J Immunol*, **166**, 3789–3796.
33. Maul, J., Loddenkemper, C., Mundt, P., Berg, E., Giese, T., Stallmach, A., Zeitz, M., and Duchmann, R. (2005) Peripheral and intestinal regulatory CD4+ CD25(high) T cells in inflammatory bowel disease. *Gastroenterology*, **128**, 1868–1878.
34. Rieger, K., Loddenkemper, C., Maul, J., Fietz, T., Wolff, D., Terpe, H., Steiner, B., Berg, E., Miehlke, S., Bornhauser, M., Schneider, T., Zeitz, M., Stein, H., Thiel, E., Duchmann, R., and Uharek, L. (2006) Mucosal FOXP3+ regulatory T cells are numerically deficient in acute and chronic GvHD. *Blood*, **107**, 1717–1723.
35. Bolwell, B., Sobecks, R., Pohlman, B., Andresen, S., Rybicki, L., Kuczkowski, E., and Kalaycio, M. (2004) A prospective randomized trial comparing cyclosporine and short course methotrexate with cyclosporine and mycophenolate mofetil for GVHD prophylaxis in myeloablative allogeneic bone marrow transplantation. *Bone Marrow Transplant*, **34**, 621–625.
36. Baron, F., Little, M.T., and Storb, R. (2005) Kinetics of engraftment following allogeneic hematopoietic cell transplantation with reduced-intensity or nonmyeloablative conditioning. *Blood Rev*, **19**, 153–164.
37. Cutler, C., Stevenson, K., Kim, H.T., Richardson, P., Ho, V.T., Linden, E., Revta, C., Ebert, R., Warren, D., Choi, S., Koreth, J., Armand, P., Alyea, E., Carter, S., Horowitz, M., Antin, J.H., and Soiffer, R. (2008) Sirolimus is associated with veno-occlusive disease of the liver after myeloablative allogeneic stem cell transplantation. *Blood*, **112**, 4425–4431.
38. Perfetti, P., Carlier, P., Strada, P., Gualandi, F., Occhini, D., Van Lint, M.T., Ibatici, A., Lamparelli, T., Bruno, B., Raiola, A.M., Dominietto, A., Di Grazia, C., Bregante, S., Zia, S., Ferrari, G.M., Stura, P., Pogliani, E., and Bacigalupo, A. (2008) Extracorporeal photopheresis for the treatment of steroid refractory acute GVHD. *Bone Marrow Transplant*, **42**, 609–617.
39. Apisarnthanarax, N., Donato, M., Korbling, M., Couriel, D., Gajewski, J., Giralt, S., Khouri, I., Hosing, C., Champlin, R., Duvic, M., and Anderlini, P. (2003) Extracorporeal photopheresis therapy in the management of steroid-refractory or steroid-dependent cutaneous chronic graft-versus-host disease after allogeneic stem cell transplantation: feasibility and results. *Bone Marrow Transplant*, **31**, 459–465.
40. Greinix, H.T., Knobler, R.M., Worel, N., Schneider, B., Schneeberger, A., Hoecker, P., Mitterbauer, M., Rabitsch, W., Schulenburg, A., and Kalhs, P. (2006) The effect of intensified extracorporeal photochemotherapy on long-term survival in patients with severe acute graft-versus-host disease. *Haematologica*, **91**, 405–408.
41. Couriel, D., Saliba, R., Hicks, K., Ippoliti, C., de Lima, M., Hosing, C., Khouri, I., Andersson, B., Gajewski, J., Donato, M., Anderlini, P., Kontoyiannis, D.P., Cohen, A., Martin, T., Giralt, S., and Champlin, R. (2004) Tumor necrosis factor-alpha blockade for the treatment of acute GVHD. *Blood*, **104**, 649–654.
42. Van Lint, M.T., Milone, G., Leotta, S., Uderzo, C., Scime, R., Dallorso, S., Locasciulli, A., Guidi, S., Mordini, N., Sica, S., Cudillo, L., Fagioli, F., Selleri, C., Bruno, B., Arcese, W., and Bacigalupo, A. (2006) Treatment of acute graft-versus-host disease with prednisolone: significant survival advantage for day +5 responders and no advantage for nonresponders receiving anti-thymocyte globulin. *Blood*, **107**, 4177–4181.
43. Berger, M., Biasin, E., Saglio, F., and Fagioli, F. (2008) Innovative approaches to treat steroid-resistant or steroid refractory GVHD. *Bone Marrow Transplant*, **42**(Suppl 2), S101–S105.
44. Le Blanc, K., Frassoni, F., Ball, L., Locatelli, F., Roelofs, H., Lewis, I., Lanino, E., Sundberg, B., Bernardo, M.E., Remberger, M., Dini, G., Egeler, R.M., Bacigalupo, A., Fibbe, W., and Ringden, O. (2008) Mesenchymal stem cells for treatment of steroid-resistant, severe, acute graft-versus-host disease: a phase II study. *Lancet*, **371**, 1579–1586.
45. Neofytos, D., Horn, D., Anaissie, E., Steinbach, W., Olyaei, A., Fishman, J., Pfaller, M., Chang, C., Webster, K., and Marr, K. (2009) Epidemiology and outcome of invasive fungal infection in adult hematopoietic stem cell transplant recipients: analysis of Multicenter Prospective Antifungal Therapy

(PATH) Alliance registry. *Clin Infect Dis*, **48**, 265–273.
46. Maertens, J., Meersseman, W., and Van Bleyenbergh, P. (2009) New therapies for fungal pneumonia. *Curr Opin Infect Dis*, **22**, 183–190.
47. Hebart, H. and Einsele, H. (2004) Clinical aspects of CMV infection after stem cell transplantation. *Hum Immunol*, **65**, 432–436.
48. Haque, T., Wilkie, G.M., Taylor, C., Amlot, P.L., Murad, P., Iley, A., Dombagoda, D., Britton, K.M., Swerdlow, A.J., and Crawford, D.H. (2002) Treatment of Epstein–Barr-virus-positive post-transplantation lymphoproliferative disease with partly HLA-matched allogeneic cytotoxic T cells. *Lancet*, **360**, 436–442.
49. Haque, T., Wilkie, G.M., Jones, M.M., Higgins, C.D., Urquhart, G., Wingate, P., Burns, D., McAulay, K., Turner, M., Bellamy, C., Amlot, P.L., Kelly, D., MacGilchrist, A., Gandhi, M.K., Swerdlow, A.J., and Crawford, D.H. (2007) Allogeneic cytotoxic T-cell therapy for EBV-positive posttransplantation lymphoproliferative disease: results of a phase 2 multicenter clinical trial. *Blood*, **110**, 1123–1131.
50. Mackinnon, S., Thomson, K., Verfuerth, S., Peggs, K., and Lowdell, M. (2008) Adoptive cellular therapy for cytomegalovirus infection following allogeneic stem cell transplantation using virus-specific T cells. *Blood Cells Mol Dis*, **40**, 63–67.
51. Einsele, H., Kapp, M., and Grigoleit, G.U. (2008) CMV-specific T cell therapy. *Blood Cells Mol Dis*, **40**, 71–75.
52. Hogan, W.J. and Deeg, H.J. (2005) Stem cell transplantation: graft-mediated antileukemia effects. *Methods Mol Med*, **109**, 421–444.
53. Weiden, P.L., Flournoy, N., Thomas, E.D., Prentice, R., Fefer, A., Buckner, C.D., and Storb, R. (1979) Antileukemic effect of graft-versus-host disease in human recipients of allogeneic-marrow grafts. *N Engl J Med*, **300**, 1068–1073.
54. Weiden, P.L., Sullivan, K.M., Flournoy, N., Storb, R., and Thomas, E.D. (1981) Antileukemic effect of chronic graft-versus-host disease: contribution to improved survival after allogeneic marrow transplantation. *N Engl J Med*, **304**, 1529–1533.
55. Ringden, O., Labopin, M., Gluckman, E., Reiffers, J., Vernant, J.P., Jouet, J.P., Harousseau, J.L., Fiere, D., Bacigalupo, A., Frassoni, F., and Gorin, N.C. (1997) Strong antileukemic effect of chronic graft-versus-host disease in allogeneic marrow transplant recipients having acute leukemia treated with methotrexate and cyclosporine. The Acute Leukemia Working Party of the European Group for Blood and Marrow Transplantation (EBMT). *Transplant Proc*, **29**, 733–734.
56. Horowitz, M.M., Gale, R.P., Sondel, P.M., Goldman, J.M., Kersey, J., Kolb, H.J., Rimm, A.A., Ringden, O., Rozman, C., Speck, B. et al. (1990) Graft-versus-leukemia reactions after bone marrow transplantation. *Blood*, **75**, 555–562.
57. Kolb, H.J., Schattenberg, A., Goldman, J.M., Hertenstein, B., Jacobsen, N., Arcese, W., Ljungman, P., Ferrant, A., Verdonck, L., Niederwieser, D., van Rhee, F., Mittermueller, J., de Witte, T., Holler, E., and Ansari, H. (1995) Graft-versus-leukemia effect of donor lymphocyte transfusions in marrow grafted patients. *Blood*, **86**, 2041–2050.
58. Bacigalupo, A., Soracco, M., Vassallo, F., Abate, M., Van Lint, M.T., Gualandi, F., Lamparelli, T., Occhini, D., Mordini, N., Bregante, S., Figari, O., Benvenuto, F., Sessarego, M., Fugazza, G., Carlier, P., and Valbonesi, M. (1997) Donor lymphocyte infusions (DLI) in patients with chronic myeloid leukemia following allogeneic bone marrow transplantation. *Bone Marrow Transplant*, **19**, 927–932.
59. Guglielmi, C., Arcese, W., Dazzi, F., Brand, R., Bunjes, D., Verdonck, L.F., Schattenberg, A., Kolb, H.J., Ljungman, P., Devergie, A., Bacigalupo, A., Gomez, M., Michallet, M., Elmaagacli, A., Gratwohl, A., Apperley, J., and Niederwieser, D. (2002) Donor lymphocyte infusion for relapsed chronic myelogenous leukemia: prognostic relevance of the initial cell dose. *Blood*, **100**, 397–405.
60. Soiffer, R.J., Alyea, E.P., Hochberg, E., Wu, C., Canning, C., Parikh, B., Zahrieh, D., Webb, I., Antin, J., and Ritz, J. (2002) Randomized trial of CD8+ T-cell depletion in the prevention of graft-versus-host disease associated with donor lymphocyte infusion. *Biol Blood Marrow Transplant*, **8**, 625–632.
61. Baron, F., Siquet, J., Schaaf-Lafontaine, N., Baudoux, E., Hermanne, J.P., Fillet, G., and Beguin, Y. (2002) Pre-emptive immunotherapy with CD8-depleted donor lymphocytes after CD34-selected allogeneic peripheral blood stem cell transplantation. *Haematologica*, **87**, 78–88.
62. Ciceri, F., Bonini, C., Gallo-Stampino, C., and Bordignon, C. (2005) Modulation of GvHD by suicide-gene transduced donor T lymphocytes: clinical applications in

mismatched transplantation. *Cytotherapy*, 7, 144–149.
63. Stauss, H.J. (1999) Immunotherapy with CTLs restricted by nonself MHC. *Immunol Today*, **20**, 180–183.
64. Bae, J., Martinson, J.A., and Klingemann, H.G. (2005) Identification of CD19 and CD20 peptides for induction of antigen-specific CTLs against B-cell malignancies. *Clin Cancer Res*, **11**, 1629–1638.
65. Moris, A., Teichgraber, V., Gauthier, L., Buhring, H.J., and Rammensee, H.G. (2001) Cutting edge: characterization of allorestricted and peptide-selective alloreactive T cells using HLA-tetramer selection. *J Immunol*, **166**, 4818–4821.
66. Bae, J., Martinson, J.A., and Klingemann, H.G. (2004) Identification of novel CD33 antigen-specific peptides for the generation of cytotoxic T lymphocytes against acute myeloid leukemia. *Cell Immunol*, **227**, 38–50.
67. Amrolia, P.J., Reid, S.D., Gao, L., Schultheis, B., Dotti, G., Brenner, M.K., Melo, J.V., Goldman, J.M., and Stauss, H.J. (2003) Allorestricted cytotoxic T cells specific for human CD45 show potent antileukemic activity. *Blood*, **101**, 1007–1014.
68. Gao, L., Yang, T.H., Tourdot, S., Sadovnikova, E., Hasserjian, R., and Stauss, H.J. (1999) Allo-major histocompatibility complex-restricted cytotoxic T lymphocytes engraft in bone marrow transplant recipients without causing graft-versus-host disease. *Blood*, **94**, 2999–3006.
69. Heemskerk, M.H., Hoogeboom, M., de Paus, R.A., Kester, M.G., van der Hoorn, M.A., Goulmy, E., Willemze, R., and Falkenburg, J.H. (2003) Redirection of antileukemic reactivity of peripheral T lymphocytes using gene transfer of minor histocompatibility antigen HA-2-specific T-cell receptor complexes expressing a conserved alpha joining region. *Blood*, **102**, 3530–3540.
70. Ljunggren, H.G. and Karre, K. (1990) In search of the 'missing self': MHC molecules and NK cell recognition. *Immunol Today*, **11**, 237–244.
71. Moretta, L., Ciccone, E., Moretta, A., Hoglund, P., Ohlen, C., and Karre, K. (1992) Allorecognition by NK cells: nonself or no self?. *Immunol Today*, **13**, 300–306.
72. Ruggeri, L., Mancusi, A., Burchielli, E., Capanni, M., Carotti, A., Aloisi, T., Aversa, F., Martelli, M.F., and Velardi, A. (2008) NK cell alloreactivity and allogeneic hematopoietic stem cell transplantation. *Blood Cells Mol Dis*, **40**, 84–90.
73. Ruggeri, L., Capanni, M., Urbani, E., Perruccio, K., Shlomchik, W.D., Tosti, A., Posati, S., Rogaia, D., Frassoni, F., Aversa, F., Martelli, M.F., and Velardi, A. (2002) Effectiveness of donor natural killer cell alloreactivity in mismatched hematopoietic transplants. *Science*, **295**, 2097–2100.
74. Dodero, A., Carniti, C., Raganato, A., Vendramin, A., Farina, L., Spina, F., Carlo-Stella, C., Di Terlizzi, S., Milanesi, M., Longoni, P., Gandola, L., Lombardo, C., and Corradini, P. (2009) Haploidentical stem cell transplantation after a reduced-intensity conditioning regimen for the treatment of advanced hematologic malignancies: posttransplantation CD8-depleted donor lymphocyte infusions contribute to improve T-cell recovery. *Blood*, **113**, 4771–4779.
75. Schaffer, M., Malmberg, K.J., Ringden, O., Ljunggren, H.G., and Remberger, M. (2004) Increased infection-related mortality in KIR-ligand-mismatched unrelated allogeneic hematopoietic stem-cell transplantation. *Transplantation*, **78**, 1081–1085.
76. Lee, G.K., Park, H.J., Macleod, M., Chandler, P., Munn, D.H., and Mellor, A.L. (2002) Tryptophan deprivation sensitizes activated T cells to apoptosis prior to cell division. *Immunology*, **107**, 452–460.
77. Munn, D.H. and Mellor, A.L. (2004) IDO and tolerance to tumors. *Trends Mol Med*, **10**, 15–18.
78. Munn, D.H., Zhou, M., Attwood, J.T., Bondarev, I., Conway, S.J., Marshall, B., Brown, C., and Mellor, A.L. (1998) Prevention of allogeneic fetal rejection by tryptophan catabolism. *Science*, **281**, 1191–1193.
79. Mellor, A.L. and Munn, D.H. (2001) Tryptophan catabolism prevents maternal T cells from activating lethal anti-fetal immune responses. *J Reprod Immunol*, **52**, 5–13.
80. Munn, D.H., Sharma, M.D., Hou, D., Baban, B., Lee, J.R., Antonia, S.J., Messina, J.L., Chandler, P., Koni, P.A., and Mellor, A.L. (2004) Expression of indoleamine 2,3-dioxygenase by plasmacytoid dendritic cells in tumor-draining lymph nodes. *J Clin Invest*, **114**, 280–290.
81. Munn, D.H. and Mellor, A.L. (2006) The tumor-draining lymph node as an immune-privileged site. *Immunol Rev*, **213**(1), 146–158.
82. Jasperson, L.K., Bucher, C., Panoskaltsis-Mortari, A., Taylor, P.A., Mellor, A.L., Munn, D.H., and Blazar, B.R. (2008) Indoleamine 2,3-dioxygenase is a critical regulator of acute graft-versus-host disease lethality. *Blood*, **111**, 3257–3265.

83. Steckel, N.K., Kuhn, U., Beelen, D.W., and Elmaagacli, A.H. (2003) Indoleamine 2,3-dioxygenase expression in patients with acute graft-versus-host disease after allogeneic stem cell transplantation and in pregnant women: association with the induction of allogeneic immune tolerance? *Scand J Immunol*, **57**, 185–191.
84. Aiuti, A., Slavin, S., Aker, M., Ficara, F., Deola, S., Mortellaro, A., Morecki, S., Andolfi, G., Tabucchi, A., Carlucci, F., Marinello, E., Cattaneo, F., Vai, S., Servida, P., Miniero, R., Roncarolo, M.G., and Bordignon, C. (2002) Correction of ADA-SCID by stem cell gene therapy combined with nonmyeloablative conditioning. *Science*, **296**, 2410–2413.
85. Aiuti, A., Brigida, I., Ferrua, F., Cappelli, B., Chiesa, R., Marktel, S., and Roncarolo, M.G. (2009) Hematopoietic stem cell gene therapy for adenosine deaminase deficient-SCID. *Immunol Res*, **44**, 150–159.
86. Aiuti, A., Cattaneo, F., Galimberti, S., Benninghoff, U., Cassani, B., Callegaro, L., Scaramuzza, S., Andolfi, G., Mirolo, M., Brigida, I., Tabucchi, A., Carlucci, F., Eibl, M., Aker, M., Slavin, S., Al-Mousa, H., Al Ghonaium, A., Ferster, A., Duppenthaler, A., Notarangelo, L., Wintergerst, U., Buckley, R.H., Bregni, M., Marktel, S., Valsecchi, M.G., Rossi, P., Ciceri, F., Miniero, R., Bordignon, C., and Roncarolo, M.G. (2009) Gene therapy for immunodeficiency due to adenosine deaminase deficiency. *N Engl J Med*, **360**, 447–458.
87. Santilli, G., Thornhill, S.I., Kinnon, C., and Thrasher, A.J. (2008) Gene therapy of inherited immunodeficiencies. *Expert Opin Biol Ther*, **8**, 397–407.
88. Howe, S.J., Mansour, M.R., Schwarzwaelder, K., Bartholomae, C., Hubank, M., Kempski, H., Brugman, M.H., Pike-Overzet, K., Chatters, S.J., de Ridder, D., Gilmour, K.C., Adams, S., Thornhill, S.I., Parsley, K.L., Staal, F.J., Gale, R.E., Linch, D.C., Bayford, J., Brown, L., Quaye, M., Kinnon, C., Ancliff, P., Webb, D.K., Schmidt, M., von Kalle, C., Gaspar, H.B., and Thrasher, A.J. (2008) Insertional mutagenesis combined with acquired somatic mutations causes leukemogenesis following gene therapy of SCID-X1 patients. *J Clin Invest*, **118**, 3143–3150.
89. Goldman, J.M. and Melo, J.V. (2003) Chronic myeloid leukemia – advances in biology and new approaches to treatment. *N Engl J Med*, **349**, 1451–1464.
90. Martinez, N., Drescher, B., Riehle, H., Cullmann, C., Vornlocher, H.P., Ganser, A., Heil, G., Nordheim, A., Krauter, J., and Heidenreich, O. (2004) The oncogenic fusion protein RUNX1-CBFA2T1 supports proliferation and inhibits senescence in t(8;21)-positive leukaemic cells. *BMC Cancer*, **4**, 44.
91. Heidenreich, O., Krauter, J., Riehle, H., Hadwiger, P., John, M., Heil, G., Vornlocher, H.P., and Nordheim, A. (2003) AML1/MTG8 oncogene suppression by small interfering RNAs supports myeloid differentiation of t(8;21)-positive leukemic cells. *Blood*, **101**, 3157–3163.
92. Scherr, M., Battmer, K., Winkler, T., Heidenreich, O., Ganser, A., and Eder, M. (2003) Specific inhibition of bcr-abl gene expression by small interfering RNA. *Blood*, **101**, 1566–1569.
93. Robbins, M.A., Li, M., Leung, I., Li, H., Boyer, D.V., Song, Y., Behlke, M.A., and Rossi, J.J. (2006) Stable expression of shRNAs in human CD34+ progenitor cells can avoid induction of interferon responses to siRNAs in vitro. *Nat Biotechnol*, **24**, 566–571.
94. Parham, P. (2005) MHC class I molecules and KIRs in human history, health and survival. *Nat Rev Immunol*, **5**, 201–214.
95. Ruggeri, L., Mancusi, A., Burchielli, E., Aversa, F., Martelli, M.F., and Velardi, A. (2007) Natural killer cell alloreactivity in allogeneic hematopoietic transplantation. *Curr Opin Oncol*, **19**, 142–147.

Subject Index

M. Sioud (ed.), *RNA Therapeutics*, Methods in Molecular Biology 629,
DOI 10.1007/978-1-60761-657-3, © Springer Science+Business Media, LLC 2010

MIX
Papier aus verantwortungsvollen Quellen
Paper from responsible sources
FSC® C105338

If you have any concerns about our products,
you can contact us on
ProductSafety@springernature.com

In case Publisher is established outside the EU,
the EU authorized representative is:
Springer Nature Customer Service Center GmbH
Europaplatz 3, 69115 Heidelberg, Germany

Printed by Libri Plureos GmbH
in Hamburg, Germany